Environmental ECONOMICS

Environmental ECONOMICS

CHARLES D. KOLSTAD

New York Oxford
OXFORD UNIVERSITY PRESS
2000

Oxford University Press

Oxford New York
Athens Auckland Bangkok Bogotá Buenos Aires Calcutta
Cape Town Chennai Dar es Salaam Delhi Florence Hong Kong Istanbul
Karachi Kuala Lumpur Madrid Melbourne Mexico City Mumbai
Nairobi Paris São Paulo Singapore Taipei Tokyo Toronto Warsaw

and associated companies in
Berlin Ibadan

Published by Oxford University Press, Inc.,
198 Madison Avenue, New York, New York 10016
http://www.oup-usa.org

Library of Congress Cataloging-in-Publication Data

Kolstad, Charles D.
Environmental economics / Charles D. Kolstad.
p. cm.
Includes bibliographical references and index.
ISBN 0-19-511954-1 (cloth : alk. paper)
1. Environmental economics. I. Title.
HC79.E5K65 1999
333.7—dc21 98-37995
CIP

9 8 7 6 5 4 3 2

Printed in the United States of America
on acid-free paper

To Kate, Jonathan, Valerie, Christine and
the memory of George Andrew Kolstad

Contents

Preface

I have been teaching environmental economics at the graduate level since the mid-1980s and at the undergraduate level since the early 1990s. A number of excellent books have been published on environmental and resource economics aimed primarily at students with a rudimentary knowledge of economics. And some books are targeted at students with no knowledge of economics. There are also a few books that are designed for use by graduate students. The gap in the literature appears to be at the level of advanced undergraduate students who have a good understanding of intermediate microeconomics. This book is written to be used by such students. Thus I assume a good working knowledge of intermediate microeconomics (at the level of Varian, 1996). Elementary calculus is useful but not essential.

In writing a book for upper level undergraduates, I was faced with the question of whether to package resource economics with environmental economics (as is done in most existing undergraduate texts) or to separate them. I chose the latter route, because the material in the two fields is somewhat distinct and because there is too much material in both environmental economics and resource economics to be covered in a single one-term course of study. At UCSB, we devote one 10-week term to environmental economics and one 10-week term to resource economics. Intermediate microeconomics and calculus are prerequisites for both courses. I have also used the book for masters-level graduate students, supplemented with calculus and some of the references mentioned in the text.

Another decision was to separate environmental policy from environmental economics, a distinction that is often not made in more introductory texts. Both policy and economics cannot be adequately treated in a single term. Thus this book is short on policy. Only one chapter (Chapter 2) treats environmental policy to any extent, and that chapter is intended to motivate the treatment of the rest of the book. There are, however, policy-relevant examples scattered throughout the text. It is my hope that the deemphasis of policy will not be considered a shortcoming of the book but rather a strength, allowing the book to focus on the key issues of environmental economics.

There is widespread worldwide interest in the field of environmental economics. Researchers in Europe in particular are very active in the field. It is a very international research community. Student interest is also international. In writing this book I have tried to make the examples international in character, so that the book seems relevant to students from countries other than the United States, not simply another transplant from the United States to other countries. I know I have not completely succeeded in this. The ideal remains—that the book should truly be an international treatment of the subject of environmental economics.

The book is divided into four primary sections. The first two chapters constitute an introduction. The goal is to define the field of environmental economics and to put it in perspective in relation to economics in general and to ecological economics and resource economics in particular. A chapter is also provided to discuss the way in which countries around the world approach environmental problems and their control.

The second section (Chapters 3–7) concerns market failure. This section covers the classic problem of externalities and public goods/bads. The goal of the section is to articulate why the market solution may in some cases be desirable and why, with environmental protection, the market often fails to work properly. This section also includes a discussion of the Coase theorem and Pigovian fees. The treatment of market failure, social welfare, and the fundamental theorems of welfare economics can be skipped by students who are well prepared in these areas of microeconomics.

The third section (Chapters 8–14) concerns the regulation of pollution. This material draws on the literature from industrial organization, though many of the applications come directly from environmental economics. Chapter 13 concerns the international dimension of environmental regulation and Chapter 14 concerns the macrodimensions of environmental regulation.

The fourth section (Chapters 15–18) concerns measuring the demand for environmental quality. This is an area of active research in the profession. It is possible to move quite quickly through this material or spend more time developing it. It is also possible to have students undertake their own empirical studies of measuring the demand for environmental goods. This is particularly easy to do for the case of stated preference methods.

All of the chapters (except for the first two) have problems at the end. These problems vary considerably in difficulty. Some require essays, some require calculus, some are quite long and some are straightforward. The most difficult are indicated with an asterix (*).

This book would not be possible without the support of a number of people, particularly my wife Valerie and children, Jonathan and Kate. The long-term encouragement of Luis Fernandez-Herlihy was instrumental in seeing the book through to fruition. I am also grateful to the University of California for facilitating my work on this book and the Stanford Institute for International Studies for providing a quiet place to work for an academic quarter.

Thanks are due to Alexander Golub (Director), Elena Strukova, and Renat Perelet of the Center for Environmental and Natural Resource Economics at the Higher School of Economics in Moscow for assistance with the section on environmental regulations in Russia. Similarly, David Cope of the U.K. Parliamentary Office of Science and Technology was most helpful in developing my understanding of UK environmental regulations. Needless to say, none of these individuals should be held responsible for my errors and omissions in the country reviews in Chapter 2.

I am also grateful to colleagues for providing comments and insights into various dimensions of the field. Professor Jim Proctor of UCSB helped me understand some of the subtleties of environmental ethics, although I readily admit to still being a neophyte on the subject. I have also appreciated the constructive comments of Linda Fernandez

(UCSB), Rick Freeman (Bowdoin), Noelwah Netusil (Reed), Charles Perrings (York), and Stef Proost (Leuven). They of course should not be blamed for the errors that undoubtedly remain. I am also grateful to a number of graduate students who have served as teaching assistants over the years, including Ken Ardon, Ed Balsdon, Jim Grefer, Glenn Mitchell and Trent Smith. In addition, Ed Balsdon has provided valuable assistance in refining the text and problems. Louise Cannell has been very helpful in manuscript preparation and in managing the various versions of this manuscript. Most of all, I am grateful to various groups of students at the University of California at Santa Barbara, the University of Illinois at Urbana, and the Catholic University of Leuven, Belgium for acting as guinea pigs in the development of this material. Thank you all.

Santa Barbara, California
March 1999

C. D. K.

Environmental ECONOMICS

1 WHAT IS ENVIRONMENTAL ECONOMICS?

Tell someone you study environmental economics and the usual response is a look of puzzlement and the query "Just what is that?" A natural reaction, particularly considering the common belief in some circles that economics is the root of the "environmental problem."

Environmental economics is concerned with the impact of the economy on the environment, the significance of the environment to the economy, and the appropriate way of regulating economic activity so that balance is achieved among environmental, economic, and other social goals. What distinguishes a morally neutral chemical such as sulfur dioxide from the pollutant sulfur dioxide is the economy. The polluters who emit sulfur dioxide do so because it is a by-product of producing some good the public wants; consumers want the good associated with sulfur dioxide but at the same time obtain disutility (damage) from the sulfur dioxide pollution. The essence of the environmental problem is the economy—producer behavior and consumer desires. Without the economy, most environmental issues are simply research questions of concern to chemists or biologists with no policy significance.

For most goods and services in a modern economy, we rely on markets to match producer costs with consumer demands to yield the "right" amount of pollution (and thus consumption). The problem with pollution is that markets do not work to yield the socially desirable amount of pollution. This illustrates the breadth of problems that need answers: What are the incentives for the generation of pollution? What are the costs of cleaning up pollution? What are the societal gains from pollution control? What is the right balance between costs of control and gains from control? What regulatory mechanisms can be designed to ensure the right balance between costs and gains? Sometimes these issues are straightforward; othertimes they are exceedingly complex.

Although the field of environmental economics probably dates to the late 1950s and early 1960s with the important contributions emerging from the "think tank" Resources for the Future,[1] the field really took off in the 1970s and has been booming ever since. In the 1990s the payoff is beginning to be seen in terms of influence on environmental policy. Marketable permits for pollution control are now widely embraced, valuation methods are an integral part of environmental prevention, environmental valuation is being used to make decisions concerning major public projects with environmental impacts, and environmental economics is playing a major role in the current climate change debate.

In the sections that follow we will develop more fully several of these dimensions of environmental economics. We first consider how environmental economics relates to environmental policy. Next, we examine how environmental economics meshes with the larger discipline of economics. We then discuss two related terms that have emerged in the academic and policy world—ecological economics and environmental economics. How do they differ? Environmental economics is also closely related to resource economics. What is the connection and what are the distinctions between them? Finally, we consider several important issues, currently the subject of much research and debate, that confront the field of environmental economics.

I. ENVIRONMENTAL ECONOMICS AND ENVIRONMENTAL POLICY

Concern with the environment is not a passing fad but a deep-seated concern, brought on in large part by the coincidence of high incomes and high population density. If there were few people in the world, the earth's environment would be very forgiving and capable of absorbing most that humans throws at it, cleansing itself automatically. The days of low population on the earth have passed; the total number of people on the planet and particularly the density of people in some parts of the planet magnify the size of environmental insults, overloading the capacity of the earth for self-cleansing. Income is also important, not only because rich people tend to consume more and thus generate more pollution, but also because the environment is often viewed as a luxury good. For the poor who struggle to keep food on the table, environmental issues often take a backseat to other more pressing needs related to survival. As people become wealthier, they turn their attention increasingly to the quality of their living environment, which ultimately is the planet. Thus if recent trends toward increasing wealth for the people of the world continue, we can expect concern for the environment to increase over time.

What are the typical issues facing most countries of the world with regard to environmental policy? In developed countries, pollution and the preservation of natural environments (e.g., wilderness areas) are major concerns. Pollution problems usually involve two major issues—what is the right amount of pollution and how can we get polluters to control their emissions?

Determining the right amount of pollution is not easy. Pollution is generated as a by-product of producing goods. To determine the costs of pollution control it is necessary to understand the structure of goods production and how costs will differ for different levels of pollution. Contrary to what most people might think, this is not an engineering question. Although it is easy to find out how much it would cost for a piece of equipment that is placed on a smokestack to reduce pollutants in the smoke (a "scrubber"), to an economist that is only the tip of the cost iceberg. Faced with the prospect of having to reduce pollution levels, the firm has many options. These include end-of-pipe treatment, modifying the production process, modifying the characteristics of the product, relocating the productive activity to reduce damage, and investing in research and development to find new ways of controlling pollution. Consumers can also reduce consumption of the polluting good. Thus characterizing costs at a conceptual level, as well as measuring these costs empirically, is a complex question without easy answers. It is also the domain of environmental economics.

Determining the right amount of pollution also involves determining damages from pollution. The words "damages from pollution" deceptively suggest that this is a natural science question, such as counting the dead fish on a polluted lake or determining the level of pollution at which people begin to get sick. This is an oversimplification of the multitude of ways pollution affects people and the relative seriousness of these effects to people. Air pollution in an urban area can cause physical irritation (itchy eyes, running nose), reduced visibility, degraded visibility (a brown pale), soiled clothes, decreased lung capacity, worry about adverse effects, increased susceptibility to illness, and of course illness itself. Some of these effects are tangible, others are very intangible. Economics is accustomed to condensing this variety of effects into a single measure—the willingness to pay to reduce pollution. If pollution is bad, people are willing to devote some of their resources to eliminating the pollution. Leaving aside the fact that most people think the polluter should pay, one way of measuring the overall magnitude or importance of pollution reduction to a person is through his or her willingness to give up something valuable in exchange for improved personal environmental quality. Measuring this willingness to pay is not easy and is the subject of much research in environmental economics.

Having characterized the importance of pollution reduction to individuals (their willingness to pay), it is possible to sum up individual preferences to obtain a societal willingness to pay to reduce pollution. It is then easy to combine this with the cost of pollution control to determine the socially optimal amount of pollution reduction. But how to obtain this? The government could tell each polluter how much to emit; but this would be analogous to central planning in the old Soviet Union—we know it works up to a point but has severe problems, particularly when there are many firms and polluters involved. It is difficult to determine the best way for the government to intervene in the economy ("regulate") to yield the right amount of pollution control without excessive administrative costs or control costs while at the same time providing the right incentives to undertake research to reduce costs for the future.

So the "simple" job of fixing the problem of pollution is not so easy at the policy level and can involve hard-to-solve problems, many in the domain of environmental economics. The examples we have used are from developed economies but a very similar analysis could apply to a developing country. Air pollution is a big problem in many cities of the developing world. Water pollution is probably the most severe environmental problem in many developing countries: water contaminated by human waste kills millions of people annually. The same issues of costs, demand for clean-up, and how to regulate apply equally to this question.

Another important type of policy question is the preservation of natural environments, broadly defined. This could involve preserving wild and scenic areas from development or protecting animal and plant species from extinction. Here the primary issue is providing balance between the forces of development that threaten these environmental resources and the social value of preservation. How can both sides of this equation be quantified to help policymakers when they are confronted with very specific decisions (such as whether to allow logging in a virgin forest)?

These examples could go on and on. The point is that environmental protection usually involves the intervention of governments in the economy and it is often difficult to decide on the proper extent and nature of that intervention. Environmental economics as it is applied to real environmental problems can be invaluable in helping make those important decisions.

II. ENVIRONMENTAL ECONOMICS AND ECONOMICS

Economics is a well-developed discipline with an extensive body of theory, a paradigm associated with how the economic world works, and a number of branches, or fields of study, associated with pieces of the economy.

The fundamental building blocks of economics are contained in microeconomics—the theory of the consumer, the theory of the firm, and the theory of market interaction. This forms the basis for nearly all of economics. Related to microeconomics is the branch of statistics applied to economic phenomena—econometrics. Microeconomics permeates all of economics and econometrics permeates all of applied economics.

Branching out from basic microeconomic theory are the several fundamental fields of economics. These would include macroeconomics (the study of aggregate as opposed to individual phenomena), public finance (the study of goods not provided by the market and the study of taxation), industrial organization (the in-depth study of how firms interact with each other and with consumers and organize themselves into industries), and international trade (concerned with how distinct and independent economies interact). Each of these major fields is concerned with major portions of economic activity and each has unique contributions to make to the overall study of economics.

There are a number of applied fields of economics that draw on all of the basic fields as well as microeconomics. These would include labor economics, health economics, monetary economics, experimental economics, development economics, international finance, law and economics, and environmental economics. Each of these applied fields draws heavily on microeconomics and the basic fields of economics. For the most part, each of these fields has contributed in some way to understanding economics outside of its own narrow set of interests. For instance, labor economics has been the source for many innovations in econometrics that have found application across economics. The primary contribution of environmental economics has been in the area of nonmarket valuation, i.e., methods for measuring demand curves for goods when there is no market. Other important components of environmental economics involve adapting tools developed in other parts of economics to questions regarding the environment.

The categorization above is by no means unequivocal. I would expect many economists in one or another of the fields mentioned above to dispute how their field has been categorized and placed in relation to other fields. To an extent they would be right: there are many different ways of summarizing the different fields of economics. The point that is being made is that environmental economics is an applied field, like many other applied fields in economics. Much of environmental economics involves adapting concepts developed in other branches of economics (particularly public finance and industrial organization) and applying them to environmental problems. Some aspects of environmental economics are unique to the field (such as valuation, mentioned above) and have potential use in economics outside of the environmental economics field.

III. ENVIRONMENTAL ECONOMICS AND ECOLOGICAL ECONOMICS

This book is concerned with environmental economics. There is another discipline that largely has grown out of systems ecology called "ecological economics." These two fields take quite different perspectives, but are ultimately concerned with helping make social decisions about environmental problems. Unfortunately, in many non-English speaking countries the distinction between the two fields is lost in the translation because of the similarities between the words "environmental" and "ecological."

A simple distinction between the two fields arises from the fact that environmental economics tends to involve economists who have extended their discipline and paradigm to consider the environment, whereas ecological economics tends to involve ecologists who have extended their discipline and paradigm to consider humans and the economy. But this is history; the appropriate question to ask is how do the two fields approach environmental problems and how do they differ?

Ecological economics (as well as conventional economics) is difficult to succinctly define. One of the leading ecological economists defines the subject as a "field of study that addresses the relationships between ecosystems and economic systems in the broadest sense."[2] The emphasis is on the very long-term health of the ecosystem, broadly defined (i.e., with humans as part of it).

One major distinction between the two fields is associated with value and thus the way in which social decisions are made that depend on measures of value of the environment. Conventional economists believe that value to society derives from the individual values held by human members of society. Ecological economists take a more biophysical view of value. For instance, some ecological economists measure value in terms of embodied energy content. Thus in comparing a typewriter to a computer, the appropriate question is which took more energy to create? Less energy is better. This is a direct extension of ecological theories that ecosystems operate to minimize the throughput of energy. To these researchers, minimizing the energy content of delivered goods and services should drive public policy. The criticism leveled at this "energy theory of value" by environmental economists is that there are many resources in short supply, including land and skilled people. Reducing the value of a good to the embodied content of any factor is an oversimplification. Environmental economists believe the value of a good stems from its embodied content of multiple scarce factors (including energy) as well as how much value individual people place on the final good. In other words, value cannot be reduced to a simple physical metric.

However, the greatest distinction between the fields emerges when considering environmental problems with very long-time horizons, such as global warming or disposal of nuclear wastes. As some environmental economists will readily admit, economics has a difficult time analyzing problems in which costs and benefits span long time horizons. For instance, storing nuclear wastes can involve potential risks that extend for a quarter of a million years. The benefits of the storage are reaped by the present consumers of nuclear power; the costs, if any, are borne by future generations that must live with the nuclear repositories. The conventional economic approach to this is to add up all of the costs and benefits, whenever they may occur, but to apply a discount factor to reduce the importance of future costs in the sum. Inevitably, this means that what happens a century from now has very little effect on the decisions that are made today. To many people, this is disquieting. Ecological economists have proposed other ways of dealing with the intertemporal decision problem, particularly the notion of sustainability. They argue that we should never undertake any action that is not sustainable in the long run. In the nuclear waste example, they would ask: Can we continue to bury waste forever and ever and be satisfied with the world that results? If the answer is no, then the action is not sustainable. It is not a matter of balancing costs and benefits. There is some intuitive appeal to such a philosophy.

To a large extent, the problem of making decisions over the very long run will not arise in this book. For the most part, we are concerned with static issues. Questions of long-run tradeoffs and dynamics are the purview of resource economics, to which we now turn.

IV. ENVIRONMENTAL ECONOMICS AND RESOURCE ECONOMICS

Nearly all textbooks combine the treatment of environmental and resource economics. In fact, most graduate programs that have a specialization in the area combine environmental and resource economics into one field. Undoubtedly this is because both concern the natural world. Environmental economics involves questions of excessive production of pollution by the market or insufficient protection of the natural world, due to market failure. Resource economics, on the other hand, is concerned with the production and use of natural resources, both renewable and exhaustible. Renewable resources would include fisheries and forests. Nonrenewables would include minerals and energy as well as natural assets such as the Alps and species of plants and animals.

So we see the distinction between the two areas but we also see the overlap. Typically, environmental economics is concerned with static questions of resource allocation. Time is not really an issue in deciding on the right amount of air pollution in London. Resource economics on the other hand is concerned with dynamics. Time is what makes renewable and exhaustible resource questions interesting. If we log a forest slowly enough, the forest can regenerate itself and we can continue logging indefinitely. How fast we extract an exhaustible resource will determine its scarcity in the future, and thus its price in the future. In both of these cases, market failure is not the essence of the problem (though poorly functioning markets can be important).

There are overlaps between environmental and resource economics. Global warming is an example of a pollution problem with a very long time frame. There are other overlaps, primarily in the preservation of natural environments. These issues involve time so they could be relegated to resource economics. On the other hand, damage to natural environments is often the incidental result of economic activity with a different primary purpose. Species loss is usually the result of conversion of habitat to human use.

Perhaps the best division between environmental and resource economics is between static issues related to the natural world and dynamic issues. For the most part this book concerns problems in resource and environmental economics that are static. Typically, we are concerned with pollution.[3]

V. POSITIVE VERSUS NORMATIVE PERSPECTIVES

There are two fundamental uses of economics. One is to try to explain what we see in the economy around us. Another is to try to explain how we would wish the economy to allocate and distribute goods and services. The terms positive and normative are used to distinguish these two perspectives. Positive economics is more value free, aiming to explain why markets and institutions have evolved as they have and how they work. Examples are understanding why the price of gasoline increases when OPEC meets to restrict output and how the spatial distribution of pollution emissions changes when a marketable permit system is established to regulate sulfur emissions.

Normative economics, on the other hand, attempts to use economic tools to design government policies to intervene in the marketplace. Inevitably the question arises as to the "best" way of intervening in the marketplace. Clearly, this requires a way of defining what is best—a much more value-laden process than merely explaining why the economy works as it does.

Unfortunately, when working with environmental problems it is not possible to restrict attention solely to positive economics. Fundamental to environmental economics is the notion of market failure. Repairing that market failure typically requires government intervention. What kind of government intervention? That is a normative question. And that is often the question that environmental economists are asked to help answer. In developing the normative theory of regulation to correct market failure or the public provision of nonmarket goods, we will try to make clear when value judgments enter the process of policy formulation. The practicing environmental economist should always be aware of the problems of venturing into the territory of normative analysis. This is one reason we turn in Chapter 3 to the question of social choice—the process of making societal decisions.

VI. IMPORTANT ISSUES IN ENVIRONMENTAL ECONOMICS

This is truly an exciting time in the field of environmental economics. There are relatively few economists who call themselves environmental economists while there are many basic problems that need addressing. We will suggest a few here.[4]

One of the most important contributions of environmental economics to economics generally has been in the area of measuring the demand for nonmarket goods. Measuring this demand has become central to many public debates over environmental quality. However, some methods for measuring demand have been the subject of great controversy. Stated preference methods involve directly asking people how they value the environment. Such methods have come under strenuous attack by some as at best biased and at worst vacuous. Others argue they are valid and of tremendous importance. A very active area of current research is the theory underlying methods for measuring the demand for environmental goods as well as empirical methods for doing so.

There is another set of issues surrounding regulation of environmental goods. Some of the unanswered questions are detailed more fully in Chapter 8. The basic problem is that economic incentives need significant refining before they can be relied on to solve many real environmental problems. These difficulties have to do with incentives, different amounts of information possessed by polluters and the government, and the role of technological change in determining future levels of pollution control. This work takes place at various levels, including institutional design, determining empircal properties of different regulatory mechanisms, and theoretical issues of regulatory design.

There are a number of international issues of environmental regulation that are not fully resolved. Some of these are discussed in Chapter 13. One major problem is in understandng how environmental regulations interact with trade restrictions (or lack thereof). Are differential environmental regulations compatible with free trade? Can environmental regulations be used effectively as barriers to trade? Does free trade tend to exploit the environments of developing countries due to their less well-developed institutions for environmental protection?

These few paragraphs are meant to suggest the menu of research questions that is being examined by environmental economists. As we travel through the book, it is my sincere hope that the reader identifies other interesting problems for further exploration.

NOTES

1. Resources for the Future (RfF), a Washington, DC environmental and resource economics research organization, was set up in the early 1950s by the Ford Foundation to address problems of materials shortages. A number of very important works emerged from researchers at RfF in the 1950s and 1960s. One contribution was the development of methods for measuring recreation demand (Clawson and Knetsch, 1966). Another was the reinvigoration of the use of emission fees by Alan Kneese. This culminated in the still-impressive study of regulation of water pollution in the Delaware River system (Kneese and Bower, 1968). Another was the analysis of scarcity of natural resources by Barnett and Morse (1963). John Krutilla forged the basis for how we view natural environments by expanding the notion of value beyond simple use (Krutilla, 1967). RfF remains today at the forefront of environmental economics research.
2. Costanza (1991), p. 3.
3. Cropper and Oates (1992), in their excellent survey of the field of environmental economics, make just this distinction between environmental and resource economics.
4. See Deacon et al. (1998) for a discussion of current research issues in environmental and resource economics.

2 ENVIRONMENTAL PROBLEMS AND POLICY SOLUTIONS

For the most part, this book concentrates on the economics of environmental problems. Policy is not a focus. However, it is useful to gain an appreciation of the scope of environmental problems, to provide a sense of the significance of our study of environmental economics. This chapter serves that purpose.

First, we will briefly discuss the recent history of environmental problems around the world, emphasizing both the relative magnitude of different problems today as well as the extent to which these problems have changed over the last few decades. We then turn to the question of how countries have approached environmental problems and their regulation, reviewing the extent to which economic incentives have played a role in regulation. Finally, we consider the costs of policies to improve environmental quality, in an effort to place environmental protection in the context of other social needs that absorb resources.

It is important to realize that this is a very quick survey of what is an enormously broad subject. The review is admittedly limited in scope. A complete review calls for volumes, not a single chapter. Furthermore, we will concentrate on problems for which data exist. (This is somewhat like looking for the keys under the lamppost because that is where the light is.) Nevertheless, some knowledge of the problems that motivate theoretical economic analyses is vital to understanding the usefulness and implications of environmental economic theory.

I. THE QUALITY OF THE ENVIRONMENT

There are many comprehensive assessments of the state of the world's environment.[1] It is unnecessary to offer another review here. What we would like to do is provide some indication of the breadth of problems that are deemed environmental and to gain an appreciation for what problems are being solved and what problems remain difficult to solve and are likely to be a focus of attention in the coming decades.

Pollution problems are not new to mankind. There are records of the Romans complaining about the "stink of smokey chimneys."[2] Pollution control laws in other parts of Europe date from the Middle Ages. Urban areas have always been problematic because of the large concentrations of people. People are associated with emissions as well as being the reason pollution is damaging. But outside of cities, historically the earth's size has been vast enough to dissipate even the most serious environmental threat. What is new is the magnitude of the problem and the fact that the world is no longer infinite compared to the ability of people to pollute it. In the 1960s and 1970s many people around the world were galvanized into doing something to curb environmental degradation. In most countries, significant movements to protect the environment date from this period.[3] To a very large extent, the enormous size of our current world's population and the high standard of living of portions of the population are what are responsible for the pressures on the environment. A larger economy generates more pollution, all other things being equal; richer citizens usually demand higher levels of environmental quality. And as long as the world becomes more populated and wealthier, the pressures will only increase. This is not to say environmental problems cannot be solved, only that it will become increasingly difficult to protect the earth's environment.

We focus on four main categories of environmental problems: air pollution, water pollution, toxic emissions, and ecosystem health. This is not to suggest that this is a comprehensive list, just that these are four major categories that encompass many of the major environmental problems faced by man today.

A. Air Pollution

Air pollution is primarily a by-product of energy consumption. Impurities in fuels lead to emissions of sulfur dioxide and particulate matter. It is a basic fact of chemistry that burning carbon-based fuels leads to emissions of oxidized carbon—carbon dioxide, a major greenhouse gas. Because our atmosphere contains significant amounts of nitrogen in addition to oxygen, burning fuels inevitably leads to emissions of nitrogen oxides. Tropospheric ozone is not directly emitted from fuel combustion but results chemically from high concentrations of nitrogen oxides (from fuel combustion) and organic vapors (from paint drying and gasoline evaporating, among other things), in the presence of sunshine. It has proved to be very difficult to control ozone in many urban areas of the world.

To a large extent, air pollutants are at their worst in urban areas, due to concentrations of people, both as sources of the pollution (directly or indirectly) and as victims of the pollution. Air pollution can lead to health problems, including sickness as well as irritation and reduced human performance. The young and those weakened by other illnesses may be particularly susceptible to the effects of urban air pollutants. Urban air pollution also damages materials (such as buildings), increases the cost of maintenance (such as increased cleaning requirements), and degrades aesthetics (no one likes to live in a brown haze).

At a regional level, air pollutants may damage crops (though some sulfur and carbon pollution can help crops). Acid deposition is a regional problem in many parts of the world. Deposition of sulfur and nitrogen-based acids can harm forest and aquatic ecosystems. Regional haze from nitrogen and sulfur pollutants is also a problem in areas with less rainfall. Carbon dioxide is a global pollutant in that overall levels in the earth's atmosphere lead to increases in the heat-trapping capacity of the atmosphere, which leads to global warming. Controlling the precursors of climate change is the subject of intense current international debate.

Some urban air pollutants have been curbed with some success, though not without cost and not in all parts of the world. Sulfur dioxide emissions in Western Europe, North America, and the rest of the developed world have been the subject of significant control over the past two decades. Aggregate emissions appear to have peaked and are now declining. Table 2.1 shows a fifteen year time series for three pollutants for three countries, one poor (China), one rich (Japan), and one in between (Iran). Note the dissimilarity among the three pollutants represented. Carbon dioxide emissions, which lead to global warming, have been increasing over the period and are higher in wealthier countries. Sulfur dioxide is almost the opposite. In fact, sulfur dioxide seems to be particularly troubling for the middle-income country, Iran. With money (income), the problem can be solved, as the Japan data demonstrate. Suspended particulates (soot) are worse in lower income countries and tend to be less of a problem as income levels rise. The story Table 2.1 tells is that not all air pollutants are the same; nor are the same pollutants always the most problematic across countries.

One of the major nonfuel air pollutants is cholorofluorocarbons (CFCs), substances that lead to depletion of the stratospheric ozone layer. These substances are now primarily used as refrigerants. CFCs have been controlled (in theory) by the Montreal Protocol, which has had some success. Worldwide production of these chemicals in 1995 was 20% of the level in 1986. One problem is a growing black market in illicitly produced CFCs

TABLE 2.1 Air Pollution Indicators for Selected Cities and Years

	1980	1990	1995
China			
GDP per capita (1995 U.S.$)[a]	907	1783	3072
SO_2 concentration (Beijing)[b]	66	107	90
Particulate concentration (Beijing)[b]	475	413	377
Per capita CO_2 emissions (tonnes)	0.4	0.6	0.7
Iran			
GDP per capita (1995 U.S.$)[a]	5377	4843	5351
SO_2 concentration (Tehran)[b]	130	165	209
Particulate concentration (Tehran)[b]	226	261	248
Per capita CO_2 emissions (tonnes)	0.8	0.9	1.1
Japan			
GDP per capita (1995 U.S.$)[a]	15538	20794	22173
SO_2 concentration (Tokyo)[b]	42	19	18
Particulate concentration (Tokyo)[b]	61	NA	49
Per capita CO_2 emissions (tonnes)	2.1	2.4	2.5

[a]*GDP per capita computed using purchasing power parities, to 1987 U.S.$. Converted to 1995 U.S.$ using US GDP implicit price deflator.*

[b]*Mean annual concentrations in city center ($\mu g/m^3$). For reference, the US primary ambient standards are 80 $\mu g/m^3$ for SO_2 and 50 $\mu g/m^3$ for particulates.*

Source: World Bank, *World Development Report* (various issues); World Bank (1992); World Bank World Development Indicators Database (World Bank, 1998); UNEP (1993); CO_2 data from US Department of Energy Climate Information Center (http://www.cdiac.esd.ornl.gov/).

as well as rapid increases in CFC use and production in the developing world. Under the Protocol, the developing world faces significantly looser restrictions than the developed world, at least for the time being.

B. Water Pollution

Water pollution has traditionally been the result of organic material deposited in waterways or lakes. Organic waste is problematic since it needs oxygen to decompose. Thus one of the measures of the quantity of pollution is biologic oxygen demand (BOD). Oxygen is of course needed for fish to survive. So if pollution has such a high BOD that the water body becomes depleted of oxygen, the water body can no longer support much in the way of life. In addition, with oxygen depleted, decomposition is now from anaerobic bacteria, which do not require oxygen. Such decomposition tends to be very odiferous. A major type of organic waste is of course human waste, which usually also involves significant human pathogens.

How serious is water pollution? The World Bank (1992) estimates that approximately one billion people are without access to safe drinking water, a fifth of the world's population. The World Bank also points out that life expectancy in cities in France went from about 35 years in 1850 to 45 years in 1900, with the timing of change closely corresponding to the introduction of modern water supply and waste water disposal. The importance of water supply and waste disposal is illustrated in Table 2.2. Note that for the ten countries shown, the child mortality is less correlated with income (GDP per capita) than with the population's access to safe drinking water and to proper sanitation services.

TABLE 2.2 GDP per Capita, Mortality, and Access to Clean Water and Sanitation for Selected Countries, 1991

Country	GDP per Capita[a]	Percentage of population with safe water[b]	Percentage of population with sanitation[c]	Child mortality rate[d]
Burkino Faso	945	65	11	159
India	1256	74	14	131
Ghana	1654	48	61	170
China	1901	74	87	43
Brazil	5534	88	73	69
Costa Rica	5758	92	96	16
Mexico	7773	83	67	39
Greece	11490	98	98	11
U.K.	17769	100	100	9
Denmark	20135	100	100	9

[a]*1995 U.S.$ per capita, using purchasing power parities to convert to 1987 U.S.$ and the US GDP deflator to convert to 1995 U.S.$.*

[b]*Percent of population with access to safe drinking water; weighted average of rural and urban.*

[c]*Percent of population with access to sanitation services; weighted average of rural and urban.*

[d]*Annual mortality per 1000 live births of children aged under five.*

Source: World Resources Institute (1994); World Bank Development Indicators Database (World Bank, 1998).

Water contaminated by pathogens is probably the primary environmental threat in much of the world.

One particularly difficult water pollution problem is pollution of groundwater. Groundwater, water in underground aquifers, is the source of drinking water for many people. Because of the cleansing ability of the earth above the aquifer, groundwater has traditionally been relatively contaminant free. However, groundwater contamination does occur, primarily from leaking storage facilities on the surface, either waste storage or storage of bulk liquids such as gasoline. In the past, chemical wastes have been dumped on the surface, finding their way into the groundwater many years later. Another source of groundwater pollution is the leaching of pesticides and fertilizers into the groundwater.

As point sources of water pollution are brought under control (something that is occurring in much of the developed world), the remaining water pollution problems are tougher to solve. Remaining pollution problems are in large part from area sources of pollution—urban runoff, agricultural runoff, as well as accidental spills.

Surface waters, such as lakes and rivers, are the ultimate repositories of much that is deposited on the land. And even the sulfur and nitrogen oxides that are emitted as air pollution are ultimately deposited on the land, either in dry form or as acidic precipitation. Lakes and rivers throughout the developed world suffer to varying degrees from excessive levels of nutrients (from agricultural runoff) and acidification. Nutrients can promote the growth of algae and phytoplankton, which increase the turbidity of the waters. Acidification has the potential of making lakes uninhabitable for most fish. Even the world's oceans are beginning to experience serious pollution, particularly smaller seas such as the Baltic and the Mediterranean.

C. Toxic Chemicals

Toxic chemicals in the environment have been a problem for decades. In fact, one of the books most important in coalescing the modern environmental movement was Rachel Carson's *Silent Spring*, a book about pesticides finding their way into the food chain and causing havoc with wild birds (among other species). The good news is that the most persistent pesticides have been banned for decades in much of the world and that many threatened species are making a comeback. The bad news is that there are still significant discharges of heavy metals and other toxics into the aquatic environment, even in developed countries.

Another serious toxic problem from the perspective of human health is lead. During this century, the most significant source of lead poisoning has been through air pollution, due to lead in motor fuels. Lead has been blamed for very serious health effects (such as mental retardation), particularly in poor urban children. Lead has been a major air pollutant because of its use as a performance enhancer in automobile fuels. Most of the developed world now significantly restricts or bans the use of lead in automobile fuel. The result of such a ban can be dramatic. According to the U.S. EPA, the average annual lead concentration in a number of U.S. cities in the late 1970's was over 1 $\mu g/m^3$. In 1990, the concentration had dropped to 0.07 $\mu g/m^3$, and the ban on lead in gasoline had still not become complete.

Aside from intentional discharges of toxics into the environment, there are two other major sources of toxics—old toxic waste sites and accidental discharges. Toxic wastes do not rapidly decay and become harmless. Some abandoned industrial sites from the nineteenth century are still considered chemically hazardous. There are many sites of nuclear and chemical wastes from the post-World War II but pre-1970s period. These can be very expensive to clean up. Accidental spills, including oil spills, and chemical spills, will always be with us. And these can be very serious sources of environmental damage.

D. Ecosystems and Natural Environments

Ecosystem health in many parts of the world has deteriorated, in large part because of the loss of habitat from an ever-expanding world population. Virgin forests are being cleared for the purpose of selling the standing timber as well as providing farm land. Wetlands are drained to obtain more land for agriculture and housing. The number of endangered species of plants and animals grows annually. The world has yet to fully come to grips with how to appreciate and protect important ecosystems.

II. ENVIRONMENTAL REGULATION

There is a surprising amount of similarity in how countries around the world have responded to environmental problems. It is beyond the scope of this book to provide a comprehensive review of international environmental regulations.[4] We will content ourselves with a very superficial examination of environmental protection regulation in the European Union, Russia, and the United States.

These reviews will focus on two issues. First, we hope to convey a sense of how environmental protection is pursued in each of these countries. Second, we will try to indicate the extent to which economic incentives have been used in these countries, particularly emission charges, in which polluters pay a fee per unit of emissions, or marketable permit systems in which polluters may buy and sell the right to pollute a given amount, thus imputing a value to reducing pollution.

A. European Union

1. Regulatory Approach. The confederation of European states that started in the 1950s as an organization to promote trade in coal and steel has developed into a federal system of independent countries that is continually ceding more and more power to the central government. As an example, monetary policy has traditionally been made on a country-by-country basis by central banks. However, beginning in 1999, most member countries surrendered monetary policy to a single European Bank, in order to unify European currencies. The balance of power between the European Union (EU) and its member states appears to be continually evolving. In this sense, the EU is very much a work in progress.

There are several basic principles that seem to permeate EU environmental policy. One is the principle of subsidiarity.[5] The principle leaves all powers to individual member states unless there is an abiding reason to take action at the European level. Thus most pollution control, at least of pollution that stays within a member state, would appear to be totally the responsibility of the member state, not a problem to be addressed at the Union level. However, because differences in pollution control regulations from one state to another can influence comparative advantage within the Union, the division of power is not quite so straightforward.[6] Another principle is the "polluter pays." This means that polluters should be required to pay for environmental damage, environmental controls, and the administration of environmental agencies.[7]

European Union policy with regard to the environment is implemented through the issuing of what are known as *directives*. A directive is an EU "law" that is binding on the governments of member countries, requiring each government to pass its own legislation to implement the directive. One of the most widely noticed directives (at least outside of the EU) was the 1988 directive that set forth sulfur emission controls for large electricity-generating stations. The impetus for EU level action was the problem of acid rain and acid deposition, which are pollution problems that do not respect national borders. The directive is very specific as to how much emissions may come from new power plants; in fact, the directive specified the pollution control technology to be used on most new power plants. An interesting part of the directive was the establishment of country caps on emissions of nitrogen oxides and sulfur oxides from pre-1987 power plants (this was to address acid rain and acid deposition). But as with all directives, member countries are required to pass laws to implement the directive. In the case of the cap, each country must determine the best way of controlling existing plants to meet the cap.

Because pollution control legislation in most member states dates from the early 1970s (and before), when there was considerably less coordination at the Union level, the approach to pollution control tends to vary from country to country. Rather than review the regulations of each country individually, we will briefly mention some of the approaches to regulating pollution used in one member country, the United Kingdom.[8]

Pollution regulation has a fairly long history in the United Kingdom. In 1273, the city of London passed a measure to control smoke.[8] Disputes regarding pollution were subject to nuisance law for centuries, and in the early nineteenth century, a parliamentary committee was convened to examine the problem of urban smoke. More problematic for many in the last century were the "noxious vapours" arising from chemical works, particularly the hydrochloric acid fumes given off in the manufacture of sodium carbonate. In 1863, the Alkali Act became law with the express purpose of controlling these emissions. The innovation of the Alkali Act was to set up the quaintly named Alkali Inspectorate, a set of experts with power to mandate emission control. The Alkali Inspectorate eventually metamorphosed into Her Majesty's Inspectorate of Pollution in the 1970s which was subsequently absorbed into the Department of the Environment, Transport and the Regions. The Alkali Inspectorate was only permitted to regulate certain "difficult" pollution processes—"scheduled processes". The list of scheduled processes started with hydrochloric acid from sodium carbonate manufacture but was slowly expanded over time. Eventually, in large part due to the deadly London smogs of 1952, nearly all potential sources of air pollution, including generators of smoke, came under the authority of the Alkali Inspectorate (in the 1956 Clean Air Act). Despite this, "routine" pollution problems were specifically delegated to local authorities. The idea was that when abatement presented no particular technical difficulties, responsibility was to be transferred from the Inspectorate to local authorities.

The Alkali Act set the stage for how pollution would be regulated in the UK. One of the important clauses in the Act as amended was that scheduled sources shall use the "best practicable means" (BPM) of pollution control. Here practicable means not only technically feasible but economical, including justifiable on the basis of the damage from the pollution. This approach has long been favored over specifying particular emission limits. The thought is that not only is BPM more flexible but that it is more likely to encourage innovation than a fixed emission limit. Thus the regulatory approach is that polluters must be registered and a requirement for registration is the use of BPM. The government has prepared "Notes" on what constitutes BPM for various industrial facilities. These guidelines are used in the source-by-source negotiating process over the required registration and re-registration process.

Although the UK is not a federal system, there are different levels of government, ranging from local to national. In the environmental area, there are clear distinctions regarding which responsibilities reside at the local level and which at the national level. As the discussion of the Alkali Inspectorate indicates, certain scheduled sources of pollution are subject to control at the national level. These are typically the sources of pollution that are the toughest to control. Small scale sources of pollution, including municipal solid waste, are typically the responsibility of local authorities.

Water pollution regulation is the responsibility of the National Rivers Authority, though this body has been merged into the Environment Agency within the Department of the Environment, Transport and the Regions. The regulatory approach is similar to that used for air pollution control in that polluters must register, and by doing so are subject to requirements regarding pollution control technology.

2. The Use of Economic Incentives. There is a longer history of the use of economic incentives in the EU than in most other parts of the world. In Europe, the emission fee is the predominant type of economic incentive.[9]

One distinction frequently made in discussing European emission charges is between revenue-raising and incentive-oriented charges. There are a great number of charges in most countries of the EU related to pollution discharges. However, most of these charges appear to be at too low a level to provide any serious incentives to reduce pollution. Their primary purpose is to generate funds to operate the administrative agencies that oversee pollution regulation.

One of the clearest economic incentives in use in the EU is the German Water Pollution charge, instituted in 1976 and implemented in 1981. The charge is established (including the level of the fee) at the federal (national) level. It is left up to the states of Germany (Lander) to implement and enforce the fee. However, it is unclear how much pollution has been reduced as a result of the fee (OECD, 1989). Germany also has a form of a marketable permit system for cities with pollution exceeding allowed limits. New firms are allowed into these cities if they can show that pollution levels will drop. In other words, new firms must find polluters and pay them to reduce their emissions in order to offset the new emissions.

Another charge that appears to have some incentive effects is the Dutch charge on the discharge of organic material into sewer systems. The problem was how to pay for the water treatment facilities necessary to improve river water quality. Somewhat logically, the parties using the system were charged a fee, and the fee was based on the load the sources would place on the treatment facility. Thus the legislative motivation appears to be simply revenue raising. However, subsequent analyses of the program suggest that the fee had a very significant effect in reducing organic material discharges from industry.[10]

B. The Russian Federation

1. Regulatory Approach. Russia is a country in transition and that applies as much to environmental regulation[11] as to any other part of the economy. Thus elements of the current Russian approach to environmental protection will undoubtedly change as the country's institutions adjust to a new economic environment. Furthermore, simply having a regulation on the books does not necessarily mean that the regulation is fully implemented and enforced.[12]

Much of the Russian approach to environmental protection has its origins in the Soviet Union. The 1970s was the beginning of significant Soviet actions to protect the environment. Both water pollution and air pollution controls were instituted in this period, along with the establishment of the country's basic framework of pollution regulation. The then Soviet and now Russian regulatory approach relies on health-based ambient standards for a wide variety of pollutants.[13] These standards are developed at the national level for use throughout the country. The standards are not absolute but depend on the nature of the use of the ambient environment. For instance, ambient standards for air depend on whether the point of application is a residential or an industrial area. Ambient standards for water depend on whether the water is used for recreation, fishing, or drinking.

Ambient standards are generic federal standards that apply throughout the country. They are used to establish source-specific emission limits. These emission limits are embodied in the pollution permits each significant source of pollution must possess. The procedure is that new sources of pollution must submit an application for a pollution permit containing a technical analysis of the effect of emissions on ambient concentrations, at the boundary of a specified zone around the source. Background pollution concentration levels must be taken into account.[14] This application is the basis for establishing an emission limit for the source. Thus this is a source-by-source procedure that is driven by ambient standards, not by a consideration of best available technology or costs. Sources that existed prior to the regulations must follow a similar procedure to obtain an emission limit as well as a permit.

As might be expected, some pollution sources have difficulty meeting the permitted emission limits. In these cases, temporary emission limits are issued, with the idea that over time, the temporary limits will become increasingly tighter until they equal the (nontemporary) emission limit. The rate at which these temporary limits are tightened is apparently somewhat negotiable.

The regulatory authority responsible for implementing these laws at the national level is the State Environmental Protection Committee.[15] In addition, there are nearly a hundred regional environmental protection committees, roughly one for every member of the Russian Federation. There are also local environmental protection committees in some of the major cities. These committees (primarily the regional committees) are responsible for implementing the environmental laws. Thus if a new source of pollution wishes a pollution permit, application is made to the regional environmental protection committee.[16]

One of the major differences between environmental regulation in Russia and the rest of the world is the use of the "environmental fund." The environmental fund is a major financial resource to be used to help clean up the environment and finance pollution control measures. Funds (pots of money) are associated with each level of environmental regulation—federal, regional, and local. The existence of such funds can probably be best understood in the context of the centrally planned Soviet Union. In the Soviet Union, the major stationary sources of pollution were state-owned enterprises. Resources were needed to pay for pollution control. Some of these resources were to come from the budgets of the enterprises but that was not considered to be sufficient. (For example, one might expect the Ministry of Steel to put environmental protection investments fairly low in its priorities for scarce investment funds.) Funds were also needed to clean up environmental accidents and other problems for which responsibility could not easily be assigned. The funds were also used to finance public environmental projects such as sewage and drinking water treatment. An interesting additional use of the funds was to pay compensation for environmental damage. In the post-Soviet era, the environmental funds have been a major source of funding for environmental protection. The funds are used for direct grants for pollution control as well as subsidized loans. An obvious question is why should a polluter pay to control pollution when a fund will pay for doing so? As might be expected, resources are limited in the funds. Polluters can make application to the fund for a direct grant or a loan or anything in between. The fund allocates resources on a case by case basis.

The focus in the above discussion is on air and water pollution from major sources. Automobile emissions are not regulated to any significant extent. Moscow bans leaded gasoline, but that is an exception. However, automobiles must undergo periodic checks on their emission rates for carbon monoxide. Nevertheless, it is not clear how strictly these limits are enforced.

2. The Use of Economic Incentives. Perhaps surprisingly, Russia makes extensive use of emission fees (as do many other countries of Eastern Europe and the former Soviet Union). In the late 1980s the Soviet Union was experimenting with emission fees and these were embraced by Russia in the early 1990s. The original intent of the emission fee was to finance the environmental funds. If the environmental funds were in turn to pay for environmental protection, it seemed logical that polluters should contribute to the fund based on their emission levels. Consequently, each source of pollution is required to pay a fee into the environmental funds per unit of emissions. Most of the payments go to local or regional funds. There are two levels of the fee, a base level, for emissions under the emission limit for the facility (described above), and a level five times greater for emissions over the limit. Although substantial revenues are collected with these fees (in 1993, nearly $100 million),[17] the individual fee levels are modest. In 1995 the sulfur dioxide base fee was 5610 Rubles (US$1.22) per ton.[18] This is significantly lower than the marginal cost of even modest pollution control, at least in the West. Several studies of the emission fee system report that the fee levels are too low to provide much of an incentive for pollution control.[19] Despite these findings, the fee system will be important to monitor as Russia's market economy evolves in the coming years.

C. United States of America

1. Regulatory Approach. The United States is a federal system with some powers vested and exercised at the national level and some powers residing at the level of the state. Although environmental protection legislation dates to the nineteenth century or before (as in many countries), substantive and major national environmental legislation was first passed in the 1960s. The post-World War II period saw rapid growth in the U.S. economy, along with increasing stress on the environment, increased per capita income, and thus increased demand for environmental quality. California took action in the early 1960s to establish emission standards for automobiles. This was quickly followed by national legislation requiring manufacturers to install certain antipollution devices on new cars and, eventually, to meet ever-tightening limits on emissions of particular pollutants per mile travelled.[20]

Several problems faced legislators in the 1960s in devising the appropriate way of controlling pollution emissions, primarily into the air and water. The problem was twofold: how to control the myriad of existing sources of pollution in a fair and equitable, but effective, manner and how to ensure that new sources of pollution (new factories, new cars) were constructed in an environmentally friendly manner. Because of the wide variety of local conditions involved in regulating existing sources of pollution, the approach taken was to let states decide how best to clean up sources of pollution that were in existence at the time environmental legislation was passed. The Federal government would promulgate goals for ambient environmental quality; states would construct "plans" to control polluters so that Federal goals could be met. These plans ("State Implementation Plans") were subject to approval by the Federal government. (Hazardous pollutants, such as chemical wastes and trace heavy metals, are regulated differently.) States are free to implement tighter standards than dictated by the Federal government.

New sources of pollution are regulated differently than old sources. One fear was that individual towns or states would relax environmental standards to attract business and jobs. To reduce this possibility, emission standards for new sources of pollution were set at the national level, without regard to location. The Environmental Protection Agency (EPA, established in 1970) would promulgate industry-by-industry emission control requirements for new sources. By doing this, concerns about environmental regulations would be eliminated from the factory location decision. The Water Quality Act of 1965 and the Clean Air Act Amendments of 1970 follow this approach to regulation. Over the years there have been some changes but the general pattern remains the same: ambient pollution goals are established at the national level, new sources of pollution are subject to nationally uniform emission standards, while local regulatory agencies oversee existing sources of pollution, with a significant amount of oversight from the national level.[21]

Land use is governed by local governments for some uses and by the Federal government for other uses. Agricultural use of land with problems of soil loss and pollution from runoff is governed by the Federal Agriculture Department. The Agriculture Department requires farmers who receive subsidies to file soil conservation plans. The conversion of wetlands to drylands is governed by a mix of Federal, state, and local laws. However, for the most part, local land use is controlled by local governments using local zoning laws, with little oversight from the Federal government, except for the Federal constitutional protection that private land may not be taken by the government without payment of compensation. For this reason, the extent to which environmental or aesthetic concerns enter into local land use decisions varies widely over the country. In recent years the Federal Endangered Species Act has occasionally had a major impact on some local land use. The Act protects species of animals and plants that the Federal government has declared endangered or threatened. In certain parts of the country this Act occasionally stands in the way of converting vacant land (habitat of endangered species) into housing tracts or commercial development. There is some debate over whether the Act's restrictions on private land constitute a "taking" that requires compensation.

Toxic wastes (also called hazardous wastes) might appear to include all pollutants; however the term is reserved for particularly hazardous pollutants that are dangerous in small doses. In the normal use of the term, pervasive pollutants such as sulfur dioxides, nitrogen dioxides, biological oxygen demand, fine airborne particulares, and tropospheric ozone are considered pollutants but not hazardous or toxic pollutants. Toxics are regulated at the Federal level in three different ways. Toxics in the workplace are regulated by the Occupational Safety and Health Administration. The generation, transportation, and disposal of toxic wastes are governed by the Environmental Protection Agency, the all-purpose Federal agency that promulgates and enforces environmental regulations. The applicable law is the Resource Conservation and Recovery Act (RCRA). A major component of the legislation is the requirement for tracking and accounting for wastes as they move through the economy, much as a registered letter would be tracked as it moves through the postal system. The Act also specifies technology standards to be used in disposing of toxic wastes. An omission from the legislation covering generation and disposal of toxic wastes is what to do when accidents occur or when old toxic waste sites are discovered. Yet another law, the Comprehensive Environmental Response, Compensation and Liability Act (CERCLA) of 1978, addressed these concerns. The law set forth liability rules apportioning responsibility for toxic waste leaks into the environment. The Act also set up a large fund (Superfund) to be used to clean up toxic waste sites immediately, without first having to seek out the responsible parties.

Pesticides are another major category of toxic substances, though strictly speaking they are not wastes but toxics that are intentionally introduced into the environment to control weeds, insects, rodents, fungus, or other pests. This does not include chemicals that are intentionally introduced into food (such as colorings), which are regulated separately by the Food and Drug Administration. Although there are several laws that regulate pesticides,[22] most pesticides (insecticides and herbicides) are controlled in a very simple and direct way by the EPA—the EPA simply certifies the label and instructions that are attached to the container before being sold. The label will be very specific—for instance, calling for use on strawberry plants at an application rate not to exceed *x* milliliters, diluted in *y* liters of water, per acre of crop. The label may also require application by certified personnel. Much of the EPA's oversight involves calculating how much of a pesticide residual will end up in the food supply, if the label directions are followed, and then how much a typical person will ingest, and whether that will consititute a health hazard.

A major gap in the previous discussion is who regulates the regulators. Specifically, when the Federal government takes actions that have environmental consequences, to whom are they accountable? In large part, it is the National Environmental Policy Act (NEPA) that applies in such cases. Whenever a project (such as building a dam) involves a Federal action, NEPA calls for the preparation of an Environmental Impact Statement (EIS), prior to the project being approved. An EIS reviews the environmental consequences of the project as well as the desirability of alternatives to the project or other steps that could ameliorate the environmental damage. Although strictly speaking this process is not a regulation, it tends to bring to public attention problems that may arise in pursuing a Federal action. If the consequences are severe enough, that is usually enough to stop the project, or cause it to be significantly modified.

2. Use of Economic Incentives. In 1990 a new market in sulfur emissions was established by legislation in an attempt to reduce sulfur emissions that lead to acid deposition (acid rain). At the time of the legislation, approximately 20 million tons of sulfur dioxide were emitted annually in the United States. The goal was to cut that in half. To accomplish that goal, a system of marketable emission permits was set up, phased in over a decade. This was to be the first major experiment in economic incentives in the United States. The market has been operating for several years now and appears to be a success. The price of sulfur (which can be equated to the marginal cost of control) has dropped from projected levels of over $500 per ton of sulfur dioxide prior to the market operating to as low as $65 per ton of sulfur dioxide after the market had been operating for a few years. Although there are many possible reasons for this price drop, it is a positive sign that control costs may be lower with the use of economic incentives. The apparent success of this experiment has led to other attempts to use market instruments for pollution control.

Although the sulfur permit market is the most significant experiment in economic incentives yet in the United States, there are other examples, most at the national level but some at the state or local level. One obvious type of economic incentive is the widespread adoption of volume-based pricing for municipal solid waste. It is now very common for households to pay by the "bag" for disposal services (see Fullerton and Kinnaman, 1996). This is in contrast to the widespread approach of a few decades ago of offering virtually unlimited household waste disposal for a fixed monthly fee. A very successful marketable permit system was that used to phase out lead in gasoline, starting in the early 1980s. The issue was how to apportion a shrinking overall cap on lead in motor fuel among different refiners. A system of tradable permits was used and appears to have worked quite well.[23] There have been a number of other emission trading schemes in the United States, most prominently the offset system whereby new polluters entering a heavily polluted area may pay existing polluters to reduce emissions. Robert Hahn (1989) has investigated many of these programs and has concluded that they have had negligible impact.

A relatively new program for trading emission in urban Los Angeles was started in 1994—the RECLAIM program.[24] This program was designed to reduce emissions from stationary sources of ozone precursors. Although the program has met with much criticism and redesign, it appears to be working quite well, at least based on the large number of trades that are taking place.

D. International Agreements

Not all environmental problems are national in scope. When more than one country is involved, the typical solution is a treaty. The two major transboundary air pollution problems are stratospheric ozone depletion and global warming due to greenhouse gases. The first of these is the subject of a treaty; the second is currently being hotly debated around the world.

The Montreal Protocol on Ozone Depleting Substances is an international agreement to control the use of refrigerants and other chemicals that tend to deplete the stratospheric ozone layer. The Protocol limits the production of such chemicals (though they may be recycled) and provides penalties for countries that violate the agreement. It appears to have been successful, at least so far. Prices of banned chemicals have risen dramatically in resale markets, suggesting they are much scarcer than prior to the Protocol.[25]

In December 1997, most countries of the world met in Kyoto, Japan, to negotiate what has become the Kyoto Protocol to control emissions of gases leading to climate change. The agreement calls for stabalizing world emissions by 2010 at roughly 5% below 1990 emissions. Whether the Protocol will be ratified by individual countries or modified prior to ratification is not known. The issue is still very much alive and unresolved.[26]

Another significant international treaty is the 1973 Convention on International Trade in Endangered Species (CITES), an agreement to limit international trade in rare animals as well as trade in animal parts such as elephant tusks or obscure parts of tigers or rhinos that are thought by some to have medicinal value.

There are truly a host of other treaties that can be classified as providing environmental protection.[27] These would include agreements to limit whaling, limit certain types of fishing practices, protect migrating birds, and protect the oceans from pollution. The list is very long. The list will undoubtedly get longer as more international environmental problems become the focus of international protection.

III. THE COSTS OF ENVIRONMENTAL PROTECTION

As with most things that are worthwhile, environmental protection costs money. Although cost information with global coverage is scarce, we do have some country-specific information on the costs of protection, particularly for the United States. Total expenditures in the United States on pollution abatement and control by industry in the United States are approximately 2% of gross domestic product (GDP), up modestly as a fraction of GDP from expenditures in the early 1970s. These are direct costs to industry from pollution control regulations, as reported to the Census Bureau. Many components of cost are potentially omitted from such a measure, such as process changes that are induced by environmental regulations; the cost of operating the bureaucracy that administers environmental laws in the United States; indirect costs such as the costs of pollution control on automobiles or the costs associated with having to use products that are environmentally friendly but of inferior quality in other respects (e.g., bans on oil-based paints); and the cost to society of the residual damage from the pollutants that are emitted, even after controls have been put in place.[28] Nevertheless, the cost of a few percent of GDP puts a correct perspective on environmental protection: it is an important cost, ranging into the hundreds of billions of dollars per year to a country such as the United States, but still relatively modest compared with the costs of other socially desirable activities (health care is estimated to be as much as 15% of GDP in the United States).

Specific industries are of course impacted harder than others. Table 2.3 indicates how pollution control costs impact specific industries in the United States. Total expenditures is not the correct measure of the impact of environmental regulations; a better measure is marginal costs or costs per unit of output. As can be seen from Table 2.3, as a percentage of the value of output, the petroleum and coal products industry is the hardest hit, but even in that case, gross abatement costs are less than 2% of the value of shipments from the industry. The chemical industry is the second hardest hit. Figures on capital expenditures tell a similar story. The petroleum and coal products industry devotes nearly 25% of its capital expenditures to pollution abatement. It is important to realize that it is not easy to measure expenditures on pollution control, and for that reason the figures in Table 2.3 are suspect. If a manufacturing process has been changed to reduce emissions, what costs are apportioned to pollution control? When a capital expenditure to remove pollutants reduces costs elsewhere in the production process, how much of the capital cost should be attributed to pollution control? The measurement problem is a difficult one in tracking the costs of environmental protection. Furthermore, the figures in Table 2.3 are for the United States; however, the figures undoubtedly are representative of the difficulty these industries have, wherever they may be located, in meeting local environmental regulations.

TABLE 2.3 Pollution Abatement Expenditures for Selected U.S. Industries, 1991[a]

Industry	Total capital expenditures	Pollution abatement capital expenditures (PACE)	PACE as percentage of capital expenditures	Total value of shipments	Abatement gross annual cost (GAC)	GAC as percentage of value of shipments
All industries	101,773	7,603	7.47	2,907,848	17,888	0.62
Industries with high abatement costs						
Paper and allied products	9,269	1,269	13.68	132,545	1,682	1.27
Chemical and allied products	16,471	2,126	12.91	300,770	4,164	1.38
Petroleum and coal products	6,066	1,505	24.81	162,642	2,931	1.80
Primary metal industries	6,049	692	11.45	136,674	2,061	1.51
Industries with moderate abatement costs						
Furniture and fixtures	750	25	3.29	41,183	140	0.34
Fabricated metal products	4,190	182	4.35	161,614	867	0.54
Electric, electronic equipment	8,356	241	2.88	203,596	857	0.42
Industries with low abatement costs						
Printing and publishing	5,187	38	0.73	161,211	235	0.15
Rubber, miscellaneous plastics products	4,337	84	1.95	103,576	454	0.44
Machinery, except electrical	7,546	132	1.75	250,512	591	0.24

[a]*Units: millions of 1992 U.S. dollars.*

Source: Jaffe et al. (1995).

TABLE 2.4 Pollution Abatement Expenditures for Selected Countries, as a Percentage of GDP

	1981	1982	1983	1984	1985	1986	1987	1988	1989	1990
United States	1.5	1.5	1.5	1.4	1.4	1.4	1.4	1.3	1.4	1.4
France	0.9	0.9	0.9	0.8	0.9	0.8	1.0	1.0	1.0	1.0
West Germany	1.5	1.5	1.4	1.4	1.5	1.5	1.6	1.6	1.6	1.6
Netherlands	—	1.2	—	—	1.3	1.5	1.5	—	1.5	—
United Kingdom	1.6	—	—	—	1.3	1.3	—	—	—	1.5

Source: Jaffe et al. (1995).

Having identified which industries are the most polluting, it becomes necessary to determine the total costs of control over all industries. Doing this is not particularly easy. Table 2.4 shows an estimate of pollution control expenditures for several European countries and the United States for the 1981–1990 period. Most importantly, the table indicates that pollution abatement is consistently between 1 and 2% of GDP in developed economies. One issue in discussing figures such as these is determining just what is included in pollution control expenditures. Do we include the cost of pollution control devices on automobiles? Typically, yes. Do we include the cost of reduced performance in automobiles due to pollution control regulations? Typically, no, though we should. Of the several estimates of the cost of pollution control, most place the figure at less than 3% of GDP. But this of course excludes the damage to the environment—these are only costs of control.

How are these costs distributed among types of environmental problems? Of the $141 billion the EPA estimates was spent on environmental protection in the United States in 1992 (2.3% of GDP), approximately 25% was expended on air pollution control, 40% on water pollution control, and the remaining 35% on land, hazardous wastes, and miscellaneous controls.[29]

It is important to realize that the reported costs of pollution control regulations reflect two things: the cost of pollution control and the unnecessary costs of inefficient regulations. Thus while pollution control regulations may cost 3% of GDP in the United States, it may be that a much higher level of environmental quality could be achieved for that same cost. It is not really known to what extent accounting costs of pollution control regulations exceed the lowest possible cost of providing the same amount of environmental quality.

IV. REMAINING PROBLEMS

Though great progress has been made worldwide over the past few decades, solving environmental problems will be a challenge that will face the world for many decades to come.

Many of tomorrow's environmental problems will arise from a significantly larger global population and a significantly richer population (per capita) than today. More people tends to be unequivocally bad for the environment, though how bad is unclear. A richer population has mixed implications for the environment. On the one hand, people will desire more of the world's resources. On the other hand, people will also demand higher levels of environmental quality and be willing to divert their incomes into environmental protection.

Global warming is a problem we will be dealing with for many decades to come. Even if an international treaty is ratified to curtail emissions of greenhouse gases, the threat of climate change will remain with us for quite some time.

Congestion associated with the motor car has yet to be adequately addressed in the world. In most cities, drivers put up with a significant waste of time, resources, and clean air, all due to road congestion. Europe has tried using extremely high taxes on automobile fuel but that has not solved the congestion problem (though it would undoubtedly be worse without the high taxes).

One of the remarkable trends of this century is the movement of more and more of the earth's natural resources from almost infinite abundance to scarcity. Vast stretches of formerly isolated forest ecosystems have been tamed by humans. Even the oceans, that almost limitless supply of water than could bounce back from any environmental insult, are being taxed by pollution and overfishing.

The cost of environmental regulation has yet to be faced seriously as a problem. Many view environmental protection as an absolute, to be provided at any cost. To the extent that such views dominate public policy, cost will take a back seat in debates over solving environmental problems. But the fact is that as environmental protection takes up more and more of a country's national income, the time will come when cost will matter. It will become necessary to ask if there are cheaper ways of providing the same level of environmental protection. If cost savings exist, they may be used to provide further environmental protection or to fund other worthy goals. When that day comes (and there is some evidence that it is here already), the efficiency of environmental regulation will be an important issue.

NOTES

1. World Resources Institute (1998) provides a good biennial review of the global environment. World Bank (1992) is a particularly good review of the environment in developing countries. The annual report of the U.S. Council on Environmental Quality (1997) provides a good picture of the environment in the United States. Information on the state of the environment in Europe is covered by Stanners and Bourdeau (1995). The information cited in this section is drawn in large part from these sources.
2. See Schoenbaum and Rosenberg (1991).
3. As always, there are exceptions. The Sierra Club, a major U.S. environmental advocacy group, was founded in California in 1892 to promote the protection of the Sierra Nevada mountains through the establishment of National parks, particularly Yosemite. In England, the National Smoke Abatement Institution was established in 1882 to lobby for pollution control in cities (Ashby and Anderson, 1981). There are undoubtedly other organizations in other parts of the world, established well before the current interest in environmental protection. Nevertheless, the bulk of environmental organizations in existence today date from recent times.
4. To my knowledge, there is no one source providing an overview of environmental regulations worldwide. The Environmental Law portion of the International Encyclopaedia of Law (Kluwer

Law International, Cambridge, MA) provides country-by-country assessments of environmental law for many countries, though it is written for lawyers rather than economists. Portney and Stavins (forthcoming) provide a good review for the United States. The World Bank website on pollution regulation (www.worldbank.org/nipr) covers many developing countries.

5. Lenaerts (1995) defines and discusses subsidiarity in the context of environmental regulation. Angelini (1993) provides a useful glossary of terms relevant to EU environmental policy.
6. Ludwig Krämer, the former head of legal affairs within the Environmental part of the European Commission, has written a very good overview (current up to the 1990s) of the EU approach to environmental protection (Krämer, 1992).
7. The "polluter pays" principle is not always followed in the EU. As Krämer (1992) points out, there are many instances of government subsidies for pollution cleanup or financing pollution control investments.
8. Linklaters and Paines Solicitors (1992) provide a detailed review of UK environmental regulations. The section draws heavily on that source as well as the excellent history of air pollution in the UK by Ashby and Anderson (1981).
9. The Organization for Economic Cooperation and Development (OECD) has conducted several surveys of the use of economic incentives among member countries (most of the developed world). Refer to OECD (1989, 1994a,b, 1995).
10. Huppes and Kagan (1989) argue that the fee had real incentive effects in addition to raising revenue.
11. There is not a great deal of information on Russian environmental regulation that is available in English, although excellent references are Averchenkov et al. (1995) and Kozeltsev and Markandya (1997). The discussion draws in part on these sources.
12. Some fairly dismal pictures have been painted of environmental quality in the former Soviet Union. See Feshbach and Friendly (1992) for one perspective.
13. There are some 1200 air pollutants for which there is an ambient standard—far wider coverage than in most if not all other countries.
14. Though there is always some latitude in conducting an analysis of the effect of emissions on ambient concentrations, the procedures for making such a calculation are indicated in some detail by federal authorities.
15. The Federal environmental agency fluctuates from being a ministerial level organization (Ministry of the Environment) and a somewhat lower level organization (State Environmental Protection Committee).
16. This organizational structure was established in the late 1980s, though there were predecessor organizations.
17. Enforcement has been a major problem with the fee system. Kozeltsev and Markandya (1997) report that in 1995 the Ministry of the Environment collected only half of the 600 billion Rubles (US$130 million) it expected from the fees.
18. Other fees are higher. The fee for lead emissions was approximately US$200 per ton in 1995 for allowed emissions and the fee for allowed emissions of benzopyrene is over US$60,000 per ton (Kozeltsev and Markandya, 1997).
19. Kozeltsev and Markandya (1997) cite several studies on the incentive effects of Russian emission fees. In one, it was estimated that charges would need to be quadrupled to provide an incentive to reduce pollution. They also note that firms may refuse to pay the charges if they become significant; many companies are in tax arrears to the government and one more debt may not make much difference.
20. An excellent and readable review of U.S. environmental regulation is Portney and Stavins (forthcoming).
21. The 1990 Clean Air Act Ammendments established a new way of regulating sulfur emissions, using tradeable emission permits (known as "allowances").
22. Historically, the two main laws covering toxic chemical substances introduced into the environment are the Toxic Substances Control Act of 1976 (TSCA) and the Federal Insecticide, Fungicide and Rotenticide Act of 1947 (FIFRA).
23. See Hahn (1989) and Hahn and Hester (1989) for information on the operation of this market.
24. Polesetsky (1995) provides an early description of how the program works.

25. Murdoch and Sandler (1997) discuss the economics of the Montreal Protocol.
26. There are numerous references to the problems of climate change. The "official" international body that is studying the problem, the Intergovernmental Panel on Climate Change (IPCC), issued a fairly comprehenisve survey report on some of the economic and social dimensions of the problem. See also Nordhaus (1994).
27. Barrett (1991) presents a detailed list of international environmental agreements.
28. Weitzman (1994) suggests that the total social costs may be two to three times these direct costs.
29. These figures are taken from the United States Environmental Protection Agency (1990).

3 SOCIAL CHOICE: HOW MUCH ENVIRONMENTAL PROTECTION?

There are two basic questions in environmental policy: "What is the right balance between environmental protection and use?" and "How do we induce economic agents to use the environment in a fashion that we have determined is desirable?" The first of these questions is the concern of this and the next few chapters.

The normative question of how much environmental protection we should have is fundamentally a social and political question: How much land should be set aside to protect endangered species of plants and animals? Should a scenic canyon be dammed to provide electricity and irrigation water? What level of urban ozone is acceptable, given the costs of control? These are similar to other fundamental questions facing society such as how much income support should be provided to the less fortunate or how much of the tax burden should be borne by the wealthy? As with many social questions, there is no unanimity within society about the best course of action. Some argue that species extinction should be avoided at all costs. Others maintain that species should be protected but not all species are equally important. Some argue that the benefits of species preservation should be balanced against the costs in terms of lost land or lumber. Still others could not care less about species preservation.

It is important to realize that when we make decisions about protecting the environment, we are making *societal* decisions. Basically, a decision to exploit or protect the environment affects a large group of people. Within this large group of people there is a variety of opinion about the best course to take vis-à-vis environmental protection/exploitation. How can we decide what action is socially desirable? Whatever action we take, some will be pleased and others disappointed. Thus the first problem we have is to translate diverse individual preferences regarding the environment into a group or societal choice.

For individuals, there are a variety of philosophical positions on environmental protection, ranging from biocentrism to sustainability to anthropocentrism and beyond. We will discuss these later in this chapter. But it is very important to distinguish between philosophical views of the role of natural environments in human existence and methods for making actual societal decisions about regulations or projects that have environmental consequences (good or bad). It is one thing to believe that no action should be taken that is not "sustainable" in the long run. It is another to try to decide whether a particular open parcel of land should be set aside as an open space for species preservation or developed for housing.

To summarize, individuals in society have widely differing opinions and views regarding the answer to our question on the right amount of environmental protection. We assume that everyone is entitled to their opinion, no matter how perverse it may appear to be. Our main problem is not in understanding why people feel the way they do. Rather, we start from individual preferences and ask how social or group decisions can be made. Our focus is on developing methods for helping to make specific societal decisions, such as whether to develop a piece of open land.

I. THREE EXAMPLES

It always helps to be concrete when discussing different ways of deciding about environmental protection. Consider three simple examples.

The first example concerns the control of air pollution in a large urban area such as Santiago, Chile. Santiago, like many urban areas around the world, suffers from severe air pollution.[1] This pollution originates from a variety of sources but particularly from diesel buses and automobiles. The pollution can be cleaned up but only at significant cost. If pollution from buses is to be reduced and bus owners are to pay for the clean-up, then the cost of transportation in the city will rise, affecting those who depend on public transportation. The wealthy are less likely to use the buses and may be particularly bothered by the pollution. On the other hand, the poor may be less concerned with pollution, having enough trouble getting food on the table. Furthermore, they are particularly likely to feel the impact of increased transportation costs. And, they may be less concerned with getting rid of the air pollution. So what should be done? If we vote, the poor are likely to have the majority and only modest pollution control will result. If we look at how much people would be willing to pay to reduce the pollution, the rich will count much more since they have more resources. In this case, the outcome may be quite different. Which is correct? Which approach should be used for making social decisions about pollution control?

Another example concerns the California gnatcatcher, a bird native to the coastal areas of Southern California and also an endangered species.[2] Coastal areas of Southern California are also favorite nesting places of people and the land is consequently highly valued. If the gnatcatcher's habitat is eliminated, the species may become extinct. To protect the bird, a significant amount of resources will have to be devoted to habitat preservation. What should be done? How to decide? If we vote, who should vote? The country? The state of California? Residents of coastal southern California? The world? Clearly the outcome of the vote depends on how the cost of habitat preservation is distributed among the population. And what about future generations? Surely they have a stake in the extinction of a species. How can we make their vote count?

A third example is the Three Gorges dam in China (Zich, 1997). This enormous dam is currently under construction. When completed, it will tame the Yangtze river, generating a reservoir over 500 km in length. On the positive side, the dam will generate a great deal of electricity and provide needed flood control. On the negative side, the dam will destroy ancient cities and inundate spectacular canyons; these are essentially irreversible consequences. Furthermore, there is always the possibility of spectacular failure of the dam. The social choice problem is whether to build the dam. Not surprisingly, opinions are mixed. But only one decision can be made. This is not an easy social choice problem.[3]

These three examples illustrate three different types of tough environmental policy problems. The first case, pollution in Santiago, is in some ways the easiest to solve. All of the affected parties presently reside in Santiago, in contrast to the other two examples in which substantial intertemporal issues are involved. What makes the Santiago example tough are the distributional issues. One person–one vote results in one decision regarding pollution; a decision based on willingness to commit resources to clean up pollution will yield a different outcome, one in which the wealthy have a larger voice. This in fact characterizes the public policy dilemma at the core of many pollution control problems in cities worldwide.

The other two examples involve decisions with long-term implications. The basic problem is that many of the people who will be affected by these decisions have not yet been born and thus are unable to voice their opinions. The example of the gnatcatcher is tough because of the moral implications (to some) of destroying a species. This is clearly an irreversible action. A decision still needs to be made, however. The example of the Three Gorges dam has very long-term implications, though it is more an issue of trading gains to the present generation for losses to future generations, which will no longer be able to experience the undamed Yangtze.

II. INDIVIDUAL PREFERENCES REGARDING ENVIRONMENTAL PROTECTION

Before examining the question of how society can make decisions about the environment, it is useful to understand the breadth of individual preferences regarding environmental protection. Understanding individual preferences is somewhat unusual in economics. Our normal approach is to assume that any preferences are possible and not be too concerned about the nature of individual preferences. Although this is certainly just as true with regard to the environment, it is useful to understand the very different views individuals hold with regard to environmental protection.

The field of environmental ethics recognizes a variety of perspectives on what is "right" regarding environmental protection.[4] This is a complex subject and we make no pretense of adequately covering the subject here. We have chosen three very different philosophical perspectives, biocentrism, sustainability, and anthropocentrism, to illustrate different ethical perspectives. We do not claim to be complete in representing environmental ethics. In fact there are many other ways of viewing inidividual attitudes regarding the environment. Our goal is to make the reader aware that intelligent and thoughtful people can have *very different beliefs* about environmental protection. Keep in mind that we are referring to how individuals think. Combining individual beliefs into societal beliefs is a separate issue.

A. Biocentrism

Biocentrism encompasses a number of different philosophical view of humans and the environment (Nash, 1989). As the word suggests, biocentrism places the biologic world (primarily nonhumans) at the center of its value system.[5] An important distinction is made in biocentrism between instrumental value and intrinsic value. Instrumental value pertains to usefulness. Something has instrumental value if it can serve as an instrument in achieving some useful objective. In contrast, intrinsic value does not pertain to usefulness. Something can be totally useless and still have intrinsic value. Consider the smallpox virus. The instrumental value of that species is generally considered to be negative—it is not useful to any living creature and for humans it is very unuseful. Yet it can be argued from a moral perspective that life itself is intrinsically valuable and thus the smallpox virus has value for that reason. To bring the example into the human sphere, consider the case of a convicted mass murderer. Most would not dispute the fact that the instrumental value of that person's life is small or nonexistent. Yet many would argue that instrumental value is not the issue; human life has intrinsic value, regardless of the deeds or actions of any particular individual.

Biocentrism is the philosophy that all living things have intrinsic value, regardless of their instrumental value. The animal rights movement is an example of biocentrism, where the intrinsic value of an animal's life is recognized, just as most people recognize the intrinsic value of human life. Interestingly, biocentrism is in conflict with some ecologically oriented views of the environment. Aldo Leopold for instance (as we discuss in the next section) viewed ecosystems as intrinsically valuable but not individual members of the ecosystem. Thus, hunting would typically be anathema to a biocentrist but for Leopold it would be perfectly acceptable, and possibly desirable to reduce overpopulation.

B. Sustainability

Sustainability, broadly defined, could be considered a recent manifestation of Aldo Leopold's "Land Ethic." Half a century ago Leopold (1949) argued that it is the health of ecosystems that is of paramount importance: an environmental policy is right if it preserves the integrity of an ecosystem and wrong if it does not. This philosophy is perfectly consistent with natural resource use, provided the use does not degrade the ecosystem. Thus fishing is acceptable but overfishing is not. Logging is acceptable, provided the long-term health of the forest ecosystem is not jeopardized. This view is the logical predecessor to what today is called sustainability—use the environment for human needs only to the extent that the long-term health of the environment is not jeopardized.

Sustainability has been an exceedingly popular word over the past few years when it comes to human interactions with the environment. Probably one reason for its popularity is that it sounds good but is difficult to define precisely. For this reason, traditional economists such as Robert Solow have embraced the notion as a concept that economists have been talking about for decades in the context of economic growth and the environment. At the same time, ecologists such as Robert Constanza and environmental activists have embraced the concept as a fresh alternative to blind economic growth (Costanza, 1991). If the term were defined more precisely, then undoubtedly more people would be taking positions on opposite sides of the issue.

One of the best known definitions of sustainability was generated by the "Brundtland Commisssion," the World Commission on Environment and Development, headed by the former Prime Minister of Norway. As quoted by Toman (1994), the commission defined sustainability as "development that meets the needs of the present without compromising the ability of future generations to meet their own needs." The debate over sustainability has focused on two key aspects: (1) the degree to which "natural capital" can be viably replaced by human capital and (2) the obligation the present generation owes to future generations.[6]

Nobel Laureate Robert Solow defines sustainability as making sure the next generation is as well off as the current generation and ensuring that this continues for all time (Solow, 1992). Key to this view is that man-made capital (machines, buildings) and knowledge are substitutes for natural capital, particularly natural resources. As we deplete the energy resources of the world, we invent ways of getting along with less energy and build machines that reduce energy use or extract energy from the sun.

There is significant justification in the historical record for this position. Barnett and Morse's classic study of a century of natural resource exploitation demonstrated that with the exception of timber, resources have been getting more plentiful (Barnett and Morse, 1963). This has occurred not because more oil has been produced by nature but because the technology of extraction and use has been advancing more rapidly than depletion.[7] Many proponents of rapid introduction of energy conservation and renewable energy sources have as their goal the substitution of man-made capital in the form of solar energy converters for natural energy resources such as oil, gas, and coal. We are also reminded of the bet between the late economist Julian Simon and the ecologist Paul Ehrlich. In the early 1980s Simon challenged Ehrlich to name several commodities he thought would be more scarce a decade hence (based on their price). When it came time to call the bet in, Simon had won hands down—all of the commodities were cheaper than they had been a decade earlier.[8]

Certainly the historic record supports Solow's position on sustainability. The past, however, may or may not be a good predictor of the future. In fact, some critics of this view of sustainability point out that the one resource that Barnett and Morse found was getting scarcer (forests) was closest to what we now call natural environments.[9]

The economist Talbot Page of Brown University defines sustainability as managing depletion, pollution, and congestion (Page, 1991). For depletion, he calls for keeping broad aggregates of resources constant from one generation to the next, for example, energy, metals, wood, soils, and water. For pollution, he sees a need to balance pollution generation with the waste assimilation capacity of the environment. For congestion, he primarily means overpopulation, but the concept goes further than that.

An ecologist's views of sustainability may differ from Page's in detail but not in spirit. The idea is that we have a God-given endowment of resources, including species and natural resources, and that we have no right to squander that endowment for short-term gain, particularly when we do not really understand the implications of our actions. Don't fool with Mother Nature!

C. Anthropocentrism

The third view of the environment is that it is there for only one purpose: to provide material gratification to humans.[10]

It is important to distinguish anthropocentrism from utilitarianism, at least as we use the terms here.[11] Although these terms are sometimes used synonymously, there is a distinction. Utilitarianism with regard to the environment emphasizes the well-being people attain from the environment, whether it be materialistic or spiritual, instrumental or intrinsic. The strictest definition of anthropocentrism places only instrumental value on the environment. If the California gnatcatcher is of no use to humans, then it has no value. Utilitarianism on the other hand would admit individual value, no matter what the source of that value. If someone obtains a warm and fuzzy feeling from knowing the gnatcatcher is alive and well, that is a legitimate utilitarian value. Many of us have an image of the "untouched" rain forests of Africa, South America, or Southeast Asia. Despite the fact that few of us will ever visit those rain forests or benefit in a material way from them, we are enriched individually by knowing they exist, at least for the time being. Thus the standing forests may have utilitarian value but little instrumental value.

III. SOCIAL CHOICE FROM INDIVIDUAL VALUES

In the previous section, we discussed many perspectives on the environment, ranging from biocentrism to anthropocentrism. Since our goal is to generate methods for making decisions about specific regulations or projects that have environmental impact, we move away from generality. In particular, we will construct different methods of making decisions based on values *individuals* place on environmental protection. These individuals may be biocentrists or selfish materialists. We make no restriction on their individual preferences.

A. The Utility Function

We start from a simple microeconomic view of the environment. Let there be N people in our society, and we will index them by $i = 1, \ldots, N$. We assume one composite material good, called x. Let $\mathbf{x} = (x_1, \ldots, x_N)$ represent individual consumption of the material good and e represent the quality of the environment (assumed to be the same for everyone). In other words, $\mathbf{x}$ is a vector representing how much of the material good each person consumes: person i consumes x_i. Define the utility or well-being obtained by individual i from a bundle of material and environmental goods, $(\mathbf{x},e)$, as $U_i(\mathbf{x},e)$.[12]

This is a standard economic perspective on utility. Individuals have preferences that are represented by their utility function. When faced with decisions about trading-off $\mathbf{x}$ and e, they will attempt to maximize utility. Pure biocentrists have utility functions that permit no substitution of $\mathbf{x}$ for e. An extreme anthropocentrist may have a utility function that allows no substitution of e for $\mathbf{x}$. Figure 3.1 shows a set of indifference curves for which the environment becomes more and more important as consumption of $\mathbf{x}$ increases. For low levels of $\mathbf{x}$, environmental quality must increase substantially to offset a loss in material goods. For higher levels of $\mathbf{x}$, a substantial increase in material goods is necessary to offset a small loss in environmental quality. Those who believe strongly in sustainability might have the utility of future generations as an explicit argument of their utility function.

In many cases of environmental protection, it is future generations who benefit from

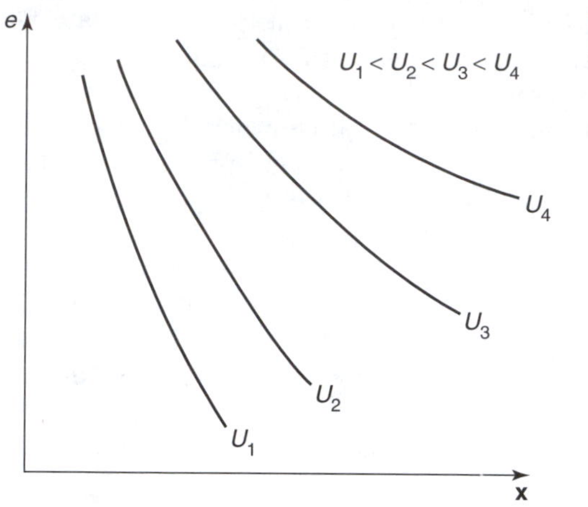

Figure 3.1 Possible sets of indifference curves.

our actions today. For instance, in controlling carbon dioxide emissions to reduce global warming, the primary benefactors are citizens of the world in 2100 and beyond. Concern about others, including future generations is known as altruism or the bequest motive—we have a desire to pass a quality environment on to the next generation. How can this be represented in our utilitarian framework? Simply by including the utility of others in the utility function: $U_i(\mathbf{x},e,U_j)$. In this case, individual i cares about his or her own consumption of $\mathbf{x}$ and e as well as the utility of person j, derived by person j's consumption of private and environmental goods, as well as the utility of people about whom j cares.

B. Social Choice Mechanisms

Having specified how strongly individual people in society feel about the environment vs. material goods, we turn to the question of how to make social decisions. Suppose we have two "bundles" of material goods and environmental goods: $(\mathbf{x}',e') = (x_1', \ldots x_N',e')$ and $(\mathbf{x}'',e'') = (x_1'', \ldots x_N'',e'')$. For instance, $(\mathbf{x}',e')$ may involve much more material goods and less environmental quality than $(\mathbf{x}'',e'')$. As is typical, material goods consumption may vary from person to person but everyone consumes the same amount of the environmental good. The question is, which should society choose—$(\mathbf{x}',e')$ or $(\mathbf{x}'',e'')$?

This is the fundamental question of social choice. We know how individuals feel about the preferability of $(\mathbf{x}',e')$ and $(\mathbf{x}'',e'')$. But which does society prefer? What this amounts to is generating a set of societal preferences over different bundles, given individual preferences over those same bundles. A "social welfare function" analogous to a utility function for an individual might also be generated. The social welfare function would give the social utility—welfare—from a particular bundle. The desirability of two bundles could be compared by comparing the welfare each gives. The primary problem is not in constructing a social welfare function to represent group preferences, but in constructing social or group preferences out of individual preferences.

We will consider several approaches to making social decisions. The first is called the Pareto criterion, which basically amounts to unanimous voting. A variant of the Pareto criterion is called the compensation principle. We then consider alternate forms of voting, such as majority rule, and the generalization of the social welfare function.

1. Pareto Criterion. The starting point for evaluating the social preference for $(\mathbf{x}',e')$ vs. $(\mathbf{x}'',e'')$ is to compare $U_i(x_i',e')$ with $U_i(x_i'',e'')$ for each individual i. If *all* individuals prefer one of these alternatives, we may conclude that society, too, should prefer that alternative. This is the Pareto criterion for social choice, named after the Italian economist Vilfredo Pareto (1848–1923). The Pareto criterion is a relatively noncontroversial way of concluding that one societal bundle is better than another. We can summarize this in a definition:

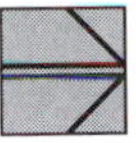

Definition. *Start with two consumption bundles,* $\mathbf{a}' = (\mathbf{x}',e')$ *and* $\mathbf{a}'' = (\mathbf{x}'',e'')$ *and a group of people* $i = 1, \ldots, N$ *with utility functions,* U_i, *defined over the consumption bundles. Then for the group as a whole,* $\mathbf{a}'$ *is* Pareto preferred *to* $\mathbf{a}''$ *if for every individual* i, $U_i(\mathbf{a}') \geq U_i(\mathbf{a}'')$, *and for at least one individual,* $\hat{i}$, $U_{\hat{i}}(\mathbf{a}') > U_{\hat{i}}(\mathbf{a}'')$.

What this says is that for an allocation $\mathbf{a}'$ to be better than $\mathbf{a}''$, everybody has to be at least as well-off with $\mathbf{a}'$ (compared to $\mathbf{a}''$) and at least one person has to be better off with $\mathbf{a}'$. Think of all members of society voting about whether they prefer $\mathbf{a}'$ to $\mathbf{a}''$, casting a "yes," "no," or "I don't care" vote. All it takes is one "no" vote to conclude that $\mathbf{a}'$ is not Pareto preferred to $\mathbf{a}''$. Thus leaving aside the individuals who are indifferent between $\mathbf{a}'$ and $\mathbf{a}''$ ("don't care"), the Pareto criterion is unanimity—everyone must prefer $\mathbf{a}'$ to $\mathbf{a}''$ for $\mathbf{a}'$ to be Pareto preferred to $\mathbf{a}''$.

We can look at this graphically. Suppose we have a society consisting of two individuals, Anna and Brewster. Figure 3.2 shows possible utility combinations that are attainable for Anna and Brewster. Resources are limited, so only the shaded area is attainable. Think of an allocation as a bundle of goods for each of our two members of society. Each allocation will yield different utility levels, indicated by different points in Figure 3.2.

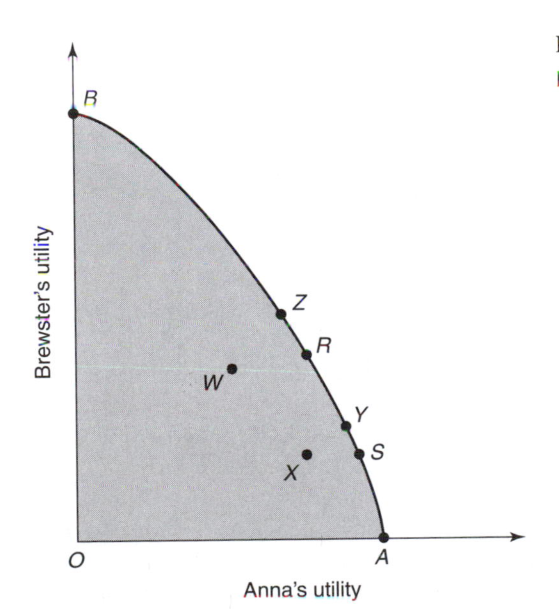

Figure 3.2 Utility possibilities for two-person society.

In Figure 3.2, compare points *W* and *Z*. *Z* is clearly Pareto preferred to *W* because the utilities of both Anna and Brewster are higher at *Z* than at *W*. The society of two will unanimously vote on *Z* being better than *W*. The same is true for allocations *X* and *Y*. Anna and Brewster will unanimously prefer *Y* to *X*; thus *Y* is Pareto preferred to *X*.

The strength of the Pareto criterion is that it is hard to fault. If *Z* is Pareto preferred to *W*, then *nobody* would rather have *W* than *Z*. In this case *Z* is unequivocally better socially than *W*, no matter what your social philosophy might be. The weakness of the Pareto criterion is that many allocations, in fact most allocations, cannot be compared. In comparing two arbitrary allocations, the most common situation is that some members of society will prefer one and some will prefer the other. In such a case, we can say nothing about which is Pareto preferred.[13]

This problem can be illustrated using Figure 3.2. Consider allocations *X* and *Z*. Although *Z* involves a little less utility for Anna and a lot more utility for Brewster, we cannot conclude which of these is better, on Pareto grounds. Similarly, we cannot compare *B* with *Z*. At allocation *B*, all of society's resources go to Brewster; poor Anna is left with nothing. Yet we cannot say that this is worse than *Z* in which both have substantial utility.

If the Pareto criterion is used to make societal decisions, then decisions may tend to be biased toward the status quo. If society is only willing to take steps that improve on the status quo for everyone, then implicitly, there is the assumption that the status quo is acceptable.[14]

2. Potential Pareto Improvement. One way around the problem of unanimity is to allow transfers of resources among individuals to increase the unanimity of opinion regarding alternatives. For instance, suppose 75% of the population prefer **a** to **b** and 25% of the population prefer **b** to **a**. Then according to the Pareto criterion, we cannot say anything about whether **a** or **b** is socially preferred. But suppose the 75% favoring **a** can pool enough resources (e.g., money) and transfer those resources to the 25% supporting **b** to make **a** more attractive to them. Presumably, if the resource transfer is large enough, unanimity can be reached on option **a**.

This idea of compensation requires a slight expansion in the definition of our economy. In addition to having a distribution of goods, $\mathbf{x}$, and a fixed amount of environmental quality, e, we now have a distribution of a tradable resource, such as money, y. Thus the allocation $(\mathbf{x},e,\mathbf{y})$ involves (x_i,e,y_i) for individual i. A *transfer among individuals* involves a vector of payments to individuals, $\mathbf{z}$, such that they sum to zero: $\Sigma_i z_i = 0$. This is essentially a redistribution of tradable resources.

Now suppose we wish to compare two allocations, $(\mathbf{a}',\mathbf{y})$ and $(\mathbf{a}'',\mathbf{y})$ where $\mathbf{a} = (\mathbf{x},e)$. Assume $(\mathbf{a}',\mathbf{y})$ is not Pareto preferred to $(\mathbf{a}'',\mathbf{y})$. But suppose $\mathbf{a}'$ is preferred by most people and in fact there is some transfer among individuals, $\mathbf{z}$, such that $(\mathbf{a}',\mathbf{y} - \mathbf{z})$ is Pareto preferred to $(\mathbf{a}'',\mathbf{y})$. In other words, if the gains to those who prefer $\mathbf{a}'$ exceed the losses to those who prefer $\mathbf{a}''$, it should be possible to choose $\mathbf{a}'$ and compensate the losers so that they are just as well off with $\mathbf{a}'$ as with $\mathbf{a}''$.

Definition. *Assume an economy of* $i = 1, \ldots, N$ *individuals. Suppose we are comparing two bundles,* $\mathbf{a}' = (\mathbf{x}',e')$ *and* $\mathbf{a}'' = (\mathbf{x}'',e'')$ *at a given distribution of tradable resources,* $\mathbf{y}$. *If there exists a vector of transfers from individuals,* $\mathbf{z}$, *which sum to zero, such that* $(\mathbf{a}',\mathbf{y} - \mathbf{z})$ *is Pareto preferred to* $(\mathbf{a}'',\mathbf{y})$, *then* $\mathbf{a}'$ *is a* potential Pareto improvement *over* $\mathbf{a}''$.

This definition is very similar to the Pareto criterion. The difference is that we are separating the tradable good from the other resources. When comparing (**a**′,**y**) and (**a**″,**y**), we ask if there is some third allocation that shuffles around **y**, (**a**′,**y** − **z**) that is clearly preferred to (**a**″,**y**). If so, we forget about **y** and simply say that **a**′ is a *potential* Pareto improvement over **a**″, potential because it is possible to find the right transfer payments to achieve a Pareto improvement.

This compensation is illustrated in Figure 3.3, which uses the same two-person society as Figure 3.2. Using the standard Pareto criterion, we cannot compare allocation *X* and *Z*. In moving from *X* to *Z*, Anna loses ΔA units of utility and Brewster gains ΔB units of utility. Although we cannot compare a unit of utility for Anna with a unit of utility for Brewster, the question is whether Brewster could transfer some resources that are not represented in this diagram (income) to Anna to make up for any losses in Anna's utility. Brewster would not do quite as well having to transfer some resources to Anna. If we start at *X*, what utility levels would result if such "side-payments" could be made? Brewster would transfer only enough to keep Anna's utility from declining. This means that *Z* is a potential Pareto improvement on *X*. The point *R* in the figure is the utility pair that would most likely result if compensation occurs.

3. The Compensation Principle. The problem is that it may not always be feasible or desirable to make such transfers. How can we identify exactly how much each individual would be willing to transfer or how much of a transfer it would take to make another individual switch from supporting **b** to supporting **a**? Furthermore, people may view such transfers as bad public policy—essentially buying support.

An alternative has been proposed, called the Kaldor–Hicks compensation principle or simply the *compensation principle*: If transfers could be made to achieve unanimity on a particular choice, then the choice is socially desirable, *even if the transfers are not actually made*. This principle is attributed to the two famous economists, Nicholas Kaldor

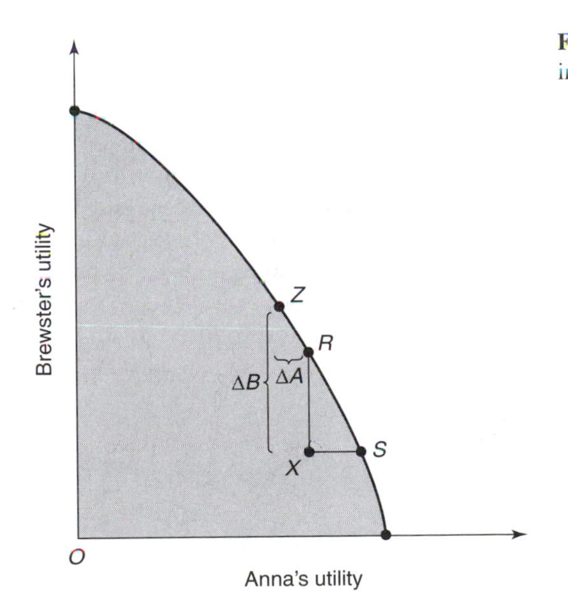

Figure 3.3 An illustration of potential Pareto improvement.

(1939) and John Hicks (1940). When a dam is built, the government asks if those who gain from the building of the dam could compensate those who lose from the building of the dam. If the answer is yes, the dam is deemed a good idea, it is built, but no compensation is paid (except as required by law for those whose land is taken). The idea is that compensation is an issue of equity, a transfer of resources from some members of society to others, and that equity can be decoupled from determining whether the dam is or is not a good idea. This idea has met with a great deal of controversy.

4. Voting. The basic problem with the Pareto criterion is the noncomparability of many bundles, which leads to incompleteness in the ordering of commodity bundles.[15] As mentioned above, the Pareto criterion is a form of unanimity: if no one opposes **a** relative to **b** then **a** is Pareto preferred to **b** (or perhaps there is indifference). Compensation solves some of this difficulty, but a bundle with compensation is really a different bundle than one without compensation.

Of course, a voting rule need not require unanimity. The typical voting mechanism is majority rule. In this case, if the majority prefers **a** to **b**, we would conclude that society prefers **a** to **b**, even though some individuals may strongly oppose **a**, relative to **b**.[16] This suggests one of the problems of voting: it cannot take into account the intensity of preferences. If the majority is barely in favor of **a** relative to **b** but the minority is violently opposed to **a**, it might be appropriate to question the social desirability of choosing **a** over **b** by simple majority rule.

Voting can be a little more complex than simple majority voting on all issues.[17] It is common in many societies to have two tiers of rules or laws. One set of laws, often embodied in a constitution, is comprised of general statements of principle. Such principles might include the right of free speech, the right to legal due process related to criminal proceedings, or the rights of private property owners to be compensated for public taking of their property. These constitutional protections are often enacted based either on general principles of what is right and wrong or on a more pragmatic basis resulting from the expected outcomes of the protection over the very long term. Free speech may be considered morally correct. Protection of private property may have no moral basis but simply be seen as necessary for a well-functioning economy. Constitutional protection of private property is necessary to allow the government to credibly commit to providing property protection in the future. Although these may be basic principles, they are decided by voting, although they may require more than majority rule.

The significance of these being constitutional protections is that they cannot be easily overridden, even if in particular cases they appear to generate bad outcomes. This is not to say that it is not possible to override these protections, only that it is difficult. In the environmental context, for example, it would be possible to prescribe a set of constitutional protections for endangered species that could be overridden but only with difficulty.

C. Social Welfare Function

For individuals, we represent individual preference orderings with a utility function. By analogy, it is logical to represent social choices with a societal utility function. We call such a societal utility function a social welfare function:[18]

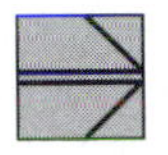

Definition. *Assume there are* i = *1, . . . ,* N *people in society, each with utility function* u_i*, defined over commodity bundles in the economy. Let* ***a*** *be a consumption bundle, indicating how much of each commodity is consumed by each individual; thus* $u_i(\mathbf{a})$ *is the utility individual i derives from* ***a****. Let* W *be a function that associates a single number with every distribution of utilities in society,* $W(u_1, u_2, \ldots, u_N)$*. If in comparing two consumption bundles,* ***a*** *and* ***b****,* $W[u_1(\mathbf{a}), \ldots, u_N(\mathbf{a})] > W[u_1(\mathbf{b}), \ldots, u_N(\mathbf{b})]$ *is equivalent to* ***a*** *being socially preferred to* ***b*** *then* W *is a Bergson–Samuelson social welfare function, or more simply a social welfare function.*

This way of writing social welfare is not the most general. We have implicitly assumed that it is the utility of different people that matters, not what is being consumed. This is not always the case. As a society, we might feel more concerned about members of society whose utility is low because of a lack of food rather than because of a lack of alcohol.

If a social welfare function exists to represent social preferences, we can draw social indifference curves, in much the same way as we draw individual indifference curves. Figure 3.4 shows a set of social indifference curves for the utility levels of Anna and Brewster. One line represents all of the utility combinations that yields social welfare W_0. The other line is the combination of utilities that yields social welfare $W_1 > W_0$. We see that point Y is Pareto preferred to point X and also yields a higher level of social welfare. Point Z is not Pareto preferred to point X but is nevertheless on a higher social indifference curve; thus Z is socially preferred to X.

There are a number of examples of social welfare functions:

Benthamite $\quad W(u_1, \ldots, u_N) = \Sigma_i\, \theta_i u_i,\ \theta_i \geq 0 \quad (3.1)$

Egalitarian $\quad W(u_1, \ldots, u_N) = \Sigma_i\, u_i - \lambda\, \Sigma_i\, [u_i - \min_i(u_i)] \quad (3.2)$

Rawlsian $\quad W(u_1, \ldots, u_N) = \min_i(u_i) \quad (3.3)$

The first of these, often called utilitarian, can be of different forms. All the weights (θ_i) may be equal or different people's utility can be weighted differently, most commonly

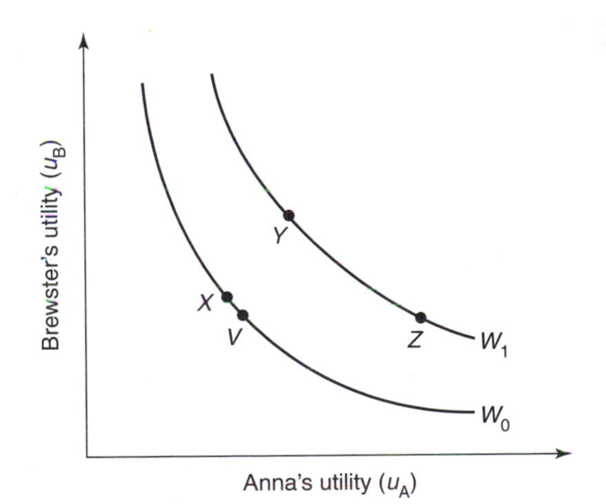

Figure 3.4 Social indifference curves.

on the basis of income.[19] With the Egalitarian representation, society cares about the total amount of utility as well as the distribution of utility. The Rawlsian welfare function, named after the Harvard philosopher, John Rawls, states that a society is only as strong as its weakest member. We see that these different welfare functions, all based on individual utility, can lead to very different views on what society considers desirable.

D. Impossibility of Perfect Choice Mechanism

Our intuition may suggest that there must be an ideal way of making social choices. Approximately 50 years ago, Kenneth Arrow set out to define the characteristics of such an ideal way. The basic question is as follows: Starting with knowledge of individual preferences over social outcomes, is there any general way of aggregating these into a social preference ordering that is reasonable? Arrow came up with six basic requirements for a social choice mechanism.[20]

A1: *Completeness.* We should be able to compare all social alternatives.

A2: *Unanimity.* If everyone in society prefers **a** to **b**, then society should prefer **a** to **b**.

A3: *Nondictatorship.* No one should always get their way, i.e., no one's preferences should be exactly the same as society's.

A4: *Transitivity.* If **a** is socially preferred to **b** and **b** is socially preferred to **c** then **a** should be socially preferred to **c**.

A5: *Independence of Irrelevant Alternatives.* Society's choice between alternatives **a** and **b** should depend only on how individuals rank **a** and **b**, without regard to other alternatives.

A6: *Universality.* Any possible individual rankings of alternatives is permissible.

The first four axioms are obvious and should not raise objections. The last two are a bit obscure and debatable. To see the significance of A5, consider the following example. Three roommates, Abel, Boris, and Curt, are trying to decide what color to paint their room: black, white, or yellow. Abel and Boris both prefer black to white with yellow coming in a close third. Curt prefers white to black. Consider the first situation where yellow is Curt's third choice. Thus it might seem appropriate to paint the room black since two out of the three prefer it. In the second situation Curt still likes white best but yellow is now his middle choice; he hates black. It might seem logical to paint the room white in this case rather than black. But A5 would prohibit our taking into account the fact that Curt hates black in making a social decision between white and black. In both situations, two prefer black to white and one prefers white to black. We cannot take into account the strength of Curt's hatred of black. The last axiom would be violated if we had somehow precluded certain types of individual choices/utility functions.

Arrow (1951) proved the following:

Arrow's Impossibility Theorem. There is no rule satisfying A1–6 for converting individual preferences into a social preference ordering.

This is a very fundamental result. It is troubling in that it implies that we should forget about having a nice neat theory of social decision making, akin to what we have in economics for individual decision making. A great deal of effort has been expended over the last half century in trying to modify these axioms so that it becomes possible to generate a social choice mechanism. And there are some relaxations that yield specific choice mechanisms. For instance, if axiom A6 is weakened to include only a certain type of "well-behaved" preferences, then majority voting emerges as a choice mechanism satisfying the other axioms.[21]

One of the problems with the Arrow axioms is that they rule out interpersonal comparisons of utility. In Figure 3.4, Z is socially desirable to X because Anna's large gain is more important than Brewster's small loss. But this involves comparing utilities of the two people, which violates axiom A5.[22]

The bottom line here is that there is no one clear unambiguous way of making social decisions. Yet social decisions must be made and are made every day. From a practical point of view, in making real social choices, the Pareto criterion or the compensation principle is often used. Arrow's impossibility theorem serves the purpose of alerting us to the fact that there may be other ways of making normative choices.

IV. CRITICISM OF THE UTILITARIAN PERSPECTIVE

Throughout the discussion in the previous section we assumed that basic individual values could be represented by utility functions and that we could then make social decisions using these utility functions along with assorted social choice mechanisms such as majority rule, supermajority voting, or the Pareto criterion. As might be expected, there are some vocal critics of this approach to making social decisions regarding the environment. This literature is quite large; we raise only a few points here.

One criticism is based on the assumption that we all have utility functions that represent our preferences and that these utility functions are immutable. Critics point out that people change their opinions over time; advertising has the express goal of doing just that. Consequently, if the environment is not highly valued by an individual, information and education programs can be used to increase the importance of the environment to that individual. However, if it is so easy to change someone's preferences, what utility function (if any) should be used to make public policy?

Another criticism is that not all affected individuals are included in many voting processes. For instance, the as-yet unborn do not vote, but are affected by actions that impact long-term environmental quality. If a local matter is put to the vote, future citizens, who may move to the locale at a later date, will be excluded (rightly or wrongly) from voting.

A third criticism of the utilitarian approach is articulated by Sagoff (1994). He questions whether public policy should be based on individual preferences rather than a concept of what is right. He says our personal desires are often dominated by what we believe is morally right or feel is our obligation to others.

V. THREE EXAMPLES REVISITED

Earlier in this chapter we presented three examples of environmental decision-making problems: air pollution in Santiago, protecting the California gnatcatcher, and the Three Gorges dam. We now return to each of these questions and examine how the outcome might differ depending on the social choice mechanism. We will also consider how the outcome might differ depending on the composition of society (e.g., if society is composed primarily of serious biocentrists).[23]

A. Air Pollution in Santiago

The issue is how much of a reduction in air pollution to pursue with polluters bearing the cost and possibly passing that cost on to customers (unless the polluters are individuals, perhaps polluting from driving cars). Other ways of bearing the cost are possible but will be ignored here.

Clearly on the basis of the Pareto criterion, the *status quo* will persist. It is unlikely that any amount of pollution control, however modest, will be supported by 100% of the population of Santiago. The poorest members of the population may be unwilling or unable to bear the increased transportation costs (due to pollution control) necessary to increase air quality. Thus under this criterion, no pollution control will be pursued.

With compensation, however, the outcome may be very different. If the wealthy feel strongly about cleaning up pollution, they can probably muster significant resources to transfer to the rest of the population to make pollution control unanimously attractive. The compensation principle would indicate that pollution control should be pursued, even though decreased pollution levels and increased transportation costs may leave many worse off.

What about majority rule voting? The outcome depends on the composition of the population. A population with a large number of poor people may not support significant pollution reduction if they have to pay more for vital services such as transportation. In these circumstances, only modest pollution reduction is likely to be supported by a majority. If, on the other hand, the population is better off or feels deeply committed to a clean environment, more substantial controls will be supported.

If a supermajority is needed to make a social decisions (such as a two-thirds majority), the pollution controls adopted will likely be more modest than under simple majority rule.

The point is that different outcomes will result from a different set of preferences on the part of the population as well as different social choice mechanisms. A corollary is that if analysis of an environmental problem concludes that only a modest level of pollution control is justified, "blame" can be placed either on the method of analysis (the social choice rule) or on the underlying preferences of the population. If many people would rather have a slightly larger TV than cleaner air, most social choice mechanisms will result in modest pollution control.

B. The California Gnatcatcher

The issue is protecting the habitat of the endangered California gnatcatcher (a bird), which nests in highly valued coastal land in Southern California. Suppose the proposal under consideration is for the government to buy all coastal land and set it aside for habitat protection. Consider two possibilities for covering this substantial cost. Option A would raise the money by a tax surcharge on the wealthiest 10% of the population; option B would involve a uniform head-tax applied to all members of the population.

Once again, the Pareto criterion is not much help. Neither option is likely to be supported by 100% of the population; thus the measure will fail. If, however, the population consists of committed biocentrists, even the richest 10% might agree that protection is a good idea. Thus option A would be favored. It is unlikely that option B would be unanimously favored because the poorest members of the population would likely find the head-tax onerous.

With compensation the two options become identical, since resources will move from those supporting habitat preservation to those opposing it. But what about the future generations of Southern California residents? They may be more than willing to compensate for setting aside habitat. However, their preferences are not known and methods are imperfect for transferring resources from the future to the present.

What about majority rule? Clearly option A would pass a majority rule test. For 90% of the population the proposal has no cost. Option B might or might not be supported by a majority, depending on how strongly the population feels about protecting the gnatcatcher. A population of biocentrists could undoubtedly muster a majority in support of option B.

But who should be voting on a proposal such as this? In the context of species preservation in Southern California, clearly the current population of Southern California is not the complete set of affected parties. Future residents of Southern California as well as people who do not currently reside in the area also have a stake in whether a species becomes extinct. Exactly who should vote is not an easy question to answer.

Consequently, we see that even a simple issue of protecting a species can be socially desirable or undesirable, depending on subtle changes in preferences as well as changes in the social choice mechanism. We cannot unequivocally conclude that such protection is socially worthwhile.

C. Three Gorges Dam

The issue is whether to flood a canyon, destroying it, for all practical purposes, for all time.[24] This is very much a problem of trading off benefits in the present with costs in the future. The dam will generate electricity and provide flood control for the coming decades. To the extent this promotes Chinese economic development, future generations will be marginally better off. However, eventually, the dam will fill with silt and from that point on provide little benefit. The loss of the river will be for all time.

Based on an intertemporal Pareto criterion, the dam should not be built. If the present generation as well as future generations are voting, unanimity will not be reached and unanimity is required to pass the Pareto test.

The compensation principle is a little more ambiguous. First, it is not possible for future generations to transfer resources to the present to compensate the present generation for not building the dam. Even if they could, it is not clear whether the benefits to the current generation would be greater or less than the compensation future generations would be willing to pay. It is likely that future generations of Chinese will be much more wealthy than the current generation and thus be able to pay considerable compensation to keep the Yangtze undammed.

What about majority rule? Clearly if we include all future generations in our list of eligible voters, the majority will vote against the dam.

Once again, we see that a complex environmental policy question will be resolved in different ways, depending on what type of social choice mechanism is adopted.

VI. CONCLUSIONS

The purpose of this chapter has been to examine different ways of making social choices. We have tried to separate the two issues of values and social choice. The individual values people place on the environment can vary widely, ranging from biocentrism to anthropocentrism. The goal of public policy is to develop ways of making social choices regarding environmental protection that are robust to any type of value system the population may possess—biocentrism, sustainability, anthropocentrism, or any combination or mixture of these.

Taking as given how people individually value the environment, we then turned to social choice mechanisms, ways of taking individual values and translating them into group values and decisions. We looked at several approaches to group decision-making, particularly the Pareto criterion and voting of one sort or another. We concluded that there is no one perfect way of making social choices.

SUMMARY

1. General principles of the importance of the environment to human life should be distinguished from principles that can be used to make decisions about specific projects or environmental regulations. The former general principles are often difficult to use to make specific decisions.

2. Three basic philosophies are considered in this chapter: biocentrism, sustainability, and anthropocentrism. Biocentrism views humans as just another species with no special claim on the world's resources. Anthropocentrism believes the environment exists only to serve people's material needs. The term sustainability suffers from ambiguity but generally prohibits actions that, if continued indefinitely, will lead to the eventual destruction of humans and the environment.

3. The basic question presented in this chapter is how to make group environmental decisions (i.e., decisions for society), starting from a disparate set of individual preferences regarding the environment. We summarize all the individual perspectives using individual utility functions, an approach that is quite general though not without detractors. We then consider several social choice mechanisms: the Pareto criterion, the compensation principle, majority-rule voting, and supermajority voting. Arrow's impossibility theorem suggests there is no perfect way of making social choices.

4. In choosing between two social alternatives, **a** and **b**, the Pareto criterion allows any member of society to veto a social choice. For **a** to be socially preferred to **b**, everyone must like **a** as well as **b** and at least one person must strictly prefer **a** to **b**. The main problem with the Pareto criterion is that it is an incomplete ordering—not all pairs of allocations can be compared using the Pareto criterion.

5. In choosing between two social alternatives, **a** and **b**, **a** is a potential Pareto improvement over **b** if resource transfers among members of society can make **a** Pareto preferred to **b**. In other words, **a** is a potential Pareto improvement over **b**, if the winners can compensate the losers.

6. The compensation principle says that if **a** is a potential Pareto improvement over **b,** then **a** is also socially preferred to **b**, even if no compensation occurs.

7. Voting is a common way of making social choices, particularly majority rule. In some cases principles or minorities are protected by more fundamental (constitutional) laws that require a supermajority to pass.

8. A social welfare function is a function that converts utility levels for all members of society into a number in such a way that the welfare function gives higher values for distributions of utility that are more socially desirable, and visa versa.

9. Arrow's impossibility theorem states that there is no perfect way of making social choices, given a set of "reasonable" assumptions. Some relaxation of these assumptions is necessary for there to exist ways of making social choices, such as majority voting.

10. The utility function-based approach to social decision making is not without its detractors. Criticisms include the malleability of preferences and the incompleteness of decision making that involves long time frames, since future generations are not represented.

PROBLEMS

1. For each of the following social choice methods, which of Arrow's axioms (A1–6) are violated, and why:
   a. the Pareto criterion
   b. the compensation principle
   c. majority-rule voting
   d. pulling a choice out of a hat (random)
2. Suppose we have a small inhabited island with three residents and a volcano that generates air pollution. Two people live upwind of the volcano and one person lives downwind. For $21,000 we can clean up the volcano with a patented "smoke guzzler." The two upwind people would pay $1000 each to get rid of the smoke whereas the downwind person would

be willing to pay $15,000. Consider two plans to finance the "smoke guzzler." Plan A calls for a head-tax of $7000. Plan B calls for the affected party (the downwind person) to pay $21,000 and everyone else nothing. Compare each plan to the status quo and indicate society's choice using (a) the Pareto criterion; (b) majority rule; (c) the compensation principle.

3. Assume a city of 1,000,000 people, 60% of whom are willing to pay $1 maximum (each) to clean up pollution. The rest of the population is better off and is willing to pay $100 each to clean up pollution. Pollution clean-up costs $2,000,000. It has been proposed that each person be taxed equally to pay for the pollution clean-up. Will that pass a majority-rule vote? Is it desirable from the point of view of the Pareto criterion? Is it a good idea, using the compensation principle? Which social choice mechanism do you think is best on intuitive grounds, and why?

4. In Figure 3.2, using the Pareto criterion, which of the labeled points are socially preferred to *X*?

5. Suppose there are three individuals, Boris, Maggie, and William. They are using majority rule to decide among themselves whether to go hiking in the wilderness this weekend (H), kayaking in the river (K), or birdwatching, looking for rare Dodos (D). Boris lives hiking best and Dodo watching worst; Maggie likes Dodos best and Kayaking worst; William likes kayaking best and hiking worst. Unfortunately, they must all do the same activity (for insurance reasons). They vote on alternatives pairwise (e.g., H vs. K; then the winner against D, etc.). Which choice will emerge as the group's choice?

6. Suppose you have a society of *n* identical individuals and the environment. Each individual likes his or her material possessions as well as access to parks and wilderness areas. In fact, the utility function of each individual is $U(x,H) = x + \mathrm{H} - 1/x - 1/H$, where x is consumption goods and H is environmental health. H attains its highest level when it is pristine: $H = 100$. Consumption of goods degrades the environment but the environment has some ability to heal itself. In fact, the environment has health $H_{\mathrm{t}} = H_{t-1} + g\ (100 - H_{t-1}) - n\ x$, where H_t is today's H and H_{t-1} is yesterday's H. The environment's ability to cleanse itself, g, is fixed.

   a. Suppose we start at $H = 100$. Assume $\mathrm{g} = 0.1$ and $n = 1$. Plot how H would evolve for $x = 0.01$, 0.1 and 1.

   b. Assume different points in time with the same population and with the same individual utility functions. Assume the society gets together to provide the desirable amount of *H*. Is this utility function consistent with (a) biocentrism, (b) sustainability, and (c) anthropocentrism? If not, what changes would need to be made in each case?

NOTES

1. Refer to World Resources Institute (1998) and O'Ryan (1996).
2. See *U.S. News and World Report*, 29 July 1996.
3. This is reminiscent of the controversies regarding dam development in the western United States (Reisner, 1986).
4. Des Jardins (1997) provides a particularly accessible survey of environmental ethics.

5. Albert Schweitzer and Paul Taylor have articulated the biocentrist philosophy (Des Jardins, 1997).
6. Toman (1994) makes this point and goes on to provide a basis for understanding sustainability within the economic paradigm.
7. Slade (1982) argues that over the life cycle of exploitation of an exhaustible resource, technological change first dominates depletion, resulting in a price decline. But later in the life cycle, depletion dominates. Thus price eventually turns up. In an empirical analysis of many resources, she argues that the turnaround point has been reached for a number of resources.
8. See Tierney (1990) for a description of this bet. The root of Simon's optimistic view of the future was his belief in continued rapid technological advance, as has occurred in the past.
9. The term natural environment is used to encompass things such as natural parks, wilderness areas, and wetlands—regions that we value for being in their natural state.
10. Nash (1989) utilizes a broader definition of anthropocentrism encompassing material and spiritual gratification from the environment. He distinguishes between biocentrism and anthropocentrism on the basis of the *rationale* for placing value on the environment. Biocentrics value the environment because that is what is morally right; anthropocentrics value the environment because it makes them feel good to protect the environment. Since these two perspectives are observationally equivalent in empirical analysis, we use the more restrictive definition of anthropocentrism to include only material gain from the environment.
11. Nash (1989) distinguishes utilitarianism from anthropocentrism by noting that early utilitarians such as Jeremy Bentham included the well-being of nonhuman animals in their calculations of the utilitarian well-being of society. Nonhumans obviously do not count to an anthropocentric except to the extent that animals give service or pleasure to humans.
12. Generally, we would expect individual i to care about $\mathbf{x}_i$, not the entire $\mathbf{x}$. However, there is no loss in making U_i a function of $\mathbf{x}$.
13. Mathematically, the problem with the Pareto criterion is that it is an incomplete ordering: many pairs of allocations cannot be ordered (one better than the other).
14. Starrett (1988) argues that the Pareto criterion is not as value free as many economists assume.
15. A complete ordering means that all pairs of bundles can be related to one another. In other words, for any two bundles, a and b, a complete ordering allows you to conclude that one of the following three alternatives is true: a is preferred to b, b is preferred to a, or a is equivalent to b. If there are any bundles for which you cannot reach one of these three conclusions, then it is an incomplete ordering.
16. One problem with voting is possible nontransitivity (Mueller, 1989). Suppose we have three alternatives, a, b, and c. It is entirely possible that a majority prefers b to a, a majority prefers c to b, and a majority prefers a to c. By sequentially voting on these alternatives, we could infinitely cycle around the three choices. This cycling was discovered by Marquis de Condorcet over two centuries ago, and further analyzed a century later by Lewis Caroll (C. Dodgson) of *Alice in Wonderland* fame.
17. Mueller (1989) lists at least eight voting alternatives to majority rule.
18. The alert reader will remark that there is no one utility function that represents an individual's preferences, but a family of utility functions, any one of which works when examining individual choice. The problem is, if we are comparing members of society, we need more than just a family of utility functions; we need a single utility function to represent a consumer's preferences. This is the distinction between ordinal utility (the utility function is useful only for ordering different bundles) and cardinal utility (the magnitude of the utility function has meaning). In general, cardinal utility is needed to make a social welfare function meaningful. This is a more restrictive assumption that we do not usually have to make when examining only individual choice. The welfare function defined here is due to Bergson (1938) and Samuelson (1947).
19. In applications, we are usually concerned with changes in social welfare. If x is the amount of some good and we assume there is a social welfare function, then the change in welfare from a small change in x can be written as $dW/dx = \Sigma_i \beta_i (\partial u_i/\partial x)/(\partial u_i/\partial y_i)$ where y_i is income of individual i and $\beta_i = (\partial W/\partial u_i)(\partial u_i/\partial y_i)$. The β_i can be thought of as welfare weights, related to individual i's income. Thus in evaluating the change in welfare from a change in x, we first

calculate the income equivalence of this change in x for each member of society and then sum these income changes over the population, weighting each element by β_i. If income differences are unimportant, then all β_i will be equal.

20. The Arrow impossibility theorem is presented in many advanced texts. See, for example, Luenberger (1995) and Mueller (1989).
21. There is a voluminous literature on these issues of "social choice." See Broadway and Wildasin (1984) for an accessible treatment of some of these issues.
22. Axiom A5 says that if we are choosing between point Z and point X in Figure 3.4, we can use only information about whether Brewster prefers X to Z and whether Anna prefers X to Z. From that point of view, the choice between X and Z appears the same as the choice between X and V, so socially we must rank X vs. V in the same order as X vs. Z. But the reason V is not as good as Z, when compared to X, is that Anna gains a lot with Z and not much with V. This is an interpersonal comparison.
23. Another example is provided by Kneese and Schulze (1985), who discuss the problem of nuclear waste disposal from several different ethical and economic perspectives. d'Arge et al. (1982) examine climate change policy from a utilitarian, egalitarian, elitist, and Paretian perspective, with and without the possibility of compensating the future.
24. A classic intertemporal analysis of the advisability of building a dam was done for the Hell's Canyon dam on the Snake River in Idaho in the United States (Krutilla and Fisher, 1985).

4 EFFICIENCY AND MARKETS

In the last chapter we took a fundamental look at how we make social decisions based on individual preferences, though at a fairly abstract level. We now turn to a more detailed examination of this issue, within the context of the economics paradigm. One of the basic normative issues in economics is for specific goods: how much should be produced and who should produce and consume—the production and allocation decisions. This is also an important issue in environmental economics. How much pollution should be produced? Clearly some pollution is necessary, because the activities that accompany much pollution are beneficial. *Totally* eliminating all automobile pollution would result in a loss for society greater than the gain obtained from cleaner air. Equally obvious is that too much pollution is undesirable. Having too much or too little pollution is viewed by economists as "inefficient."

Having determined how much pollution is desirable ("efficient"), we must determine how to achieve that desirable level of pollution. For instance, suppose we determine that ozone levels in a metropolitan area should be half what they currently are; we must then determine which polluters should reduce pollution and by how much. Each polluter could reduce emissions by 50%; this would probably achieve the overall pollution goal. Alternatively, those polluters for which pollution control is easy could reduce pollution more than those polluters for which pollution control is difficult. Obtaining the right amount of pollution, but doing it in a way that costs more than necessary, is also viewed by economists as "inefficient."

These are two concepts of efficiency: efficiency in obtaining the right overall amount of pollution control and efficiency in allocating pollution control responsibility to specific polluters. In this chapter we will develop both of these concepts. We first define efficiency, using the concept of Pareto optimality defined in the previous chapter. We then show the relationship between efficiency and a market equilibrium: a competitive market equilibrium is generally efficient. We also show that market failures, including pollution, can result in inefficiencies in a market equilibrium that should be corrected by government intervention.[1] We also develop some of the tools of cost–benefit analysis, the method often used to determine efficient outcomes in the absence of a competitive market.[2]

I. WHAT IS EFFICIENCY?

Many people in developed economies have heard the pronouncement that competition is desirable (at least up to a point) and that competition in an economy is somehow good for the country. There are really two reasons why competition may be desirable. One is that competition helps a market economy work smoothly. Competition usually makes an economy more agile and responsive in a changing world. The second reason is somewhat more controversial. The issue is whether the outcome achieved by competition in a market economy (including the division of the "pie") is "best" for society. It is important to be somewhat cautious in jumping to such a conclusion, keeping separate criteria for determining an economic allocation (production and distribution) that is socially desirable versus one that is achieved by competitive markets. Whether competitive markets result in the best social outcome depends on how social welfare is defined.

In the previous chapter we examined different ways of deciding what is socially "best." For our definition of best, we adopt the weakest criterion, that of Pareto. The Pareto criterion is an important one in economics. Despite the fact that it is not useful in comparing many social allocations, it is important for the notion of efficiency. To illustrate this, we return to the two-person economy developed in the Chapter 3. Figure 4.1 shows attainable utility levels for the two members of our economy, Anna and Brewster. Note that for every allocation that is *inside* the feasible set *ABO*, it is possible to find a Pareto preferred allocation (a *Pareto improvement*). Basically, for any interior allocation, any allocation to the upper right (i.e., in the direction that improves both Anna's and Brewster's utility) is a Pareto improvement. Note in the figure that for *X*, all allocations in the shaded region are Pareto preferred to *X*. This is because all allocations in that region give both Anna and Brewster just as much utility as at *X*, and for at least one of them, more util-

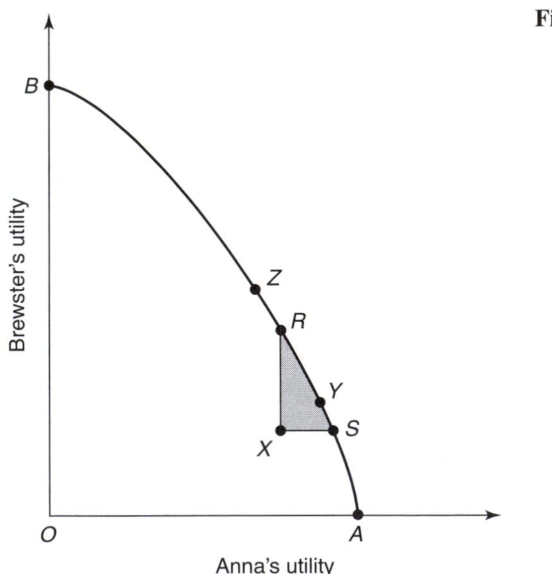

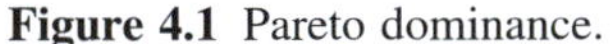

Figure 4.1 Pareto dominance.

ity. What allocations cannot be improved? All allocations on the line *AB* are as good as possible, in a Pareto sense. This leads to the notion of Pareto frontier.

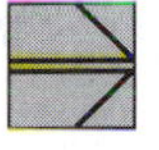

Definition *The* Pareto frontier *consists of all allocations for which there are no allocations that are Pareto preferred. Alternatively, the Pareto frontier consists of all allocations that cannot be improved in the Pareto sense.*

In other words, if *Z* is on the Pareto frontier, it is not possible to find another allocation that is preferred to *Z*. This is our notion of economic efficiency. It is also referred to as *Pareto optimality*.

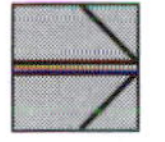

Definition *An allocation is* efficient, *or* Pareto optimal, *if it lies on the Pareto frontier; an allocation is* inefficient *if it is not on the Pareto frontier.*

This is an important concept that we will return to time and again. An allocation is inefficient if it is interior to the Pareto frontier. The obvious reason inefficiency is undesirable is that resources are "left on the table." In other words, it would be possible to move to an allocation where everybody does better and has more. Thus in Figure 4.1, *X* is inefficient because any allocation in the stippled region is a Pareto improvement over *X*. Nobody would choose *X* over these alternative allocations.

This seems like a straightforward concept. Where are the problems, then? The problem is that we cannot compare different allocations that lie along the Pareto frontier. In Figure 4.1, the curved line *AB* is the Pareto frontier. We cannot say anything about the relative desirability of any points along the frontier. Points *A* and *B* have all of the utility going to one of the two individuals (which is not too equitable). We cannot say whether *A* or *B* is more or less socially desirable than an allocation such as *Z*.

This is the basic source of controversy when economists call for actions that promote efficiency. Suppose society is at point *X* in Figure 4.1. Economists might promote a move to *S* because it is a Pareto improvement. However, some might view Anna as undeserving and might prefer *Z*, where the pie is divided a little more evenly. A movement to *S* seems counterproductive. The bottom line is that efficiency may not be the only goal in dividing the social pie.

II. EFFICIENCY AND COMPETITIVE MARKETS

Efficiency is one of the most important words in the vocabulary of an economist. To an economist, efficiency means Pareto optimality; inefficiency means a lack of Pareto optimality. If an economy is operating on the Pareto frontier, it is efficient; if not, it is inefficient. Throughout this book we will use Pareto optimality and efficiency synonymously.

Typically, there are two types of inefficiencies that exist in an economy: inefficiencies in exchange and inefficiencies in production. Inefficiency in exchange means that resources could be moved around among individuals without making anyone worse off. For instance, take the case of municipal solid waste from a large city. Suppose a law prohibits the transport of wastes across state or national boundaries.[3] This means that wastes must be disposed of locally, possibly in an area of high population concentration. There might be a rural region that would be willing to take the wastes for an appropriate fee (and at lower disposal costs), making everyone better off. Efficiency would be promoted by removing such a restriction on waste shipments.

The other type of inefficiency occurs in production. For instance, suppose an environmental regulation calls for the use of a "scrubber" on the smoke generated at a factory. The factory could achieve the same level of emissions reduction by changing its production process, but regulations do not allow using process changes to achieve emissions targets, only the scrubber. In this example, resources are wasted in production with no apparent gain. Everyone could be better off by using a method of pollution control that costs less. Providing too much or too little pollution overall would also be inefficient.

One of the most important results in economics is that competitive markets are efficient. That means that the allocation achieved when a perfectly competitive market operates is on the Pareto frontier. The goal of this section is to demonstrate this. What we will do is develop a simple model of an exchange economy (with two goods and two members of the economy), develop the Pareto frontier and the competitive equilibrium, and then show that these two are the same. We will confine ourselves to an exchange economy because it is simple and easy to grasp. We will briefly cover the extension to an economy with production but leave a full development of that to another course.[4]

A. Efficiency in Exchange: Goods

In the previous section we compared the attainable utility of two people, Anna and Brewster. We were not concerned with exactly how they achieved their levels of utility, nor how the utility of each depended on the actions of the other. Our goal was to develop the concept of Pareto optimality. We now expand this example somewhat and assume that Anna and Brewster have some resources and wish to enter into trade. Initially we will consider the example of two conventional goods—cheese and wine. After becoming comfortable with the case of two goods, we will switch to the case of one good and one bad—wine and garbage.

Figure 4.2 shows indifference curves for cheese and wine, reflecting Anna's preferences. These are typical indifference curves. Note that they are concave upward by assumption. There are two ways of explaining the shape of these indifference curves. One explanation is that mixtures are better than extremes. Consider the dashed line, *CD*. Would we expect more utility from consuming one of the extremes (all wine, no cheese or all cheese, no wine) or a mixture of the two (point X^*)? Typically, we assume the mixture is better—thus the utility level is higher at X^* than at the points *C* or *D*. Another explanation of the curvature of the indifference curves is in terms of the marginal rate of substitution between cheese and wine. Recall that the marginal rate of substitution between cheese and wine at any point (like X^*) is the slope of the tangency to the indifference curve at that point. The marginal rate of substitution (MRS) of wine for cheese indicates how much cheese consumption would need to increase to offset a one unit drop in wine consumption to keep utility constant. At a point such as *A* in Figure 4.2, where there is more cheese and less wine relative to point X^*, it is reasonable to think that more cheese

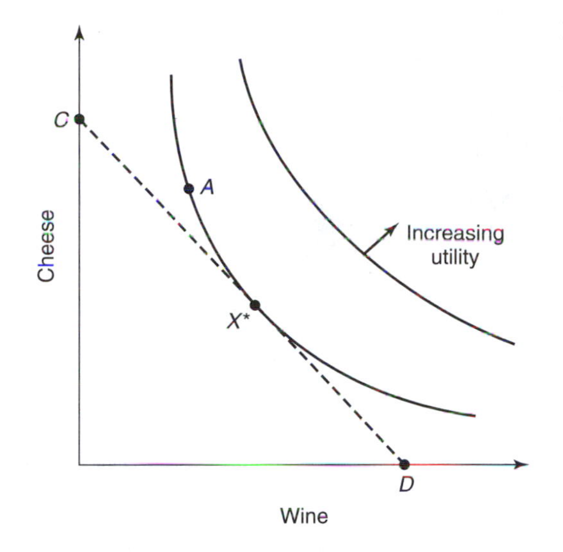

Figure 4.2 Utility maximizing choice of wine and cheese. CD, Mixtures of the basket *C* (all cheese) and the basket *D* (all wine); X*, choice point.

would be required to compensate for a one unit drop in wine—the MRS of wine for cheese increases as cheese consumption increases and wine consumption decreases. (The more cheese you have, the less valuable one more unit is and the less wine you have, the more valuable one more unit is.) If this is the case, then the indifference curves will have the shape shown in Figure 4.2.

Suppose Anna and Brewster are endowed with certain quantities of cheese and wine. We are interested in how these people might trade. We have no prices—this is pure barter. Assume Anna has a kilogram of cheese (no wine) and Brewster has a bottle of wine (no cheese). Both would prefer a little of each good.

The Edgeworth box is a useful graphic way of representing exchange of two commodities between two individuals. Figure 4.3 shows such an Edgeworth box for Anna and Brewster and their wine and cheese. Essentially what we do is take Anna's indifference curves (Figure 4.2) and superimpose them on a similar figure for Brewster, so that the origin of Brewster's set of indifference curves is in the upper right of Figure 4.3. The total amount of wine to trade is the length of the horizontal axis; the total amount of cheese to trade is the height of the vertical axis. Any *point* in Figure 4.3 represents a division of the cheese and wine between the two people. In particular, point *A* in the upper left represents the initial endowment: Anna with all the cheese and Brewster with all the wine.

Any division of the cheese and wine that lies between the lines *AB* and *AC* will result in increased utility for both individuals and thus will be a Pareto improvement. That is because all of these points (stippled in Figure 4.2) lie on higher indifference curves for both people. But this region is very large—many divisions are possible. Can we narrow the outcomes further? One way would be to focus on the Pareto frontier of the region *ABC*—the divisions of the "pie" for which one party cannot do better without making the other worse off. It turns out that any point in the *ABC* region at which the indifference curves of Anna and Brewster are tangent to one another lies on the Pareto frontier—also called the contract curve. To see this, consider a point at which tangency does not apply, such as point *A*. There are clearly allocations that are Pareto preferred to *A*—in fact, any

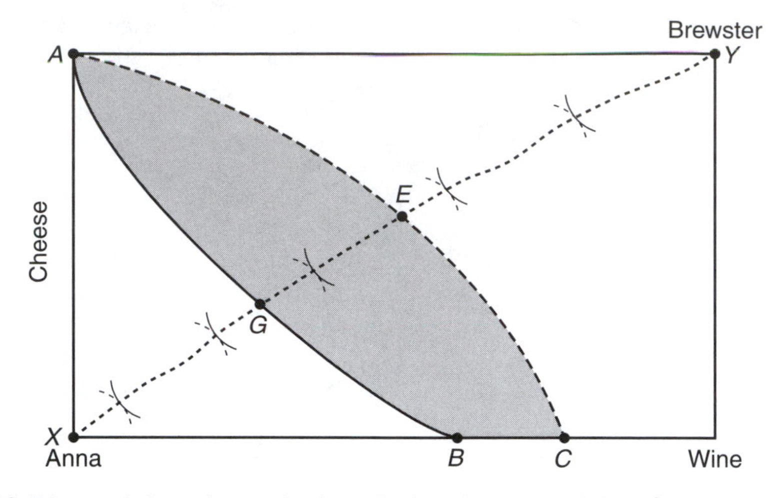

Figure 4.3 Edgeworth box characterization of wine–cheese bargaining. A, Initial endowment; AGBCE, points of Pareto impovement over *A*; XGEY, contract curve; GE, portion of contract curve that improves on *A*.

allocation in the shaded lens *ACB*. Now consider those allocations on the dotted line *XY* corresponding to allocations where one of Brewster's indifference curves is tangent to one of Anna's. Along this contract curve (*XY*), one person can improve utility only by decreasing the other's utility. This is precisely the criterion necessary for the Pareto frontier. So whatever the initial endowment of the two individuals, we know the result of barter will be somewhere along the contract curve. In our case, starting at point *A*, the result of bargaining must lie on the segment of the contract curve, *EG*.

Tangency between the two indifference curves gives us a necessary condition for a Pareto optimum. What is the interpretation of this tangency? Basically, tangency is the same as the slopes of the indifference curves being simultaneously the same for the two people at a single point. The slope of the indifference curve gives us the rate at which each person can substitute wine for cheese while keeping utility constant. For instance in Figure 4.3, if Anna's slope at the point of tangency is 2, that means that if cheese is decreased and wine increased small amounts in the ratio of 2:1, then utility will not change (or change only very slightly). This slope is, of course, the marginal rate of substitution of wine for cheese. Efficiency requires that the marginal rate of substitution of wine for cheese be the same for both Anna and Brewster. More generally, we have the following result: at a Pareto optimum, the marginal rate of substitution between any two goods should be the same for all individuals.

So how will the cheese and wine be divided? All we can say is that Anna and Brewster will end up somewhere along the *EG* portion of the contract curve, i.e., that portion of the contract curve that results in an improvement in utility (a higher indifference curve) for both individuals, in comparison to their initial position, *A*. Therefore, Anna will pay Brewster in cheese for some of his wine. Furthermore, they will trade until the gains from trade are exhausted, ending up at a Pareto optimal division of the cheese and wine.

We have seen how to represent exchange graphically. But more importantly, we have shown that in this simple model, free exchange leads to a Pareto optimal division of the goods. In other words, efficiency is achieved.

B. Markets and Exchange

Now suppose we move away from barter and introduce prices for the two commodities. We will allow trade to occur, but trade must be accompanied by payment. Thus the value of any bundle that Anna ends up with must have the same value as Anna's initial endowment. And the same for Brewster. Neither can generate income except through trade. To be more concrete, let C_0 and W_0 be Anna's initial endowment of cheese and wine respectively. Let C^* and W^* be the quantities of the two commodities Anna eventually ends up with. If we let prices of the two commodities be p_C and p_W, then Anna's budget balancing requires the following to hold for any (C,W) consumed:

$$p_C C_0 + p_W W_0 = p_C C + p_W W \tag{4.1}$$

Clearly if Anna has the same amount of money before and after trade, Brewster will too, since there is nowhere else for the money to go. Note that for a given set of prices, Eq. (4.1) basically described a linear relationship between C and W with the left-hand side of the equation a constant. It may be easier to visualize Eq. (4.1) if we write it with C on the left-hand side and W on the right-hand side:

$$C = [C_0 + (p_W/p_C)W_0] - (p_W/p_C)\,W \tag{4.2}$$

In Eq. (4.2), the portion in brackets is a constant. Equation (4.2) describes a straight line with slope $-(p_W/p_C)$. Equation (4.2) defines the budget constraint—combinations of C and W that are affordable. Obviously, the initial endowment is affordable, which can be checked by substituting (C_0,W_0) for (C,W) into Eq. (4.2).

This can be represented in an Edgeworth box. Figure 4.4 shows a simplified version of the Edgeworth box of Figure 4.3. In addition, Eq. (4.2), the budget constraint, has been drawn. As it should, the line passes through (C_0, W_0), the initial endowment, labeled point A. For an allocation X^* to be a market equilibrium, both consumers must be simultaneously satisfied with X^*, given the prices. In other words, both individuals, assumed to be price-takers, must have indifference curves tangent to the budget line at X^*.[5] This of course is the same condition for points along the contract curve—that indifference curves be tangent. Thus market equilibrium will occur where the budget line (the straight line from A in Figure 4.4) intersects the contract curve (X^* in Figure 4.4).

It is important to remember that the prices are unknown. In fact, the prices are what we are interested in determining. All we know is that whatever they are, they must satisfy two requirements:

1. The budget line (Eq. 4.2) must pass through the initial endowment.
2. There must be some allocation at which the indifference curves of both parties are tangent to the budget line.

If we are lucky only one line satisfies both of these conditions. It can be found by passing a straight line through the initial endowment (point A) and rotating the line (mentally)

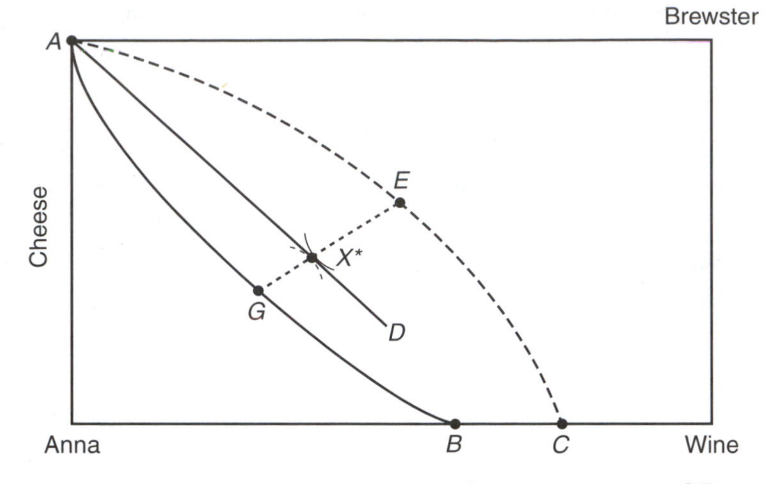

Figure 4.4 Market equilibrium in cheese–wine bargain. A, Initial endowment; GE, portion of contract curve that improves on *A*; X*, market outcome with initial allocation at *A*; AD, budget constraint.

around *A* until the line intersects the contract curve in such a way as to be tangent to both indifference curves at the same time. This is the line *AD* in Figure 4.4.

As can be seen from Eq. (4.2), the slope of the budget line in Figure 4.4 is the negative of the price ratio ($-p_W/p_C$). This is a negative number since the prices are positive. This price ratio is also equal to the marginal rate of substitution at X^* between cheese and wine for both Anna and Brewster.

So what? What we have shown, albeit in a somewhat simplified example, is that a market equilibrium results in a Pareto optimal allocation. Markets are efficient. Either Anna and Brewster can negotiate directly, leading to a set of possible efficient allocations along the contract curve or the market can operate, leading to one particular efficient allocation along the contract curve. Do not fail to notice that there are other allocations along the contract curve that are also Pareto optimal but will not result from the operation of a market as just described. Furthermore, if the initial endowment (point *A* in our example) is different, a different allocation may result. For instance, if Brewster has a little cheese and not quite so much wine, he may buy less cheese and agree to sell less wine.

In fact, note that by changing the initial distribution of cheese and wine, any point on the contract curve (Pareto frontier) can be obtained using the price system.

C. Efficiency in Exchange: Bads

The previous section was designed to be as familiar as possible—two goods in exchange, an example you have probably encountered before in an intermediate microeconomics course. But this is a book on environmental economics, so now we will consider exchange involving one good and one bad—wine and garbage. Think of garbage as bags of trash rather than the more esoteric water or air pollution. Consequently, garbage consumption really consists of storing the bags of trash in an out-of-the-way place until they decay into harmless compounds. Ignore the fact that the garbage may smell and bother the neighbors or that rainwater may wash it into streams or groundwater. Garbage is simply something that is undesirable but that can be stored so that it affects no one besides the person doing the storing.

Figure 4.5 describes Anna's preferences regarding wine and garbage. It shows her indifference curves, although they are a little different than normal because we have one good and one bad. The indifference curves slope upward, away from the origin, because as wine consumption increases, garbage "consumption" must also increase if we are to keep utility constant (a requirement for an indifference curve). Can we say anything about the shape of these indifference curves? They are shown as being concave downward, but is that necessary? It is not strictly necessary, though it is the most reasonable assumption. As the indifference curve is drawn in Figure 4.5, the marginal rate of substitution of garbage for wine at a point such as *A* is higher than at a point such as X^*. (Recall that the MRS is the slope of the tangency to the indifference curve.) At point *A*, Anna has no garbage and less wine than at X^*. Thus if we give her a bottle of wine, we will have to give her quite a bit of garbage to keep utility constant. At point X^*, she has more wine, so another bottle is not quite as valuable; furthermore, she has more garbage than at *A* so one more unit of garbage is more damaging. Thus if we give her one more bottle of wine, we will not have to give as much garbage to keep her utility constant. This is equivalent to a declining MRS, which is all that is necessary to yield indifference curves that are shaped as the ones in Figure 4.5.

How might trade occur in this example? We start with an endowment of wine and garbage. Assume Anna has all of the wine and all of the garbage. Poor Brewster has none. Because we are looking at exchange, the quantities of wine and garbage are fixed—there is no production.

The Edgeworth box can be used in this case to examine what trade might take place. Figure 4.6 shows such an Edgeworth box for Anna and Brewster and their garbage and wine. This is constructed in exactly the same way as Figure 4.3, except that the indifference curves have positive rather than negative slopes, due to the fact that garbage is a bad. The length of the horizontal axis is the total amount of wine available and the height of the vertical axis is the total amount of garbage available. The point *A* in Figure 4.6 is the initial endowment—everything to Anna and nothing to Brewster.

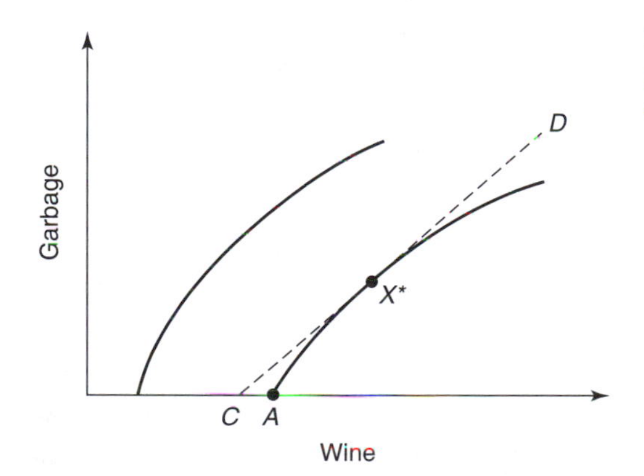

Figure 4.5 Anna's preferences for wine and garbage. CD, Slope is marginal rate of substitution of wine for garbage at X^*.

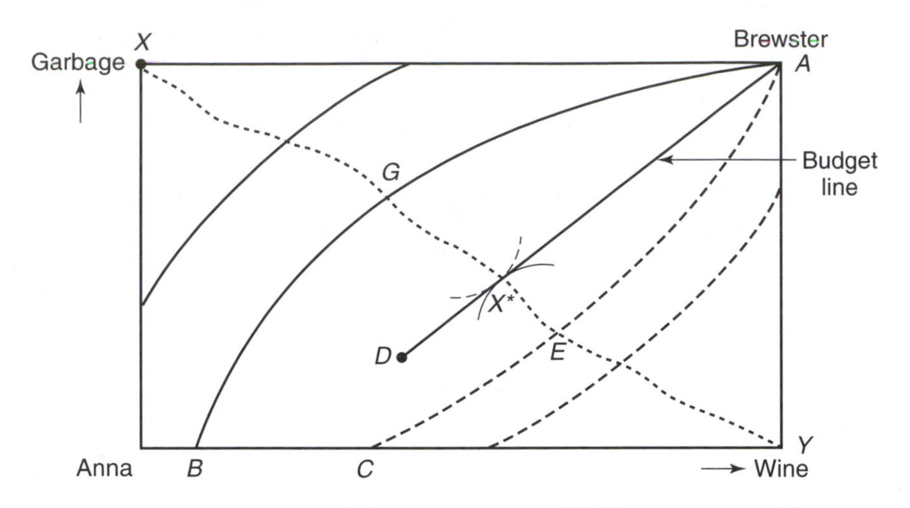

Figure 4.6 Garbage–wine exchange. A, Initial endowment; XGEY, contract curve; X*, market equilibirum with *A* as initial endowment; AD, budget constraint, AGBCE, region of Pareto improvement over *A*.

Any division of the garbage and wine that lies between the curved lines *AB* and *AC* will result in increased utility for both individuals and thus will be a Pareto improvement. That is because all of these points lie on higher indifference curves for both people. But we are really interested in the set of Pareto optimal divisions of the wine and garbage, i.e., what is the Pareto frontier/contract curve? As before, that is the collection of tangencies between Anna's indifference curves and Brewster's indifference curves, shown as the dotted line in Figure 4.6, *XY*. However, only a segment of that will be better than the initial endowment for both Anna and Brewster—the line segment *EG*. So barter will eventually end up with some mixture of wine and garbage for each person, represented by one of the points along *EG*. We cannot say which one.

Now suppose we move away from pure barter and introduce prices for the two commodities. A price for wine is of course quite familiar. In fact a price for garbage should also be familiar. The price of garbage is negative. When you take a bag of garbage you pay a negative price for it—you are paid to take it. This is what waste disposal firms do. However, restricting ourselves to the two-person exchange, the value of any bundle of wine and garbage that Anna ends up with (or Brewster) must be exactly equal to the value of the initial endowment of each person. Money cannot be created or destroyed. Let G_0 and W_0 be the initial endowment of garbage and wine possessed by Anna. Let G and W be the amount of garbage and wine Anna ends up with. Budget balancing requires

$$p_G G_0 + p_W W_0 = p_G G + p_W W$$

or

$$W = [(p_G/p_W)\, G_0 + p_W W_0] - (p_G/p_W)\, G \tag{4.3}$$

This is Anna's budget constraint. Any (W,G) combination that satisfies Eq. (4.3) will be affordable and only such (W,G) combinations. Notice that it is a straight line. The bracketed term is a constant and the slope of the line is $-(p_G/p_W)$, which is positive since the price of garbage, p_G, is negative. A quick check will see that the initial endowment (G_0, W_0) satisfies Eq. (4.3), which means the line passes through the initial endowment.

Furthermore, as a budget line, any choice made by Anna must be where the budget line is tangent to an indifference curve. And the same for Brewster. The line *AD* is just such a budget line. It passes through the initial endowment and is tangent to an indifference curve for each person at X^*. Thus X^* will be the amount of wine and garbage consumed by the two people and the exchange price of garbage, relative to wine, will be equal to the slope of the line *AD*.

The bottom line is that we obtain exactly the same result as we did with exchange of two goods. A market equilibrium results in a Pareto optimal allocation. Markets are efficient. Either Anna and Brewster can barter directly leading to a set of possible efficient allocations along the contract curve or the market can operate, leading to one particular efficient allocation along the contract curve.

D. Efficiency in Production

In exchange, we are concerned with moving goods and bads around among individuals to improve everyone's welfare. Basically, we were looking at different ways of dividing pies. We now turn to the question of combining resources to produce new goods and bads. We are concerned with efficiency in production. Suppose we have a firm—Chateau Mull—that makes wine with garbage as a by-product.

We start with a few assumptions, continuing to focus on the production of wine and garbage. Assume we have a fixed amount of resources to produce these two items—Anna's and Brewster's labor (L). If we put all of our resources into wine production, we will produce the maximum amount of wine, as well as the maximum amount of garbage. We can redirect some of our labor into garbage prevention, reducing the amount of garbage generated. With less labor available for wine production, we will inevitably generate less wine. The question is how to split labor between these two activities. Alternatively, how much garbage and wine is to be produced? Contrast this to the previous example which there was a fixed amount of garbage and wine and the question was how to divide it up between Anna and Brewster.

Figure 4.7 shows the possible production combinations of wine and garbage (the shaded region). Note that to produce the maximum amount of wine, a maximum amount of garbage will also be produced. Analogously, less garbage means less wine. The points along the line *AB* are the efficient combinations of wine and garbage—the Pareto frontier, also called the production possibilities frontier. Note that we could conceivably be in the interior of the figure *ABC*—the labor in wine production or in garbage prevention could be wasted. In such a case, we might obtain a point such as *D* in Figure 4.7. The point *D* is in the interior of the production possibilities set and is inefficient because it is possible to produce more wine and less garbage (moving from point *D* to point *E*), which has to be desirable to everyone in the economy. So the first condition for efficiency is that the economy operates on the production possibilities frontier.

The inputs into production in our example (labor) are fixed resulting in costs C. But the producer does have a choice about how much wine and garbage to produce. Profits for the producer are given by

$$\pi = p_W W + p_G G - C \tag{4.4}$$

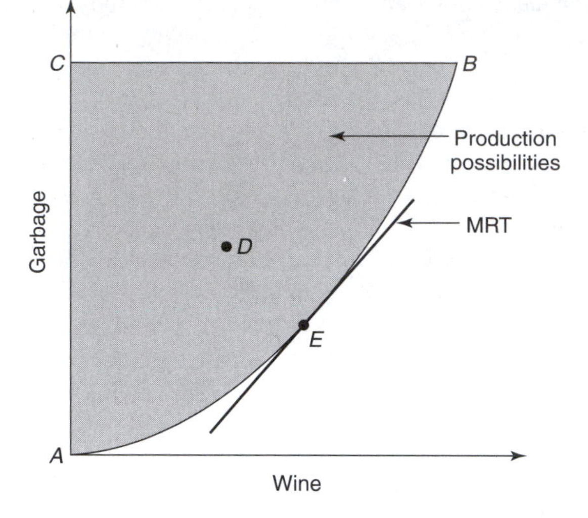

Figure 4.7 Efficiency in production.

which can be rewritten with G on the left-hand side and W on the right-hand side:

$$G = (\pi - C)\, p_G - p_W /\, p_G\, W \tag{4.5}$$

This is obviously a straight line with intercept $(\pi - C)/\, p_G$ and slope $-p_W/\, p_G$, which is positive because p_G is negative. Profits are highest when the intercept is as low as possible (since p_G is negative), which means the line shifted as far to the right as possible, while still having a point in common with the production set. The profit-maximizing production combination is shown as point E in Figure 4.7.

The slope of the production possibilities frontier at any particular point is called the *marginal rate of transformation* (MRT) between the two goods (a good and a bad in our case). This MRT is shown in Figure 4.7 for the point E on the production possibilities frontier. In general, the MRT shows the rate at which output of one good has to be sacrificed to increase output of the other good. In our case of a good and a bad, the MRT of wine for garbage shows how much garbage production goes up when we try to increase wine production by one unit (at point D). For instance, if an additional 2 kg of garbage needs to be generated to produce one more bottle of wine, at point E, then the marginal rate of transformation of wine for garbage is 2 (kilos/bottle) at E. With a market, the marginal rate of transformation is equal to the price ratio.

One condition that is not obvious from Figure 4.7 is that if there are multiple producers of wine and garbage, for efficiency to hold, the marginal rate of transformation between wine and garbage must be the same for all producers. Suppose there are two such producers. If producer A has a larger MRT between garbage and wine than producer B and if we reduce wine production from A by one unit and increase wine production from B by one unit, then wine production stays constant and overall garbage production declines. This is obviously a Pareto improvement. Consequently, a necessary condition for efficiency is that the marginal rate of transformation between any two outputs be the same for all producers in the economy.

This is an important point that in fact often fails to hold for most pollution regulations. If several firms generate pollution along with a useful good, what we have just shown is that the marginal rate of transformation between pollution and the good should be the same for all firms. In simple terms, the marginal cost of pollution control should be the same for all firms. This is what is known as the equimarginal principle and will receive significant attention in later chapters on regulation. Unfortunately, this principle rarely holds for actual environmental regulations. More often, firms will be told to reduce emissions by some percentage. For some firms this will be cheap and for other firms this will be expensive.

All that remains is to put production and exchange together. The basic necessary condition for efficiency is that the marginal rate of transformation (in production) between any two commodities be the same as the marginal rate of substitution between these two commodities for any individual consuming both commodities.

E. Efficiency with and without Markets

In the previous sections we examined conditions that need to hold for economic efficiency. These conditions had nothing to do with prices; in fact, the conditions can be applied in an economy without prices. One basic requirement is that the marginal rate of substitution (in consumption) between any two commodities be equated to the marginal rate of transformation (in production) between those two commodities. Secondarily, we saw that for any two firms producing both commodities, the marginal rate of transformation between the two commodities must be the same for each firm. Similarly, any two consumers consuming both commodities must have the same marginal rate of substitution between the two commodities.

We also saw in a simple exchange economy that a competitive market will generate prices so that the marginal rate of substitution is the same for all consumers and equal to the ratio of prices of the two commodities. We also showed that with a competitive market, the marginal rate of transformation will be equal to the price ratio.

These results can be generalized into two basic theorems, the first and second theorems of welfare economics.

> **First Theorem of Welfare Economics.** In a competitive economy, a market equilibrium is Pareto optimal.

This first theorem is a restatement of the results of this section. If we let a competitive market operate, a Pareto optimum will prevail.

The second theorem states that a market can be relied on to give *any* Pareto optimal allocation. As we saw earlier in the simple exchange economy between Anna and Brewster, the initial endowment of wine and garbage was important in determining the eventual split of garbage and wine between the two people. In fact, any point on the contract curve could be obtained using a market, simply by changing the initial endowment of wine and garbage. This can be extended to the case involving production:

> **Second Theorem of Welfare Economics.** In a competitive economy, any Pareto optimum can be achieved by market forces, provided the resources of the economy are appropriately distributed before the market is allowed to operate.

This all sounds very easy: let the market operate. Get out of the way and all problems will be solved. Then why is pollution a problem in a market economy? The reason is that the generation of most pollution violates one of the assumptions of a competitive market.

What are those assumptions? What conditions must prevail in a market before we deem it competitive and thus subject to these two theorems of welfare economics? Four conditions must hold before we declare a market competitive:[6]

1. *Complete property rights.* A well-defined, transferable, and secure set of property rights must exist for all goods and bads in the economy so that these commodities can be freely exchanged. All the benefits or costs must accrue to the agent holding the property right for the good or bad.
2. *Atomistic participants.* Producers and consumers are small relative to the market and thus cannot influence prices. Instead, they maximize profits or utility taking prices as given.
3. *Complete information.* Consumers and producers have full knowledge of current and future prices.
4. *No transaction costs.* It must be costless to attach prices to goods traded.

Clearly pollution violates condition (1). If I own a bottle of wine, I am the sole beneficiary. If I sell the bottle of wine, I transfer all of the benefits to the person who buys the wine. Many forms of pollution cannot be restricted in that fashion. If I have a factory that generates smoke, everyone in the neighborhood is forced to consume the smoke.

In the next chapter we examine more closely just what it is about environmental goods that violates the assumptions of a competitive market. We will see that markets typically fail when pollution is present. We will also see how those market failures can be corrected through government intervention so that we may rely on the market to achieve a Pareto optimal outcome.

III. SUPPLY, DEMAND, AND EFFICIENCY

In the last section we took a fairly abstract look at the way markets operate. This was useful because it gives a fundamental understanding of exchange, production, and efficiency. We can also look at the question of efficiency and markets through a much more familiar lens, that of supply and demand. In doing so, we assume a market setting and ask if the market will give us efficiency (a Pareto optimum).

A. A Market Equilibrium is Pareto Optimal

The first welfare theorem states that for a competitive market, a market equilibrium is Pareto optimal. We can also see this in a supply and demand context. If we take a commodity such as wine, we know that we can represent a consumer's preferences regarding wine through a demand curve, an expression that gives the quantity of wine demanded as a function of the price of wine relative to other goods. Alternatively, the demand curve may be viewed as indicating the marginal willingness to pay for one more unit of the good, as a function of the total quantity of the good being consumed. Similarly, a supply curve represents the amount of wine that will be supplied at various prices. Figure 4.8 shows such demand and supply curves for wine. The market equilibrium is the point where the two curves cross at w^* quantity of wine. To see that this is Pareto optimal, consider other quantities of wine such as a little less, w_1, or a little more, w_2. At w_1, additional wine could be supplied to our consumer at a cost less than the willingness-to-pay for the wine. Clearly allowing such consumption would be a Pareto improvement. Similarly, at w_2, the last bit of wine (between w^* and w_2) is worth less to the consumer than it costs to produce it. By reducing wine consumption, we save the producer more than the consumer loses; transfers can make everyone better off. In this sense a market equilibrium is Pareto optimal.

B. Consumer and Producer Surplus

We know how to move from measures of utility (indifference curves) to a demand curve. This is the familiar maximization of utility subject to a budget constraint, which ultimately generates a demand curve. We can also move in the other direction. Starting with a demand curve, we can obtain a measure of the overall welfare gain associated with consuming some amount of good. The welfare gain is in monetary terms, since utility is not directly measurable. This is consumer surplus—the difference between what a consumer would be willing to pay for a bundle of goods and the amount the consumer actually has to pay. You have been walking through the desert all day without water and come upon a shop selling bottled water for $2 per bottle. You buy a bottle. What is the value of that purchase to you? Unless you are very poor, you would be willing to pay more than $2 for the bottle, perhaps very much more. The surplus associated with that transaction is the difference between what you would have been willing to pay and what you did pay. Where do we obtain information on how much you would pay? From the demand curve.

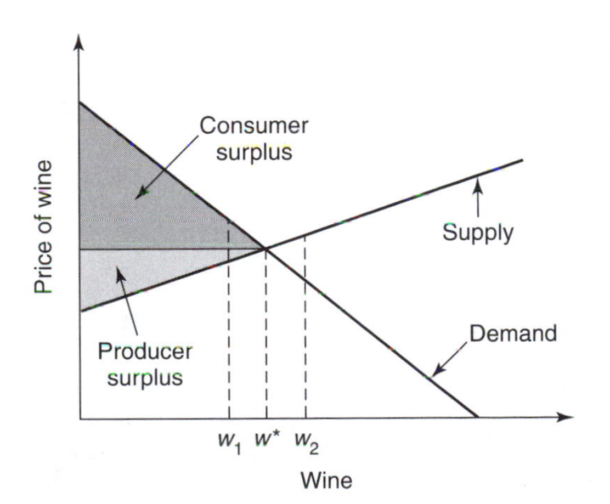

Figure 4.8 Supply, demand, and surplus measures in the market for wine. w^*, market equilibrium quantity of wine; w_1, too little wine; w_2, too much wine.

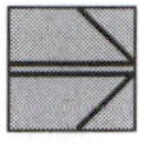

Definition *The consumer surplus associated with consuming* q* *units of a commodity for a consumer with demand* Q(p) *is the area between the demand curve and the horizontal axis and between the vertical lines* q = *0 and* q = q*. *The surplus is reduced by any payment the consumer makes for the commodity.*

We can also carry this concept over to the producer. When supply curves are upward sloping, the interpretation is that the first units supplied are less expensive than subsequent units. This may simply be the result of using some scarce factor in production whose value is bid up as the industry expands. Whatever the reason, there may be a difference between revenue received and the cost of production—this is the producer surplus.

Definition *Producer surplus is the difference between revenue received for sale of* q* *units of a commodity and the cost of providing that commodity,* C(q*). *This difference is generally equal to the area between the supply curve and a horizontal line through the price, and between the vertical lines* q = *0 and* q = q*.

These concepts are also shown in Figure 4.8, which is a representation of supply and demand curves for wine. Shown in the Figure are consumer surplus, producer surplus, and total surplus (the sum of producer and consumer surplus). Total surplus at q^* is the area under the demand curve less the area under the supply curve, measured between $q = 0$ and $q = q^*$. This is the shaded area.

Several qualifications are in order. One is that total surplus may not exist. If, for instance, the demand curve were asymptotic to the vertical axis, the area under it might be infinite. If this is the case, the best we can do is talk about the difference in surplus between consuming two different quantities, say four bottles of wine and two bottles of wine.

A second qualification is that the surplus measure depends on prices of other goods and, possibly, income. For instance if the price of beer drops, we might expect the demand for wine to shift. Or if incomes were to drop, demand for wine might decline and the demand for beer might increase.

One important thing to infer from Figure 4.8 is that total surplus is maximized at the market equilibrium. It is easy to see that for any quantities of wine less that w^*, surplus (the shaded area) is less than at w^*. But for quantities of wine greater than w^*, note that the additional surplus from another bottle of wine is negative. The area under the demand curve less the area under the supply curve is negative, to the right of w^*. Thus, at least graphically, we have shown the equivalence among a market equilibrium, a Pareto optimum, and a surplus maximum.

C. Supply and Demand for Bads

Supply, demand, and surplus measures for goods should be familiar concepts. This framework can easily be extended to bads. Consider our example of garbage. Viewing garbage as a commodity, it has a negative price. To be willing to consume (i.e., store) garbage, you must pay a negative price (i.e., receive a positive price—compensation). Figure 4.9a

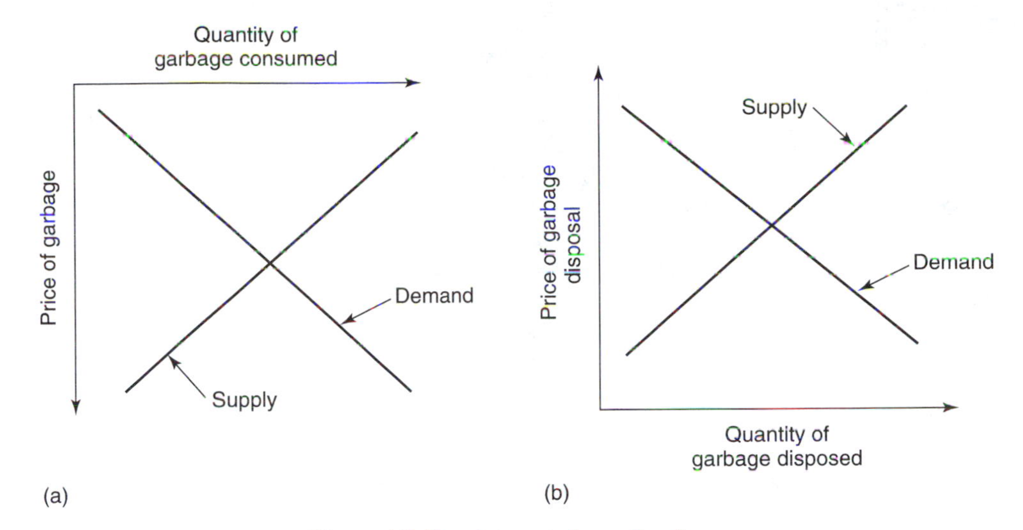

Figure 4.9 Two interpretations of garbage.

shows what supply and demand might look like for garbage. Demand is downward sloping. If the price (compensation) is low, people will not be willing to take much; if the price is high, people will take more. Similarly, generators of garbage will provide little if they must pay a large price; they will provide more at prices closer to zero.

This may be a little confusing. Thinking in terms of negative prices is not common in economics. Usually, however, we can change how we view the problem to make it more consistent with the standard model of supply and demand. Instead of thinking in terms of garbage (a bad), think in terms of garbage disposal (a good). A consumer of garbage is a supplier of garbage disposal. A producer of garbage is a consumer of garbage disposal. Figure 4.9b provides the same information as Figure 4.9a except that the information is given in terms of garbage disposal rather than garbage.

Where do these supply and demand curves come from? This is where we connect supply and demand to the Edgeworth box. If a consumer is faced with an income A and prices for wine and garbage so that their budget constraint is AB in Figure 4.10b, she would choose to consume point b in the left panel of Figure 4.10b. This is simply consumer choice: utility maximization subject to a budget constraint. If we now raise the price of garbage (moving it closer to zero), and give our consumer enough income to keep utility constant (compensated demand), then the slope of the budget line will become steeper and the consumer will move to a choice point such as a in Figure 4.10b. As the price of garbage is changed, consumer choice changes, tracing out a demand curve for garbage as shown in the right panel of Figure 4.10b. As the "law of demand" requires, this demand curve is downward sloping: raise the price and the amount demanded declines.

The supply side can be developed similarly. Figure 4.10a, left panel, shows the production possibility set for producing wine and garbage. As the price of garbage changes, keeping the price of wine constant, choice points A, B, and C result. Translating this into the right panel of Figure 4.10a results in a supply curve for garbage. It has the expected slope as well—a lower (higher in absolute value terms) price (higher compensation must be paid) results in less garbage being generated.

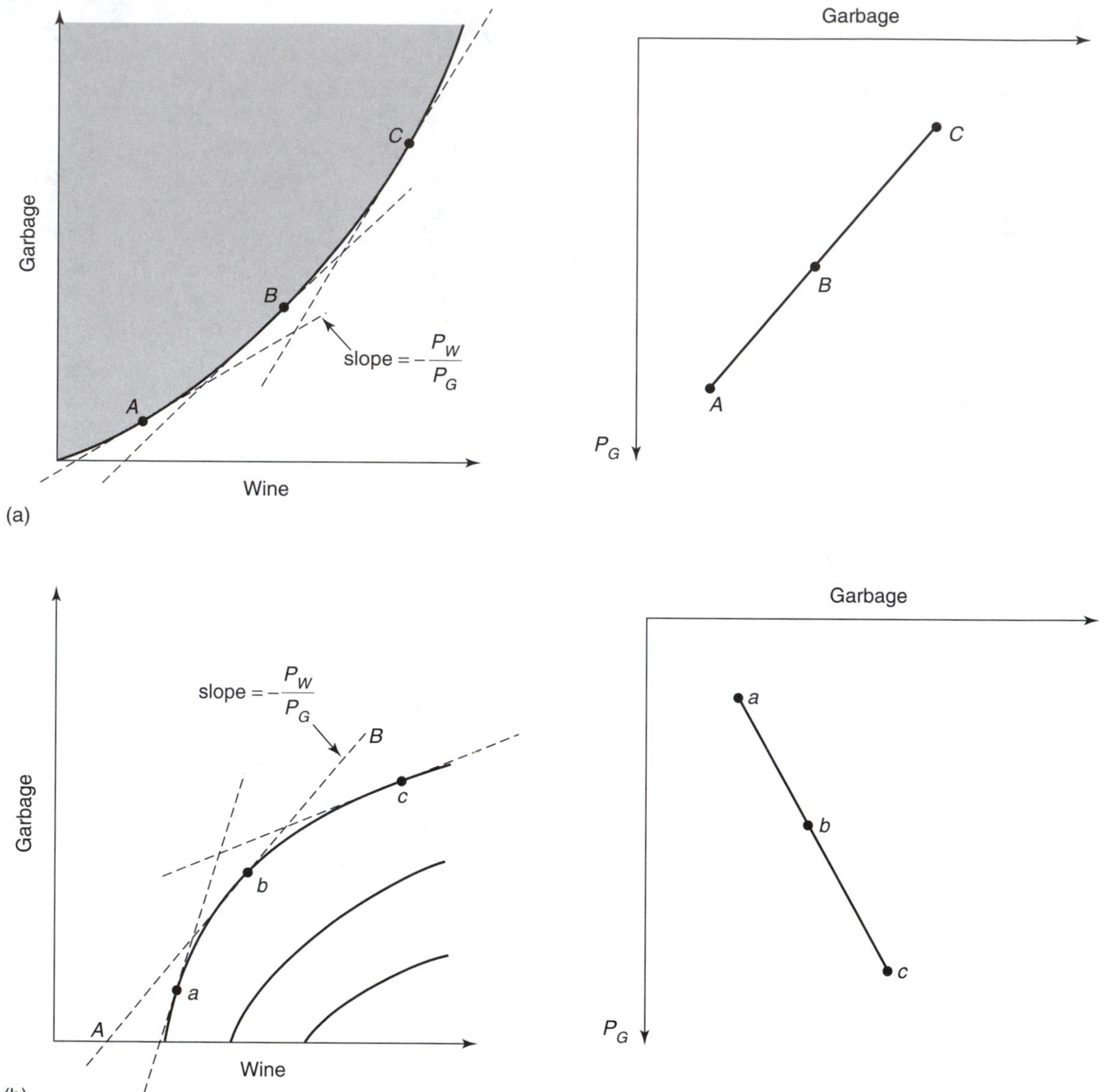

Figure 4.10 Obtaining supply (a) and demand (b) for garbage, holding the price of wine constant.

It is important to realize that we have been talking about individual supply and demand curves—for individual consumers or producers. Such demand and supply curves follow directly from utility and profit maximization. Often, in the marketplace, we are dealing with aggregate demand and supply—the sum of many individual demand and supply curves. Such aggregate supply and demand curves may look like those for individuals, but they need not share many of the same properties since they do not result directly from utility or profit maximization.

D. Surplus Measures for Bads

We can also obtain measures of producer and consumer surplus for bads. Focus more closely on the demand for garbage disposal, obtainable directly from the supply of garbage. Figure 4.11 shows Chateau Mull's demand curve for garbage disposal, along with some specific hypothetical quantities and prices. How much would Chateau Mull be willing to pay to dispose of five bags of garbage? To dispose of its first bag, it would be willing to pay up to \$6.10. To get rid of the second bag, it would be willing to pay up to \$5. Consequently, to get rid of two bags, it would be willing to pay up to \$11.10 = \$6.10 + \$5. Obviously if it does not have to pay that much it would be happy. The point is, the \$11.10 is its maximum willingness to pay for garbage disposal. Continuing, it would pay up to \$4 to get rid of the third bag, \$3.50 for the fourth bag, and \$2.80 for the fifth bag. Its total willingness to pay for the five bags is \$21.40 = \$6.10 + \$5 + \$4 + \$3.50 + \$2.80. This is a measure of the value of disposing of five bags of garbage—the consumer's surplus. Chateau Mull is willing to pay \$21.40—but not a penny more—to dispose of the five bags. This is why the \$21.40 is the value of getting rid of five bags of garbage. Chateau Mull is indifferent between generating no garbage (and no wine) and generating five bags of garbage, paying \$21.40 to get rid of it.

Note that what we are actually measuring is the area under the demand curve between the vertical axis and a vertical line through five bags of garbage. In Figure 4.11, the hatched and shaded area is the consumer surplus from consuming five bags of garbage disposal.

Now suppose Chateau Mull has to pay \$2.80 per bag for disposal or 5 × \$2.80 = \$14.00 total. The total surplus would then be reduced by payments, from \$21.40 to \$7.40 (\$21.40 − \$14.00). This is where the concept of surplus comes in. We do not buy if something is worth less than the price; we often buy when something is worth more. Our surplus is the difference between what the good is worth to us (willingness to pay) and what it costs. This is the single-hatched shaded area (above the horizontal line through \$2.80).

We can also carry this concept over to the producer. Think of Brewster as storing garbage bags in his back yard. Figure 4.11 also shows a hypothetical supply curve for

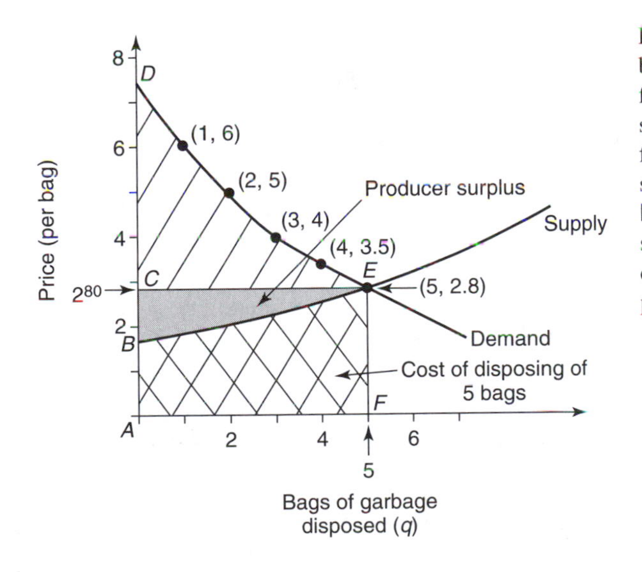

Figure 4.11 Chateau Mull's possible demand and Brewster's supply for garbage disposal. ABDEF, Consumer benefit (excluding payments) for disposal services; CDE, Consumer surplus for disposal services; BCE, Producer surplus for disposal services; ABEF, Cost of providing disposal services; BCDE, Total surplus for providing disposal services.

garbage disposal services. The first bag of garbage can be disposed of relatively easily (cheaply). The cost per bag goes up as the number of bags increases. Let $C(q)$ be the cost of disposing of q bags of garbage. The total cost of disposing of 5 bags of garbage is the area under the supply curve up to $q = 5$ the double hatched area of the figure. Brewster would never provide any services without payment. Suppose Brewster receives $2.80 per bag disposed. Then his total profit is revenue less costs. This is producer surplus shown as the shaded area in Figure 4.11.

Another interpretation of total surplus is the area under the demand curve less the area under the supply curve. As can be seen from Figure 4.11, this is also total surplus. The area under the demand curve between $q = 0$ and $q = 5$ is the total benefit of providing 5 units of garbage disposal services and the area under the supply curve is the total cost of providing these garbage disposal services. Thus a second interpretation of total surplus is as the benefits less the costs of providing the good. This leads us to the subject of benefit–cost analysis.

IV. BENEFIT–COST ANALYSIS

In the last section, we saw how measures of surplus could be used as a welfare measure. For a competitive market, consumer plus producer surplus is maximized at a market equilibrium and at a Pareto optimum. But environmental protection often requires government intervention in the market to correct market failures. The unfettered market produces too much pollution (for reasons we will see in subsequent chapters). Some public projects (such as National Parks and dams) are inadequately provided for by private markets. In these cases, we still desire a Pareto optimum quantity of goods and services, but we cannot rely on the market to provide them. How can we use the tools of economics to help decide if a public project or regulation is a good idea?

One of the primary tools for deciding on the appropriateness of a government intervention in the economy is benefit–cost analysis. The basic idea is simple: find the project that gives the largest surplus. As we saw in the previous section, total surplus (consumer plus producer surplus) is equivalent to total benefits less total costs. We know that in general a surplus maximum is equivalent to a Pareto optimum. The idea may be simple but implementing the idea is far from simple. The usual problem is difficulty in quantifying some of the benefits and/or some of the costs.

In the next few sections we will take a closer look at issues in benefit–cost analysis. In the next section we will continue our simple example of garbage disposal. We will look at how to evaluate the net benefits (benefits less costs) of several governmental programs for regulating garbage generation and disposal. We will then turn to issues of how to treat surplus in secondary markets and how to treat benefits and costs that occur at different points in time. These are important issues in implementing benefit–cost analysis.

A. Measuring the Effects of Government Programs

Consider an example. Chateau Mull is generating garbage which is being consumed by Brewster. For a variety of reasons, no market has emerged to solve the garbage disposal problem. The government must step in to decide what the Pareto optimal amount of garbage might be. Or a less ambitious goal would be for the government to propose several levels of garbage and try to choose which generates the most surplus. In our example, we let Chateau Mull have a demand curve for garbage disposal services (we could have also examined Anna's demand for garbage disposal services). We let Brewster act as the provider of garbage disposal services (we could have fabricated a firm to do this). The government is concerned because too much garbage is being generated. The government wishes to solve the problem and we will assume it has two options: directly dictate to Brewster and Chateau Mull how much garbage to generate and store or discourage garbage production by penalizing Chateau Mull's garbage-generating activities, but at the same time encourage the garbage market to operate. In particular, for every bag of garbage Chateau Mull generates, it would have to pay a fee to the government, in addition to any payments paid to Brewster. We are interested in measuring the welfare impacts of these policies.

Figure 4.12 shows the demand and supply curves for garbage disposal. If a market were to operate, we would expect the price of garbage disposal to be p^* and the quantity of garbage disposed of to be q^*, corresponding to the point at which supply and demand intersect in Figure 4.12, point *C*. The surplus associated with such an outcome is the hatched area of Figure 4.12—*ABCDE*. This is the area between the demand and supply curves and between $q = 0$ and $q = q^*$. Note that the market equilibrium maximizes the total surplus, since additions to surplus for $q > q^*$ are negative.

1. Quantity Control. Now suppose instead of a market for garbage disposal, the government steps in to provide the service. The government decides that the amount of garbage generated and disposed of should be q_1. The surplus associated with this level of service is the shaded area in Figure 4.12—*ABDE*. The loss associated with this incorrect level of garbage disposal service is the difference between these two areas: $ABCDE - ABDE = BCD$.

There are two ways of interpreting this result. If there had been no market for disposal services prior to government action, government action has netted society the surplus *ABDE*. Alternatively, there is a loss compared to what could be achieved, and that loss is *BCD*.

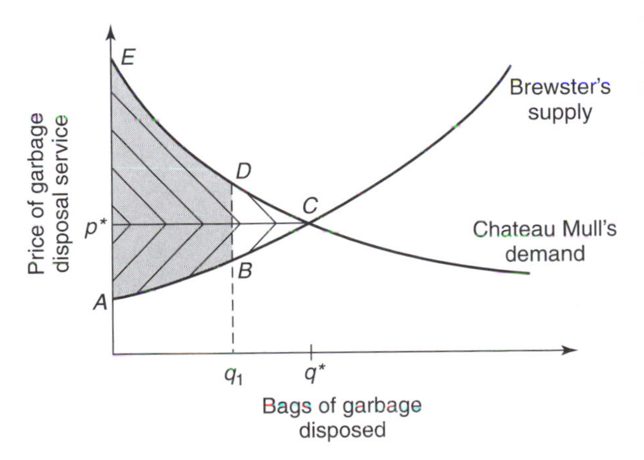

Figure 4.12 Total surplus from garbage. (q^*,p^*), Market equilibrium amount of garbage disposal; ABCDE, surplus from market equilibrium; ABDE, surplus from quantity restriction to q_1.

2. Tax. Now let us evaluate the operation of the garbage market in conjunction with a tax, paid by Chateau Mull, for every bag of garbage generated. What will happen to demand for garbage disposal services? Suppose the tax is \$1 per bag. If Chateau Mull formerly were willing to dispose of five bags if it paid \$5 per bag, it is now willing to pay Brewster only \$4 to dispose of the same quantity, since an additional dollar is being paid to the government. Thus Chateau Mull's demand curve drops uniformly by \$1 per bag. This has the effect of shrinking garbage disposal (and generation) as shown in Figure 4.13. Services shrink to q_1. How do we evaluate the welfare consequences of this?

The true demand and supply for garbage services have not really changed by the imposition of a tax. Chateau Mull still values garbage disposal in the same way and Brewster still has the same disutility from garbage disposal. Thus we would measure the consumer and producer surplus using the true supply and demand curves. Because of this, we find that the tax has had the same effect as the quantity control. The loss relative to the ideal amount of garbage disposal services (q^*) is the area *BCD* in Figure 4.13.

To evaluate the consequences of a tax or subsidy it is first necessary to determine what quantity of the good (or bad) will be supplied in response to the tax or subsidy. Having determined that, the undistorted supply and demand curves should be used to measure the surplus effects.

B. The Role of Secondary Markets

Often, a government action in one market will impact other markets. The obvious question is, how many markets do we have to examine in measuring the surplus associated with the government action? For example, suppose the government tightens regulations on sulfur dioxide emissions from industry. We would expect this to increase the demand for scrubbers (abatement equipment), which in turn might increase the demand for steel, which might increase the demand for iron. Where do we stop?

It turns out that secondary markets can be ignored so long as prices do not change in those markets.[7] Consider an example drawn from an actual empirical analysis of air pollution control The case is that of a large American city, Baltimore. The question is how much particulate matter should be emitted into the air in the city—what should be

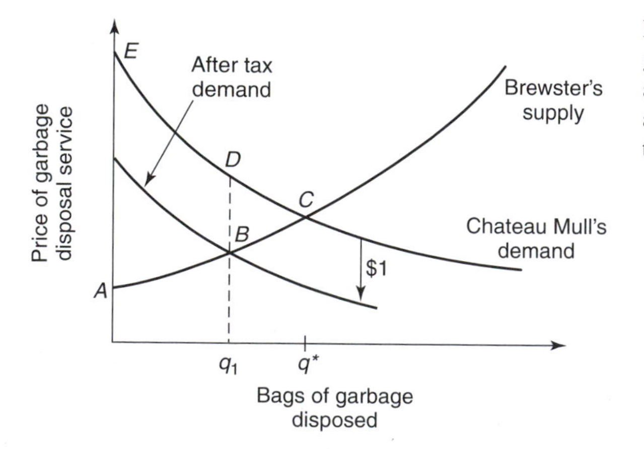

Figure 4.13 Effect of a tax on garbage production. q^*, Efficient amount of garbage disposal; q_1, amount of garbage disposal with \$1 tax.

the concentration limit on total suspended particulates (TSP)? This question was examined by several economists at the University of Maryland (Oates et al., 1989). The top panel of Figure 4.14 shows their estimate of the marginal cost to firms of different levels of the TSP standard (supply of pollution) and the marginal benefit of pollution (demand for pollution) in Baltimore.[8] Note that both are negative, consistent with pollution being a bad. The demand for pollution is roughly horizontal—constant marginal benefits. The bottom panel of Figure 4.14 shows a hypothetical market for electrostatic precipitators, the primary means of controlling emissions of TSP.[9] The supply of precipitators is shown as a horizontal line, since it is assumed to be a constant cost industry. Also shown in Figure 4.14 are two demand curves for precipitators, one with no TSP control in Baltimore and one (further to the right) with significant TSP control. Note that there will be no price effect in the precipitator market. Thus we can ignore it.

Returning to the supply and demand for TSP concentrations (top panel, Figure 4.14), note that 87 μg m^{-3} is the efficient level at which to set TSP limits. The benefits of moving from 120 μg m^{-3} to this efficient level is the shaded area in the top panel of Figure 4.14, equal to approximately \$45 million per year.

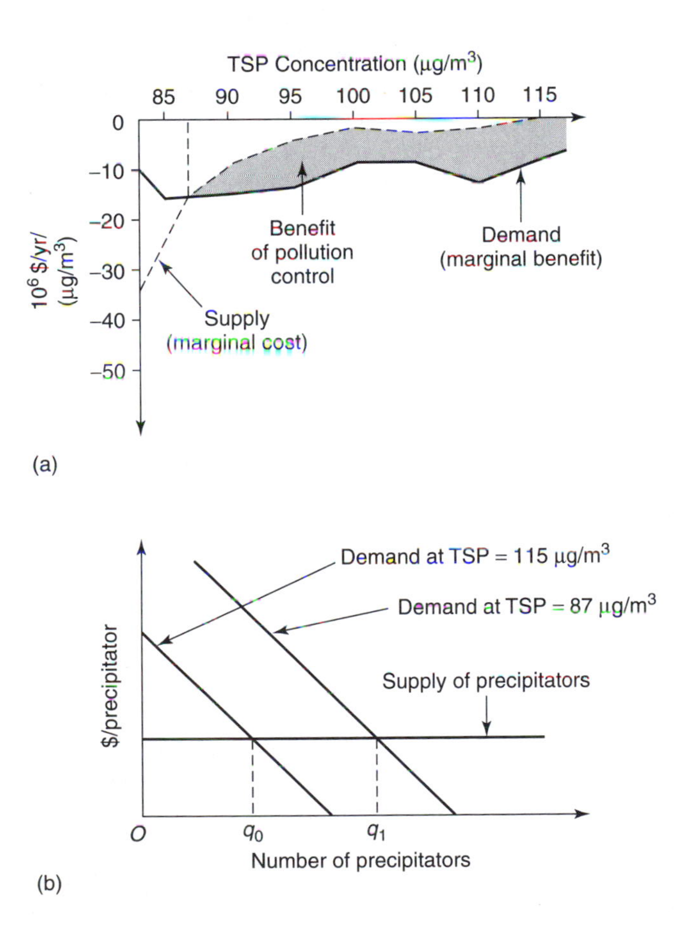

Figure 4.14 Cost–benefit analysis of pollution (TSP) reduction in Baltimore. Supply and demand for (a) pollution (adapted from Oates et al., 1989) and (b) pollution control (electrostatic precipitators).

C. Multiple Time Periods

Costs and benefits often occur at various points in time. For instance, we may spend money to reduce emissions of carbon dioxide today but see the benefits of this action only decades or centuries into the future. How do we evaluate such an intertemporal mixture of costs and benefits? The standard approach is to write a future cost (or benefit) as an equivalent current cost (or benefit), using a *discount factor*. The discount factor for time period t, β_t, is used to convert costs in time period t, C_t, into costs today, $\beta_t C_t$. For example, what would be the *current* equivalent of $1 paid 2 years from now? We would use a discount factor, some number less than 1. Suppose that discount factor is $\beta = 0.9$. That would mean that we would be indifferent between receiving $0.90 today and $1 two years from now. In general, if we are faced with a stream of costs $\{C_0, C_1, C_2, \ldots, C_T\}$ and a stream of benefits $(B_0, B_1, B_2, \ldots, B_T\}$, we can convert these streams into a net present value (NPV) of benefits less costs using the appropriate discount factors:

$$\text{NPV} = \sum_{t=0}^{T} \beta_t \{B_t - C_t\} \quad (4.6)$$

The next logical question is where do the values of β_t come from? There is some controversy over what the appropriate value of β_t is to use for environmental decision making—the social discount factor. The simplest answer is that the discount factors are market determined, based on how consumers and producers trade off the present with the future. Figure 4.15 shows how society might view decisions between consumption today (c_0) and consumption tomorrow (c_1). The line TXT' in Figure 4.15 is the production possibilities for our society between consumption today and tomorrow. If we consume less today and invest it (for example, plant trees to harvest tomorrow), we should have more to consume tomorrow. That is why the line segment OT is slightly smaller than the line segment OT'. The point T represents "consume everything today and nothing tomorrow." The point T' represents "consume nothing today and everything tomorrow." Which point along this production possibilities frontier will society choose? We can represent societal preferences by drawing indifference curves in Figure 4.15 based on a representative consumer. The indifference curve marked U_0 represents the highest level of utility attainable

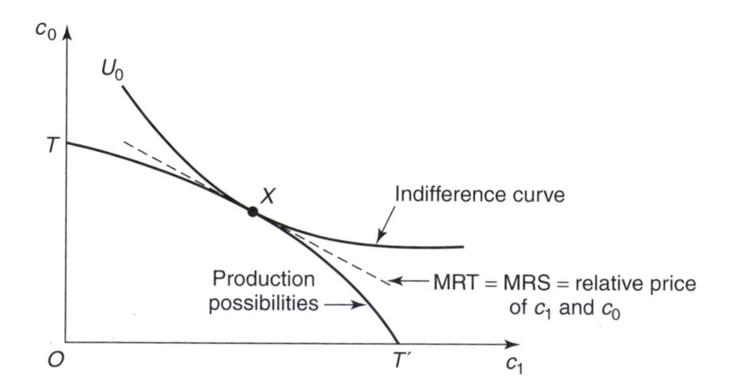

Figure 4.15 Market determination of discount factor.

that is still feasible (intersects with TXT'). The point X is the optimal consumption point. This is the point at which the marginal rate of transformation of c_1 for c_0 is equal to the marginal rate of substitution of c_1 for c_0. The slope of the tangent to either curve at this point represents the price ratio between consumption today and consumption tomorrow. Suppose this slope is β. That means that if I increased consumption of c_1 by \$1 and reduced consumption of c_0 by β, my utility should stay the same. But this is the same definition we used for the discount factor. If the discount factor is β, you should be indifferent between one dollar tomorrow and β dollars today.

Thus we see that the discount factor is determined by two things: the consumer's rate of time preference (willingness to defer consumption) and how productive investments are (the marginal productivity of capital).[10] The typical assumption regarding the discount factor is that discounting is exponential. In other words, β_t is defined by

$$\beta_t = \beta^t = (1 + r)^{-t} \tag{4.7}$$

where r is the discount rate. One characteristic of exponential discounting is that the future does not matter very much unless r is very close to zero. In Eq. (4.7), if r is 5%, then β_t for $t = 100$ would be 0.0076. This means you would be indifferent between \$1 a century from now and less than 1¢ today. This is somewhat troubling when trying to justify controlling carbon dioxide emissions today in order to avoid significant damage to the planet a century from now.

There are two ways out of the troubling conclusion that discounting trivializes what happens a century from now. One is to use a very low discount rate. A discount rate of 1% results in a discount factor of 0.36 for $t = 100$ years. Another approach is to question exponential discounting. There is some empirical evidence that an individual's discount factor is not exponential as in Eq. (4.7) but "slower" so that the medium future is discounted to the same extent as the distant future.[11] There is one fundamental problem with either using a discount rate that is lower than the market rate or using nonexponential discounting.[12] The problem can best be illustrated with an example. Suppose investing \$1 today in a risk-free investment (perhaps planting trees) can be expected to return \$100 in 50 years (compond return at the risk-free market rate). In contrast, an environmental policy (for instance, a reduction in carbon emissions) involving spending \$1 today will avoid \$30 of environmental damage in 50 years. Which policy would the residents of the earth in the year 2050 prefer us to undertake?

D. Using Benefit–Cost Analysis

What we have shown is that measures of demand and supply can be used to make normative decisions even if markets do not exist. In particular, demand and supply can give us monetary measures of the benefits of pursuing various governmental interventions on behalf of the environment. To be honest, we have painted a relatively rosy picture of benefit–cost analysis. Though potentially powerful, it is not easy to implement. Further, there are subtle assumptions built into the approach.

One assumption regards the distribution of income. As mentioned earlier, if incomes change, supply and demand curves will probably change, and thus so will measures of surplus associated with a government policy. Thus if one does not consider the income distribution "correct" or "just," then there may be reason to question the policy prescriptions arising from benefit–cost analysis.

Another assumption is that the compensation principle applies. Suppose cost–benefit analysis concludes that building a dam in a scenic canyon is a good idea because the benefits exceed the costs. This means that the gains to some from having the dam exceed the losses to others from having the dam. For there to be a Pareto improvement, the winners should compensate the losers so that everyone is better off with the dam. However, this compensation rarely occurs. This is the controversial Kaldor–Hicks compensation principle at work.

Another problem, not insurmountable, is that measuring the damage to the environment from pollution is no easy task. We will consider that later in this book and show that great strides have been made. It would be foolish to pretend that it is easy to quantify the marginal willingness to pay for all environmental goods, since many are very intangible. This is not a fatal problem for benefit–cost analysis, only a factor that makes its application difficult in some cases.

Another implementation problem involves what economists call the theory of second best. If there are prior distortions in the economy, such as taxes or regulations, perhaps unrelated to the environmental problem at hand, supply and demand curves will be distorted. And distorted supply and demand will give distorted measures of surplus.

This list of problems with benefit–cost analysis should not be taken as an indication of fatal flaws in the method. Benefit–cost analysis remains a vital part of quantifying and evaluating the desirability of government intervention in the economy on behalf of the environment. One should simply be aware of its shortcomings.

SUMMARY

1. The Pareto frontier consists of all allocations that cannot be improved on in the Pareto sense. In other words, all other allocations make at least one person worse off.

2. Efficiency to an economist means being on the Pareto frontier—no redistribution of resources can be undertaken without making someone worse off.

3. In a competitive exchange involving a good and a bad, the bad will be negatively priced and the good will be positively priced. Trade will generally take place.

4. In a two-good barter economy, exchange will take place where the marginal rate of substitution between the two goods is the same for all participants. Suppose consumption of one good is increased a small amount and consumption of the other good is decreased so that utility stays constant. The ratio of these two changes is the marginal rate of substitution.

5. In the context of trade and bargaining, the contract curve is another name for the Pareto frontier.

6. When prices are used for exchange in a two-good exchange economy, the equilibrium quantities of goods will generally be the point on the contract curve at which the budget line is equal to the marginal rate of substitution for each consumer and at which the budget line passes through the initial endowment.

7. Efficiency in production requires that the marginal rate of transformation for a producer be equal to the price ratio. Suppose production of one good is increased a small amount and production of another good is decreased so that profit stays constant. The ratio of these two changes is the marginal rate of transformation.

8. Efficiency in a competitive economy with production and trade requires that the marginal rate of substitution between two goods be the same for all consumers consuming both goods, that the marginal rate of transformation between two goods be the same for all producers producing both goods, and that the marginal rate of substitution equal the marginal rate of transformation, which in turn equals the price ratio.

9. The first theorem of welfare economics states that a competitive market equilibrium is Pareto optimal.

10. The second theorem of welfare economics says that any Pareto optimum can be supported by a competitive market equilibrium simply by adjusting the initial distribution of resources in the economy.

11. Markets usually fail when pollution is present. Thus we cannot rely on markets to give us a Pareto optimum. It is usually necessary for the government to intervene to achieve efficiency. Benefit–cost analysis is a method for determining if such government actions are efficient.

12. Consumer surplus is a measure of the value of goods to consumers, over and above the amount paid for the goods.

13. Producer surplus is a measure of the gains from selling a good, after the cost of production has been taken into account.

14. Total surplus of increasing the quantity supplied from q^* to q^{**} is the area bounded above by the demand curve, below by the supply curve, to the left by the vertical line $q = q^*$, and to the right by the vertical line $q = q^{**}$.

15. Benefits or costs at a future point in time can be converted into equivalent benefits or costs today using a discount factor. The discount factor is an equilibrium concept, equal to the marginal rate of substitution of consumption tomorrow for consumption today, also equal to the marginal rate of transformation in the economy between consumption tomorrow and consumption today.

PROBLEMS

1. In Figure 4.5, describe the conditions necessary for the indifference curves to be concave upward.
2. In Figure 4.6, show what trade will take place if the initial endowment is that Brewster has all the garbage and Anna has all the wine (i.e., start at the point in the lower right corner). Explain.
3. In Figure 4.7, the slope of production possibility frontier (the MRT of wine for garbage) is increasing as garbage and wine increase. Does this make sense? Why?

4. Combine Figures 4.5 and 4.7 together into a single figure showing how much garbage and wine will be produced in an economy consisting of Anna and Chateau Mull. Show the relative price of garbage and wine in the figure. Assume Anna's income is fixed.

5. In the context of Problem 4, suppose you know that Chateau Mull's MRT of wine for garbage was 2:1 (it would require the creation of two additional bags of garbage to produce one more bottle of wine) and Anna's MRS of wine for garbage was 3:1 (she would take three more bags of trash to get one more bottle of wine). For this society would you recommend greater expenditures on garbage control, less expenditures on garbage control, or the status quo? Explain your answer.

6. Suppose an environmental regulation requires all polluters to reduce emissions by 50%, even though the cost of pollution control differs widely from one polluter to another. Why might this be inefficient, using the definition of efficiency developed in this chapter?

***7.** Suppose Humphrey and Matilda live together. Humphrey currently smokes 20 packs of cigarettes per month; Matilda hates the smoke. They currently have no agreement restricting smoking. Their only joint expense is monthly rent, which they split 50:50. Draw an Edgeworth box with two goods—smoke and rental payments. Make up some reasonable indifference curves. Show the initial endowment. What Pareto efficient points might result from bargaining to restrict smoke? How does the graph show what price per pack Matilda might pay to buy down Humphrey's smoking (i.e., show the relative prices on your figure)? How would your answers change if the status quo is that the two have an agreement for no smoking and Humphrey would *like to smoke* as much as 20 packs per month? He must seek Matilda's permission to do so. (Hint: For Matilda, redefine Humphrey's smoking as smoke reduction.)

NOTES

1. Government failure is also possible but not considered here. See Weimer and Vining (1992).
2. These ideas are developed in more detail in most microeconomics texts (such as Varian, 1996). See also texts on applied welfare economics, such as Just et al. (1982) or texts on cost–benefit analysis, such as Gramlich (1981).
3. Although there are few prohibitions on the interstate transport of waste within the United States, there are differential regulations that discriminate against placing out-of-state waste in local waste disposal facilities. Waste shipments from Germany to Eastern European countries have generated public criticism within Germany and elsewhere. There are international treaties prohibiting certain international waste shipments.
4. See Varian (1996) for a full development of a two-agent (Robinson Crusoe) economy with production and exchange.
5. The purist will correctly point out that this is not always the case—corner solutions can occur. We are ignoring those difficulties for the time being.
6. These conditions for a market to operate are discussed by many authors, including Bator (1958). It is possibe to relax some conditons (such as complete information) and still have a functioning market. An exploration of that issue would lead us too far astray.
7. It is also important that the secondary market not be distorted by some other government program.
8. The marginal costs and marginal benefits are in 1980 US$. It is assumed that these are annual marginal costs and benefits, though that is not clear from Oates et al. (1989).

9. The precipitator market was not examined by Oates et al. (1989). It is used here for illustration purposes only.
10. When we observe discount factors in the marketplace, they usually have additional components added in, particularly uncertainty. Uncertainty about the future is risk that will inevitably cause one to discount further a promise of a dollar to be received in the future. Tax policy also drives a wedge in this market making observations of the discount rate difficult. Pearce and Turner (1990) and Nordhaus (1994) provide good discussions of the use of the rate of time preference and the marginal productivity of capital to deduce a social discount rate for use in environmental policy formation. See also the more technical volume by Lind et al. (1982).
11. Cropper et al. (1992, 1994) provide some experimental results on discount factors being less than exponential in the case of public projects intended to save lives. Harvey (1994) refers to this as "slow discounting"—discounting the future at a rate slower than exponential. Hyperbolic discounting is a particular type of slow discounting.
12. Schelling (1995) discusses some other problems with abandoning discounting in evaluating climate change policy.

5 MARKET FAILURE: PUBLIC BADS AND EXTERNALITIES

In this chapter we consider the fundamental difference between conventional goods (e.g., bread), which the market allocates and rations efficiently, and environmental commodities (e.g., pollution), which the market fails to allocate efficiently. It is the very nature of pollution that leads to this market failure. Economics has developed the concepts of public bads and externalities to describe the characteristics of these environmental commodities that lead to market failure. A bad is of course the opposite of a good. A good is individually beneficial and people want it; a bad does harm and people do not want it. Quite simple, really.

I. PUBLIC GOODS AND BADS

Despite the virtues of a price system for making decisions on the production and consumption of goods, the price system does not always work nor is it always desirable to rely on it. This breakdown can occur either on the consumption or production side of the market. Production complications can involve scale economies and the existence of a natural monopoly. This leads to the desirability of having a few firms, which may lead to price manipulation by firms and issues of budget balancing. Market failure on the consumption side usually involves goods that have characteristics of "publicness" or involve externalities, concepts that we explore in this chapter.

Economics has defined two fundamental characteristics of goods: excludability and rivalry. Excludability has to do with whether it is *possible* to use prices to ration individual use of the good. Rivalry has to do with whether it is *desirable* to ration individual use, through prices or any other means.[1]

A. Excludability

To be able to use prices to allocate a good, it is necessary to ensure that consumers do not consume a good unless an appropriate price has been paid. Thus it must be possible to keep the consumer from the good. This is excludability. Not all goods are excludable. For instance, a regular broadcast television signal can be received by anyone with a television set. It is not possible to selectively determine who will be able to receive the signal. Another example is the fishery on the high seas. Keeping consumers (in this case fishermen) from consuming this resource is very difficult (i.e., costly). As an example of a bad, air pollution is also nonexcludable. Air, of whatever level of cleanliness, is all around us. We cannot exclude certain people from consuming air pollution.

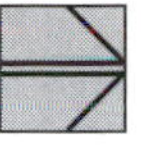

Definition *A good is excludable if it is feasible and practical to selectively allow consumers to consume the good. A bad is excludable if it is feasible and practical to selectively allow consumers to avoid consumption of the bad.*

Why is excludability important? Because to attach a price to the consumption of a good or bad, we have to be able to deny that consumption if the price is not paid. This is how market allocation works. For a good, this is a straightforward concept. If I produce hamburgers, I must be able to deny you the right to consume a hamburger unless you pay for it. Some goods are nonexcludable. A city park without a fence and entrance control is not excludable. I cannot charge for admission because anyone can use the park, whether or not they have paid admission. Thus no one will pay admission. A fence and entrance control could be added, but this would be so expensive that its cost would probably outweigh any benefits associated with restricting entry to the park. Generally, we would expect to see exclusion only when the benefits of exclusion outweigh the costs of exclusion. Exclusion must be not only technologically possible but also not too costly, relative to the benefits of exclusion (if any). This is the reason the word "practical" is used in the definition of exclusion.

For a bad, the concept is a little more subtle. Suppose I have produced garbage and I want someone to take it off my hands—"consume" it. With excludability, I can pay someone to store it for me and I am rid of it. A price can be attached to the act of consumption. Without excludability, I can pay someone to consume the bad, but that person can give it right back to me—there is nothing to force that person to actually consume the bad. Thus prices do not work.

Household garbage is excludable with the right laws on littering and trespass; it is not excludable without those laws. Suppose I have generated some household trash and wish to pay someone to consume it (i.e., store it safely). Without laws on trash disposal, whomever I pay to dispose of my garbage could simply leave it by the side of the road or throw it back into my yard. With laws preventing such actions, when I pay someone to consume my trash, that person must retain control of the trash until someone else voluntarily (i.e., with pay) agrees to take it.

Urban air pollution is not excludable. It is not possible to selectively target who is to consume the air pollution (i.e., breathe it). Everyone consumes it to the same degree. If it were possible to exclude air pollution, only those people who agree to be compensated for the pollution would consume it. If someone is not compensated, they would not consume.

Two factors play a major role in excludability. One is the cost of exclusion and the other is the technology of exclusion and how it changes over time. Consider the Great Plains of the central United States. In the last century, it was open to everyone for grazing (for everyone's animals, that is), since the cost of fencing was prohibitively expensive compared to the gains from exclusion (prevention of overuse). Over time, the payoff to exclusion increased but, more importantly, the technology of exclusion changed (the invention of barbed wire), making fencing cheaper. A similar case applies to television, particularly those programs disseminated via satellite and cable. Historically, it has been too expensive to exclude consumers. However, with the development of low-cost signal scramblers and unscramblers, exclusion became economically feasible, particularly for high value programs such as recent films. A third example is local parks. A local park could be made excludable by building a fence around it and installing a gatekeeper to control access. However, compared to the value generated by the park, such costly measures are rarely warranted. Only when there are low cost ways of excluding, such as parking controls for parks that can be accessed only by car, will exclusion exist.

Thus exclusion must not only be physically possible but also must be a good idea, given the cost of exclusion compared to the benefits. In general, the benefits involve the additional social value of having limited as opposed to unlimited access to the resource. If the good is being provided privately, the benefit is the revenue that can be gained by charging admission.

Environmental goods have undergone some change in excludability over the past few decades, mostly through legal rather than technological changes. The simplest example is garbage (like the normal household type). Without institutions, garbage is not excludable. Without laws against littering, garbage will simply be dumped wherever convenient, much as it was in the Middle Ages in Europe when people would throw it out their window onto the street. This, however, has been defined as littering or creating a nuisance and is illegal in most locales. Consequently, garbage is now excludable. People can selectively choose to trade money and garbage. This is the economic transaction of the trash collector and the household. This legal aspect of excludability of course could also apply to ordinary goods. Without laws protecting property and outlawing stealing, all goods would be community property and exclusion would not be possible.

Space plays a particularly important and, in some ways, confusing role with regard to exclusion. Most nonexcludable goods and bads are provided locally—city parks, television, air pollution. A consumer can effectively be excluded through location. If a consumer can be prevented from living in a community then the consumer will live elsewhere and effectively be excluded from the community's locally nonexcludable goods. It will be too costly to travel to consume those goods. The basic point is that goods and bads may be locally nonexcludable though globally excludable.

Why is excludability important? For a price system to work it must be possible to take possession of the good or bad for which the price is being paid. Without excludability, a price system cannot work.

B. Rivalry

A second important characteristic of goods is rivalry. This is a slightly more subtle concept than excludability. Rivalry pertains to the manner in which a good is consumed. As an example, compare the consumption of a hamburger with that of a flower garden. For the hamburger, the act of consumption destroys the good and makes it unavailable for anyone else to consume. In contrast, the act of consumption of a flower garden involves light bouncing off flowers and being transmitted to the eyes of the consumer. This is fundamentally nondestructive to the flower garden and in no way diminishes the ability of someone else to "consume" the flower garden in precisely the same way.

These examples best illustrate the concept of rivalry. A good is rival in consumption if the act of consumption reduces the amount of the good that might be available for other consumers. A good is nonrival if consumption does not diminish what is available for others. Note that in contrast to exclusion, this is not a characteristic of a good that might change with technology or costs. A hamburger will always be rival since the nature of consumption will not change. Thus rivalry is a more fundamental characteristic of a good or bad than is exclusion.

Garbage (the household variety) is an example of a rival bad. When I "consume" a bag of garbage, I am taking control of the bag, perhaps storing it in my backyard. When I consume that bag, it is unavailable for others to consume. Thus standard garbage is a rival bad.[2] Air pollution on the other hand is nonrival. When I take a deep breath in downtown London, I in no way diminish the ability of others to "enjoy" dirty London air.[3] My act of consumption does not diminish the quantity of the bad available to others. The global climate (threatened by greenhouse gases) is the purest example of a nonrival good. One person's enjoyment of the climate in no way impinges on the ability of others to enjoy the climate.

One way to view the idea of rivalry is through opportunity cost. When I consume a rival good such as a hamburger, I am reducing the number of hamburgers available for others, or perhaps necessitating that another hamburger be manufactured to return us to our starting point. In either case, there is an opportunity cost for others associated with my consumption. In contrast, when I consume a flower garden, there is no social opportunity cost (for others) of that consumption. The same amount of flower garden is available for others. No additional flowers need be planted. When I breathe the air in downtown Mexico City, I am not changing the amount of pollution available to others. Thus I am neither benefitting nor imposing additional costs on others. There is no opportunity cost (positive or negative) imposed on others from my breathing. If, on the other hand, my breathing reduced the amount of pollution, then there would be a negative social opportunity cost (an "opportunity benefit") associated with my breathing polluted air. This is the case for garbage consumption. When I consume a bag of garbage (i.e., store it in my back yard), I diminish the amount of garbage everyone else has to tolerate.

Definition *A bad (good) is* rival *if one person's consumption of a unit of the bad (good) diminishes the amount of the bad (good) available for others to consume, i.e., there is a negative (positive) social opportunity cost to others associated with consumption. A bad (good) is* nonrival *otherwise.*[4]

One complicating factor that applies to goods more than bads is congestion. A sparsely populated rural highway is nonrival in that there is no opportunity cost associated with one additional person using the road. We all know, however, that once congestion sets in, there is an opportunity cost of that extra driver and the road is no longer nonrival. A road is a congestible good—nonrival for low levels of consumption and rival for high levels of consumption. The typical reason for congestibility is an indivisibility in production that is closely related to economies of scale. We cannot produce a one-person road. Roads by their very nature can handle a certain amount of traffic without becoming congested. Alternatively, even if we could produce a one person road, it would be almost the same price as a many-person road. It would be less desirable to provide many one-person roads rather than a single multiperson road.

Why is rivalry an important issue? The key is efficiency. If there is no cost associated with incremental use, and if price equals marginal cost, the price should be zero. But with a price of zero, how can revenues balance costs so that the good or bad is efficiently provided? We return to this issue later.

C. Pure Public and Pure Private Bads

We can now put these two concepts of excludability and rivalry together and classify goods on the basis of both criteria. Figure 5.1 shows these two criteria together. The degree of rivalry is shown horizontally and the degree of excludability is shown vertically. To the right, bads are rival and to the left, bads are nonrival. The top represents bads that are excludable and the bottom is reserved for bads that are nonexcludable. Although Figure 5.1 is drawn so that bads are either rival or nonrival, or excludable or nonexcludable, there is really a continuum in both cases. For instance, because noise disipates with distance, noise can be made excludable by prohibiting people from being within a certain distance of the noise. That is why noise could be both excludable and nonexcludable—it depends on the nature of the noise. Garbage is shown as excludable and rival. Greenhouse gases, leading to climate change, are shown as nonrival and nonexcludable. No one can easily be prevented from being subject to the global climate and any one person's consumption of the global climate imposes no costs on others.[5]

Bads in the lower left-hand corner of the figure (nonrival, nonexcludable) are termed public bads. Bads in the upper right are termed private bads. There is no particular terminology for the land between these two extremes. The significance is that private goods and bads will typically be provided by the market, and provided efficiently. In many communities in the United States, private companies have spontaneously developed to haul and dispose of garbage; individuals contract with these companies to dispose of their trash. Pure public bads will not be provided by the market, primarily because nonexclusion makes it impossible to charge or compensate for use. The nonrivalness of these goods makes such prices undesirable anyway, as we shall see later.

There is a certain amount of redundancy between rivalry and excludability, despite the fact that they are different concepts. It is difficult to generate examples of nonex-

	Nonrival	Rival
Excludable	• Water pollution in small lake • Indoor air pollution	• Household garbage today
Nonexcludable	• Noise • Greenhouse gases	• Household garbage in Middle Ages

Figure 5.1 Excludability and rivalry for bads.

cludable rival goods or bads. This is because rivalry involves physical possession and destruction for consumption (e.g., the hamburger). If this is possible, it is probably also relatively easy to exclude.

Note that this entire discussion pertains to the characteristics of goods in consumption, not production. There are of course characteristics of the production process (such as natural monopoly) that create difficulties for using markets to govern production. But for the most part, public goods and bads can be privately produced and sold for a price like any other good or bad. The difficulty is typically in consumption. It is the characteristics of consumption of pollution that cause the market failure. We cannot keep pollutants contained and one person's consumption does not reduce pollution levels.

II. OPTIMAL PROVISION OF PUBLIC GOODS AND BADS

The problem now is to determine how much of a public good or bad should be provided. This is one of the basic questions in economics. The standard response in the case of private goods is that they should be provided up to the point at which the marginal cost of production equals the price and where the quantity produced equals the amount demanded at that price. Quite simply, produce and consume where demand and supply curves intersect.

The conventional approach to demand for goods is, for a given price, to sum individual quantity demanded to obtain market demands. At a given price, we see how many loaves of bread are demanded by each individual and sum these quantities to obtain the total number of loaves of bread demanded at that price. This is commonly referred to as horizontal summation of individual demands and is illustrated in Figure 5.2. Figure 5.2a shows the individual demands for some good by the two members of our society, Anna and Brewster. The horizontal axis shows the quantity and the vertical axis shows price or marginal willingness to pay (MWTP). The MWTP is how much one more unit of the good is worth to the person. In a market, the price is also equal to the value of one more unit of the good to the person. If the good is a rival good, then the aggregate demand is obtained by summing individual demands horizontally, as shown in Figure 5.2b. For any given price, we see the total amount of the good the consumers wish to consume.

If the good is nonrival then we proceed differently. We are interested in the total marginal willingness to pay for specific amounts of the nonrival good. In other words, for a particular price, how much of the good will consumers wish to consume? The difference here is that all consumers consume the same amount of the good. So the price paid to producers for the good is the sum of the prices paid by the individuals. For example, if Anna is willing to pay $1.90 for the fifth public park and Brewster is willing to pay $2.80 for the fifth public park then together they are willing to pay $4.70 for the fifth public park. To capture this, we sum demand vertically, as shown in Figure 5.2c.

The good is produced in the same way whether or not it is rival in consumption. Thus the same supply or marginal cost curve applies to its production. Supply is shown together with demand in Figure 5.2d. If the good is considered rival, consumption is Q_R, whereas if the good is considered nonrival, consumption is Q_N. Note that more of the good will be produced if it is nonrival, but this need not always be the case.

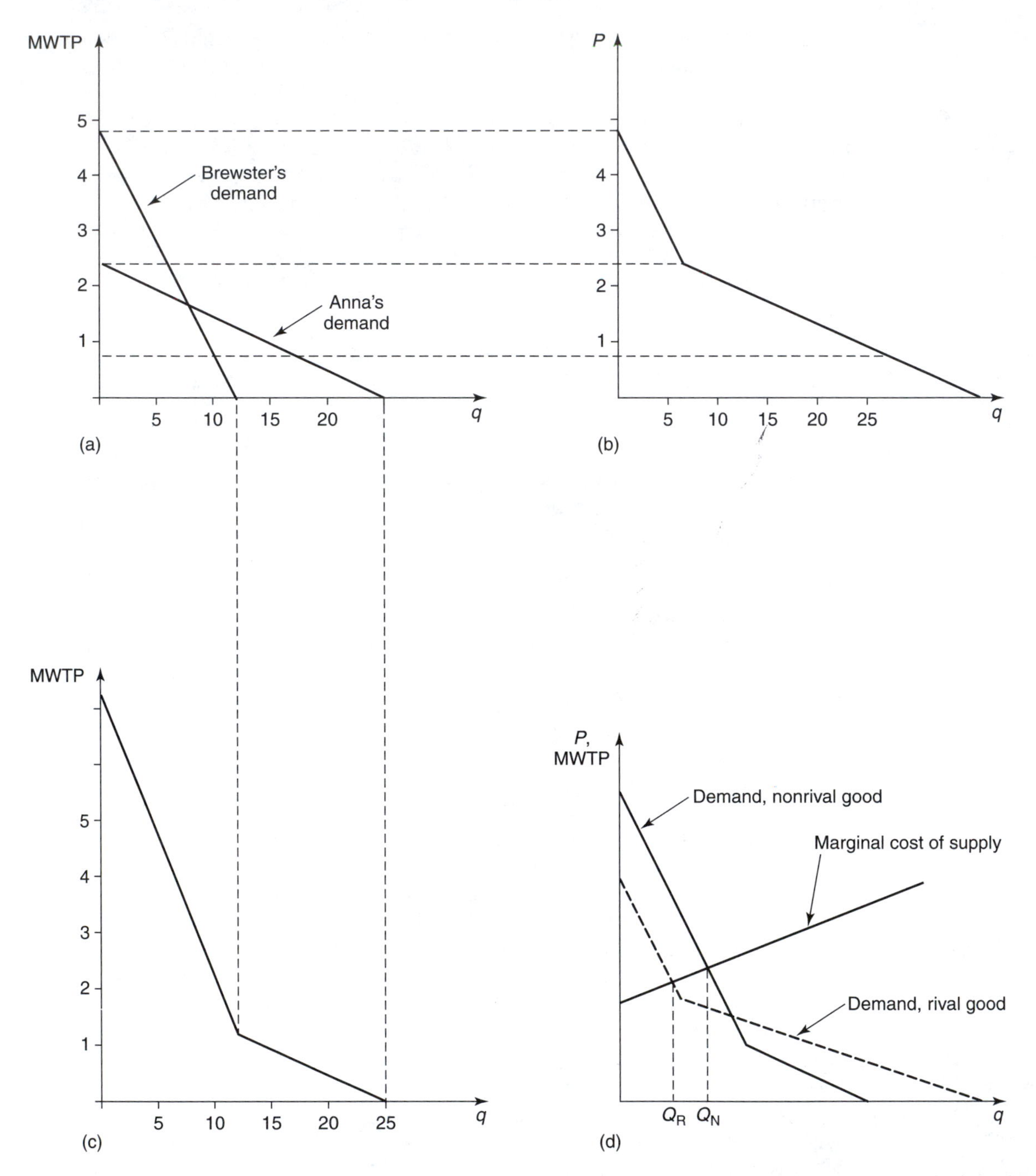

Figure 5.2 Supply and demand, rival vs. nonrival goods. (a) Individual demands; (b) aggregate demand, rival good; (c) aggregate demand, nonrival good; (d) efficient provision. Q_R, Quantity produced of rival good; Q_N, quantity produced of nonrival good.

Also note that although we can infer the market price for a rival good from the intersection of supply and demand, we cannot do this for a nonrival good. At the price at which supply and demand intersect for the nonrival good, individual demand will be quite low, since aggregate demand is the *sum* of individual willingnesses to pay. It is not easy to induce the optimal provision of the nonrival good shown in Figure 5.2d.

This approach to supply and demand for a public good is known more generally as the Samuelson condition for optimal provision of a public good. Those conditions are that for efficiency, the marginal rate of transformation of a public good for a numeraire private good should be equal to the *sum* over all individuals of their individual marginal rates of substitution of the public good for the private good.[6]

It is worth emphasizing that supply and demand curves exist for any commodity, independent of whether there is a market for that commodity. Thus despite the fact that markets rarely exist for public goods or bads, demand curves still exist, though they may not be well known.

III. PRICING PUBLIC GOODS AND BADS

We have detailed how public goods and bads differ from private goods and bads and have intimated that markets will not work for public commodities. We have also examined how much of the public good or bad is desirable. In this section, we take this question further, examining what happens if markets are used. We will see that the basic problem is a mismatch between what is optimal for the producer and what is optimal for the consumer. We will assume excludability is possible; otherwise there is no hope of using prices as a rationing device. The first question is what happens when markets are left to supply the public good?

A. Market Provision of Public Goods

What happens if markets are used? We sort of know what the answer is but it is important to understand why: if we let the market operate without intervention, public goods generally will be underprovided.[7] To see this most simply, assume we have a society with N identical individuals. Each individual can consume two goods, a rival, excludable private good, x, and a nonrival, nonexcludable public good, G. To clarify things, the public good is nonexcludable in consumption (everyone gets to consume it) but it is of course excludable in production—it is produced like any other good at a cost and can be sold at a price to anyone. The price of the private good will be set to one and we will adjust the units with which we measure the public good, so that its price is also one. Assume that each person has income w available to them. The utility of any of the identical people is given by

$$\text{Utility} = u(x,G) \tag{5.1}$$

We might think that G is out of the control of an individual, because it is a public good. However, individuals are free to purchase some of the public good, if that is in their own best interest (they are unlikely to take anyone else's interests into account). Let g be the amount of the good that one individual will purchase and let $\overline{G}$ be what is provided by everyone else: $G = \overline{G} + g$. This of course means that if our individual provides g of the public good, only $w - g$ of money is available for the private good. Thus we can rewrite the individual's utility as

$$\text{Utility} = u(w - g, g + \overline{G}) \tag{5.2}$$

Now our individual must decide how much g to choose to consume. The choice depends, of course, on how much of the public good is being provided by everyone else. Figure 5.3 shows indifference curves for our individual with the two goods being g and $\overline{G}$. Remember wealth or income (w) is fixed. If we assume that preferences are such that for any utility level it will be desirable for our consumer to spend his money on both private goods and public goods, then indifference curves must be U-shaped. More $\overline{G}$ is needed to compensate for the extremes of all private good consumption or all public good consumption, in order to keep utility constant. However, at high levels of public goods provision by others, $\overline{G}$, it is entirely plausible that no privately provided public goods are necessary. In this case, $\overline{G}$ needs to increase to keep utility constant as g rises from zero and indifference curves will be upward sloping in g.

Note that if $\overline{G}$ is fixed, the best choice of g is the point at which one of the indifference curves just touches the horizontal line at the fixed $\overline{G}$. As $\overline{G}$ increases (other people are providing more of the public good), the best choice g falls. If others are providing the public good, then we do not need to provide as much ourselves. The "best response" line in Figure 5.3 indicates the choice of g that will be made in response to the amount of the public good selected by others, $\overline{G}$. For a high enough $\overline{G}$, no g will be provided.

Now how much will be provided by our little society? Since everyone is the same, everyone will make the same choice about g. Thus $\overline{G} = (n - 1)g$. This is a straight line of slope $n - 1$, also drawn on Figure 5.1. The point at which it intersects the best response line (point N) provides the amount of g and $\overline{G} = (n - 1)g$ that will be generated in this economy: g_N, $\overline{G}_N$. So some amount of the public good is provided privately.

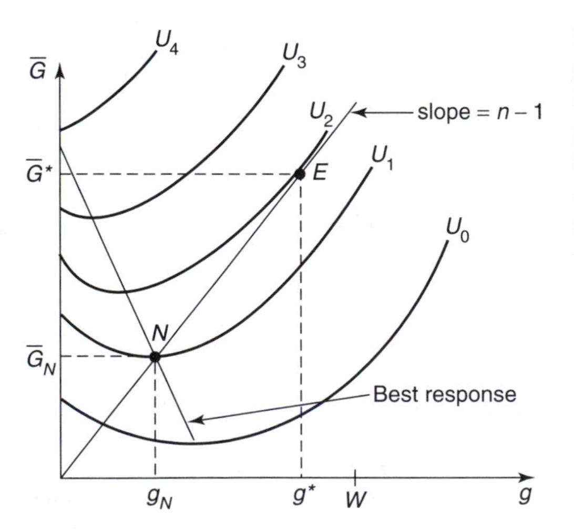

Figure 5.3 Market provision of and efficient provision of a public good. n, Number of identical people in society; g, individual's provision of the public good; $\overline{G}$, provision of the public good by everyone else; w, wealth of each individual; N, g_N, $\overline{G}_N$, market provision of public good; E, g^*, $\overline{G}^*$, efficient provision of public good; U_i, indifference curves.

How much should be generated in this society? In other words, what is the efficient level of g and $\overline{G}$? The reason we would expect the public good to be underprovided is that if consumers raise their g, they benefit individually but everyone else also benefits and they do not take that into account. Suppose they could be sure that when they raise their g, everyone else matches them by raising their gs? Then consumers see the full benefit of raising g. We can see this in Figure 5.3. If we assume that $\overline{G}$ stays at $(n - 1)g$ as consumers raise and lower g, then consumers see all of the advantage of raising and lowering g. In other words, if consumers know that g and $\overline{G}$ will lie along the upward sloping straight line in Figure 5.2, the best choice of g is the point at which an indifference curve is just tangent to the straight line, labeled E in Figure 5.3, resulting in public good provision g^* and $\overline{G}^*$.

Note that the noncooperative provision of the public good is lower than the efficient provision, though still positive. The basic reason for this is that when each person chooses g noncooperatively, he or she sees some benefits from buying g (thus $g_N > 0$), but most of the benefits occur to others and are ignored in deciding how much to consume (thus $g_N < g^*$).[8]

We know that there is a great deal of private contribution to protect the environment—such contributions support international organizations such as Greenpeace, the Worldwide Fund for Animals/World Wildlife Fund (WWF), and many national organizations such as the Sierra Club. The Nature Conservancy is specifically set up to accept private donations and use these to purchase important natural areas and set them aside for prosperity—a very clear example of private provision of a public good.[9]

The public bad version of this involves a number of polluters. They are injured somewhat by polluting, but most of the damage accrues to someone else. Thus they will control pollution somewhat, reflecting the damage they incur, but they will not control pollution as much as is socially desirable.

The bottom line for both goods and bads is that the market may very well provide some public goods or public bads. However, left to its own devices, the market typically under-provides public goods and over-provides public bads.

B. Efficient Public Good Pricing: Consumers

Now suppose that somehow we could set prices for public goods and bads. Can we do this and obtain efficiency? We approach the case of generating an efficient price for the nonrival bad by first considering the nonrival good. Suppose we have a park with a fence around it and thus can control access. Suppose the park is large so that congestion is never a problem and suppose the cost of operating it is all fixed costs with no costs depending on visitation. What then should be the efficient admission price (one price for all consumers)? To answer this, we should determine first what is the efficient amount of use of the park. Since there is no cost associated with letting one more person use the park, efficient use clearly calls for everyone having access to the park. If we were to exclude anyone, then a Pareto improvement could be attained by allowing whomever has been excluded to use the park—costs would not change, current users would not have their utility reduced, and those newly admitted would have their utility increased.

What admission price would support this Pareto optimum? Clearly only a price of zero. Figure 5.4 shows a typical demand curve for park visits. At any price $p_1 > 0$, there will be a certain number of people (q_1) who will pay the admission price and visit the park. There will also be a number of people ($q_0 - q_1$) who will not visit the park because the price is too high. We would argue that a lower price would be Pareto preferred. This is because if the price is lowered, all previous attendees would be just as well off, some people for whom the price was previously too high would be able to enjoy the park at the lower price, and the cost of providing the park services is unchanged. This process continues until the price drops to zero. The consequence is that for a nonrival excludable good, the optimal price from the consumer's point of view is *zero*.

What is the analog for a nonrival bad? For the sake of simplicity, suppose we have a factory in the middle of nowhere generating smoke that fills a circular area around the factory 2 km in radius. There are no effects outside this circle. Furthermore, there is no good reason for anyone to live inside the circle—assume that workers at the plant are housed at a location 15 km away and are provided free transportation to the plant. The following question is before us: If people decide to live in the polluted area around the plant, what compensation should be paid for the damage they incur?

First it is necessary to determine the Pareto optimal location of consumers. Clearly it is Pareto optimal for nobody to live inside the polluted area. Anyone inside the polluted area can be made better off by moving outside the area, since we assumed there are no advantages to living close to the factory. What level of compensation to people living with the pollution would support this efficient location of consumers? The answer is zero compensation. Any positive level of compensation will attract those consumers for whom damage is less than the compensation offered. It is a Pareto improvement to lower the compensation, inducing those people to move outside the circle (or not move there to begin with). In fact, the optimal compensation is zero since compensation induces people to inefficiently move to the pollution.[10]

This example is a bit stylized since there is no reason for people to live near the pollution. Consider a slightly more realistic example of an airport located within a city. Noise bothers people and the further they are from the airport, the less the noise and the annoyance. People live near the airport because they work near there so commuting is shorter. What compensation should the airport pay to consumers? The same argument ap-

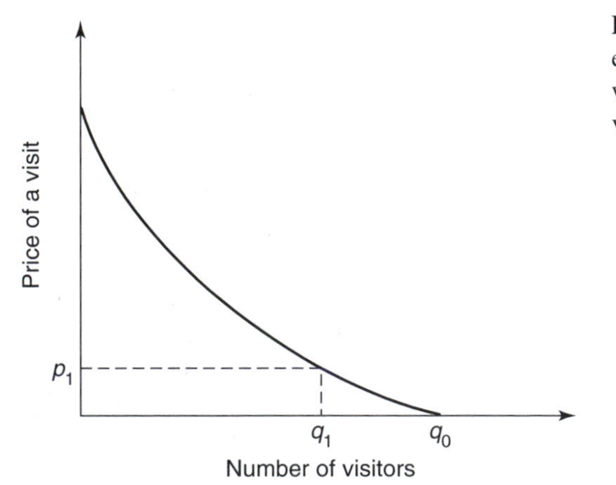

Figure 5.4 Demand for visits to nonrival, excludable parks. q_0, Number of visitors with no admission charge; q_1, number of visitors when price is p_1.

plies. Without compensation, consumers optimally trade-off the noise with convenience in finding a place to live. The consumer lives where the marginal convenience cost of moving a little further away just equals the marginal gain to moving in reduced noise. Any positive compensation that is offered will diminish the net damage from the noise and cause the consumer to locate slightly closer to the airport.[11] This is not socially desirable. Key here is the nonrivalry. If the noise were rival (i.e., my consumption of it reduces the amount available to others), then my consumption provides a community service for which I could be paid. Because my consumption of the nonrival bad (noise) imposes no costs, positive or negative, on anyone else, there is no justification for my receiving payment for consumption. In fact, to the extent that such compensation distorts my decisions about where to locate, it is undesirable. However, if compensation does not distort decisions (e.g., lump-sum compensation to land owners is nondistortionary since land cannot be moved), compensation is neutral and perhaps desirable on equity grounds.

Another way of looking at this is that there are actions producers can take to reduce damage (e.g., emit less smoke or noise) and there are actions "victims" can take to reduce damage (e.g., change location or install household noise insulation). The pollution the victim suffers provides sufficient incentive to the victim to take action to reduce damage. Any compensation that distorts those incentives is undesirable. Nondistortionary compensation is fine for equity reasons but of no consequence to efficiency.

This point about the undesirability of compensation for consumption of nonrival bads is difficult to grasp intuitively.

C. Budgets and the Optimal Producer Price

Having discussed consumption we now examine production. We have seen that the optimal provision of nonrival goods involves a price to the consumers of zero. Production of nonrival commodities is similar to production of any other good or bad. There are costs associated with production that are positive for the case of goods and negative for the case of bads. Thus to efficiently supply a nonrival good, the private producer must have revenue sufficient to cover costs. Unfortunately, if the producer must price at zero, which we have just indicated is the efficient consumer price, then the producer cannot afford to provide the right amount of the nonrival good. Budgets will not balance and a deficit will result. If the producer raises prices to generate revenue for provision, demand will be depressed. There will be too little of the public good provided.

For a nonrival bad, a similar thing happens. The cost of production of a bad such as smoke is negative, i.e., the producer saves money by producing the smoke. Since the price is zero (no compensation should be paid), the producer of smoke runs a budget surplus. In other words, the producer saves money by generating smoke and pays no compensation. Smoke is good for the producer; without intervention the smoke producer will ignore damage to the consumer. Now suppose we require a balanced budget, with the producer's savings paid out as compensation to consumers. This would be analogous to the private provision of a public good where the budget must balance. If people are compensated, they desire more smoke than if they were not compensated. Thus in this case there is an incentive to overprovide the public bad. Too much smoke will be generated.

D. Lindahl Prices and Free-Riding

A theoretical but impractical solution is available for this problem using prices for public goods and bads. First decide how much of the public good is to be provided by determining where aggregate demand and supply intersect. Then charge each consumer his or her individual willingness to pay for that amount of the good. By construction, the total marginal cost of provision will be the sum of the individual willingnesses to pay, thus we should have budget balancing. If Anna loves to hike and really values public parks, she should pay her marginal willingness to pay for those parks. If Brewster couldn't care less, perhaps because he has an allergy attack when he goes outdoors, Brewster should pay little toward the public parks. If each pays according to his or her marginal willingness to pay, not only will the right amount of the public good be provided but the budget will balance—the amount collected will equal the amount necessary for provision. This is what is known as a Lindahl equilibrium, and the prices everyone pays are known as Lindahl prices. These are named after the Norwegian economist Erik Lindahl.

One of the main failings of the Lindahl equilibrium is that information is required on individual demand curves. Those demand curves must be provided voluntarily by individuals and they have every incentive to distort their articulated demand. Since the greater your demand, the more you will pay, you have an incentive to understate your demand. This is what is known as *free-riding*. Even if you have a large willingness to pay, you know that if you fib and indicate a low willingness to pay, you will still get the benefits from the good (assuming nonrivalry). You free-ride on everyone else's willingness to provide the public good. Everyone of course has this incentive so that all of the articulated individual demand curves can be assumed to understate demand, although by how much we cannot know.[12]

If we were to seek contributions from consumers to reduce pollution, we would expect some people to claim they were damaged only slightly since their contribution would be based on their announced damage. This is free-riding on the provision by others.

IV. EXTERNALITIES

An externality occurs when one person's or firm's actions affect another entity without permission.[13] If I like to play my stereo loudly, my neighbors must listen as well. If a laundry is located next to a steel mill, the act of making steel increases costs for the laundry because of all the dirt and smoke generated in making steel. One more additional driver on a congested highway generates externalities for all the other drivers. Of course, there can also be positive externalities. In a classic example of externalities,[14] the owner of an apple orchard provides a positive externality for a neighboring apiary (in terms of the quantity and sweetness of the honey). And the apiary provides a positive externality to the orchard in terms of the bee pollinating the apple blossoms. The generation of knowledge also provides positive externalities in the sense that its benefits are rarely confined to those who generate the knowledge.

Suppose an individual's utility function is given by $U(x,y)$ where x and y are quantities of two goods (or bads) consumed. The individual chooses how much of x to consume but has no control over consumption of y. How much y will be consumed is chosen by others. This is an externality.

Definition *An* externality *exists when the consumption or production choices of one person or firm enters the utility or production function of another entity without that entity's permission or compensation.*

It is important to exclude from this concept actions between two agents that are mutually agreed and for which payment occurs. When iron ore is sold to a steel mill, the mill's profits are improved, but that is hardly an externality; rather it is a simple transaction. It is unstated, but we also preclude from our definition of externality intentional harm to another (e.g., murder) or intentional good to another (altruism).

Figure 5.5 illustrates a production externality. Shown are production possibilities for steel and laundry—the maximum amounts of the two that can be produced. Note that with no externality, both can produce some maximum amount and the two firms are essentially independent of one another. With a modest externality involved (e.g., smoke), increased steel production diminishes laundry output. If the externality is strong enough,[15] the possible output combinations can even be nonconvex, leading to potential pricing problems that will not be explored here.[16] What would influence the strength of the externality? Clearly location is important. As we move two firms away from each other, the strength of the externality is bound to decline.

A production externality exists when one firm's profits are involuntarily affected by another's. More specifically, suppose the technology of producing laundry is given by

$$L = f_L(x_1, \ldots, x_n, e) \tag{5.3}$$

where L is output of laundry and $x_1, \ldots, x_n$ are n different inputs into laundry production. The variable e is smoke emissions from steel manufacturing, which is of course cho-

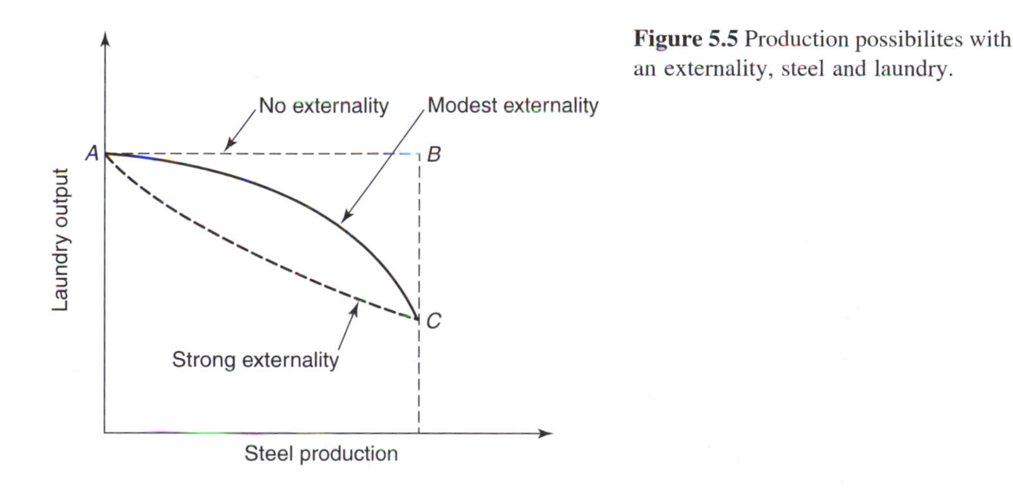

Figure 5.5 Production possibilites with an externality, steel and laundry.

sen by a steel mill. The steel mill produces steel (S) and smoke emissions (e) using inputs $z_1, \ldots, z_m$, according to

$$S = f_s(z_1, \ldots, z_m) \tag{5.4a}$$

$$e = f_e(z_1, \ldots, z_m) \tag{5.4b}$$

Note that e is chosen by the steel mill [Eq. (5.4b)] but e enters into the production function of the laundry [Eq. 5.3)]. The laundry has no say over what the level of e is nor how e enters its production function. Thus the smoke is an externality.

A consumption externality is very analogous. The difference is that we deal with utility functions instead of producion functions [as in Eq. (5.3)]. Consider the utility function $U(w,e)$, where w is just a basket of goods chosen for consumption and e is levels of pollution, which enters the utility function through no choice of the consumer. This is a consumption externality.

Figure 5.6 shows an example of a consumption externality. The externality is generated by a paper mill that discharges pollution into a river that is used for swimming. For illustrative purposes, we will assume the water pollution is undesirable only to the extent that it degrades the swimming experience; for instance, the river may totally clean itself (naturally) within a few miles of the paper mill. Shown in Figure 5.6 is a set of indifference curves between pollution and other goods (for some typical consumer). Note the two distinct regions in Figure 5.6. To the left (low levels of pollution), swimming and pollution are inversely related. More pollution means the swimming experience is degraded. Thus additional quantities of other goods are needed to keep utility constant. At some point, however, swimming is just not fun anymore; the river has become so polluted that swimming ceases. This is the point of the kink in the indifference curves in Figure 5.6. To the right of this point, increased pollution has no effect on utility. One problem with these indifference curves is that they are nonconvex.[17] As mentioned in the context of Figure 5.5, this can lead to serious problems with pricing, which will not be explored here.[18]

So why should we be worried about externalities? The basic problem is that the generator of the externality is deciding how much of the externality to produce but is not taking into account the effects of the externality on others. Figure 5.7 illustrates the case of

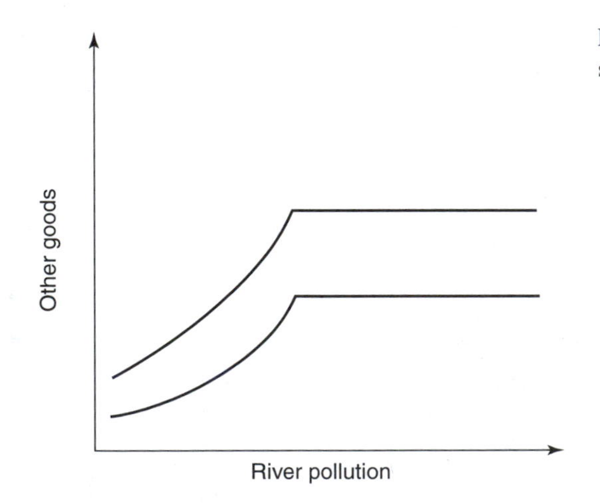

Figure 5.6 Indifference curves with consumption externality.

the steel mill and the laundry with a modest externality in the form of smoke from the steel mill. Shown in Figure 5.7 are possible output combinations of laundry and steel given a fixed amount of inputs (production possibilities). The maximum amount of steel that can be produced with these resources is S_0. With no intervention, the steel mill will ignore the effects it might have on the laundry and produce S_0 of steel; the laundry will produce as much as it can, L_0. Assuming prices for steel and laundry (p_S and p_L), the total value of output can be read off the vertical axis as shown in Figure 5.7.[19] Now eliminate the externality by merging the laundry and steel mill into one firm. Steel production will adjust, taking into account the negative effects of the smoke, finding a point at which the total *value* of output (Y) is greatest, (S_1,L_1). Reading output off the vertical axis, we can see that indeed $Y_1 > Y_0$. Note that some smoke may still be generated, but the amount is chosen so as to balance the interests of the steel mill with those of the laundry. If there is a transfer from the laundry to the steel mill, this is a Pareto improvement (both the laundry and steel mill are made better off). If no transfer takes place, (S_1,L_1) is a potential Pareto improvement over (S_0,L_0). This is because the pie is bigger when the externality is corrected.

There is a question of semantics about whether we have eliminated the externality at (S_1,L_1). There is still smoke being produced so there is still an external effect. The convention is to distinguish between Pareto-relevant and Pareto-irrelevant externalities.[20] It is socially desirable to correct a Pareto relevant externality; after the correction, the externality becomes Pareto irrelevant. Thus smoke at (S_0,L_0) is Pareto relevant whereas smoke levels at (S_1,L_1) are Pareto irrelevant.

There is another type of externality that can best be illustrated by an example. Suppose you have been a swordfish eater for many years. Recently you have seen the price of swordfish rise dramatically due to the increasing popularity of fish in a healthy diet and depletion of swordfish stocks. The actions of others have driven up the price of fish so that now you have to pay more so you consume less swordfish. This is what is known as a pecuniary externality.[21] Basically, your prices change due to the actions of others. Your utility changes because your income remains fixed. While this is certainly an external effect, it is not something that we will be concerned about. The basic reason is that

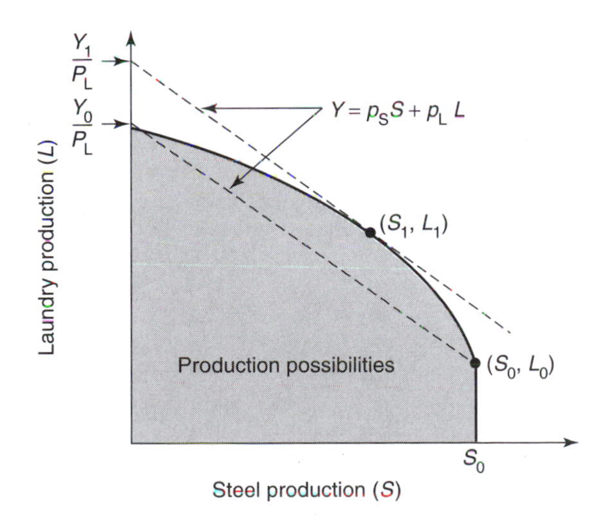

Figure 5.7 Pareto relevance of externality. (S_0,L_0), Steel and laundry production without coordination; (S_1,L_1), steel and laundry production with firms merged; Y_1, value of output of merged firm; Y_0, value of output without coordination; (p_S,p_L), prices of steel and laundry.

there is no inefficiency involved in introducing a pecuniary externality. Prices change, but we have efficiency before and after the price change. We have just moved from one point on the Pareto frontier to another. Although this may have distributional consequences, it does not have the efficiency effects of a conventional externality which involves the inappropriate physical amount of an externality.

V. EXTERNALITIES AS PUBLIC BADS

After having developed the notion of externality, we now wish to argue that as a term it is somewhat redundant with the concept of public bads developed earlier, a concept that is more consistent with conventional consumer and producer theory. An externality involves a good or bad whose level enters the utility functions of several people or firms but is chosen by only one person or firm. This is exactly what happens with a nonrival, nonexcludable good or bad—a public good or bad. When the steel mill generates smoke, it is doing so not intentionally to hurt the laundry but rather because the smoke is nonexcludable. (The fact that smoke is nonrival is not relevant to this example.) When I choose to enjoy my blasting stereo, you also listen to it because the sound is nonexcludable. I am a producer of the nonexcludable noise; you are the consumer. Prior to laws prohibiting the dumping of household garbage, my creation of garbage was an externality because I simply dumped it into my neighbor's yard or into common areas. With laws making garbage excludable the externality disappears because my garbage no longer enters your utility function without permission.

The notion of externality has arisen because of its intuitive appeal. I do something for my own benefit, ignoring the fact that my action also affects you. However, modern economics starts from the basis of preferences represented by utility functions, with physical quantities of commodities (goods or bads) as arguments, and production functions, also with physical quantities of commodities as arguments. Most externalities can be viewed as a nonexcludable good or bad being produced by one agent and being consumed by one or more agents. The producer chooses how much to produce based on his or her own calculus. The consumer has no choices since the good or bad is nonexcludable. Air pollution fits into this framework, as does noise, and even positive externalities such as knowledge.

The point is that we could just as well dispense with the term externality and fit environmental problems into conventional theory using the concept of public bad or excludability. However, the term externality is so common that students should be aware of it. It is sometimes intuitively appealing to discuss an action resulting in an externality rather than convert the action to a good or bad.

SUMMARY

1. Two basic characteristics of a good or bad are the degree of rivalry in consumption (termed rivalry) and the ability of producers to exclude consumers, preventing consumption (termed excludability).

2. Excludability is determined by the availability of a technology of exclusion. Furthermore, the costs of exclusion must be low enough that the benefits of exclusion exceed the costs.

3. Rivalry is a more fundamental characteristic of consumption of a good or bad. A good or bad is nonrival if its consumption involves no opportunity costs for others. In other words, one person's consumption does not change the amount of the good or bad available for others to consume.

4. A good or bad that is both rival and excludable is a *private* good or bad. A good or bad that is both nonrival and nonexcludable is a *public* good or bad.

5. Aggregate demand for rival goods is determined by summing individual demand curves horizontally; aggregate demand for nonrival goods is obtained by summing individual demand curves vertically.

6. Public goods may be privately provided by individuals in the market. In such a case, the market provision will be lower than is efficient. For public bads, market provision will be higher than is efficient.

7. Nonrival goods or bads should have a price of zero, which makes it difficult for revenues to equal costs. This also implies that no compensation be paid to victims of pollution, in general.

8. An externality exists when the consumption or production choices of one person or firm enter the utility or production function of another person or firm without permission or compensation.

9. A pecuniary externality occurs when one person's actions affect the prices paid by another person. Since prices do not enter into utility or production functions, this is not a conventional externality and in fact does not involve an inefficiency.

10. The fact that a public bad is nonrival means that everyone consumes the same quantity of the public bad. This is equivalent to the amount of consumption being chosen by someone else, the generator of the public bad. This is an externality. Thus there is an intimate connection, and in fact some redundancy, between the concepts of public bad and externality.

PROBLEMS

1. Consider an air basin with only two consumers, Huck and Matilda. Suppose Huck's demand for air quality is given by $q_H = 1 - p$ where p is Huck's marginal willingness to pay for air quality. Similarly, Matilda's demand is given by $q_M = 2 - 2p$. Air quality can be supplied according to $q = p$ where p is the marginal cost of supply.

a. Graph the aggregate demand for air quality along with individual demands.

b. What is the efficient amount of air quality?

2. Consider an airport that produces noise that decays as the distance (d), in kilometers, from the airport increases: $N(d) = 1/d^2$. Fritz works at the airport. Fritz's damage from noise is \$1 per unit of noise and is associated with where Fritz lives. His costs of commuting are \$1 per kilometer (each way). The closest he can live to the airport is $d = 0.1$ km.

a. Write an expression for Fritz's total costs (noise and transportation).

b. What is the distance Fritz will live from the airport in the absence of compensation for the noise? What are his total costs?

c. Suppose Fritz is compensated for his damage, wherever he may live. How close to the airport will he choose to live? How much will he be compensated? (*Hint:* Solve graphically or using calculus.)

3. When sulfur dioxide is emitted into the air it is transported over long distances and is converted to sulfuric acid. This gradually falls to the ground, either as rain or snow or simply by settling out of the air. This is called acid deposition. In what way could acid deposition be consider a rival bad? [*Hint*: See the paper by Freeman, 1984.]

4. Why is lump sum compensation for an externality nondistortionary?

5. Two types of consumers (workers and retirees) share a community with a polluting cheese factory. The pollution is nonrival and nonexcludable. The total damage to workers is p^2 where p is the amount of pollution and the total damage to retirees is $3p^2$. Thus marginal damage to workers is $2p$ and marginal damage to retirees is $6p$. According to an analysis by consulting engineers, the cheese factory saves $20p - p^2$ by polluting p, for a marginal savings of $20 - 2p$.

a. Find the aggregate (including both types of consumers) marginal damage for the public bad.

b. Graph the marginal savings and aggregate marginal damage curves with pollution on the horizontal axis.

c. How much will the cheese factory pollute in the absence of any regulation or bargaining? What is this society's optimal level of pollution?

d. Starting from the uncontrolled level of pollution calculated in part (c), find the marginal willingness to pay for pollution abatement, A, for each consumer class. (Abatement is reduction is pollution; zero abatement would be associated with the uncontrolled level of pollution.) Find the aggregate marginal willingness to pay for abatement.

e. Again starting from the uncontrolled level of pollution, what is the firm's marginal cost of pollution abatement? What is the optimal level of A?

f. Are the problems of optimal provision of the public bad (pollution) and the public good (abatement) equivalent? Explain why or why not.

6. Consider the problem of carbon dioxide emissions. We will abstract away from the problem slightly, assuming there are polluters and consumers in two regions, the OECD (O) and the rest of the world (R). Suppose the marginal cost of controlling CO_2 emissions is \$10 per ton of emissions. Let the marginal willingness to pay for pollution reduction be $13 - Q$ for region O and $12 - 2Q$ for region R, where Q is the amount of pollution reduction. The United Nations is considering two proposed methods for conrolling CO_2 emissions, both involving polluters paying for the damage they cause. Proposal A involves the polluters paying damages to each region for the pollution generated. Proposal B involves the polluters in each region independently negotiating pollution reductions, assuming the other region is not undertaking pollution reduction.

a. Graph the marginal abatement cost and the total marginal willingness-to-pay schedules. What is the socially efficient level of emission reductions, Q?

b. How much total pollution reduction will occur under proposal A and what will be the total compensation received by regions O and R? If those payments were instead placed in the general coffers of the UN, would the outcome be any different from an efficiency point of view? Why or why not?

c. How much pollution would be generated under proposal B? Explain any differences between this answer and the answer to parts (a) and (b).

7. In the manner of Figure 5.3, draw a set of indifference curves such that an individual's private provision of the public good increases as the number of people in the society increases. Explain your result. Are these indifference curves plausible?

NOTES

1. This taxonomy is attributed to Head (1962). Randall (1983) provides an illuminating discussion of rivalry and excludability and related classifications. See also Musgrave (1959). See Laffont (1989) or Starrett (1988) for more advanced treatments. The economics community is not totally united in this classification of goods, however. Laffont (1989) provides a classification scheme that builds on rivalry and excludability, but adds the issue of jurisdiction (local vs. national) and the extent to which consumption is voluntary or not. Hershleifer (1983) argues that the nature of consumption provides an additional classification scheme in which the total amount of a public good may or may not equal the sum of what is provided by government and individuals. Congestion also causes additional problems and has motivated the study of club goods (see Cornes and Sandler, 1996). These various distinctions are left to a more detailed examination of the theory of public goods.
2. When the garbage in the dump leaks into the groundwater it becomes both nonexcludable and nonrival.
3. Whenever I use this example in class, someone always points out that some of the pollution will stay in my lungs and diminish what is available for others. This is of course true but the effect is so minuscule that we ignore it.
4. See Starrett (1988) for further discussion of the notion of rivalry. Baumol and Oates (1988) use the term depletability to mean rivalry for the case of a bad.
5. Consumption of the climate does not include injecting CO_2 into it. This involves the production side of the global climate, which is not the issue here.
6. We have glossed over a theoretical issue here: that the marginal rates of substitution usually depend on the consumption of the private good. This implies that if the distribution of the private good among the population changes, the marginal rates of substitution between the public and private good may change and thus the efficient amount of the public good may change.
7. There is a large literature on the private provision of public goods, with many important issues. For instance: How does the size of a society influence public good provision? How does government provision affect private provision? How close to efficient is the private provision? Myles (1995) provides a good treatment of some of these issues.
8. The reason everyone provides a little of the public good is that everyone is identical. If one person provided none of the public good, everyone could be assumed to do the same and there would be no provision. If our society consisted of heterogeneous individuals then we would probably see some people providing some amount of the public good and other people providing nothing (see Bergstrom et al., 1986, or Myles, 1995).
9. One of the issues in private provision is the extent to which government provision will "crowd out" private provision. One empirical study of charitable giving in the United States estimated that every dollar of government support for charities crowded out (eliminated) private giving of almost \$0.30 (Abrams and Schmitz, 1978). Theoretical studies indicate the effect could be even greater (Bergstrom et al., 1986).
10. Baumol and Oates (1988) discuss this issue of compensation in the case of public bads.

11. In law and economics, this is referred to as "coming to the nuisance" (see Cooter and Ulen, 1997).
12. There is a fairly advanced literature on what are known as "demand-revealing mechanisms." These are complex ways of calculating payments for public goods based on articulated willingness to pay. It is possible to structure these so that people have an incentive to tell the truth about demand; unfortunately, for these mechanisms, we cannot always be assured that we will collect the right amount of money to pay for the public good. See Tideman and Tullock (1976).
13. There is a large and sometimes confusing literature on externalities. Some highlights are Heller and Starrett (1976), Baumol and Oates (1988), Freeman (1984), Buchanan and Stubblebine (1962), Mishan (1971), Plott (1983), Bator (1958), and Meade (1952, 1973).
14. See Meade (1952).
15. This example involving the "strength" of an externality is motivated by Baumol and Bradford (1972) and Baumol and Oates (1988).
16. A convex set is a set in which a straight line segment connecting any two points in the set lies entirely within the set. In Figure 5.5, a line connecting points *A* and *C* will not be entirely within the set of production possibilities for the case of the strong externality.
17. Nonconvexity for an indifference curve means a straight line segment connecting any two points on the curve does not lie entirely above the curve. A line connecting a point just to the right of the kink in any indifference curve in Figure 5.7 with a point just to the left of the kink will be entirely below the indifference curve, implying nonconvexity.
18. The problem with nonconvexities in either production or consumption is that a market equilibrium, supported by prices, generally cannot be relied on to exist when production sets (Figure 5.4) are nonconvex or when indifference curves (Figure 5.6) are nonconvex. The problem may be that for all possible prices, consumers will want one bundle of goods to consume and producers will want to produce some other bundle. Prices may never bring consumers and producers together, as is usually the case. See Starrett and Zeckhauser (1974). Cooter (1980) argues that this is a nonissue in practice.
19. If the total value of output (S_0, L_0) is $Y_0 = p_S S_0 + p_L L_0$, we want to solve for Y_0. If we plot $Y = p_S S + p_L L$ and pass it through (S_0, L_0), then that line will intersect the vertical axis at $L = Y_0/p_L$.
20. See Buchanan and Stubblebine (1962).
21. See Viner (1931) and Baumol and Oates (1988).

6 PROPERTY RIGHTS

In the last two chapters we took a close look at why markets are desirable when conditions are right ("competitive") and why markets can fail, particularly in the context of environmental goods and bads. We now complete this examination by seeing what changes can be made to correct market failure. Although we will examine this question in detail when we consider regulation later in the book, it is instructive to examine the two generic ways in which market failure can be corrected. One involves correcting imperfect property rights; recall that well-defined property rights were an important requirement for a market to be competitive. We consider property rights in this chapter. In the subsequent chapter we will consider the introduction of administered prices for pollution—Pigovian fees.

I. INTRODUCTION

As the discussion of exclusion in the last chapter should have made clear, institutions play an important role in allowing markets to function to allocate resources. If stealing is not illegal, then nearly all goods are nonexcludable and thus cannot be rationed using prices. A simple institutional change, establishing enforceable property rights, makes goods excludable and thus allows a market system to operate. In the case of garbage, if no laws exist prohibiting littering, garbage is nonexcludable; with enforced prohibitions on littering, garbage becomes excludable and the price system works to ensure proper disposal of trash.

The basic point is that properly defined property rights can make a big difference in whether a market will allocate goods and bads efficiently. In this chapter we explore this issue further, asking how property rights should be initially allocated by government. Should we all have the right to clean air, and if pollution occurs be entitled to elimination of the pollution or compensation for pollution damage? Or should polluters have the right to pollute? This seems like a simple question with an obvious answer. The "polluter pays" principle is enshrined in public policy regarding pollution control in many parts of the world. On the other hand, most countries have implemented subsidies, particularly through the tax system, to encourage pollution control. This indicates at least some acknowledgment of polluter rights.

This is not quite as simple an issue as might first appear. Consider two people, a polluter and a victim. The conventional view of the problem is that the polluter is the source of the problem and that blame must fall on the polluter's shoulders. However, leaving aside any preconceptions of right and wrong, the victim could also be blamed for being next to the polluter. Without the victim, the pollution would not be a problem. Morally, it seems that responsibility for cleaning up the pollution should fall on the factory. Can we use economic efficiency arguments to reach this same conclusion?

This subtle issue of who should have rights, the polluter or the victim, was raised in 1960 by Ronald Coase in an important paper that contributed to his winning the Nobel Memorial prize in economics. Coase (1960) asks whether it is logical to assign rights to one or the other of these parties—the victim or the polluter. One of Coase's conclusions is that under some conditions, it makes no difference to efficiency whether the polluter has a right to pollute or the victim has a right to clean air (although it will make a great deal of difference to each of the two parties). Since the right to pollute is a property right that has value, if trade is allowed in those rights, efficiency should prevail, no matter how they were initially allocated. If the right is worth more to the victim than the polluter, the victim will end up with the right, no matter how it was initially distributed. Of course, the initial distribution of rights does matter to questions of equity. Rights can be valuable. Vesting someone with a right is like giving them money and resources.

Coase's arguments regarding property rights are often dubbed the Coase Theorem, falsely suggesting precision.[1] There are two parts of the Coase Theorem, at least insofar as we consider it here. The first concerns the case in which there are no impediments to buying and selling pollution rights. In this case, the initial distribution of rights does not matter. The second involves impediments to trading property rights. Not surprisingly, in this case the initial distribution of rights does matter.

To most effectively present Coase's results, we rely on an example of a polluter (a steel mill) and a victim (a laundry). We first demonstrate Coase's result in this example, before stating his result more generally.

II. THE POLLUTER AND THE VICTIM: WHO SHOULD HAVE RIGHTS?

To explore the Coase Theorem, we return to our friends, the steel mill and the laundry. We ask if it makes any difference whether the steel mill has the right to pollute or the laundry has the right to clean air. And to have a benchmark for comparison, we also determine what is the "correct" amount of pollution, as well as steel and laundry output.

To structure the problem, the steel mill produces steel, S, and pollution emissions at a cost $C_S(S)$. The laundry produces clean clothes, L, at a cost $C_L(L,S)$, which depends on the amount of steel produced because steel and pollution go together. This is a true externality, since the laundry does not choose S yet S enters into the laundry's production function—S depresses laundry output, holding inputs fixed. The price of steel is p_S

and the price of laundry is p_L. Both these prices are determined by a larger market and thus can be assumed to stay fixed throughout our analysis.

A. What Is Efficient?

First we need a benchmark: How much would be produced if we could "internalize" the externality? The simplest way to do this is to merge the firms. If a single firm produces steel and laundry, then there is no externality. Thus the profit-maximizing choices for this merged firm will be efficient, socially desirable. The production function for this merged firm contains S and L. Both of these variables are chosen by the merged firm. Profits for the merged firm are

$$\Pi_M (S,L) = p_S S + p_L L - C_S(S) - C_L(L,S) \tag{6.1}$$

The merged firm wishes to choose S and L to maximize Π_M. Recall that profit is maximized with respect to a variable such as S when small changes in S generate no change in Π_M. The following optimality conditions should hold for the profit-maximizing output quantities, which we will denote S^* and L^*:

$$\begin{aligned}\frac{\Delta\Pi_M}{\Delta S} &= p_S - \frac{\Delta C_S(S + \Delta S)}{\Delta S} - \frac{\Delta C_L(L,S + \Delta S)}{\Delta S} \\ &= p_S - MC_S(S) - MD_L(L,S) = 0\end{aligned} \tag{6.2a}$$

$$\frac{\Delta\Pi_M}{\Delta L} = p_L - \frac{\Delta C_L(L + \Delta L,S)}{\Delta L} = p_L - MC_L(L,S) = 0 \tag{6.2b}$$

Note in Eq. (6.2a) that there are two marginal costs for the steel mill portion of our firm. One, the marginal cost for steel, MC_S, indicates how steel mill costs change when steel output changes. The other, marginal damages to the laundry, MD_L, indicates how laundry costs change when steel output changes. We would expect both MC_S and MD_L to be positive. Equation (6.2a) states that the price of steel should be equal to the marginal cost to both the laundry and the steel mill, $MC_S + MD_L$. Figure 6.1 shows how S^* is chosen. The curve labeled private costs is simply the marginal costs of steel production to the steel mill, MC_S. The curve labeled social costs is the marginal costs of the steel mill plus the marginal damage to the laundry from steel production, $MC_S + MD_L$. This is labeled social costs because it is the cost of producing steel, including the external costs to the laundry. However, in our case, the costs are all within one firm. Distinguishing between the two costs is still useful. Also shown in Figure 6.1 is a demand curve for steel from our firm, shown as horizontal, reflecting the fact that price must be taken as given to the individual firm. The point at which the demand curve intersects the marginal social cost curve is the profit-maximizing amount of steel production, S^*.

Equation (6.2b) states that the price of laundry should be equal to the marginal cost for the laundry, MC_L. This condition for the laundry is standard.

It is important to remember that it is not always just the marginal conditions that matter. It may be desirable to totally shut down the steel mill or the laundry. This decision will not be apparent from Eq. (6.2), except that there may be no solution to Eq. (6.2) for which $L^* > 0$ and $S^* > 0$. Fixed costs do not enter into the marginal conditions yet

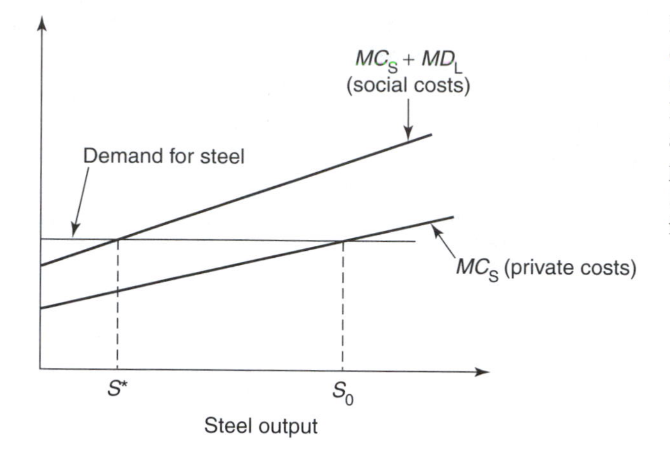

Figure 6.1 Social and private costs of producing steel. S_0, The steel mill's choice, ignoring the laundry; S^*, socially optimal amount of steel; MC_S, marginal private cost of producing steel; MD_L, marginal damage to laundry from steel production.

may determine whether the revenue of a firm is sufficient to cover costs. If they are not the firm has no choice but to cease production.

Should the laundry close, there is no damage from smoke; so steel production will be chosen so that marginal costs equal the price of steel:

$$p_S = MC_S(S) \tag{6.3}$$

Call this amount of steel S_0. This results in profits Π_M $(S_0,0)$. Similarly let the optimal amount of laundry when the steel mill is shut down be L_0 and the associated total profits, $\Pi_M(0,L_0)$. If there is a solution to Eq. (6.2), (S^*,L^*), we should compare $\Pi_M(S^*,L^*)$, $\Pi_M(S_0,0)$, and $\Pi_M(0,L_0)$, choosing whichever is greatest and letting that dictate our shutdown decisions.

B. The Coase Theorem: Marginal Conditions

Having determined the conditions for an efficient amount of an externality, we now turn to the more interesting case of the two firms operating separately. We are interested in two property rights regimes, one in which the steel mill has the right to pollute and the other in which the laundry has a right to clean air. We are also interested in what levels of steel and laundry will be produced in each of these property rights regimes.

1. Laundry Has Rights to Clean Air. First consider the case in which the laundry has a right to a smoke-free environment. This means that the steel mill must compensate the laundry for any damage it does.

Whatever the steel mill produces, the laundry maximizes profits by setting the price of laundry equal to the marginal cost of producing laundry:

$$\frac{\Delta\Pi_L}{\Delta L} = p_L - \frac{\Delta C_L(L + \Delta L,S)}{\Delta L} = p_L - MC_L(L,S) = 0 \tag{6.4a}$$

or

$$p_L = MC_L(L,S) \tag{6.4b}$$

Given S, Eq. (6.4b) can in principle be solved for L, which defines $L(S)$, the amount of laundry that will be produced when the steel mill produces S.

If the steel mill produces S, damage would be defined as the difference in profits for the laundry between the case of no steel production (and no smoke) and the case of steel production at the level S. Thus profits for the steel mill will be

$$\Pi_S(S) = p_S\, S - C_S(S) - \{\, \Pi_L[L(0),0] - \Pi_L[L(S),S]\} \tag{6.5}$$

where $\Pi_L(L,S)$ is the profit of the laundry when it produces L while the steel mill is producing S. $L(S)$ is the amount of laundry that will be produced when the steel mill produces S. The term in braces in Eq. (6.5) is the damage to the laundry from producing steel in the amount S. This damage must be paid to the laundry; consequently, it decreases the profits of the steel mill.

Now, we are interested in how much steel will be produced. In other words, we should find the marginal profit (the ratio of a change in profit to a small change in S), set it equal to zero, and solve for the S that satisfies that equation. From Eq. (6.5),

$$\frac{\Delta\Pi_S(S)}{\Delta S} = p_S - MC_S(S) + \frac{\Delta\Pi_L[L(S),S]}{\Delta S} = 0 \tag{6.6a}$$

where

$$\frac{\Delta\Pi_L[L(S),S]}{\Delta S} = \frac{\Delta\{p_L L - C_L[L(S),S]\}}{\Delta S} \tag{6.6b}$$

$$= p_L\frac{\Delta L}{\Delta S} - \frac{\Delta C_L[L(S) + \Delta L,S]}{\Delta L}\frac{\Delta L}{\Delta S} - \frac{\Delta C_L[L(S),S + \Delta S]}{\Delta S} \tag{6.6c}$$

$$= [p_L - MC_L(L,S)]\ \Delta L/\Delta S - MD_L[L(S),S] \tag{6.6d}$$

$$= -MD_L[L(S),S] \tag{6.6e}$$

Note two things in the derivation of Eq. (6.6). The first term in braces in Eq. (6.5), laundry profit at $S = 0$, is a constant, so it does not change when S changes; thus it does not enter in Eq. (6.6a). Second, the term in brackets in Eq. (6.6d) is exactly zero, according to Eq. (6.4b). Thus, Eq. (6.6e) states that the marginal change in profits for the laundry from extra steel is exactly equal to the marginal cost to the laundry of extra steel, MD_L. Revenue changes play no role in determining the marginal damage to the laundry of extra steel.

Now we can combine Eq. (6.6a) with (6.6e) to obtain the conditions for how much steel will be produced:

$$p_S = MC_S(S) + MD_L[L(S),S] \tag{6.7}$$

How far away from efficiency is this? In the previous section we determined the amount of laundry and steel that would be produced if we internalized the externality by merging the two firms. In that case, Eq. (6.2) determined the amount of laundry and steel. In the case here, the amounts of laundry and steel are determined by Eq. (6.4b) and Eq. (6.7). A quick inspection will reveal that these conditions are the same. If the steel mill is liable for damage, it will pay those damages and the amount of steel and laundry will be efficient.

2. Steel Mill Has Right to Pollute. Now suppose the laundry has no rights to clean air. We basically have a free-for-all. We ask the same question: How much steel will be produced? An initial reaction would be an amount such that the price of steel equals $MC_S(S)$. Let that quantity of steel be S_0. But let us ask ourselves whether the laundry might not pay the steel mill to reduce steel output somewhat. How much would the laundry be willing to pay to reduce steel output to S? Income to the laundry is greater at S than at S_0 (for $S < S_0$) and, *potentially*, the laundry would be willing to pay nearly all of that increased income $\{\Pi_L[L(S),S] - \Pi_L[L(S_0),S_0]\}$ to the steel mill to reduce steel output.[2] This will reduce the costs of the steel mill so that its total profits from producing S are

$$\Pi_S(S) = p_S\, S - C_S(S) + \{\ \Pi_L[L(S),S] - \Pi_L[L(S_0),S_0]\ \} \tag{6.8}$$

which can be rewritten as

$$\begin{aligned}\Pi_S(S) = p_S\, S - C_S(S) &- \{\Pi_L[L(0),0] - \Pi_L[L(S),S]\ \} \\ &+ \{\Pi_L[L(0),0] - \Pi_L[L(S_0),S_0]\ \}\end{aligned} \tag{6.9}$$

Note that in Eq. (6.9), the terms in the last set of braces equal a constant—the difference in laundry profit with no steel and laundry profits with the maximum amount of steel. Consequently, it can be ignored when looking at marginal profits. But without that last term, Eq. (6.9) is exactly the same as Eq. (6.5), profits when the laundry has a right to clean air. Consequently, the marginal conditions for choice of S and L must be the same as in that case.

We have seen that the marginal conditions for choosing S and L are identical under either property rights regime. Furthermore, those conditions are the same as for a merged firm, with the externality eliminated. This is the fundamental result of Coase, that the initial assignment of rights vis-à-vis pollution does not matter for efficiency.

Note that the *distribution* of resources certainly does depend on the assignment of property rights. Although the marginal conditions for the steel mill were the same for the two different property rights regimes, the total costs and thus profits of the steel mill were quite different. Compare Eqs. (6.5) and (6.9). When the steel mill has the property rights, its costs are significantly lower, by the amount the laundry paid to reduce pollution, possibly as much as $\{\Pi_L[L(0),0] - \Pi_L[L(S_0),S_0]\}$. This is essentially the value of the property right to clean air and, in the "no rights" case, it is being purchased by the laundry from the steel mill. After the purchase, the two property rights regimes become identical.[3]

Because the total costs differ under the two property rights regimes, there are other outcomes that might emerge from the different assignment of rights. One would be that the two firms might merge, in which case the externality would be eliminated since the costs and benefits of pollution control would be within the same firm. Equivalent to a merger would be if one of the firms bought the other. So instead of paying the laundry for pollution damage, the steel mill buys the laundry from its owners.

It might also be that one of the firms cannot afford to pay the other for the right to clean air, in which case that firm would go out of business. For instance, if the laundry has rights to clean air, the steel mill may go out of business rather than pay pollution damage. Alternatively, if the steel mill has rights to clean air, the laundry may go out of business rather than bribe the steel mill to clean up its act. In this case, the steel mill will pollute without concern for damage since without the laundry, there is no pollution damage (at least given our assumptions).

Because the total costs differ in the two property rights regimes, we might expect differences to emerge in the long run. When the laundry has rights, the average cost of laundry will be lower than when the steel mill has rights. Similarly when the laundry has rights, the average cost of steel will be higher than when the steel mill has rights. Higher costs inevitably mean higher prices and decreased output. Thus in the long run, if property rights reside with the polluter, we would expect to see more of the product associated with the pollution and less of the victim's product.[4]

C. The Coase Theorem: A Numerical Example

We can take the analysis of the previous section a little further and at the same time make it more concrete by assuming some specific numerical values for costs and prices. Suppose the costs of producing steel are $C_S(S) = S^2 + 8$ and the costs of producing laundry are $C_L(L,S) = L^2 + L\,S + 4$. These cost functions are shown in Figure 6.2. Thus each of these cost functions involves a fixed cost, which we assume is not sunk.[5] These fixed costs are real costs that do not depend on output levels. However, should a firm decide to shut down and produce nothing, that cost is not incurred. In other words, $C_S(0) = 0$. Marginal costs and damage can be easily calculated:[6] $MC_S(S) = 2S$; $MC_L(L,S) = 2L + S$; $MD_L(L,S) = L$. We will assume the price of laundry is fixed throughout our analysis at $p_L = 10$ but we will vary the price of steel, letting it take on three values: $p_S = 8$, 11, and 14.

How much should be produced (if the firms merged) and how much will be produced under the two property rights regimes? We know the marginal conditions for choice of L and S are given by Eq. (6.2) [which we have argued is the same as Eqs. (6.5b) and (6.7)]. These marginal conditions are the same, regardless of the allocation of initial property rights. What will differ is the profits of each of the two firms.

There are several situations that must be examined in determining how the two firms will operate. The efficiency standard is of course the two firms operating as one merged

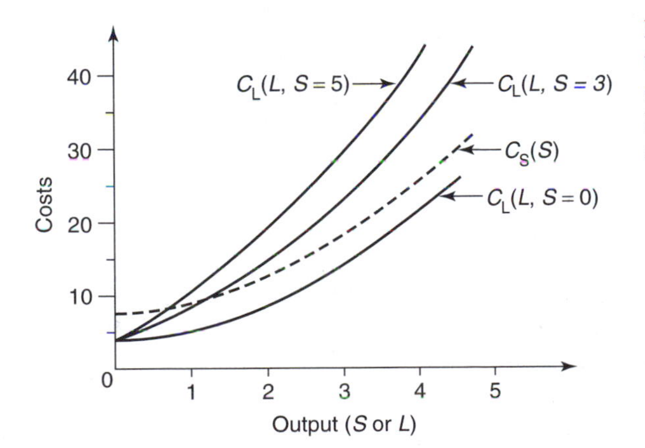

Figure 6.2 Steel and laundry production costs. $C_S(S)$, costs of producing steel (S); $C_L(L,S)$, costs of producing laundry (L) with steel production (S) externality.

firm. This involves the application of Eq. (6.2) to determine output. But as was discussed earlier, we must also separately consider the possibility of shutting down either the laundry or the steel mill and comparing the resulting profit with profit when both firms operate. Table 6.1 shows the numerical results of just such a computation. Column (D) shows how the merged firms would operate. Note that when the price of steel is low ($p_S = 8$), it pays to shut the steel mill down, resulting in a profit of 21; for an intermediate price ($p_S = 11$), it pays to have both firms operating; and for a high price of steel ($p_S = 14$), it pays to shut the laundry down. Also shown in Table 6.1 (columns B and C) is what happens when either of the two firms is closed down, with the open firm choosing output based on Eq. (6.2). Shown in column (A) is what happens if the steel mill ignores the laundry and the laundry does the best it can, based on the pollution it must endure. No transfers or financial lures are involved—the firms act independently and noncooperatively.[7]

Now what about our two property rights regimes? First consider the case of the steel mill having rights. This means that the steel mill is free to pollute without concern for the damage caused to the laundry. The laundry may, however pay the mill to reduce output. The numerical results are shown in column (E) in Table 6.1. Note the result we theoretically developed earlier—the output levels are exactly the same as the efficient output levels (column D). For instance, when $p_S = 8$, it pays to shut down the steel mill. How will

TABLE 6.1 Illustration of Coase Theorem[a,b]

	(A) No transfers/ no bargaining	(B) Close laundry	(C) Close steel mill	(D) Merged firm	(E) Steel mill has rights	(F) Laundry has rights
I $p_s = 8, p_L = 10$						
Steel mill output	4	4	0	0	0	0
Laundry output	2	0	5	5	5	5
Steel mill profits	8	8	0	0	8	0
Laundry profits	4	0	21	21	13	21
Transfer to laundry	0	0	0	0	−8	0
Total profits	12	8	21	21	21	21
II $p_s = 11, p_L = 10$						
Steel output	5.5	5.5	0	4	4	4
Laundry output	2.25	0	5	3	3	3
Steel mill output	22.3	22.3	0	20	22.3	4
Laundry profits	1.1	0	21	5	2.7	21
Transfer to laundry	0	0	0	0	−2.3	16
Total profits	23.4	22.3	21	25	25	25
III $p_s = 14, p_L = 10$						
Steel mill output	7	7	0	7	7	7
Laundry output	0	0	5	0	0	0
Steel mill profits	41	41	0	41	41	20
Laundry profits	0	0	21	0	0	21
Transfer to laundry	0	0	0	0	0	21
Total profits	41	41	21	41	41	41

[a]*See cost conditions in text.*

[b]*In column (E), any transfers are assumed to flow to the steel mill; in column (F) any transfers are assumed to flow to the laundry. Alternate cases are considered in Problem 4.*

this occur? Either the steel mill will buy the laundry or the laundry will buy the steel mill. Once they have merged, the outcome in column D results. How much would the laundry be willing to pay for the steel mill? On its own, the steel mill can make 8 (column A); thus the laundry should be able to buy the steel mill for 8, close it down, and gross a profit of 21, for a net profit of 21 − 8 = 13. Of course, the steel mill could also buy the laundry. In the case of $p_S = 11$, it is efficient for both firms to operate. When the steel mill has rights, the laundry must "bribe" it to reduce steel output from 5.5 (column A) to the efficient level of 4. The laundry has to compensate the steel mill for its lost profit, 2.3. For the high steel price, $p_S = 14$, it is best if the laundry closes down. If the steel mill acts independently, producing 7 units of steel (coumn A), the laundry cannot make a profit and closes down. Thus the steel mill need do nothing regarding the laundry and that is what is shown in column E.

Now suppose the laundry has rights. These results are shown in column F. Once again, the efficient output will be achieved. With $p_S = 8$, it is best for the steel mill to close down. If the steel mill must pay damages to the laundry, those damages will be potential profits of the laundry (21—column C) less laundry profits with the steel mill operating. Thus whatever amount of steel is produced, the laundry must be compensated so that its profits are at least 21. Clearly the steel mill cannot afford this and shuts down. With the intermediate price of steel, $p_S = 11$, both firms operate. The laundry could make 21 (column C) and at the efficient level of production is making only 5; thus it must be compensated 16 by the steel mill. This reduces steel mill profits from 20 to 4. Finally, with the high price of steel, $p_S = 14$, we know it is efficient for the laundry to be shut down. The laundry can make as much as 21 in profit (column C). Thus the steel mill could buy the laundry for 21 and shut it down, netting the steel mill profit of 20. The laundry could also buy the steel mill.

The bottom line is that in two different cases, one in which both firms operate and one in which one firm shuts down, the initial assignment of rights to clean air is irrelevant to the outcome in terms of the amount of pollution and the amount of goods produced. Furthermore, in both cases the outcome is efficient. The initial assignment of rights is purely a distributional issue. And as Table 6.1 shows, there are large distributional differences.

III. THE COASE THEOREM

The preceding discussion illustrates that bargaining between a polluter and a victim can solve pollution problems without worrying about the legal assignment of rights to a clean environment—you end up at the same spot in both cases. A key assumption in the discussion was that bargaining between the parties was easy. This of course is not always the case. If there had been 1000 laundries and one steel mill in our example, having all the laundries agree to pay the steel mill to reduce pollution would be difficult. Or the next time you are in a restaurant and someone at the next table starts smoking a cigar (presumably to your consternation), try paying the person to stop smoking. Bargaining costs usually do exist. The results of the previous section assumed that bargaining was costless.

More generally, the previous section assumed *no transaction costs*. Just what do we mean by transaction costs? Transaction costs are the costs of entering into a transaction, over and above the exchange of money for a good. Transaction costs cover a multitude of different costs associated with agreeing to transact. At the simplest level, a transaction cost is an extra payment associated with consummating a transaction. For instance, in buying an automobile, there is the cost of the automobile itself but there are also a host of costs associated with the transaction: sales tax, registration, transfer tax, V.A.T., etc. In purchasing a house, the transaction costs can be even more significant.[8]

A more subtle kind of transaction cost is involved in bargaining. Paying smokers at neighboring tables in a restaurant not to smoke involves significant effort and perhaps some trauma. Many of us are shy about asking a stranger not to smoke. Even though there may be no monetary transaction costs, there are psychic costs associated with striking bargains with neighboring diners. The costs are just as real as monetary costs. That is why we also think of this impediment as a transaction cost, even though it is more subtle than a transfer tax paid when acquiring an automobile.

We are now in a position to state the Coase Theorem, one version of which was illustrated in the previous section with the steel mill and the laundry. (It is called a theorem though its statement and its proof are considerably looser than most mathematical theorems.)[9]

> **Coase Theorem.** Assume a world in which some producers or consumers are subject to externalities generated by other producers or consumers. Further, assume (1) everyone has perfect information, (2) consumers and producers are price-takers, (3) there is a costless court system for enforcing agreements, (4) producers maximize profits and consumers maximize utility, (5) there are no income or wealth effects, and (6) there are no transaction costs. In this case, the initial assignment of property rights regarding the externalities does not matter for efficiency. If any of these conditions does not hold, the initial assignment of rights does matter.

What this theorem states is that if there are no barriers to reaching agreement (such as the laundry paying the steel mill not to pollute), we get efficiency, regardless of how property rights are initially distributed.[10] In fact, the theorem goes beyond externalities broadly to all market failures. With a monopoly and no barriers to agreement, consumers can simply pay monopolists to produce at the efficient level, sharing the gains in efficiency. The most significant condition in the Coase Theorem is probably the zero transaction costs assumption. In most real-world situations, there are significant transactions costs, which limits the practical application of the Coase Theorem. In fact, Coase has stated that the case of zero transaction costs is of limited interest:

> We do not do well to devote ourselves to a detailed study of the world of zero transactions costs, like augurs divining the future by the minute inspection of the entrails of a goose.[11]

What the Coase Theorem does is direct our attention away from efficiency arguments to explain why property rights should be allocated in a particular way.

Cooter (1987) argues that even when there do not appear to be transaction costs, there are impediments to reaching bargains (such impediments are basically the same as transaction costs). He points out that in a bilateral negotiation with gains from consummating an agreement, the bargaining may break down because of an inability to decide on the precise division of the gains. Another instance in which there appears to be no transaction costs when in fact there are involves public bads in which many affected parties must bargain with the polluter. Because this example is important, we will consider it in more detail in the next section.

Another important clause in the Coase Theorem concerns wealth effects. If I am endowed with a right that has value, I am richer. If I do not have that right, I am poorer. We know that the demand for goods depends on our income; consequently, the existence of wealth effects would generate differences in the final bargain, depending on how rights were initially allocated. This is the reason for condition (5) in our statement of the Coase Theorem.

It is important to keep in mind that Coase is not saying property rights are not important. As we saw in the previous chapter, a complete set of property rights is necessary for a market to support a Pareto-optimal allocation. What Coase is saying is that for efficiency it does not matter how these rights are distributed. If the polluter has the right to use the air for smoke, that is a well-defined right, even though some of us might think it unfair. If citizens have the right to clean air, that is also a well-defined right. Coase simply says that if trading in these rights is easy and costless, it does not matter how they are initially distributed; they will subsequently be traded so that they end up in the right hands.

Let us now focus on the second part of the Coase Theorem, when transaction costs exist. Consider the example of the steel mill and the laundry: suppose there is a transaction fee of $10 for any transaction associated with the externality (payment of damages, buying out the laundry, paying the steel mill to reduce output). Then only if the payoff from any of these transactions exceeded $10 would the transaction take place. Consider the case of p_S = $11 with the steel mill having the right to pollute. We saw that the laundry could pay $2.30 to induce the steel mill to reduce output from 5.5 tons to 4 tons when the price of steel was $11 (Table 6.1, column E). Laundry profits increased from $1.10 to $2.70. If a transaction cost of $10 is associated with this, there is not enough profit to pay the transaction cost to induce the steel mill to reduce output. Thus the status quo will persist—excessive steel production and excessive smoke.

So we see that when transaction costs are present, it does matter where rights are initially vested. With transaction costs, the status quo tends to prevail. This means that when rights are initially established by the legal system, it is important to vest rights in those parties that have the greatest need for those rights. One should not rely on trade to redistribute rights to pollute. Of course another conclusion may be that attention should be paid to reducing the transaction costs associated with trading rights to pollute.

IV. PROBLEMS OF PUBLIC BADS AND BARGAINING

The problem of reaching an agreement over a public bad is fundamental. Let us modify our steel mill/laundry example slightly to one polluting power plant and a number of people surrounding the power plant. If the people own the right to pollute, the power plant will have to compensate people for damage. Although this can lead to problems as we saw in the previous chapter, some lump-sum compensation is possible. On the other hand, if the right to pollute is vested in the power plant, the people surrounding the plant will have to get together to pay the plant not to pollute. The problems of reaching agreement among the people surrounding the plant are significant. As we shall see, no matter whom the property right is vested with, the basic problem is free-riding and truthful revelation of demand.

Expanding our example, suppose there are 20 people surrounding the power plant, the cost of cleaning up the plant completely is $91, and the damage to each person from the pollution is $5. Thus the aggregate damage is $100 so it is socially desirable to undertake pollution control, for a net gain of $9. Although we have stated the damage is $5 per person, in reality, the damage is private information to the individuals. This is where the problems arise.

Suppose first that the right to pollute is vested with the power plant. The Coase Theorem suggests that efficiency (pollution control) can be attained via payments from the individuals to the plant. Indeed, one possibility is that all 20 people get together, each contributes $4.55 (for a total of $91), and the plant is paid to clean-up. However, it is entirely possible that one of the individuals pretends to not really mind the pollution and is not really willing to contribute anything. In this case the other 19 individuals can each contribute $4.79 (collecting a total of $91.01). One person free-rides but we are still able to obtain a Pareto-optimal outcome. However, if two people get the idea to free-ride, there is no way the other 18 people can pool money to raise $91 while individually paying no more than the $5 of damage. Consequently, the problems of free-riding combined with private information on damages make it very difficult to reach a Coasian solution when rights to pollute are vested in the power plant.[12]

Is it any better if the individuals have the right to clean air? The problem of not encouraging people to move to the plant should pollution continue can be solved by paying each person their damage, assuming they are behaving optimally. But the first question the plant must address is how much damage there is and whether it exceeds the cost of pollution control. Once again, the power plant must ask the citizens what their damage is; however there are incentives for individuals to overstate this damage, since compensation may be based on their response.[13] In this example, the result will be installation of pollution control equipment, the desired outcome. However, suppose damages are more modest and it is optimal to compensate. Then the tendency to overstate damage could lead to excessive compensation or overzealous pollution control measures.

The reason these problems arise is because of the nonrival aspect of the pollution. This leads to free-riding and difficulties in striking bargains. If the pollution were a rival good, such as garbage, the plant would offer the pollution at a price up to its marginal cost of control (as in the steel mill/laundry example earlier). If someone takes the pollution, there is that much less for others to consume. This is a characteristic of rivalry. In such a case, we do not have a problem.

V. BARGAINING WITHIN GROUPS[14]

The Coase Theorem is most applicable when two individual firms or people are bargaining. We do not have the problems of free-riding and negotiating costs we saw in the previous section. There is, however, a more fundamental problem when more than two people are bargaining: there may be no bargain that is agreeable to all three people simultaneously, even when transaction costs are not present. If this is the case, it presents additional problems for the Coase Theorem. We show this with an example. Before doing so, it is useful to introduce the economic concept of the "core."

One of the fundamental concepts of exchange involving multiple agents is the "core." In Chapter 4 we introduced the concept of the contract curve as the set of trades with tangency between indifference curves. The idea was that both agents were operating under the same marginal rates of substitution between the two goods. The core in those exchange examples was the portion of the contract curve that yielded utility for both parties higher than their initial situation. For two people, this is the core of the exchange example—the set of allocations to which nobody will object (individuals will object if they can do better on their own).

With three or more people, we have the same concept, although we have to generalize it a little. The core is still a piece of the contract curve but now we say that for an allocation to be in the core nobody individually will object to it (compared to their initial situation) and, furthermore, no grouping of agents (smaller than the whole or the whole itself) will object to the allocation. These groupings are called coalitions. A coalition can be one person, all people involved, or anything in between. The appeal of the core is that it is an allocation with which everyone is happy. No under-the-table (or above-board) collusion between a few of the agents changes that—they cannot do any better.

So what is the relevance of the core to the Coase Theorem? The point is that if there is an empty core, there is no allocation that everyone can agree is better than the *status quo*. If under one set of property rights we have an empty core, the status quo will prevail. If that is different than under another set of property rights, we have an apparent violation of the Coase Theorem. Actually, it is not a violation since the Coase Theorem is predicated on there being no transactions costs. And problems of bargaining are transactions costs. The example is a useful one to further illustrate problems with the Coase Theorem.

Suppose we have our old friend the steel mill and the laundry. This time, however, we will add a railroad, which also generates smoke. Use the letters S, L, and R to denote these three agents. The possible cooperative arrangements are $A_1 = \{(R),(S),(L)\}$, $A_2 = \{(R),(S,L)\}$, $A_3 = \{(R,S),(L)\}$, $A_4 = \{(R,L),(S)\}$, and $A_5 = \{(R,S,L)\}$. These are the five possible coalition structures. The first of these, $A_1 = \{(R),(S),(L)\}$, simply means that each of the three firms operates on its own. The second of these, $A_2 = \{(R),(S,L)\}$, means that the railroad operates by itself, but that the steel mill and laundry merge into a single firm and act for their joint good. The profits for individual coalitions are given in Table 6.2. Note that we have specified above what a coalition may get but not how that coalition would divide its profit among its members.

Think of the following physical arrangements to go along with these profit levels. Whenever one of the polluters gets together with the laundry, the polluter stops producing and laundry output goes up dramatically. If both polluters get together with the laundry, both polluters shut down, generating the largest laundry output. Efficiency calls for this to happen, as can be seen by inspecting the payoffs. Adding up the total payoff to all of the coalitions in any particular partitition demonstrates that A_3 yields the highest total, \$40,000. This is the case in which both polluters shut down and the laundry cleans up.

Now let us examine what will happen under two property rights regimes: the laun-

TABLE 6.2 Payoffs to Coalitions: Example of Railroad, Steel Mill and Laundry

Partition[a]	Payoff[b]	Description
$A_1\{(R),(S),(L)\}$	{\$3,\$8,\$24}	Each firm acts independently
$A_2\{(R,S),(L)\}$	{\$15,\$24}	Railroad and steel mill merge
$A_3\{(R,S,L)\}$	{\$40}	All three firms merge
$A_4\{(R),(S,L)\}$	{\$3,\$36}	Steel mill and laundry merge
$A_5\{(R,L),(S)\}$	{\$31,\$8}	Railroad and laundry merge

[a]*A partition is a collection of coalitions; i.e., a particular way of dividing all of the firms into coalitions.*
[b]*Payoffs in thousands of dollars.*

dry has a right to clean air and the laundry has no rights to clean air. Suppose first that there is a right to clean air. The railroad and the steel mill will have to pay damages. Without the pollution, the laundry makes \$40,000 [since the grand coalition of (R,S,L) involves stopping production of R and S]. If the laundry operates independently, it makes \$24,000 for a loss of \$16,000. The railroad and steel mill cannot afford the \$16,000 damages so they will shut down. All other coalitional structures run into a similar problem. Consequently, the only outcome that will result is no pollution and the laundry producing \$40,000 of output.

Now suppose there is no right to clean air. Now the laundry must induce the polluters to stop polluting. There are two issues. First, will the laundry have enough money to induce the polluters to shut down? Second, will any arrangement the laundry sets up with the polluters be stable in the sense of there being no incentives for the participants to break the deal? The answers to these questions will become clearer as we delve into the problem.

The laundry stands to gain \$16,000 by shutting down the polluters (compare A_3 to A_1 or A_2). The two polluters together can make \$15,000 (A_2), so there should be enough extra money to induce them to shut down. This would leave the laundry with an extra \$1000 from striking the deal. How will the railroad and steel mill divide the \$15,000? The mill must get at least \$8000 since that is what it makes on its own (A_1). That would leave the railroad with \$7000. Thus the railroad makes \$7000, the steel mill makes \$8000, and the laundry makes \$25,000. Call this the "grand bargain." Clearly, everyone prefers this arrangement to independent action (A_1). Would any smaller group of the three agents prefer to get together without the third? If the mill were to get together with the laundry (A_4), they stand to make \$36,000, which is \$3,000 greater than the \$33,000 the laundry and steel mill netted in the grand bargain. So the coalition (S,L) "blocks" the grand bargain. One could continue like this, selecting a coalition and a split of profits that looks better than an alternative, only to find some other coalition and split that dominates it. The point is, no matter what payments you propose from the laundry to the two polluters (to shut down), one of the other coalitions will be preferred by the alternative coalition's members. For this reason, there is no coalition and division of the pie that is not dominated by another. This means that the core of this "economy" is empty. One interpretation of this is that no bargain will be struck. They will continue trying to strike a deal for all time.

The significance of this result is that striking a "deal" to eliminate externalities may not be as easy as suggested by the Coase Theorem. The outcome may indeed depend on who initially is allocated the rights to pollute. Another interpretation of this is that the problem of reaching an agreement over pollution is just another form of transactions cost. Consequently, the example we have given is not counter to the Coase Theorem. Although that is a valid interpretation, if difficulties in reaching agreement are excluded from the result, the Coase Theorem becomes weaker and even less applicable to real world problems.

VI. THE POLICY SIGNIFICANCE OF THE COASE THEOREM

The Coase Theorem is remarkable because it is counterintuitive, at least when first encountered. But is it anything more than a curiosity of academic economics? Is the Coase Theorem relevant to environmental policy?

There are several clear policy implications of the Coase Theorem. One is that transaction costs matter to the efficient distribution of rights to a clean environment. This means that in designing a set of property rights for the environment, it is important both to distribute them approximately efficiently and/or work to reduce the costs associated with trading those rights.

As markets for rights to pollute become increasingly common, it is important to heed this advice. In the case of tradable emission rights in many urban areas in the United States, a great deal of paper work is associated with every trade in rights. As a result, the number of trades is often disappointing.[15] More attention to reducing transaction costs would help in such situations.

Another policy implication of the Coase Theorem is that victims of pollution should not be loathe to pay polluters to reduce pollution. Often the political power of polluters is such that it is impossible to force them to reduce pollution and get them to pay for it. This often results in a stalemate that can last for years. Rarely is the stalemate resolved by acknowledging reality with the victims paying for cleanup. The desirable outcome is the absence of pollution. Who pays is (to a certain extent) a secondary issue. This is a conclusion of the Coase Theorem that has yet to find wide acceptance in environmental policy.

There are a few cases in which the victims of pollution pay to eliminate or reduce the pollution. One case occurred in Santa Maria, California in 1996. A cattle-fattening operation ("feedlot") had been in existence for many years, generating pungent smells but located well away from houses, at least initially. Over time, as the town grew, houses were built closer and closer to the feedlot. Rather than try to force the feedlot to control its pollution, the City Council in Santa Maria voted to tax the residents around the feedlot and use the revenue to pay the feedlot owner to cease operations.[16] Efficiency was attained even though the victims paid to attain it.

SUMMARY

1. Property rights are important to a well-functioning market. Without property rights, even the most ordinary market transactions are difficult.

2. Transaction costs are those costs associated with consummating a trade, over and above the trade itself (i.e., the money exchanged for the good). A simple example is the commission paid to an agent to sell a house. Transaction costs can also involve non-monetary costs such as the difficulty in striking a bargain.

3. The Coase Theorem has two parts: one with transaction costs and one without. In achieving the efficient level of an externality (such as pollution), the initial assignment of property rights (right to pollute, right to clean air) is irrelevant if there are no transaction costs. With transaction costs, the initial assignment matters.

4. When transaction costs are present, it is important for the government to distribute property rights in an efficient way and to try to reduce transaction costs associated with trading rights.

5. With public bads, where there are a few generators of pollution and many consumers, bargaining implicitly involves costs. Thus the Coase Theorem states that the free trade of property rights cannot be relied on for efficiency.

PROBLEMS

1. In Section V, for each partition, propose a division of the profits that makes each firm at least as well off relative to independent action. Then show how one or more firms can actually do better in a coalition in another partition.

2. Consider a pollution problem involving a paper mill located on a river and a commercial salmon fishery operating on the same river. The fishery can operate at one of two locations: upstream (above the mill) or downstream (in the polluted part of the river). Pollution lowers profits for the fishery: without pollution, profits are $300 upstream and $500 downstream; with pollution, profits are $200 upstream and $100 downstream. The mill earns $500 in profit, and the technology exists for it to build a treatment plant at the site that completely eliminates the pollution, but at a cost of $200. There are two possible assignments of property rights: (i) the fishery has the right to a clean river and (ii) the mill has the right to pollute the river.

a. What is the efficient outcome (the maximum of total joint profit)?

b. What are the outcomes under the two different property rights regimes, when there is no possibility of bargaining?

c. How does your answer to (b) change when the two firms can bargain costlessly?

3. Discuss the following statement: Most pollution problems can be traced to disputes over property rights, imperfectly defined property rights, or ambiguously defined property rights. Give examples to support your comments.

4. Suppose in column (E) of Table 6.1, the laundry decided to purchase the steel mill and merge operations. What is the minimum the laundry would have to pay for the steel mill under the three assumptions regarding the price of steel?

5. Brussels is currently considering two plans to control pollution caused by heavy trucks. Plan A calls for the institution and enforcement of tight emission standards on commercial trucks with fines for noncompliance. Plan B involves tax incentives for firms operating commercial trucks that achieve lower emission levels. You have analyzed these plans and conclude that each plan has identical effects on emissions from trucks and resulting pollution concentrations. What are the potential differences between these two plans, on the basis of efficiency or other criteria?

6. A beekeeper and a farmer with an apple orchard are neighbors. This is convenient for the orchard owner since the bees pollinate the apple trees: one beehive pollinates one acre of orchard. Unfortunately, there are not enough bees next door to pollinate the whole orchard and pollination costs are \$10 per acre. The beekeeper has total costs of $TC = H^2 + 10H + 10$ and marginal costs $MC = 10 + 2H$ where H is the number of hives. Each hive yields \$20 worth of honey.

a. How many hives would the beekeeper maintain if operating independently of the farmer?

b. What is the socially efficient number of hives?

c. In the absence of transaction costs, what outcome do you expect to arise from bargaining between the beekeeper and the farmer?

d. How high would total transaction costs have to be to erase all gains from bargaining?

7. Foster and Hahn (1995) have examined trading of air pollutant emission rights within the Los Angeles air basin in 1985–1991. They hypothesized that two factors would lead to high transaction costs: exchanges of thinly traded pollutants (few trades) and exchanges involving small quantities of pollution. Transaction costs for thinly traded pollutants would be high because of the difficulty in finding someone with whom to trade. The following is a table showing the number of trades for five pollutants during the period in question in Los Angeles. Shown are trades from one factory to another within the same company ("internal"), trades involving brokers or other intermediaries ("brokered"), and trades between companies without assistance of intermediaries ("external").[17]

Pollutant	Internal	Brokered	External	Total
CO	3	6	1	7
NO_X	4	21	4	29
PM	6	4	1	11
ROG	32	55	61	148
SO_X	3	3	0	6

a. How would you expect the transaction costs to differ among the three categories of trade: internal, brokered or external?

b. Do the data in the table support the hypothesis that transaction costs might be higher for thinly traded pollutants? Why or why not?

c. What effect would you expect the size of a trade (number of tons of pollution) to have on transaction costs (per ton of pollution)? How would you expect the size of a trade to influence whether a pollutant is more likely to be internally traded, brokered, or externally traded?

NOTES

1. There has been a large literature on the Coase Theorem. For a sampling, see Dahlman (1979), Hoffman and Spitzer (1982), Regan (1972), Schulze and d'Arge (1974), and Calabresi (1968). Cooter (1987) provides a nice summary.
2. If the laundry is losing money at S_0, it will choose not to produce any laundry, resulting in a profit of zero. The same applies at S.

3. One might ask whether the laundry has the resources to purchase the property rights: Can it still have nonnegative profits if it purchases the rights? We will consider that question in the numerical example that follows.
4. This point is made by Schulze and d'Arge (1974).
5. A sunk cost is a cost that cannot be recovered. If you dig a hole in the ground to install a swimming pool but half-way through construction you change your mind, then the cost of digging the hole cannot be recovered. If on the other hand, you buy some equipment to use in your pool, you may return that to the store for a refund; the equipment can be used by another. Both are examples of fixed costs; only the first is sunk.
6. These marginal cost and marginal damage functions can be obtained by examining the ratio of changes in costs to changes in S or L. Alternatively, they may be obtained by simple calculus.
7. This outcome is referred to as the Nash equilibrium, using game theory terminology. In the case of bargaining, it is referred to as the threat point—the point to which both parties will revert if bargaining breaks down.
8. High transaction costs have been blamed for the failure of several pollution trading systems. See Hahn (1989) and Stavins (1995).
9. The characterization of the Coase Theorem presented here is drawn from Hoffman and Spitzer (1982).
10. Baumol and Oates (1988) cite a Swedish example of a refinery damaging an auto manufacturing plant. Negotiation between these parties took place to eliminate the Pareto-relevant externality.
11. Coase (1981), p. 187.
12. Even if the damage figure were public knowledge so that everyone knew the individual damage was \$5, the potential for holdout is still there. By this we mean that one or more of the people can refuse to pay, even though the payment would make them better off. Provided there is only one holdout, the mill can still be paid and Pareto optimality achieved.
13. Here the holdout problem (mentioned in note 12) does not exist. If the plant knows the damages of each, it simply offers compensation to everyone in the amount of \$5 or, alternatively, installs pollution control equipment.
14. This section is based in part on Aivazion and Callen (1981).
15. Hahn (1989) provides a very accessible treatment of this issue.
16. See Bedell (1996) for an account of this transaction.
17. For pedagogical reasons, the data shown in the table are a slight oversimplification of the data in Foster and Hahn (1995).

7 PIGOVIAN FEES

In the past few chapters we explored the reasons that markets cannot be relied on to provide the right amount of pollution—markets fail when it comes to public bads. Normally, consumer preferences are communicated to producers through the price system. The basic problem in the case of pollution is that without a price system, polluters do not "see" the damage caused by the pollution they emit. In the last chapter, we explored how liability and negotiation could solve the problem. We now consider the alternate approach of the government intervening and sending a signal to polluters—the government prices pollution. Since pollution is a bad, the price is negative. In other words, polluters pay a price for every unit of pollution they generate. This corrects the market failure, at least in theory.

I. PIGOVIAN FEES: SINGLE POLLUTER

Early in the twentieth century the English economist Arthur C. Pigou argued for the imposition of taxes on generators of pollution.[1] Since the social cost of pollution is in excess of the private cost to the polluter (actually, polluters have a negative cost since they save money by polluting), the government should intervene with a tax, making pollution more costly to the polluter. If the pollution is more costly to produce, the polluter will produce less pollution. This tax has come to be called a Pigovian fee or Pigovian tax.

Suppose we have a factory generating pollution in the amount x and goods output in the amount y. Production costs for the factory depend on x and y (as well as input prices) and can be written $C(x,y)$, assuming input prices are constant. For the most part, we can also assume y will be produced where the marginal cost of production of goods is equal to the price of goods. For that reason, and to keep things simple, we will suppress writing y as a determinant of costs. This does not always simplify things, as is demonstrated in Problem 5 at the end of the chapter. The elimination of y and the suppression of input prices results in costs which depend on x only, $C(x)$, with marginal costs $MC(x)$—the additional costs from producing one more unit of pollution, x. Since costs decline as x is increased, marginal costs are actually negative. Another way of thinking about this is in terms of marginal savings—the savings from emitting one more unit of pollution. Of course the marginal savings is the negative of the marginal costs: $MS(x) = -MC(x)$.

Further assume there are N people surrounding the factory and that pollution causes damage. For the time being, assume that people cannot use locational choice to change the amount of pollution they face. Thus there is nothing a person can do to reduce his or her exposure, short of getting the factory to cut back. For person i, the damage from pollution is $D_i(x)$, which is positive and increases in x. There are several other ways of interpreting this damage. We could also say that person i benefits from the pollution in the amount $B_i(x)$ with benefits negative and decreasing in x. Or, we could say that $D_i(x)$ is the willingness to pay to eliminate the pollution. Total damages are given by

$$D(x) = \Sigma_i D_i(x) \tag{7.1}$$

The right amount of pollution is the amount that minimizes total costs and damages:

$$x^* \text{ minimizes } \{C(x) + D(x)\} \tag{7.2a}$$

We know that something is minimized when its marginal is zero. Further, the marginal of a sum is equal to the sum of the marginals. Thus we can set the marginal of the quantity in braces in Eq. (7.2a) to zero:

$$MC(x^*) + MD(x^*) = 0 \tag{7.2b}$$

Substituting the marginal version of Eq. (7.1) into Eq. (7.2b) and recognizing that marginal savings is the negative of marginal cost, we obtain

$$MS(x^*) = \Sigma_i MD_i(x^*) \tag{7.2c}$$

In other words, we seek a level of pollution such that the marginal savings to the firm from pollution ($-MC$) is equal to the marginal damage from pollution over the entire population. Since pollution is a public bad, the aggregate marginal damage (MD) is the vertical sum of the individual marginal damages (MD_i).

A. The Fee Level

We have seen that a market will not spontaneously emerge to supply the right amount of pollution. If it did, each individual would receive compensation equal to $MD_i(x^*)$ per unit of pollution (Lindahl prices) and the firm would pay $MS(x^*)$ per unit of pollution. This would induce the firm to generate the correct amount of pollution. Budgets would balance because of the relationship in Eq. (7.2b). Suppose the firm paid $-MC(x^*)$ per unit of pollution but paid it to the government instead of to the consumers of the pollution. This would result in the correct amount of pollution being generated and, since consumers cannot affect their exposure, the fact that they are not compensated is irrelevant for efficiency. This is the concept of a Pigovian fee.

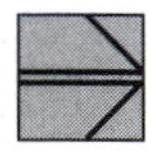

Definition A Pigovian fee *is a fee paid by the polluter per unit of pollution exactly equal to the aggregate marginal damage caused by the pollution when evaluated at the efficient level of pollution. The fee is generally paid to the government.*

This situation is illustrated in Figure 7.1 for the case of one polluter and two victims of the pollution. Shown in the lower half of the figure is the marginal cost of pollution. Note that this is negative since every extra unit of pollution the factory is allowed to emit lowers total costs for the factory (up to a limit of course). The marginal savings to the factory is the negative of this and is shown in the first (upper) quadrant. As the factory increases pollution from no emissions at all, savings are initially quite high. When emissions are relatively large, the savings from emitting a little more are much smaller. Thus $MS(x)$ is downward sloping.

Also shown in Figure 7.1 are the marginal damage functions for the two victims of the pollution: $MD_i(x)$. Marginal damage is the negative of the demand function for pollution for each of the individuals. Each of the marginal damage schedules is upward sloping. When pollution levels are small, one more unit of pollution causes little damage. When pollution levels are higher, that extra unit causes more damage. Since the pollution is a public bad, aggregate marginal damage, like aggregate demand, is the vertical sum of individual marginal damages. This is also shown in the figure [$MD(x)$]. The optimal amount of pollution is the x for which $MD(x) = MS(x)$, shown as x^* in Figure 7.1. Also shown in Figure 7.1 is the Pigovian fee, p^*. If the polluter is charged p^* per unit of pollution, the polluter basically sees pollution as priced at p^*. Thinking of pollution as an output of the firm, the firm receives $-p^*$ of revenue for each unit of pollution it generates. We know that the firm will produce so that price equals marginal cost:

$$MC(x^*) = -p^* \qquad \text{or} \qquad MS(x^*) = p^* \tag{7.3}$$

The total amount of money the firm pays for the pollution is p^*x^*.

Another way of viewing this problem is that without the Pigovian fee and without any other markets or regulations to restrict pollution, the firm basically sees a price of zero for pollution. It is optimal (from the firm's point of view) for the firm to respond to a zero price by producing where marginal cost is zero, $\hat{x}$ (see Figure 7.1). To reduce pollution generation, we must increase the cost of pollution by raising the fee. As the fee is

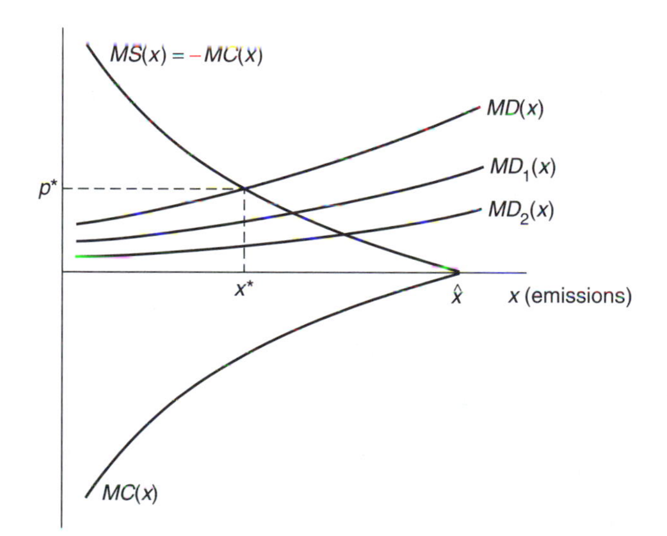

Figure 7.1 Optimal Pigovian fee on pollutant emissions with two victims of pollution. $MD_1(x)$, Marginal damage to victim 1; $MD_2(x)$, marginal damage to victim 2; $MD(x)$, aggregate marginal damage; $MC(x)$, marginal cost of emitting for the polluter; $MS(x)$, marginal savings from emitting for the polluter; $\hat{x}$, pollution levels with no regulation; x^*, efficient amount of emissions; p^*, Pigovian fee.

raised from zero, pollution generation gradually declines until we reach x^* when the fee rises to p^*.

Note that the Pigovian fee is defined as the marginal savings from pollution generation *at the optimal level of pollution*. If we are not at the optimum, the Pigovian fee will be neither the current marginal cost of pollution control nor the marginal damage from pollution. Thus the Pigovian fee is not any emission fee; it is the marginal savings from pollution at the optimal pollution level.

B. Should Victims be Compensated?

One feature of the Pigovian fee is that it is paid to the government and the government keeps it. It is not necessary to pay it out as compensation to the victims of the pollution, although it is not undesirable to do so. Why? Remember that the only thing that matters for efficiency is that the actions of consumers and producers be correct. How much is paid is irrelevant except to the extent that it induces optimal behavior. Since consumers can do nothing to influence how much pollution they are exposed to, payment of compensation will not change their behavior vis-à-vis pollution. Such payment would be only an income transfer. On the other hand, if consumers can change their location, we saw in Chapter 5 that consumers should not be compensated. If they are, they will tend to move toward the pollution. In this case payment of compensation is worse than neutral—it reduces efficiency. Thus in either case, as long as we are dealing with a public bad, the Pigovian fee should be levied and collected but need not be paid out to victims.

The astute reader may ask the following: If the Coase Theorem fixes the problem of externalities, why do we need Pigovian fees? Further, won't Pigovian fees exacerbate the problem if the Coase Theorem is also operating?

Consider the case of Anna and Brewster, who this time are neighbors. Brewster generates a lot of garbage and Anna does not. To start with, we have no property rights, so Brewster gets rid of his excess garbage by tossing it over the fence into his neighbor's yard. This situation is represented graphically in Figure 7.2.

Shown in Figure 7.2 is the marginal savings to Brewster from throwing garbage over the fence into Anna's yard (MS_B). Also shown is the marginal damage Anna suffers from this unneighborly activity (MD_A). Brewster is initially inclined to dump $\hat{g}$, the point

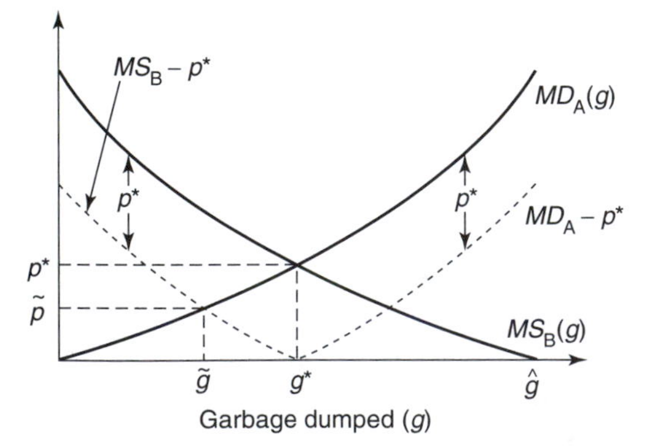

Figure 7.2 Pigovian vs. Coasian solutions. $MD_A(g)$, Marginal damage from garbage to Anna; $MS_B(g)$, marginal savings to Brewster from garbage; g^*, socially efficient level of garbage; p^*, Pigovian fee; $\tilde{g}$, garbage level from bargaining with Pigovian fee; $\tilde{p}$, Coasian payment to Brewster, with Pigovian fee.

at which the marginal cost of dumping (zero) equals the marginal savings from dumping. Obviously, society would like g^* to be dumped. If Anna and Brewster have a polite discussion of this matter, they could reach an agreement whereby Anna pays Brewster p^* for every bag of garbage he does not throw over the fence. This is worthwhile to Anna since the marginal damage of those bags is in excess of p^*. It is also a good deal for Brewster since his marginal savings from the dumping is less than p^*. This is as Coase would have predicted.

Now suppose society institutes a Pigovian tax of p^* on Brewster. Brewster's marginal savings curve is reduced by p^* as shown in the figure ($MS_B - p^*$). Without anything else happening, Brewster will generate g^* of garbage. But suppose Anna and Brewster have their little discussion and Anna offers to pay Brewster $\tilde{p}$ for each bag of garbage Brewster does not throw over the fence. We will end up at $\tilde{g}$ with too little garbage going over the fence. On the other hand, if the Pigovian fee is paid to Anna, Anna's damage is reduced, as is also shown in Figure 7.2 ($MD_A - p^*$). Then Anna will not pay anything to reduce garbage below g^* and Brewster will not wish to produce garbage in excess of g^*.

The point is that it is important to consider both the consumer and producer side of this equation. If both are considered, there is no conflict between the Coasian and Pigovian solutions.

Unfortunately, the case of a public or nonrival bad is more complex. We have already discussed the fact that compensation is undesirable or at least unnecessary when a bad is nonrival. Thus the appropriate Pigovian fee is levied on the generator of the public bad and the receipts are not distributed to the victims of the pollution. When the fee is levied, the source of the pollution cuts back until the marginal savings from the pollution equals the Pigovian fee. However, if all the consumers get together and agree to pay the factory to reduce pollution further, that will be done and pollution will be reduced below its optimal level. The saving grace is that we have also argued that the Coasian solution is unlikely to work for the case of a public bad because of the problems of bargaining and free-riding.

II. MULTIPLE POLLUTERS: THE EQUIMARGINAL PRINCIPLE

We have seen how a Pigovian fee can generate the correct amount of a public bad. The case we have looked at involves a single polluter. Suppose we have more than one polluter. For the time being, assume we have two polluters. Figure 7.3 illustrates the case for two polluters in which the marginal damage function (MD) is the aggregate damage to all consumers. Shown is the marginal savings to each of two firms from generating pollution. How much pollution should each firm generate and how should the Pigovian fee be set to support that level of pollution?

An aggregate marginal savings function for a group of polluters indicates what the marginal savings will be if the total amount of pollution increases by one unit. This depends, of course, on what assumptions are being made about how the total amount of pollution is distributed among the individual polluters. If one polluter is doing all of the pollution control and all other polluters are doing none, the marginal savings will be higher

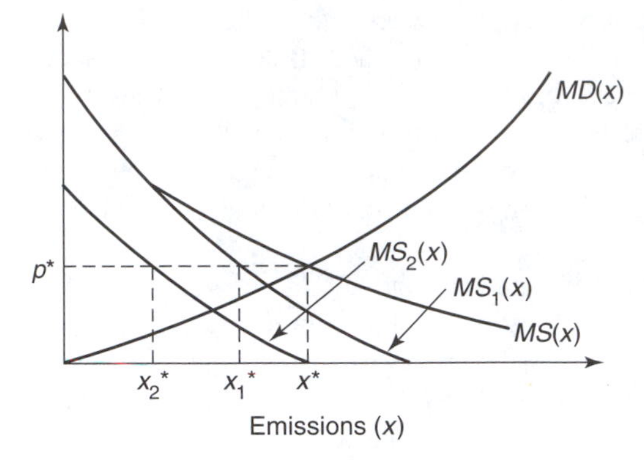

Figure 7.3 The case of two polluters. $MS_1(x)$, Marginal savings from emitting firm 1; $MS_2(x)$, marginal savings from emitting firm 2; $MS(x)$, aggregate marginal savings from emitting; $MD(x)$, marginal damage from emitting; p^*, Pigovian fee; x^*, total amount of emissions with Pigovian fee; x_1^*, emissions from firm 1 with Pigovian fee; x_2^*, emissions from firm 2 with Pigovian fee.

than if the pollution control were more "evenly" distributed among the polluters. One way of avoiding this ambiguity is to assume the polluters are sharing the obligation of pollution abatement in an efficient or least-cost manner.

The efficient way of distributing a pollution control obligation among various polluters is to do so in such a way that the marginal control costs are equalized among them. This is really quite logical. Suppose we have two polluters and we wish to emit in total the quantity S of pollution. Further, suppose we divide S among the two polluters in such a way that one polluter has a higher marginal cost of pollution control than the other. We can reduce costs without changing the total amount of pollution by increasing emissions from the low-cost polluter by one unit and decreasing emissions from the high-cost polluter by one unit. Control costs will decline while the total amount of pollution remains constant. We can continue this thought experiment until the marginal control costs for the two firms are equalized.

This notion that efficiency requires marginal pollution control costs (or, equivalently, marginal savings from polluting) to be equalized among polluters has become known as the *equimarginal principle*:

Definition *In controlling emissions from several polluters whose emissions all contribute to damage in the same way, the* equimarginal principle *requires that the marginal cost of control be equated across polluters to achieve an emission reduction at the lowest possible cost.*

Clearly a Pigovian fee generates pollution in such a way that the equimarginal principle is satisfied, since all polluters set their marginal savings function to the same number, the Pigovian fee.

Now that we know how an aggregate quantity of pollution control will be distributed among several polluters, we can return to our question of how to construct an aggregate marginal savings function. The simplest way to construct such an aggregate is to do it backward. Since we know all firms will have the same marginal savings from emitting if we are distributing emissions efficiently, we can start with a particular number for marginal savings and ask how much each firm would emit to yield that marginal savings. We then add all of these answers and obtain the aggregate emissions that would be associated with that marginal savings. If we continue this process for several different levels of marginal savings, we will soon have constructed a marginal savings curve.

Returning to Figure 7.3, how do we aggregate the two marginal savings functions to obtain an aggregate? The aggregate marginal savings function is found by horizontally summing the marginal savings for the two firms.[2] The aggregate marginal savings curve is the curve that indicates for a particular pollution price, how much pollution each firm would generate. This is the standard way of generating an industry marginal cost function from firm-level marginal cost functions. Thus for any level of the fee, $MS(x)$ tells us how much x in total will be emitted and each $MS_i(x)$ tells us how much each firm will contribute to that total. We have constructed $MS(x)$ in such a way that the amount of pollution from each firm will sum to the total.

In Figure 7.3, to determine the optimal amount of pollution, we note where the marginal savings curve (MS) intersects the marginal damage curve (MD). That determines the optimal amount of pollution (x^*) and the marginal savings to polluters at x^* : p^*. Thus p^* is the correct Pigovian fee. This is shown in Figure 7.3. At that fee level, firm 1 will generate x_1^* and firm 2 will generate x_2^*. Note that each firm operates so that marginal savings from polluting are set equal to the Pigovian fee:

$$MS_i(x_i^*) = p^* \tag{7.4}$$

Furthermore, by the way in which $MS(x)$ was constructed,

$$MS(x^*) = p^* \tag{7.5}$$

Equation (7.4) illustrates one of the primary virtues of the Pigovian fee: all firms will control pollution at the same level of marginal costs. Marginal costs of pollution control will be equated across all polluters. Firms with different pollution control costs receive the correct signals regarding how much pollution to generate. Those with high control costs will control relatively less than firms with lower control costs.

Example: In December 1997, most of the countries of the world met in Kyoto, Japan to try to reach an agreement on controlling emissions of gases that lead to global warming. The Kyoto Protocol emerged from that meeting. The Protocol calls for countries to reduce emissions by a certain percentage below 1990 levels and to achieve this by the year 2012. The European Union (EU) would be expected to reduce emissions to 7% below 1990 levels, Japan 6%, Australia 8%, and the United States, 7%. Other countries have different targets. Overall, if all countries abide by the Protocol, global emissions should be 5% below 1990 levels (Anderson, 1998). One of the requirements the Europeans pushed for was the ability to pool European countries, reducing emissions more in some countries than others, simply requiring the 7% target be met for the EU as a whole.

Why would the EU want to pool emissions when determining compliance with the Protocol? Figure 7.4 shows one estimate of the marginal costs of controlling carbon dioxide emissions in various European countries. Although there is considerable debate regarding the true magnitude of these costs, Figure 7.4 illustrates the potential differences in control costs among different countries.[3] It is easy to see that a 7% reduction in emissions in each country of the EU would definitely not satisfy the equimarginal principle.

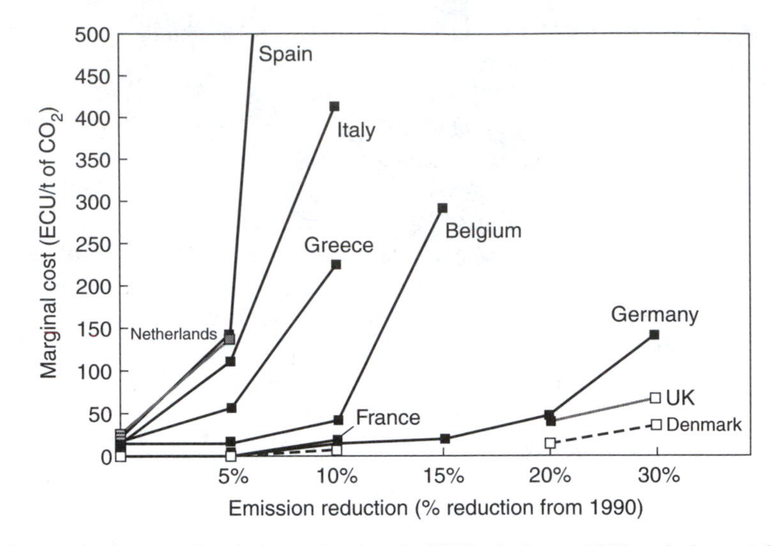

Figure 7.4 Marginal costs of emission reductions in 2010 relative to 1990 emissions. Adapted from Hourcade et al. (1996), p. 318.

Italy's marginal costs would be in the neighborhood of 300 ECU per ton of CO_2 while France and Germany's marginal costs would be significantly less than 50 ECU per ton. With an EU-wide emission reduction target, Figure 7.4 suggests that the reductions would most likely be concentrated in Germany, the United Kingdom, Denmark, France and Belgium.

III. FEES VERSUS SUBSIDIES

The Coase Theorem suggests that it makes no difference whether the polluter must compensate the victim of the pollution or the victim must pay the polluter not to pollute. There is an analogous issue in the context of Pigovian fees. Is it possible to obtain the same outcome by subsidizing firms to reduce pollution? In the "real" world, is there any danger in providing tax breaks and other subsidies for pollution control, rather than making polluters pay for the pollution they generate? Is it possible to obtain efficiency using a subsidy instead of a fee? This is an important question since subsidies are usually much more politically popular than taxes or fees.

The answer to this is that these two approaches yield different outcomes.[4] The tax is efficient, whereas the subsidy can result in too many firms in the industry and thus an inefficient amount of both pollution and the good associated with the pollution. We will consider two cases. The first is for the short run—there is no time for new firms to enter the industry. The second is for the long run—there is time for entry and exit (although exit in the sense of shut-down is always possible, even in the short run).

A. The Short Run

Let us consider a competitive industry, producing some good in conjunction with pollution. Initially, assume all of the firms in the industry are identical. Under a pollution tax (t), the production costs of a typical firm would be

$$C_T(y,e) = V(y,e) + t\,e + FC \tag{7.6}$$

where y is the amount of the good being produced, e is emissions, $V(y,e)$ represents variable production costs for producing e and y, and FC is the fixed cost of production. To simplify things, suppose there is a fixed relation between output and emissions—the more you produce, the more you pollute. In particular, suppose emissions are related to output by $e = ay$ where a is a constant. We can rewrite Eq. (7.6) as

$$C_T(y,ay) = V(y,ay) + tay + FC \tag{7.7a}$$

Recognizing that V and C_T are now functions of y only, we let $TC(y) = C_T(y,ay)$ and $VC(y) = V(y,ay)$ and rewrite Eq. (7.7a) as

$$TC(y) = VC(y) + tay + FC \tag{7.7b}$$

This means that marginal production costs are

$$MC(y) = MVC(y) + at \tag{7.8}$$

So basically, marginal costs are increased by at.

Now consider what happens with a subsidy, s. With no attention to pollution control, a firm might pollute at the level $\hat{e}$. With a subsidy, the firm will be paid to reduce emissions. If the firm reduces emissions to e, the subsidy payment will be $s(\hat{e} - e)$. This means that costs will be

$$\begin{aligned} TC(y) &= VC(y) + FC - s(\hat{e} - e) \\ &= VC(y) + say + \{FC - s\hat{e}\} \end{aligned} \tag{7.9}$$

Note that the term in braces is a fixed cost, consisting of the standard fixed cost plus a lump-sum transfer of $s\hat{e}$ that is independent of the firm's choice of y or e. Thus the variable costs in both cases [Eqs. (7.7b) and (7.9)] are exactly the same. Only the fixed costs are different. Consequently the short-run marginal costs of production will be identical in the two cases and the firm will produce exactly the same amount of pollution and the good. In fact, the marginal costs from Eq. (7.9) are

$$MC(y) = MVC(y) + as \tag{7.10}$$

which is exactly the same as Eq. (7.8) except we have an s here instead of a t.

Consequently, our first result is that in the case of identical firms in the short run, Pigovian fees and subsidies yield exactly the same outcome. We should note that this result applies even if there is a more complex relationship between output and emissions than the fixed ratio assumed here. Showing that is more cumbersome mathematically so we omit it here.

We now turn to an industry with heterogeneous firms. This may be because of different technologies used by different firms due, for instance, to their different ages. This case is best understood graphically. Suppose we have an industry composed of two classes of firms, old firms and new firms. Newer firms may have higher fixed costs but lower variable costs. We are concerned with industry behavior in the short run under Pigovian fees and subsidies. Since this is the short run, no new firms may enter. Any firm may, however, choose to produce nothing, shutting down. If a firm produces nothing, the subsidy disappears. In other words, we only pay firms to produce less pollution. We do not continue to pay firms if they decide to go out of business.

Figure 7.5 shows the marginal cost curves and average variable costs for these two types of firms. Since this is a short-run analysis, we are not concerned with total costs. The issue is whether prices cover average variable costs and, if they do, production will be at marginal cost equals price. As we saw above, the effect of a tax on marginal cost is identical to the effect of a subsidy on marginal cost—both raise marginal costs relative to the unregulated case. The U subscripts in the figure correspond to average variable cost (*AVC*) and marginal cost (*MC*) in the unregulated, pretax, or presubsidy case. The T and S subscripts refer to the case of a Pigovian tax or subsidy, respectively.

Note in Figure 7.5 that although the tax and the subsidy raise the marginal costs by

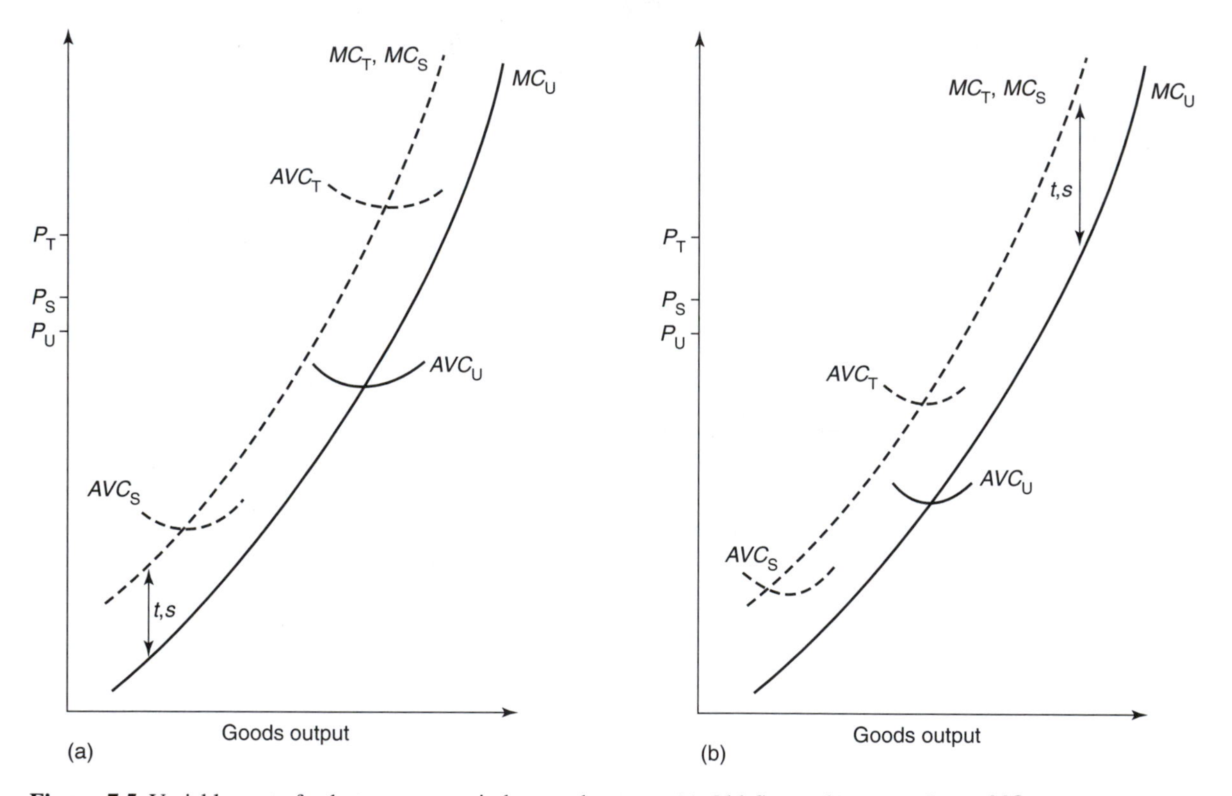

Figure 7.5 Variable costs for heterogeneous industry, short run. (a) Old firms; (b) newer firms. MC_U, Marginal cost unregulated case; MC_S, marginal cost, with emission control subsidy; MC_T, marginal cost, with emission fee; AVC_U, average variable cost, unregulated case; AVC_S, average variable cost, with emission control subsidy; AVC_T, average variable cost, with emission fee; p_U, goods price, unregulated case; p_S, goods price, with emission control subsidy; p_T, goods price, with emission fee.

the same amount, the subsidy lowers average variable cost whereas the tax raises average variable cost. The reason is that we have assumed the subsidy applies only if the firm is operating. The lump sum $s\,\hat{e}$ in Eq. (7.9) goes away if the firm shuts down. Thus it counts as a variable cost. Fixed costs (FC) are incurred whether or not the firm shuts down (in the short run).

Now we turn to determining what the market price of the good might be under these three regimes. Figure 7.6 traces out the short-run supply functions for this industry for the three cases, unregulated, Pigovian taxes, and subsidies. Also shown is a typical demand function for the good. Recall that firms will operate on the portion of their marginal cost curve that lies above the average variable cost. We can see that with both the unregulated and subsidy cases, both types of firms are operating, yielding product prices of p_U and p_S, respectively. In the case of the Pigovian tax, only the newer firms operate, yielding product price p_T. These prices are shown on Figure 7.5. Note that all prices are above average variable costs for the newer firms, which is why they operate in all three cases. However, for the old firms, p_T is below AVC_T, the average variable cost for the Pigovian tax case. This is why the old firms shut down in this case.

Our conclusions are that taxes and subsidies have different effects in the short run. A subsidy may allow firms to continue operating that would not continue in the case of a tax. Which is efficient? The subsidy requires a lump-sum transfer, which has to be obtained from somewhere. Even more important, the subsidy involves the operation of firms that are really losing money (negative profits). This is not efficient.

B. The Long Run

We now turn to the case of the long-run effects of Pigovian taxes and subsidies. We saw in Eq. (7.7)–(7.10) that the effect of a tax or subsidy was to raise marginal costs, but that a subsidy lowered average costs while a tax raised average costs. This applies in the short run as well as in the long run. If we assume a constant-cost industry, all firms will operate at the bottom of their average total cost curve, in long-run equilibrium. Thus the supply schedules for the industry will be horizontal and as shown in Figure 7.7b. The result

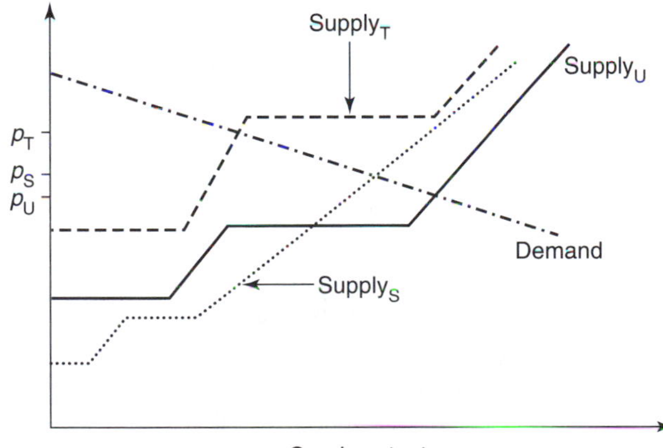

Figure 7.6 Short-run supply and demand, heterogeneous industry, with and without taxes and subsidies. Supply$_U$, Goods supply, unregulated case; Supply$_S$, goods supply, with emission control subsidy; Supply$_T$, goods supply, with emission fee; Demand, goods demand; p_U, goods price, unregulated case; p_S, goods price, with emission control subsidy; p_T, goods price, with emission fee.

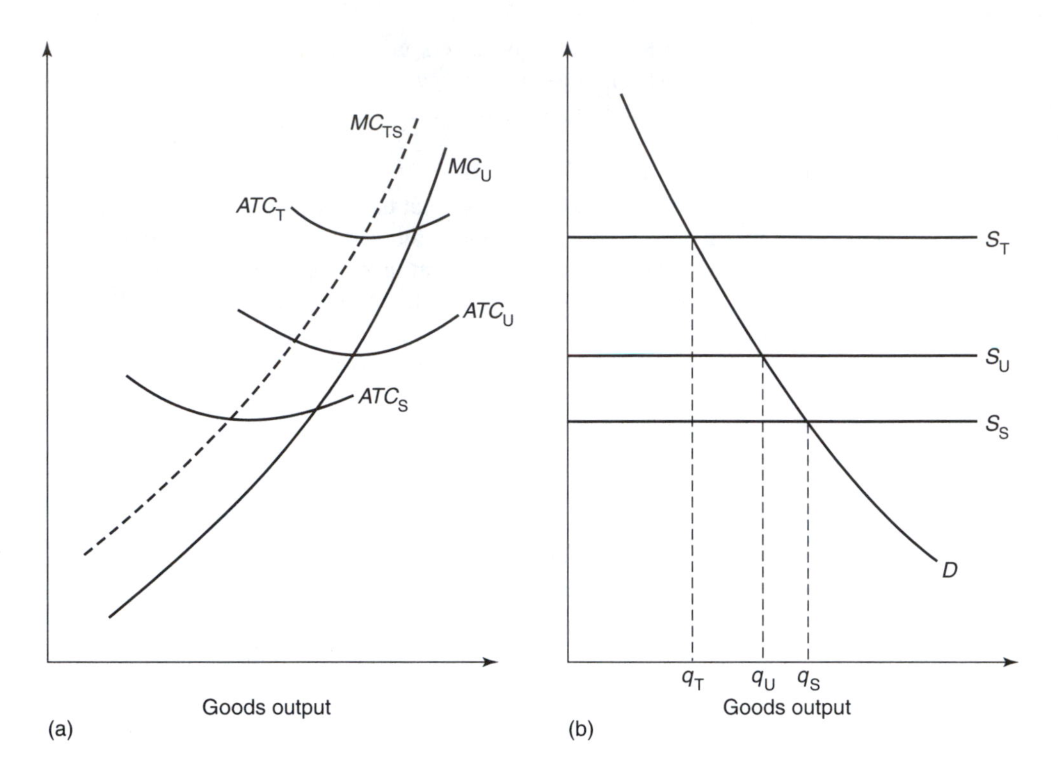

Figure 7.7 Long-run supply and demand, constant cost industry, with and without taxes and subsidies. MC_U, Marginal cost, unregulated case; MC_{TS}, marginal cost, with emission control subsidy or emission fee; ATC_U, average total cost, unregulated case; ATC_S, average total cost, with emission control subsidy; ATC_T, average total cost, with emisison fee; D, goods demand; S_U, long-run supply of good, unregulated case; S_S, long-run supply of goods, with emission control subsidy; S_T, long-run supply of goods, with emission fee; q_U, equilibrium goods output, unregulated case; q_S, equilibrium goods output, with emission control subsidy; q_T, equilibrium goods output, with emission fee.

of this is that goods prices will be higher with a Pigovian tax than with a subsidy. Furthermore, there will be more firms in the industry with a subsidy than with a Pigovian tax.

Other than the fact that a subsidy has to come from somewhere, a subsidy is undesirable because it does not allow the market to communicate the true costs of the product being consumed to the consumer. To be quite concrete, suppose we are dealing with paper mills producing paper from trees and polluting rivers at the same time (virgin mills). Other mills produce paper from recycled products and we will assume they are pollution free (which is not actually the case). A pollution subsidy to clean up the virgin mills will make virgin paper more attractive (compared to recycled paper) than if the virgin paper manufacturer had to pay a pollution tax. The result is that a subsidy results in the overuse of trees and underuse of recycled paper, compared to the case of a Pigovian tax. If a product generates pollution, we want consumers to see the full costs associated with producing that product when the consumers decide what to buy and how much to buy.

IV. IMPERFECT COMPETITION

When markets are not competitive, a host of efficiency problems generally arise, and controlling pollution is no exception. There are two cases we will consider. One concerns a monopolist in some goods market who is also a polluter. For instance, the monopolist may dominate the steel market while also generating smoke. The second case is more likely: a firm that is the only producer of smoke in some appropriate region. In this case we have a monopoly but a monopoly, in the provision of the bad.

The general conclusion that results from the analysis is that when there is market power, a Pigovian fee can make matters worse. This is a result in what economists call the theory of "second best." When there is a distortion in an economy, such as monopoly, levels of output and prices will be distorted—not at their efficient levels. In such a case, the best way to correct inefficiencies such as pollution will not be to blindly use prescriptions from theory developed for efficiency. Other, "second best" methods must be used to correct the problems. This is a large area of study in economics. We will consider only the example of monopolists and monopsonists generating pollution.

A. Monopolist in the Goods Market

Assume the steel industry is a monopoly and it produces smoke.[5] There are many smoke producers but only one producer of steel. For simplicity we will assume, as in the previous section, that smoke and steel output are proportional. We know from intermediate microeconomics what the steel mill should do. This is illustrated in Figure 7.8. Shown in Figure 7.8 is a typical demand curve for steel and the associated marginal revenue function. Also shown is the marginal cost of producing steel, ignoring pollution, MC_U, and the marginal social cost of producing steel, including pollution, MC_T. The producer will produce where marginal costs equal marginal revenue, at S_U, and price on the demand curve. Conventional analysis would call for computing the inefficiency (the deadweight loss) as the sum of the hatched and solid portions of Figure 7.8. But pollution is being generated. Tak-

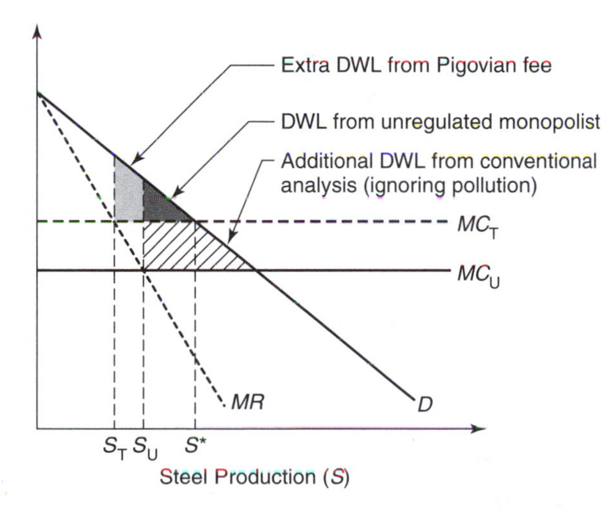

Figure 7.8 Imposing a Pigovian fee on a goods monopolist. MC_U, Marginal private cost of producing steel; MC_T, marginal social cost of producing steel; D, demand for steel; MR, marginal revenue for steel; S_U, steel output level, unregulated; S_T, steel output level with Pigovian tax; S^*, socially optimal output of steel.

ing into account the social costs of steel production, the optimal amount of steel production would be S^*. So, in actual fact, the deadweight loss (inefficiency) associated with the unregulated output of steel is just the solid area in Figure 7.8. This is the area between the true marginal cost function (MC_T) and the demand curve, bounded by S_U and S^*. This is smaller than when we used MC_U instead of MC_T. The reason is that when an unregulated monopolist reduces output to increase profits, it also reduces pollution.

Suppose we impose a Pigovian tax on the smoke generation, raising the private marginal cost of producing steel to MC_T. The monopolist will not keep producing at S_U, but will produce where the new marginal cost function equals marginal revenue, at S_T. Now the inefficiency is the sum of the shaded and solid areas in Figure 7.8. The imposition of the Pigovian fee has made matters worse!

The intuition behind this is quite simple. Efficiency is not served if there is too much or too little pollution. Monopoly tends to reduce output and emissions, assuming there is no intervention. If a Pigovian fee is also imposed, output is restricted even further. Now we have much too little goods output and somewhat smaller smoke output than desirable. The Pigovian fee has overdone it.

B. Monopolist in Bads Production

The second case we consider is that of a firm that is competitive in its goods output market but is the sole supplier of pollution. This is a remarkably common situation. Consider the company town with one large producer of pollution. Since the pollution has relatively local effects, not spilling over to the next town, this firm is the monopolistic provider of pollution.

This situation is illustrated in Figure 7.9. We focus on the output of smoke since the goods market is competitive. Shown in Figure 7.9 is the marginal savings (*MS*) to the firm from producing smoke levels s. Also shown is the marginal damage to the populace (*MD*) from smoke. The enlightened population of the town where the polluter is decides to impose a pollution tax equal to marginal damage from the smoke. But rather than set the tax equal to marginal damage at s^*, they set the tax to $MD(s)$, at whatever level of s prevails. The tax payment by the firm is then

$$T(s) = s\ MD(s) \tag{7.11}$$

As is always the case, the firm wants to produce where marginal cost (i.e., marginal tax payments) equals marginal revenue (i.e., marginal savings from polluting). To compute marginal tax payments for some level of smoke, s, we start with total tax payments and purturb s a bit, by Δs. Total tax payments will also change:

$$T(s + \Delta s) = (s + \Delta s)\ MD(s + \Delta s) \tag{7.12}$$

The marginal tax payments will be the ratio of the change in taxes to the change in smoke levels:

$$\begin{aligned} MT(s) = \frac{T(s + \Delta s) - T(s)}{\Delta s} &= \frac{(s + \Delta s)\ MD\ (s + \Delta s) - s\ MD\ (s)}{\Delta s} \\ &= s\{[MD(s+ \Delta s) - MD(s)]/\Delta s\} + MD(s+\Delta s) \\ &= s\ MMD(s) + MD\ (s+\Delta s) \\ &\approx s\ MMD(s) + MD(s) \end{aligned} \tag{7.13}$$

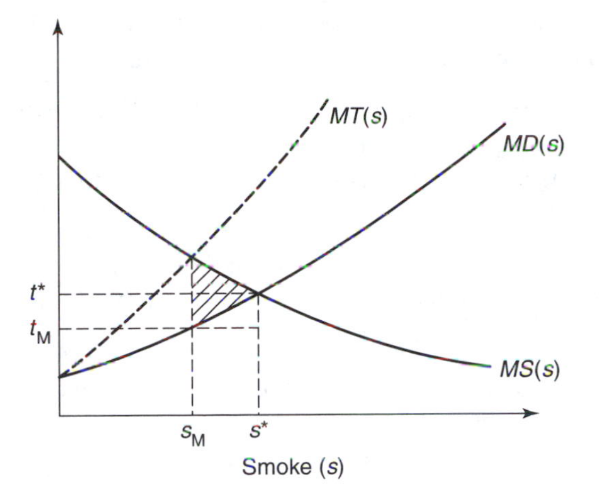

Figure 7.9 Monopolist in provision of pollution. *MS*, Marginal savings from polluting; *MD*, marginal damage from polluting; *MT*, marginal emission fee payments; t^*, Pigovian fee; t_M, monopoly emission fee level; s^*, efficient amount of smoke; s_M, monopoly emissions of smoke.

The $MMD(s)$ in Eq. (7.13) is the *marginal*marginal damage of smoke, i.e., how much marginal damage changes for a small change in smoke levels. Since $MD(s)$ is getting larger as s gets larger in Figure 7.9, $MMD(s)$ is positive. Also, the last approximate equality ($\approx$) in Eq. (7.13) applies if Δs is very small so that s and $s + \Delta s$ are virtually the same.[6]

Note from Eq. (7.13) that the marginal tax payment is in excess of the marginal damage. If the firm generates one more unit of smoke, tax payments go up because the firm has to pay $MD(s)$ for that extra unit of smoke. But also, the tax rate goes up by a little, which means payments for all smoke, not just the last unit, go up a bit. Consequently the total tax bill goes up by more that just $MD(s)$. The discussion is exactly equivalent to why marginal revenue for a monopolist in a goods market is lower than price. As monopolists expand output by one unit, they gain the revenue for that unit (which is the price) but the price goes up a bit for all the units they are still selling. This is why marginal revenue is less than price.

Now the rational smoke monopolist will produce where the marginal tax payments equal the marginal savings from polluting, namely s_M in Figure 7.9. Efficiency would call for smoke levels of s^*. The firm is underpolluting. At s^* the emission fee would be t^*. However, the smoke monopolist has managed to drive the emission fee down to t_M. The deadweight loss associated with this is the shaded area in Figure 7.9.

To summarize, the monopolist provider of pollution can manipulate the emission tax by reducing pollution below the efficient level. It is tempting to say that this outcome is fine. We have less pollution than otherwise, which cannot be too bad. However, it is important to realize that too little pollution may mean too many resources are devoted to pollution control. Those resources are being diverted from other socially desirable purposes.

SUMMARY

1. A Pigovian fee is a charge per unit of pollution generated, set equal to the marginal damage of pollution, at the efficient level of pollution generation.

2. A Pigovian fee is levied by the government, which collects the fee revenue. The fee generally induces provision of an efficient amount of pollution.

3. With multiple polluters, efficiency in pollution control requires that the marginal cost of control be the same for all polluters, provided the emissions from each polluter contribute to damage in the same way. This is the equimarginal principle.

4. In comparing a subsidy for pollution control with a tax on pollutant emissions, both result in the same marginal conditions for pollution emissions. However, the subsidy results in excess production in the polluting industry, in both the short and long run.

5. If a monopolist is the sole producer of a good in a market and also pollutes as a by-product of goods production, a Pigovian fee can make matters worse. A Pigovian fee will raise costs of production and thus reduce output of the monopolized good even more, which increases the inefficiency associated with monopoly.

6. With a monopolist in pollution production, if an emission fee is set equal to marginal damage, the monopolist will drive the fee down below the level of a Pigovian fee and reduce emissions below the efficient level.

PROBLEMS

1. Assume an economy of two firms and two consumers. The two firms pollute. Firm one has a marginal savings function of $MS_1(e) = 5 - e$ where e is the quantity of emissions from the firm. Firm two has a marginal savings function of $MS_2(e) = 8 - 2e$. Each of the two consumers has marginal damage $MD(e) = e$, where e is this case is the total amount of emissions the consumer is exposed to.

a. Graph the firm-level and aggregate marginal savings functions.

b. Graph the aggregate marginal damage function.

c. What is the optimal level of pollution, the appropriate Pigovian fee, and emissions from each firm?

2. In Section IV,B does the problem of monopoly provision of a bad arise with a true Pigovian fee? Why?

3. Consider the market for electricity. Suppose demand (in megawatt hours) is given by $Q = 50 - P$ and that the marginal private cost of generating electricity is \$10 per megawatt hour (P is in the same units). Suppose further that smoke is generated in the production of electricity in direct proportion to the amount of electricity generated. The health damage from the smoke is \$15 per megawatt hour generated.

a. Suppose the electricity is produced by an unregulated monopolist. What price will be charged, and how much electricity will be produced?

b. In part (a), what is the consumer surplus from the electricity generation? What is the net surplus, taking into account the pollution damage?

4. Consider the case of a rival bad. Would efficiency require that a Pigovian fee be levied on the producer of the bad and the receipts given to the consumers as compensation? Does it matter if the bad is excludable or nonexcludable?

5. The Fireyear and Goodstone Rubber Companies are two firms located in the rubber capital of the world. These factories produce finished rubber and sell that rubber into a highly competitive world market at the fixed price of £60 per ton. The process of producing a ton of rubber also results in a ton of air pollution that affects the rubber capital of the world. This 1:1 relationship between rubber output and pollution is fixed and immutable at both factories. Consider the following information regarding the costs (in £) of producing rubber at the two factories (Q_F and Q_G):

Fireyear	Costs: $300 + 2Q_F^2$	Marginal costs: $4Q_F$
Goodstone	Costs: $500 + Q_G^2$	Marginal costs: $2Q_G$

Total pollution emissions generated are $E_F + E_G = Q_F + Q_G$. Marginal damage from pollution is equal to £12 per ton of pollution.

a. In the absence of regulation, how much rubber would be produced by each firm? What is the profit for each firm?

b. The local government decides to impose a Pigovian tax on pollution in the community. What is the proper amount of such a tax per unit of emissions? What are the postregulation levels of rubber output and profits for each firm?

c. Suppose instead of the emission tax, the government observes the outcome in part (a) and decides to offer a subsidy to each firm for each unit of pollution abated. What is the efficient per unit amount of such a subsidy? Again calculate the levels of output and profit for each firm.

d. Compare the output and profits for the two firms in parts (a) through (c). Comment on the differences, if any, and the possibility of one or both of the firms dropping out of the market.

6. The Rocky Mashed Potato Factory produces output at costs $C = Q^2$ (marginal costs $2Q$), where Q is the quantity of mashed potatoes produced, in tons. In addition, 2 units of emissions are produced for each ton of mashed potatoes ($E = 2Q$). Pollution damage is \$2 for each unit of emissions, which leads the government to charge \$2 per unit of emissions as a Pigovian fee. The firm's output sells competitively for \$10 per ton.

a. How many tons of mashed potatoes will the Rocky Mashed Potato Factory produce? How much does it pay in emission fees? What are its profits?

b. A device is invented that would reduce the firm's emissions to one unit for each ton of output ($E = Q$). How much would the firm be willing to pay for such a device?

c. How would your answer to part (b) change if there were no government regulation of pollution emissions? What does this lead you to say about the relationship between government regulation and the market for pollution abatement equipment?

7. In Figure 7.8, the unregulated monopolist produces less than the efficient amount of steel. Does this always have to be the case when external costs are present? Can you redraw Figure 7.8 such that the profit-maximizing monopolist produces more than the socially efficient amount of steel? Could it ever be the case that the unregulated monopolist produces exactly the efficient amount?

NOTES

1. See Pigou (1962). This rambling treatise covers many "imperfections" in a market economy and discusses how they can be repaired. The first edition was published in 1920.

2. Horizontal summation means adding up the pollution generated by the different firms for a given marginal savings. Mechanically, this means rewriting $MS_1(x_1)$ and $MS_2(x_2)$ as $x_1(MS_1)$ and $x_2(MS_2)$. We are then interested in $x(MS) = x_1(MS) + x_2(MS)$, which can be rewritten as $MS(x)$. This is illustrated in Problem 1.
3. Hourcade et al. (1996) report a variety of cost figures for reductions in carbon emissions. What is clear is that there is a great deal of disagreement over the cost of abating carbon emissions.
4. Polinsky (1979) considers this in a manner similar to the approach here. See also Spulber (1985) and Carlton and Loury (1980).
5. This problem was first identified by Buchanan (1969). See also Barnett (1980) and Martin (1986).
6. Eq. (7.13) is quite simple, using calculus: $MT(s) = \frac{dT(s)}{ds} = \frac{d[s\,MD(s)]}{ds} = s\frac{d\,MD(s)}{ds} + MD(s)$.

8 REGULATING POLLUTION

In previous chapters, we discussed the virtues of markets for allocating goods and services but also the problems markets have in allocating environmental goods, particularly pollution. In the last chapter, we took a first step toward correcting the market failure associated with pollution by introducing the concept of a Pigovian fee, which generally works to correct the efficiency problems associated with pollution. But a Pigovian fee does not spontaneously emerge as markets do for conventional goods. It needs a central authority—a government—to implement the fee.

In this chapter we introduce the government as an active player in solving the problems associated with pollution. In some cases, the government will play a modest role, simply laying down the ground rules for a quasimarket to operate to solve the pollution problem. In other cases, the government will play a much more visible role, directing specific polluters as to what emissions are allowed. Although governments can solve problems that decentralized markets cannot, governments can also fail. Government failure is not a focus here, but we should be aware that government intervention to cure market failure is not always successful.[1] Some ways of intervening to solve pollution problems are better than others.

I. RATIONALE FOR REGULATION

The theory of economic regulation goes far beyond the issues of concern in environmental regulation. In fact, environmental regulation is a special, and relatively recent, example of economic regulation. For that reason, it is appropriate to place it in a larger context.

Economic regulation involves the government intervening, in a variety of ways, in the private actions of firms and individuals. There are two basic theories of regulation, the public interest theory and the interest group theory.[2] The public interest theory of regulation views the purpose of regulation as the promotion of the public interest. In this context there are three general reasons why regulation might exist: imperfect competition, imperfect information, and externalities. The interest group theory of regulation views the purpose of regulation as promoting the narrow interests of particular groups in society, such as individual industries.[3] We consider these in turn.

To a certain extent, the public interest theory of regulation is a normative theory. Recall that normative theory seeks to explain what should happen in an ideal world. In contrast, the interest group theory is a positive theory, attempting to explain why the world works as it does.

Imperfect competition, particularly natural monopoly, is the traditional normative justification for government regulation.[4] In the case of natural monopoly (such as an electricity distribution company), economic efficiency calls for a single firm. It is not a good idea to have multiple sets of poles and wires traveling down streets, connecting residences to sources of power.[5] The role of government is to guarantee a monopoly to a particular firm (restrict the entry of new firms) and, in addition, to control prices in order to protect consumers from monopoly pricing.

A related role of government is to prevent undue concentration of power in markets in which multiple competing firms represent the best organizational structure. In this case, the government attempts to prevent collusion and restricts mergers that will create excessive market power. The array of U.S. antitrust laws, starting with the Sherman Act of 1890, is designed to preserve a competitive environment in the United States, outlawing practices that are deemed anticompetitive. Many other countries have a more laissez-faire attitude toward anticompetitive activities, since such activities often serve to bolster domestic industries in the international marketplace.

The second major rationale for government regulation is the case of imperfect information. Acquiring information is costly. As a consequence, when consumers are about to enter into a transaction, they may not always have complete information on items such as product quality. Furthermore, because of the cost of acquiring information, it may not even be desirable for consumers to acquire complete information—the costs may far exceed the benefits. Imagine that each time we entered a grocery store we had to conduct extensive tests on the safety of each food item. This is a justification for the government to step in to compensate for incomplete information.[6] The role may be a relatively "hands-off" type of intervention such as establishing a set of liability rules to encourage the provision of safety-related quality. If problems occur in consuming a product, a firm can be held liable. Properly designed, such liability rules can induce firms to provide an efficient level of safety-related quality. Of course, government can also more directly intervene in the market, specifying acceptable levels of quality, such as is generally the case with regulations on food additives.

A third rationale for government regulation is in the area of the provision of public goods and bads. Public bads and externalities are of course our focus. As we know from earlier chapters, when there are elements of "publicness" (nonrivalry or nonexcludability), private markets are inefficient. Government intervenes to try to correct the problem. Government may step in to directly provide these goods or bads at efficient levels, effectively eliminating the private market. This is frequently the case with public goods (e.g., national defense) and rarely the case with public bads. In the case of public bads, the usual approach is for government to define a set of institutions and regulations to govern the provision of these public bads, e.g., the government establishes a set of regulations to restrict the production of pollution.

The interest group theory of regulation maintains that rent seeking is the primary rationale for regulation.[7] As will become clear, rent seeking is less a justification for regulation than an explanation of why some regulations exist—a positive theory of regulation. What is meant by rent seeking? Rent seeking involves private individuals or firms using the government to guarantee extra profits (rents) through government-mandated restrictions on economic activity. For instance, U.S. requirements that a certain fraction of clean-fuel gasoline additives be from renewable sources is fundamentally a subsidy to producers of ethanol in the midwest United States (the requirement is largely unrelated to air quality). Groups that benefit from such regulation lobby government for regulations that provide them with rents that would not exist in a competitive market.

II. A POLITICAL ECONOMY MODEL OF REGULATION

The basic problem of environmental regulation involves the government trying to induce a polluter to take socially desirable actions, which ostensibly are not in the best interest of the polluter. But the government may not always be able to precisely control the polluter. To further complicate matters, the government faces a complex problem of determining exactly what level of pollution is best for society. In reality, the government faces pressures from consumers and polluters. Although a full development of this interaction is beyond the scope of this discussion, Figure 8.1 captures some of this complexity. Shown in Figure 8.1 is a highly stylized schematic of the interactions among government, polluting firms, and consumer citizens.

The government is shown in Figure 8.1 as consisting of three branches, the legislature, the judiciary, and the regulators. Although the nomenclature may vary from country to country, this is a reasonably general representation of basic government. In the United States, the legislature would be Congress, the regulators would be the EPA, and the judiciary would be the Federal courts. In the United Kingdom, Parliament would be the legislature. The regulator would be the Department of Environment, Transport and the Regions, although other branches of national and local government also oversee pollution control (as in the United States). The courts and the House of Lords would be the equivalent of the judiciary, though the regulatory process tends to be less litigious in the EU than in the United States. The legislature passes laws defining what the regulators are to do in controlling pollution. The regulators are charged with the detailed implementation of the legislature's laws. The regulator's actions are tempered by the judiciary. Note that the legislature may have difficulty achieving its goals since it does not directly influence the polluter but must act through another part of government. This other part of government—the regulator—may have different goals. For instance, regulators may be interested in job security or the size of the bureaucracy they control as well as in reducing pollution levels. Furthermore, the judiciary may temper the actions of the regulators.

The firm, as shown in Figure 8.1, consists of several pieces. The Board of Directors is the core of the firm, although it is subject to oversight by the owners—the stock and bond holders. The Board issues directives to the managers; the managers issue directives to employees who produce the product of the firm as well as pollution. The key point is that regulators direct the Board to take certain actions, but the Board is removed,

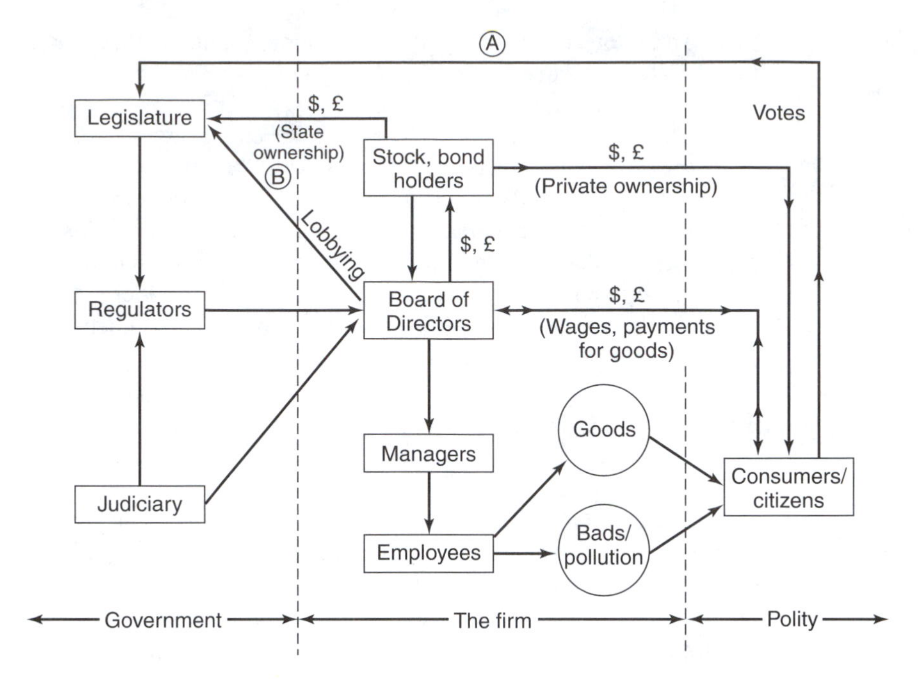

Figure 8.1 Schematic of interactions among government, polluting firms, and consumer citizens.

by several steps, from the employees who actually generate the pollution. Furthermore, the Board has other objectives in addition to pleasing the regulators (such as pleasing the stockholders). Generally, this is termed the principal-agent problem because of the inability of the EPA (the principal) to completely control the polluter (the agent). Note that the firm may not be a passive entity but may in fact influence legislation, through lobbying or financial incentives. These are shown as line B in Figure 8.1.

Finally, it is the consumers who consume the goods produced by the firm as well as the pollution generated by the firm. Consumers in turn are the citizenry whom the legislature is supposedly serving. Thus consumers direct votes and other influence to the legislature (line A), while at the same time sending money to the firm in exchange for the goods consumed.

There is obviously much detail missing from Figure 8.1. We could have employees or stockholders lobbying the legislature, for example, or firms lobbying the citizenry through advertising. Nevertheless, the essence of the process is represented in Figure 8.1.

There are really two important lessons that should be taken from Figure 8.1. The first is that there are many imperfect links between the legislature and the pollution-generating process. Since the legislature cannot physically control pollution directly, it must rely on indirect means to obtain its ends, and often these indirect means may be less than perfect. Regulation may be excessively costly, may result in considerable cheating, may result in excessive time delays, and may result in too much pollution.

The second point to take from Figure 8.1 is that the legislature does not necessarily act as an efficient benevolent maximizer of social well-being. Although we may think it is "best" if the legislature is influenced only by citizen welfare, as represented by line

A in Figure 8.1, the legislature will likely be influenced by other interests, such as the well-being of the polluter, as reflected in line B. This is the distinction between normative theory, what *should* happen (perhaps maximizing citizen welfare with some concern for equity), and positive theory, what *will* happen (a balancing of interests, some "legitimate," some not). This distinction is important. Sometimes we are trying to design a best regulation, in which case we want to know what *should* happen. In other cases, we are trying to understand why a certain regulation is on the books, in which case we should be interested in what *will* happen or why something has happened. Often the inclusion of line B is termed regulation with *endogenous* politics; the omission of line B is termed regulation with *exogenous* politics.

We can also view exogenous politics as coinciding with the public interest theory of regulation. Though it may be quite difficult to induce the firm to do the right thing, the fundamental objective of the regulatory agency is to maximize social welfare (however defined), subject to the regulatory constraints it faces. Typically, the objective of the regulator is viewed as maximizing the sum of producer and consumer surplus, perhaps with a slight bias toward consumer surplus.

The positive interest group theory mentioned earlier is consistent with the endogenous politics model of regulation. With endogenous politics, we are conscious of the ability of interest groups (boxes in Figure 8.1) to influence the regulators, legislators, and perhaps the judiciary. Environmental regulation is particularly susceptible to interest group influence. One interest group is clearly the firms potentially subject to environmental regulation—the polluters, or "browns." Another interest group, however, is the "greens," or environmental interest groups. Understanding how the browns and greens interact with the legislature and regulatory agencies can explain why we see the environmental regulations we do.

III. BASIC REGULATORY INSTRUMENTS

Having just discussed the prevalence of conflicting incentives and private information, this is an appropriate time to introduce the two major types of environmental regulation: economic incentives and command and control.

A. Command and Control

Command-and-control regulation is the dominant form of environmental regulation in the world today. Although it can take many forms, the basic concept of command and control is for the regulator to specify the steps individual polluters must take to solve a pollution problem. The essence of command and control is that the regulator collects the information necessary to decide the physical actions to control pollution; the regulator then commands the polluter to take specific physical steps to control the pollution. The regulator is generally quite specific as to what steps must be taken.

Command-and-control regulations can take many forms. One way of conveying the nature of command and control is by example. For instance, the Clean Air Act in the

United States requires the EPA to determine the minimum pollution control "performance" of new sources of pollution. The EPA is required to specify for each new category of source (e.g., new power plants or new tire factories) what pollution controls and emission rates are deemed acceptable. This means that the EPA must investigate the production process for literally every type of plant or factory in the United States, at least if significant pollution is involved. For instance, in tire manufacture, the EPA has hired engineers to examine the process of tire manufacture, generating a "Control Technology Guideline," which indicates what kinds of pollution control are appropriate for new tire manufacturing plants. In industries that apply surface coatings (e.g., furniture manufacture), regulations may impose limits on the types of coatings that may be used or firms may be required to use a certain type of vent system to recapture vapors that are emitted during painting. Furthermore, products may be banned altogether as is the case with many oil-based paints in urban areas of the United States most severely affected by photochemical smog. Power plants (producing electricity) may be required to use certain technologies to reduce emissions of sulfur dioxide. At one time in the United States, flue-gas desulfurization was required on all new coal-fired power plants, regardless of the uncontrolled emission rate of the plant.

Command-and-control regulations take many specific forms. Specific pollution-control equipment requirements can be specified as in the above example. Alternatively, the regulation may specify an emission limit for particular types of plants and particular pollutants. An example might be the U.S. standards for new automobiles: every new car may emit no more than x grams of carbon monoxide per mile driven.[8] Furthermore, all new cars are required to have a specifically defined system for capturing vapor that might escape from the gasoline tank during refueling. (In the case of automobiles, a major factor that bears on pollution is how much the car is used. There are virtually no regulations in the United States addressing distance driven except for modest taxes on gasoline.) In the case of power plants, fuel quality may be limited (e.g., sulfur content cannot exceed 1%) or emissions per unit of fuel use may be limited (e.g., no more that 1.2 pounds of sulfur dioxide may be emitted per million Btu of fuel used).

Command and control may in fact be combined with significant fines and penalties associated with noncompliance. Such incentives to comply with a command-and-control regulation should not be confused with an economic incentive to abate pollution. Though command and control may take many forms, there are two key features that distinguish command and control from economic incentives: (1) restricted choice for the polluter as to what means will be used to achieve an appropriate environmental target; and (2) a lack of mechanisms for equalizing marginal control costs among several different polluters. For example, if a polluter is told by a regulator that it must use a particular type of equipment, the polluter has little choice in determining emissions. If there is a cheaper way to attain the same level of emissions, such cost savings cannot be pursued. Some command-and-control regulations may afford some discretion to the polluter. Polluters may, for instance, be told that they can emit up to a certain amount per unit of goods output (e.g., grams of SO_2 per kwh sold). This gives more discretion to the polluter, although the polluter is not free to adjust emissions among sources that have different pollution control costs. The one thing that characterizes all command-and-control regulation is a centralization of some of the pollution control decisions that could be made by the polluter.

The best analogy for command and control is the system of central planning that existed in the former Soviet Union to manage its economy. Rather than let prices signal relative scarcity and thus direct goods around the economy, nearly all decisions on production, investment, and even interplant trade were made centrally, by central planners. In theory this can work as well as any system. The problem is that the informational requirements are enormous. In actuality, planners operate with very incomplete information, which leads to serious inefficiencies. Despite these problems, this is more or less how pollution generation is managed in most countries. Although this approach to running an economy proved too burdensome for the Soviet Union, environmental regulation in most market economies is unlikely to have such a dramatic effect since pollution control is only a modest part of modern economies. On the other hand, one might expect it to be possible to significantly improve on the current system.

What are the pros and cons of command and control? Command-and-control regulations have one major advantage: more flexibility in regulating complex environmental processes and thus much greater certainty in how much pollution will result from regulations. In an urban area with factories at different locations contributing differently to overall levels of urban pollution, it can be difficult to fashion a workable set of emission taxes or other incentives to ensure a certain level of pollution. Furthermore, in an atmosphere of uncertainty, where it is unclear how a polluter might respond to an economic incentive, command and control gives greater certainty on how much pollution will actually be emitted.[9] Another advantage of command and control is in simplifying monitoring of compliance with a regulation. If a regulation states that a particular piece of pollution control equipment must be used, monitoring simply involves seeing whether that equipment has been installed. This is easier than measuring pollution emissions.

There are of course disadvantages to command and control. Because the informational costs are high for command and control, such a regulatory system can be very costly to administer. Each plant, or at least each industry, must be analyzed in detail to determine the appropriate level of emission control. This is very costly, not to mention fraught with errors. There is also the potential for fundamental information problems. The regulator often needs to rely on information from the polluter, either in terms of emissions or costs of control. Because of this, the polluter has an incentive to distort information provided to the regulator. (Though this problem often applies to other forms of regulation as well.)

A very significant problem with command and control is reduced incentives to find better ways to control pollution. This may be purely a static issue, finding process changes or other means to reduce pollution. Or it might involve investing in research into better ways of controlling pollution. In either case, many types of command and control provide weak incentives for innovation (see Jaffe and Stavins, 1995).

Perhaps one of the biggest problems with command and control is difficulty in satisfying the equimarginal principle. It is almost impossible for command-and-control regulations to ensure that the marginal costs of pollution control are equalized among different polluters generating the same pollution. This could occur only if regulators are completely correct in their assessment of each firm's control costs. If the marginal costs of pollution are not equalized, costs of pollution control will be unnecessarily high.

This failure of the equimarginal principle is illustrated in Table 8.1, which shows the marginal cost of controlling biological oxygen demand (BOD), a measure of water pollution in rivers and lakes. Table 8.1 is drawn from an analysis (Magat et al., 1986) of

TABLE 8.1 Marginal Treatment Costs of BOD Removal, U.S. Regulations

Industry	Subcategory	Marginal cost[a]
Poultry	Duck—small plants	3.15
Meat packing	Simple slaughterhouse	2.19
	Low processing packinghouse	1.65
Cane sugar	Crystalline refining	1.40
Leather tanning	Hair previously removed/chromium	1.40
Poultry	Duck—large plants	1.04
Leather	Save hair/vegetable	1.02
	Hair previously removed	1.02
Meat packing	High processing packinghouse	0.92
	Complex slaughterhouse	0.90
Paper	Unbleached kraft	0.86
Leather tanning	Save hair/chromium	0.75
	Pulp hair/chromium	0.63
Poultry	Turkey	0.60
Cane sugar	Liquid refining	0.51
Paper	Paperboard	0.50
	Kraft—NSSC	0.42
Leather	Pulp or save hair/no finish	0.39
Poultry	Further processing only—large plants	0.35
	Chicken—small plants	0.25
Paper	NSSC—ammonia process	0.22
Raw sugar processing	Louisiana	0.21
Poultry	Fowl—small plants	0.20
	Chicken—medium plants	0.16
Raw sugar processing	Puerto Rico	0.16
Paper	NSSC—sodium process	0.12
Poultry	Fowl—large plants	0.10
	Chicken—large plants	0.10

[a]*Units: U.S. dollars per kilogram of BOD removed.*
Source: Magat et al. (1986), p. 136.

how the USEPA translates legislative mandates regarding water quality into regulations to apply to individual firms. Clearly, the equimarginal principle is violated. It would be much more cost effective, without hurting the environment, to relax regulations on some of the high cost industries and tighten regulations on some of the lower cost industries.[10]

As another example, Table 8.2 is a listing of a number of U.S. command-and-control regulations that are designed to protect health and save lives. Interpreting lives saved as the only goal of the regulation, the equimarginal principle also fails to hold. Some regulations save lives at very low cost, whereas some are extremely costly (e.g., formaldehyde in the workplace, at $72 billion per life saved).

A final problem is that with command-and-control, the polluter pays only for pollution control, not residual damage from the pollution that is still emitted even after controls are in place. This effectively provides a subsidy to the polluter, which may create a variety of distortions. As an example of this consider paper manufacture with pulp (the raw material) either coming from recycled paper (assumed pollution free) or virgin wood

TABLE 8.2 Average Cost of US Regulations to Reduce Risk of Death

Regulation	Initial annual risk	Expected annual lives saved	Cost per expected life saved (millions of 1984 $)
Unvented space heaters	2.7 in 10^5	63.000	0.10
Airplane cabin fire protection	6.5 in 10^8	15.000	0.20
Auto passive restraints/belts	9.1 in 10^5	1,850.000	0.30
Underground construction	1.6 in 10^3	8.100	0.30
Servicing wheel rims	1.4 in 10^5	2.300	0.50
Aircraft seat cushion flammability	1.6 in 10^7	37.000	0.60
Aircraft floor emergency lighting	2.2 in 10^8	5.000	0.70
Crane suspended personnel platform	1.8 in 10^3	5.000	1.20
Concrete and masonry construction	1.4 in 10^5	6.500	1.40
Benzene/fugitive emissions	2.1 in 10^5	0.310	2.80
Grain dust	2.1 in 10^4	4.000	5.30
Radionuclides/uranium mines	1.4 in 10^4	1.100	6.90
Benzene in workplace	8.8 in 10^4	3.800	17.10
Ethylene oxide in workplace	4.4 in 10^5	2.800	25.60
Arsenic/copper smelter	9.0 in 10^4	0.060	26.50
Uranium mill tailings, active	4.3 in 10^4	2.100	53.00
Asbestos in workplace	6.7 in 10^5	74.700	89.30
Arsenic/glass manufacturing	3.8 in 10^5	0.250	142.00
Radionuclides/DOE facilities	4.3 in 10^6	0.001	210.00
Benzene/ethylbenzenol styrene	2.0 in 10^6	0.006	483.00
Formaldehyde in workplace	6.8 in 10^7	0.010	72,000.00

Source: Viscusi (1996), pp. 124–125.

(with associated pollution). Efficiently regulating the manufacture of virgin wood pulp involves seeing that the proper pollution control is achieved and that the damage from any remaining pollution is included in the price of pulp. This way the price of pulp reflects pollution control costs and residual damage. However, if command and control regulations require only pollution control and not the payment of residual damage, the resulting price of virgin pulp will be lower. Why is this undesirable? It is easy to see that with a lower price of virgin pulp, paper manufacturers are more likely to choose virgin pulp over recycled pulp, because it is cheaper. As a result, there will be more pollution than is efficient.

B. Economic Incentives

Economic incentives, in contrast to command-and-control regulation, provide rewards for polluters to do what is perceived to be in the public interest. We are all familiar with incentives. Instead of closely monitoring a child's homework activity, a parent may provide a substantial reward for good grades (or a disincentive for bad grades). The idea is to align public and private incentives. In the context of pollution, there are three basic types of economic incentives: fees, marketable permits, and liability.

Pollution fees involve the payment of a charge per unit of pollution emitted. If the fee is at the right level, this would be an example of a Pigovian fee. When a polluter must pay for every unit of pollution emitted, it becomes in the polluter's interest to reduce emissions.

A marketable permit allows polluters to buy and sell the right to pollute. Thus what starts as something akin to command and control (a permit to pollute) turns into an economic incentive by allowing trading. Trading induces a price or value on a permit to pollute, thus causing firms to see polluting as an expensive activity; less pollution means fewer permits need be bought. There is an opportunity cost of emitting; by not emitting, the firm can sell more permits.

This is illustrated in Figure 8.2, which corresponds to the situation in which there are two polluters. We are interested in allowing 100 units of pollution in total. We start by giving each firm 50 permits. The marginal savings from polluting functions are shown in Figure 8.2 for the two firms. They have been drawn so that any point along the horizontal axis gives the number of permits each firm holds: read from the left for firm 1 and from the right for firm 2. Note that trading will occur until firm 1 holds e^* and firm 2 holds $100 - e^*$. The equilibrium price of a permit is p^*.

It is important to note that an emission fee of p^* would achieve exactly the same outcome. Suppose there is a little uncertainty. With an emission fee, we know precisely what the marginal cost of control will be; we are less sure about the quantity of pollution. With a marketable permit, we know exactly how much pollution there will be; we are less sure of the marginal cost of control.

Liability is a third type of economic incentive. The basic idea is that if you harm someone, you must compensate that person for damage. In theory, this means that when you undertake a risky activity (such as polluting or storing hazardous wastes), you will take all potential damage from your activity into account when deciding how carefully to perform your activity. The important issue is that the government is not telling you what to do, just that you will be responsible for any consequences. This creates an incentive to be careful when undertaking risky activity and, in fact, to take the socially desirable amount of precaution in undertaking such risky activity.

To illustrate liability, suppose we have a hazardous waste storage facility (a "dump"). The dump can do things to minimize the risk of hazardous wastes leaking into the envi-

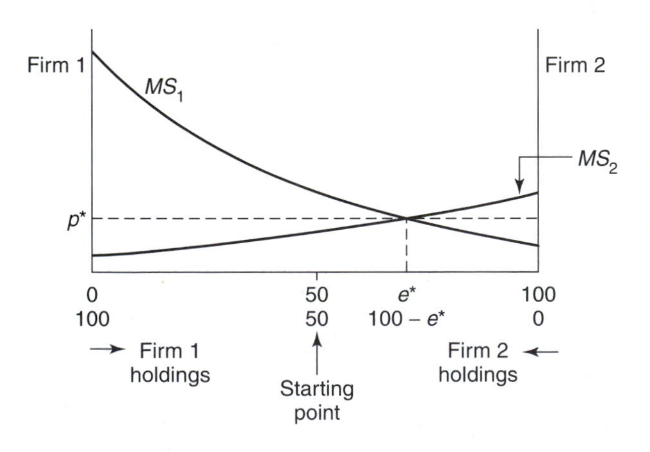

Figure 8.2 Marginal savings from polluting functions for two firms. MS_1, Marginal savings from emitting, firm 1; MS_2, marginal savings from emitting, firm 2; e^*, equilibrium holding of permits; p^*, equilibrium price of permits.

ronment. We can lump all of those risk-reducing activities into one term, "precaution." If the dump takes a great deal of precaution, the risk of a leak will be low. If the dump takes little precaution, the risk will be high. Precaution is of course expensive for the dump to undertake. All other things being equal, the dump would prefer to take little precaution. Damage to society also depends on the level of precaution. This is illustrated in Figure 8.3, which shows both costs to the dump and damage to society as functions of the level of precaution. There is some socially desirable level of precaution, x^*, at which the marginal costs of taking more precaution are just offset by the reduction in marginal damage from taking more precaution. Liability works by saying to the dump: "Do whatever you wish but should an accident occur, we will find the socially desirable level of precaution; if you were not taking that level of precaution, you will be responsible for all of the environmental damage from the accident." This is how negligence liability works, although other types of liability work in a similar fashion. This threat of being held responsible for accident damages is often a sufficient incentive for firms to take the socially desirable amount of precaution.

One of the dominant questions in environmental economics and regulation over the past three decades is why command and control dominates environmental regulation worldwide when most economists believe economic incentives are much better.

Clearly, economic incentives have a number of advantages over command and control. First, informational requirements are less significant. It is not necessary to know what it going on within a firm to use an emission fee. Furthermore, economic incentives will provide an incentive for a polluter to innovate, finding cheaper ways of controlling pollution.[11] Also, in contrast to command and control, economic incentives involve the polluter paying for control costs as well as pollution damage. Thus there is no implicit subsidy to the industry. As we will see, the fee payments approximate the damage associated with the pollution. Consequently the cost (and thus the price) of the product manufactured in association with pollution will reflect control costs as well as residual pollution damage. In the previous section, in the example of paper manufacture, we saw the importance of reflecting environmental damage in product price, something that command and control fails to achieve.

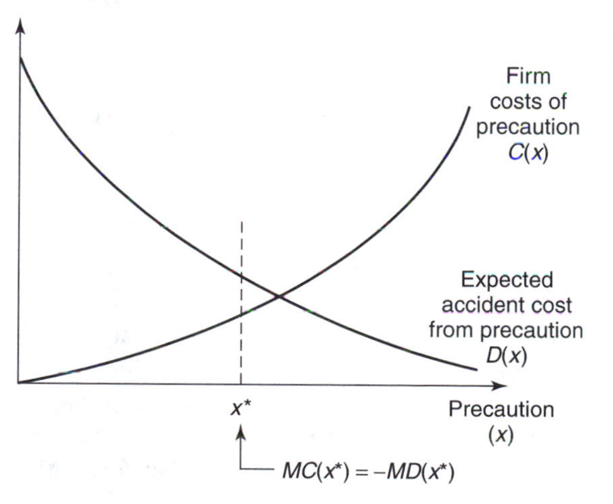

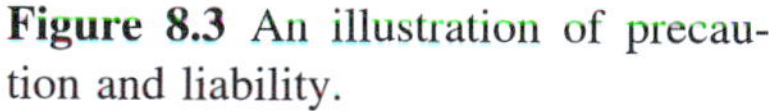
Figure 8.3 An illustration of precaution and liability.

Probably the biggest advantage of economic incentives is that the equimarginal principle will automatically hold for most types of economic incentives. For instance, with an emission fee, all firms set their marginal cost of pollution control equal to the fee. The equimarginal principle is trivially satisfied. When firms trade pollution permits, the price of the permit that will be determined by the permit market sends the same signal to all polluters regarding the opportunity cost of emitting. The equimarginal principle will hold. As we will see in Chapter 12, liability effectively has the polluter setting its marginal abatement cost equal to marginal damage. Thus the equimarginal principle will also hold. Why is meeting the equimarginal principle so important? Basically because the costs of regulation will be higher if the equimarginal principle does not hold, perhaps much higher. Thus economic incentives have this cost-saving advantage.

There are disadvantages to economic incentives. One problem is forging a set of economic incentives that can accommodate the complexities of environmental transformation without being excessively complex and impractical. Just think of urban air pollution in which the damage from a unit of emissions can vary considerably in both space and time. Developing an economic incentive that efficiently and perfectly takes these complexities into account can be very difficult.

A second problem with economic incentives is largely political. If there is a great deal of uncertainty associated with the environmental problem being controlled, it may be necessary to adjust the level of the incentive (level of the fee, number of marketable permits issued) over time, as information becomes available. This may be very difficult in many practical situations. For instance such an adjustment in the United States might require Congressional action. It took over 10 years of debate in Congress before the U.S. Clean Air Act was amended in 1990 to include acid rain control.

A third problem, also political, is that many economic incentives involve massive transfers from firms to the government. An emissions tax generates a tremendous amount of revenue for the government administering the tax. This may be good for the government, but instituting such a tax may be very difficult politically, precisely because of these wealth transfers. This explains in large part why substantial emission fees have gone nowhere in most market economies. In contrast, emission fees have been widely used in the former Soviet Union and Eastern Europe because of the traditional dedication of the fee revenue to investment in pollution control in state enterprises.[12]

IV. COMPLICATIONS FOR ENVIRONMENTAL REGULATION

A. Space and Time

In contrast to most types of regulation, pollution regulation is complicated by the physical environment that interposes itself between polluters and consumers. This is illustrated in Figure 8.4. A polluter generates *emissions*. These emissions are transformed, possibly in a complex fashion, to *ambient concentrations* of pollution. Emissions cause no damage; it is ambient concentrations that cause damage. This is an important distinction. The word ambient refers to the world around us. That is its use here. Ambient concentrations are the concentrations of pollution in the air around us or in the water we drink. It is ambient con-

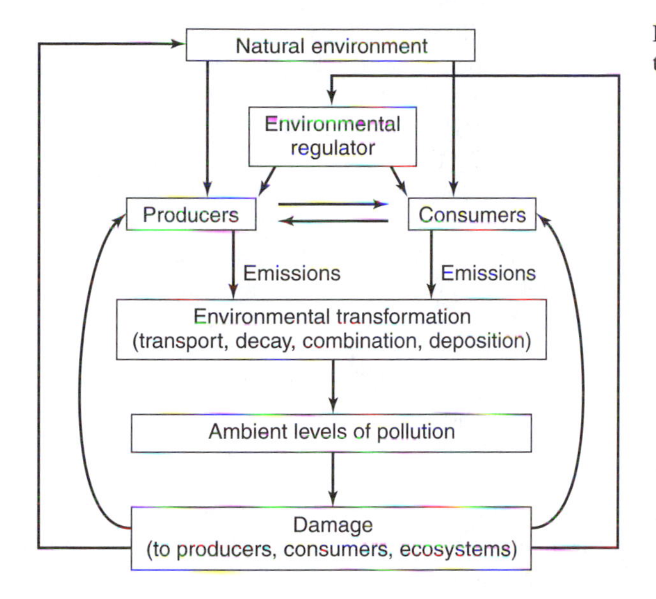

Figure 8.4 Environmental regulators face a complex task.

centrations, not emissions, that should be of concern when we discuss environmental damage. The regulator, too, is interested in ambient concentrations because that is the source of damage. However, ambient concentrations are imperfectly connected to emissions and emissions are what firms emit and what needs to be regulated. This creates a complexity that is not present in other forms of regulation, such as monopoly or occupational safety.

Space is a major player in environmental transformation. If we are interested in pollution levels in central Tokyo, sources nearby will generate more damage than sources located in distant suburbs. Or if we are interested in acid rain damage in the Black Forest, emissions from German power plants will be more damaging than emissions from more distant English power plants. We take this issue up in more detail in the next chapter.

Time is also a significant factor in environmental transformation, though probably less important than space. Urban photochemical smog (primarily ozone) involves sunlight and a mixture of nitrogen oxides and volatile organic compounds. Without sunlight, the problem is significantly less severe. Furthermore, the residence time in the atmosphere of some of these compounds is less than a day. Thus emissions in the evening or at night will be less damaging than emissions in the morning or mid-day. Seasonal variation is also important. In some areas, winter involves less heat and sunshine, which tends to reduce the problem of smog. Thus emissions in the winter may be less damaging than emissions in the summer. Designing regulations to reflect this hour-to-hour, day-to-day, and season-to-season variation is not easy.

B. Efficiency versus Cost Effectiveness

The complex and imperfectly understood relationship between emissions and ambient concentrations and, in turn, environmental damage, presents a conundrum for regulators. Ideally, regulations would target damage and ambient concentrations, but that is often too difficult. Often, regulations will have as a goal some overall level of emissions, with no clear connection to ambient concentrations. For instance, the 1990 U.S. acid rain legislation calls for a reduction in annual sulfur dioxide emissions of 10 million tons, without regard for where that reduction takes place. But emissions close to sensitive lakes and forests are much more damaging than emissions that, for instance, largely blow out to sea. However, it is too complicated to try to reduce acid deposition based on variable emission reductions at specific locations.

Efficiency calls for emissions that balance the costs of emissions control with the damage from ambient pollution, fully taking into account the complexities relating emissions to damage. When this is not practical (as is often the case), goals or targets will be established regarding desired levels of ambient concentrations or desired levels of emissions. These goals may be only imperfectly related to efficient levels of pollution, since efficient levels of pollution many vary through time and over space.

As another example, efficiency may call for the level of ozone in the Los Angeles basin to vary spatially (depending on population density, for example, and control costs) as well as temporally (depending on the time of day). But given the imperfect knowledge of ozone damage, formation, and transport, a second-best approach is to establish a target upper bound on concentration of ozone in the entire basin, averaged over an hour. But even such a simplified target may be impractically complex. Because of the imperfect knowledge of the ozone problem, regulations may establish targets for emissions that may be only imperfectly related to ambient concentrations. For instance, goals or targets are established for overall emissions of precursors of ozone (those pollutants that lead to ozone). Regulations apply to individual polluters and are designed to achieve approximately the emission targets for the basin.

This example illustrates the compromises that are often (though not always) necessary in actually regulating polluters. If regulators are unable to control emissions from individual polluters to balance costs and damage, a fall-back position may be to establish ambient pollution targets and regulate polluters in such a way as to achieve the ambient pollution target in the best way possible. But even this may not always be possible, in which case a target is established for emissions and this target may be only approximately related to the target for ambient concentrations. The goal of regulation then is to control individual polluters in such as way as to achieve the emission target in the best way possible.

Establishing emission targets or ambient targets is usually a compromise that sacrifices efficiency in pollution control. But even with an emission target, there are both good ways and less desirable ways of regulating emissions to achieve the target. Different polluters have different costs of pollution control. The least-costly way of achieving a given emission target involves controlling pollution from various sources in a way that reflects different costs of pollution control. If a set of environmental regulations achieves the emission target at least cost, we say the regulation is *cost effective*. If the regulation is cumbersome and poorly matches emission cutbacks with control costs, then it is likely the regulation will not be cost effective.

This is an important concept. Even though efficiency is not attainable for many regulations, cost effectiveness is attainable. Unfortunately many environmental regulations around the world are far from cost effective in achieving either emission targets or ambient targets. In some cases this lack of cost effectiveness is for very real and understandable reasons, particularly concerns about equity. Cost effectiveness may call for most of the damages or costs of emission control to fall on a few individuals or firms.[13] Society may deem this inequitable and consciously move away from cost effectiveness. However, it is just as common that regulations are poorly designed, leading to excessive costs for no particular reason.

C. Ambient-Differentiated versus Emission-Differentiated Regulation

We have already concluded that targets may be ambient targets or emission targets. Regulations, too, can apply to polluters based on their emissions or on their contribution to ambient concentrations. This is a bit subtle but actually relatively simple. Suppose we have an ambient target. Typically, a unit of pollution from one polluter will have a different effect on ambient concentrations than a unit of pollution from another polluter. An ambient-differentiated regulation will control these two polluters differently, based on their different contributions to ambient concentrations. An emissions-differentiated regulation will ignore the differences between the two polluters, though the overall level of emissions will still be controlled in such a way as to achieve the ambient target.

Using the example of ozone in Los Angeles again, regulation is driven by ambient targets on maximum concentrations of ozone at any point in the air basin. However, sources are regulated without much regard for their contribution to ambient concentrations. A source of a particular type (e.g., furniture manufacture) will be subject to the same emissions regulation regardless of its location in the basin. This is an example of emission-differentiated regulation. Contrast this with regulations applying to new sources of pollution in the Los Angeles basin. A new source seeking to locate in the basin must arrange for emission reductions from existing sources and demonstrate that the ambient concentrations resulting from the new source combined with the arranged reductions in existing sources will be better than the previous status quo. This is an example of an ambient-differentiated regulation.

This concept of ambient-differentiated regulations will be explored more fully in the next chapter.

V. BASIC ISSUES IN ENVIRONMENTAL REGULATION

This chapter has served as an introduction to environmental regulation. Lest the reader be seduced into thinking that designing good environmental regulation involves a simple application of economic theory, it may be instructive to briefly mention some of the major issues in environmental regulation. In the succeeding chapters, we will examine many of these issues in more detail.

The debate over command-and-control regulations vs. economic incentives is still as lively as ever. Although there have been some recent major inroads of economic incentives into actual environmental regulation, the verdict is still out on how well these incentives work. Furthermore, some of the more vexing environmental problems, such as urban air pollution, are so complex that no complete package of incentives has been suggested that might achieve efficient control of the problem. Furthermore, although command and control is theoretically undesirable, empirical evidence is sparse on exactly how much money is being wasted in command and control.

Public sources of pollution are very significant, particularly for some media and in some countries. Sewage plants, for instance, are often owned and operated by local governments or other quasigovernmental entities. It is not known how well economic incentives will work on a government agency that has a soft budget constraint.[14] Are there incentive alternatives to command and control that will work with public enterprises?[15]

One important issue in environmental regulation concerns information, particularly private information polluters may have that regulators need. How best can regulations be structured to give polluters incentives to act efficiently and truthfully divulge information necessary for regulation? How can enforcement policies be designed to give polluters incentives to obey regulations? How can regulations be structured for the case in which we cannot observe emissions, such as soil erosion from multiple farms in a watershed?

Risk is becoming more important for many pollution problems. A hazardous waste dump may be well designed, but it still poses a risk to nearby residents, since failure of the dump is always possible. Toxins in the environment may not directly cause cancer but may increase the probability of getting cancer. How do we deal with these problems? What sort of regulations are appropriate?

Another issue concerns competition between jurisdictions vis-à-vis environmental regulations. With localities competing to attract industry, to what extent will environmental regulations be weakened as an incentive to attract a new employer? If pollution is coming from outside one's jurisdiction, can anything be done to try to reduce that pollution, such as countervailing tariffs? This issue applies to competition among states in a federation (e.g., states of the United States, countries of the EU, or republics of Russia) as well as competition among countries.

A related issue concerns the appropriate political level to institute environmental regulations. Also known as environmental federalism, the question is whether it is better to institute regulations at the Federal level or at the state level. In the case of the EU, should a regulation be promulgated from Brussels, applying to all member states, or should each state be allowed to develop its own regulations? Put differently, what kinds of pollution problems are best regulated at the Federal level and which problems should be left to member states?

Another issue concerns the incidence of regulations. When a regulation is instituted, who bears the burden (either in costs or pollution damage)? Who reaps the benefits? This is not as simple as it might seem. Consider a poor urban area with high levels of pollution. As is often the case, many people in the urban area rent their apartment or house, and the landlord/owner typically lives in a more desirable location, perhaps a different town. A regulation is proposed to reduce pollution in the poor urban area. Who benefits from this pollution reduction? It might seem at first that the poor residents benefit, but this is not correct. When pollution levels decline, property values rise. Thus the residents end up with cleaner air and higher rents. Property owners are the primary beneficiaries. Who pays for the regulation? If the polluters primarily serve the local market, they will most likely be able to pass the costs of pollution control on to local residents in increased product prices; if they serve a large market, the polluters will probably have to bear the costs. The question of who benefits and who pays for an environmental regulation is not easy to answer. And sometimes the answers are not what we expect.

Another major issue in environmental regulation concerns innovation and technical change. Theoretically, economic incentives are assumed to prompt more innovation and cost reductions in pollution control, relative to command and control. The magnitude of this effect is an empirical question. One of the issues in controlling the emission of greenhouse gases is the extent to which a pollution tax will decrease the cost of pollution control in the future, due to effects on innovation. If the effect is substantial, this is a significant justification for instituting controls now rather than later, when the problem becomes more severe.

SUMMARY

1. There are two basic theories to explain the presence of economic regulation. One is the public interest theory, under which government regulates to correct a market failure and increase societal well-being. The other theory is the interest group theory. Under this theory, interest groups influence government to promote their own agenda. Since regulation tends to benefit some and hurt others, interest groups will lobby for those regulations that are in their self-interest.

2. Under the public interest theory of regulation, there are three primary reasons for regulation: imperfect competition, imperfect information, and externalities. The case of externalities is of most interest to us. Externalities generate market failure that government regulation seeks to correct.

3. Any market economy is a complex web of government agencies, firms with hierarchical structures, and consumers who wear multiple hats (consumers of goods, victims of pollution, stockholders in corporations, voters). Under the public interest theory of regulation, politics is exogenous to this complex structure. In such a case, the regulator designs regulations that maximize a weighted sum of consumer and producer surplus. However, the regulator may have incomplete control over the individuals who actually make the pollution cleanup decisions. This imperfect connection is termed the principal-agent problem.

4. There are two basic types of environmental regulations: command and control and economic incentives. Command and control involves the regulator making many of the pollution control decisions for the firm, reducing firm choice. Economic incentives give more choice to the firm, simply providing an incentive for the firm to find the best way to reduce pollution.

5. The advantage of command and control is that there is more flexibility in regulating complex pollution problems. Command and control may also give greater certainty as to how much pollution will be emitted. On the negative side, command and control typically is much more costly than efficient alternatives. One of the primary reasons it is more costly is that it is difficult for the equimarginal principle to hold.

6. Economic incentives fall into three main categories: emission fees, marketable emission permits, and liability. Emission fees involve the polluter paying a fee per unit of pollution to the government. Marketable emission permits involve the government issuing (selling or giving away) permits to pollute, which may be traded among polluting firms. Liability involves holding polluters liable for any damage they may cause. Liability is particularly useful for regulating risky activities that lead to occasional accidents.

7. Economic incentives have the advantage of placing decision making for pollution control in the hands of those most familiar with pollution control options—the polluter. Economic incentives work to keep costs low and also to spur innovation. The equimarginal principle automatically holds with most economic incentives. A disadvantage of economic incentives is that it can be very difficult to design a set of economic incentives for complex pollution control problems, such as urban ozone.

8. One complication of environmental regulation is how to deal with space and time. We distinguish between emissions (which come directly from a polluter) and ambient pollution, the levels of pollution in the air or water around us. Ambient pollution causes damage, but emissions are what is regulated.

9. Regulations, too, may make a similar distinction. Emission-differentiated regulations treat all emissions as the same, no matter where they come from. Ambient-differentiated regulations treat emissions from different sources differently, based on their relative contribution to ambient pollution levels.

10. Because of the difficulty in designing regulations that achieve efficiency, regulations sometimes take the approach of specifying environmental targets (e.g., maximum pollutant concentration levels) and then ask what is the cheapest way of reaching these targets. This is termed cost effectiveness, as opposed to efficiency.

PROBLEMS

1. Explain how the public interest theory of regulation might come to a different conclusion regarding emission fees vs. marketable permits than the interest group theory.

2. The Kyoto Protocol, drafted in December 1997, calls for each developed country to cut its emissions of carbon dioxide by the year 2012 to approximately 8% below 1990 levels. Some countries are concerned about the cost of making such a cut in carbon emissions. Expectations are that the cost of pollution control will be $5–10 per ton of carbon emitted. However, there really is a great deal of uncertainty about what the costs might be; some say it will be easy to reduce carbon emissions, others say it will be very expensive. In one developed country, two proposals are being considered to achieve the reduction. Proposal A involves the issuance of permits for carbon emissions every year, with the number of permits equaling the targeted "8% less than 1990 emissions." Each permit will be good for emitting one ton of carbon. Proposal B is the same as Proposal A except that the government agrees to sell an unlimited number of additional permits for $10 each. Explain how each of these proposals works and what the price of permits might end up being under each proposal. Discuss the pros and cons of each proposal.

3. Suppose we have one hundred different firms emitting pollution. The Ministry of the Environment (MOE) currently regulates these firms by requiring specific pollution control equipment to be installed by each firm. The MOE is considering issuing tradable permits to each firm indicating how much pollution the firm may emit. Permits would be issued in

such a way as to result in the same amount of pollution as the technology regulations. Discuss the implications of switching to tradable permits on the likelihood of seeing research and development to reduce the cost of pollution control.

4. A classic way of introducing pollution regulation into an urban area is the "rollback" method. First you determine the maximum tolerable level of a particular pollutant, x_s. Next you find the maximum current concentration of that pollutant in the urban area, x_m. Presumably $x_m > x_s$ (otherwise, there is no problem). Regulations then require all polluters in the city to reduce emissions from their current level of e_m to $e_s = e_m\ (x_s/x_m)$. Explain how this would work to bring pollution levels down to an acceptable point. What might be the efficiency problems with such an approach?

NOTES

1. See any good policy analysis book for a discussion of government failure (e.g., Weimer and Vining, 1992).
2. See Laffont and Tirole (1993) for a good discussion of the basic theories of operation of regulatory agencies.
3. The interest group theory can be traced to Stigler (1971) and Peltzman (1976), though the ideas certainly predate these individuals.
4. This subject is treated in depth in the field of industrial organization and regulation. For example, see Carlton and Perloff (1994), Viscusi et al. (1992), or Spulber (1989).
5. The electricity industry is being deregulated worldwide. However, local retail distribution always (to my knowledge) remains a regulated monopoly or state-owned enterprise.
6. Although incomplete information is a problem, it need not always be solved by government intervention. In the case of food safety, private firms could test products and give their "seal of approval" to products meeting certain standards.
7. In general, rent is defined as a payment for a good in excess of the cost of provision. Not all rent is "bad." Scarcity rent is the payment for a scarce good (e.g., the premium that ocean-front property often obtains relative to more plentiful inland property). The problems with monopoly are due to the monopolist seeking monopoly rent. As we know, monopoly is usually undesirable because of the inefficiencies.
8. The 1990 Clean Air Act in the United States specifies the following "tailpipe" emission limits in grams per mile: nonmethane hydrocarbons, 0.25; carbon monoxide, 3.4; nitrogen oxides, 0.4. [Bresnahan and Yao (1985) estimate uncontrolled emission rates to be 8.7, 87, and 4.4, respectively.]
9. The view that command and control gives greater certainty as to pollution levels is sometimes misleading. Urban areas of the United States have been using command and control for decades to try to bring ambient air pollution levels below statutory limits. In general, this effort has failed (though pollution levels have been reduced in many areas).
10. Assuming that the USEPA is rational, Magat et al. (1986) attempt to explain the failure of the equimarginal principle to apply in this case. They statistically analyze how industry-level BOD standards are affected by variables such as marginal contol costs (these are significant—higher costs lead to weaker standards), industry concentration, which is associated with greater economic power of the industry (this is significant—more concentrated indstries get weaker standards), quality of information on control costs (also significant—less uncertainty leads to tighter standards), and the volume of comments sent to the EPA by industry on the proposed regulations (this is insignificant).
11. It is widely accepted that providing an economic incentive to reduce pollution emissions will also provide an incentive to innovate and invent better and cheaper ways of controlling pollution. Laffont and Tirole (1996) suggested that economic incentives are not all equal when it

comes to providing an incentive to innovate. In particular, they point out that if there are a fixed number of permits in a market, innovation may drive down the price and reduce the gains from innovation. They also suggest ways to avoid this potential problem.

12. Emission fees were used in the former Soviet Union. The fees were basically used to generate a pool of funds that could be used to purchase pollution control equipment for the state-owned polluting enterprises. Thus the fee was more of a way of raising an investment fund, charging firms in proportion to their emissions. There is no evidence that these fees provided any incentive to reduce pollution in order to reduce fee payments. In Europe, emission fees have been levied for many years, particularly on sources of water pollution. These fees were generally set at levels necessary to finance the operation of pollution control agencies rather than at levels associated with marginal damage (Bower, 1983). The general consensus is that such fees were set at too low a level to provide any meaningful incentive to reduce pollution (Howe, 1994). See Chapter 2.
13. For instance, in many urban areas, a good portion of auto emissions comes from older cars. Cost effectiveness usually requires substantial efforts to control these older cars Yet it is often the poor who own older vehicles. Thus cost effectiveness would impose a burden on the poor that may be considered unacceptable politically.
14. A soft budget constraint means that an enterprise need not show a profit. Private firms will go out of business if they consistently fail to make a profit—this is a hard budget constraint.
15. Most approaches to using economic incentives for governmental polluters focus on determining what objectives bureaucrats have (akin to profit maximization). Oates and Strassman (1978) examine this question in more detail.

9 EMISSION FEES AND MARKETABLE PERMITS

By now you should be relatively familiar with the concepts of emission fees and marketable emission permits as economic incentives for controlling pollution. Our goal in this chapter is to discuss this topic in more detail, looking at how these incentives might work and exploring some of the potential problems.

Several issues complicate using incentives to control pollution. One is space—pollution transport is often highly dependent on location. Another is time—primarily accumulation over time, such as the multidecade accumulation of greenhouse gases (but also including daily and seasonal variation). A third is imperfect competition. We consider these issues here. We will also examine the question of just how much more efficient economic incentives are than the alternative—command and control.

I. SPACE

A. Sources, Receptors, and Transfer Coefficients

The problem of dealing with spatial effects is very real for pollution control. At the simplest level, Figure 9.1 shows a river with two factories discharging organic waste (such as sewage) into the river and a municipal water supply taking water out of the river (yuk!). For the time being, ignore the fact that the river may be useful for many purposes (such as fishing, swimming, and ecosystem services) and focus on its use as a water source for the town.

The problem of course is that the two factories are upstream of the municipal water supply. Consequently, what these factories do is of importance to the municipality. Fortunately, the river is able to clean itself somewhat. As the river flows, bacteria will work on organic material that the factories have discharged into the river. So the further one goes downstream, the smaller the effects from the pollution. Although this is good for the environment, unfortunately it complicates our analysis. To correctly regulate the two factories, we must take into account their individual effects on the municipality.[1]

To take space into account, we will introduce two terms: *sources* and *receptors*. A source is a point of discharge of pollution. Each factory in our example is a source. A re-

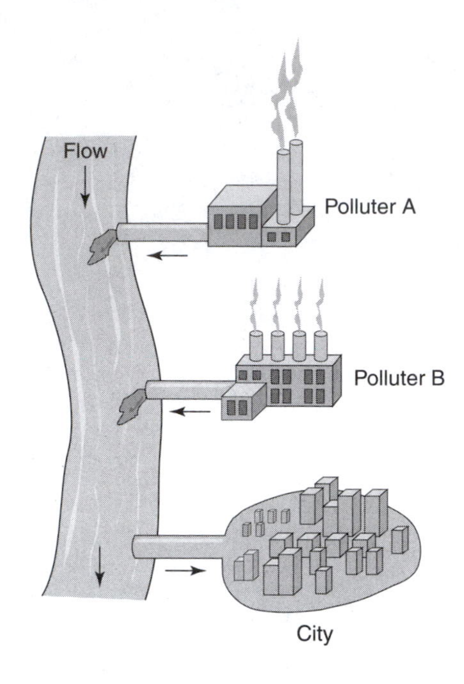

Figure 9.1 River with two factories discharging organic waste into it and a municipal water supply taking water out of it.

ceptor is a point at which we care about the level of ambient pollution. (Recall that ambient pollution refers to the level of pollution in the surrounding environment.) In our example, a logical receptor is the intake point for the municipal water supply. If we are concerned about the health of the river at other points, we might have several receptors located at different points along the river. It might seem that we care about pollution everywhere, not just at a few receptors. While this may be true, in practice we usually identify a small set of receptors where pollution levels will be measured. Typically, these receptors are scattered over space and serve as good proxies for the overall level of pollution.

Generally speaking, there is some relationship between emissions at the various sources, $e_1, e_2, \ldots, e_I$ (where I is the number of sources) and concentrations of pollution at any receptor j:

$$p_j = f_j(e_1, e_2, \ldots, e_I) + B_j \tag{9.1}$$

where B_j is the background level of pollution at j (perhaps zero). Fortunately, in many environmental problems the physical environment is linear[2]; that is, Eq. (9.1) can be written as

$$p_j = \Sigma_i\, a_{ij}\, e_i + B_j \tag{9.2}$$

The coefficient a_{ij} is called the *transfer coefficient*. Typically, we assume $B_j = 0$. Note that in Eq. (9.2), if we change emissions at some source i by a little (Δe_i), pollution will change by $a_{ij}\, \Delta e_i$. This brings us to a definition of the transfer coefficient:

Definition *Suppose a change in emissions from source* i *(Δ e_i) results in a change in pollution at receptor* j *(Δp_j). The transfer coefficient between the source* i *and the receptor* j *is defined as the ratio of the change in pollution at* j *to the change in emissions at* i*:*

$$a_{ij} = \Delta p_j \,/\, \Delta e_i \tag{9.3}$$

The concept of a transfer coefficient is really useful only when the relationship between emissions and pollution is linear, such as in Eq. (9.2). In general, for our analysis, we will make the linearity assumption. Equation (9.3) in essence gives the conversion rate for emissions to ambient concentrations. For instance, if a_{ij} is equal to 2, then every unit of emissions at i yields two units of ambient pollution at j.

B. How Much Pollution Do We Want?

We first turn to the question of what is the efficient amount of pollution. As we know, efficiency involves equating marginal damage with the marginal savings to the firm from pollution generation. But marginal savings is relative to emissions and marginal damage is relative to ambient pollution. To link these, we must either convert a firm's marginal savings function to marginal savings *per unit of ambient pollution* or express marginal damages as marginal damage *per unit of emissions*. We take the second course. Let us term marginal damages per unit of emissions from source i as the function $MDE_i\ (e_i)$, which is in contrast to marginal damages per unit of ambient pollution, $MD(p)$. For now we will work with one receptor. We know that MDE_i is the ratio of the change in damages $[D(p)]$ to the change in emissions at source i:

$$\begin{aligned} MDE_i\ (e_i) &= \{D(p+\Delta p) - D(p)\} / \Delta e_i \\ &= MD(p)\ \Delta p / \Delta e_i \\ &= a_i\ MD(p) \end{aligned} \tag{9.4}$$

This relationship between MDE and MD is intuitive. If, as was mentioned in the previous paragraph, a_i is 2, one more unit of emissions yields two units of pollution and thus twice as much damage as one more unit of ambient pollution. Thus MDE is twice as big as MD.

What is the efficient amount of pollution? As before, efficiency calls for equating the marginal savings from emissions with the marginal damage, and this must apply to all sources (from the equimarginal principle). So if there are $i = 1, \ldots, I$ sources, the following must hold:

$$-MC_i\ (e_i) = MDE_i\ (e_i) = a_i\ MD(p), \qquad \text{for all } i = 1, \ldots, I \tag{9.5}$$

Eq. (9.5) is really a set of I equations, one for each source, setting that source's marginal savings from emissions equal to marginal damage. Since all of the $MD(p)$ terms in Eq. (9.5) are the same, this implies that for any two sources, m and n,

$$MC_m\ (e_m) / a_m = MC_n\ (e_n) / a_n = -MD(p) \tag{9.6}$$

The MC/a terms can be interpreted as the marginal cost per unit of ambient pollution. Thus Eq. (9.6) says that marginal cost in terms of ambient pollution must be equal to the negative of marginal damage. In other words, efficiency calls for all sources to have the same marginal costs of emissions, normalized by the source's transfer coefficient. Thus if a source has a larger impact on ambient pollution (a is larger), its marginal cost of pollution control (marginal cost of emissions) must be larger. This is the equimarginal principle, modified for ambient pollution.

To summarize, two conditions are necessary for efficiency. First, the marginal cost of emissions, normalized by the transfer coefficient, must be equalized for all sources. Second, that normalized marginal cost must equal the negative of marginal damage.

Example: Suppose that in our problem with the river and the municipality in Figure 9.1 the basic conditions are

$$\begin{aligned} a_1 &= 2 \\ a_2 &= 3 \\ MC_i\,(e_i) &= -14 + 7e_i,\ i = 1, 2 \\ MD(p) &= \mathrm{p} \end{aligned} \tag{9.7}$$

Also assume the background pollution level is zero. How much should each source efficiently emit?

Note that we have assumed that both sources share the same marginal cost function. This of course need not be the case. Note also that marginal costs of emissions are negative. Recall that firm costs decrease as emissions increase, yielding negative marginal costs. Marginal savings from emissions are, in contrast, positive.

Applying Eq. (9.6) to this example yields the following efficiency conditions:

$$MC_1\,(e_1)\ /\ a_1 = (-14 + 7e_1)/2 = MC_2\,(e_2)\ /\ a_2 = (-14 + 7e_2)/3 \tag{9.8a}$$

$$\begin{aligned} MC_1\,(e_1)\ /\ a_1 &= (-14 + 7e_1)/2 = -MD(p) \\ &= -MD(2e_1 + 3e_2) = -(2e_1 + 3e_2) \end{aligned} \tag{9.8b}$$

We have two equations in two unknowns that we know can be solved. First, simplify, rewriting Eq. (9.8) as

$$-42 + 21e_1 = -28 + 14e_2 \tag{9.9a}$$

$$-14 + 7e_1 = -4e_1 - 6e_2 \tag{9.9b}$$

which can be solved for $e_1 = 1$ and $e_2 = 0.5$. The marginal cost for firm 1 is -7 and the marginal cost for firm 2 is -10.5. The total ambient pollution is 3.5, which yields marginal damage of 3.5. Check that the marginal cost, normalized by the transfer coefficient, is equal to the marginal damage.

C. Emission Fees

Having set up the framework for dealing with space, it is relatively straightforward to see how emission fees should be structured to yield efficiency. We seek emission fees, t_i, one for each firm i, that yield efficiency, defined by Eq. (9.6). These will be ambient-differentiated emission fees. We know that whatever fee is used, the firms will respond by minimizing direct costs plus fee payments. This is equivalent to setting marginal cost equal to the negative of the fee:

$$MC_i\,(e_i) = -t_i, \qquad \text{for all sources } i \tag{9.10}$$

This means we can rewrite the conditions for efficiency in Eq. (9.6) as

$$t_n / a_n = t_m / a_m, \qquad \text{for all firms } n \text{ and } m \tag{9.11a}$$

and

$$t_n = a_n\, MD(p), \qquad \text{for any firm } n \tag{9.11b}$$

Thus the emission fees levied on the firms must be equal, after normalizing by the transfer coefficient. The second condition says that for any firm, marginal damage in emission units (*MDE*) must be equal to the emission fee. The first condition equalizes control costs across firms; the second condition equalizes marginal pollution damage and control costs. An alternative interpretation is that all firms face the same emission fee per unit of ambient pollution, but to convert it to emission units, it must be multiplied by the appropriate transfer coefficient.

Thus we see that ambient-differentiated emission fees can achieve efficiency. Sometimes, however, it is too complicated to let emission fees vary from location to location. For instance, there may be too much uncertainty about the level of the transfer coefficients; or it may be considered unfair to levy different fees on different firms. Most emission fees do not depend on location, even though damages do. How inefficient is applying a uniform emission fee when damages depend on location? Figure 9.2a is a graphic representation of Eq. (9.5). Shown is the marginal savings curve, assumed the same for each firm as well as marginal damages, normalized by the transfer coefficient. We know that efficient taxes are set so that Eq. (9.11a) holds. Thus for efficient taxes (t_1^*, t_2^*), emissions from the two firms will be as shown in Figure 9.2 (e_1^* and e_2^*). But suppose we use

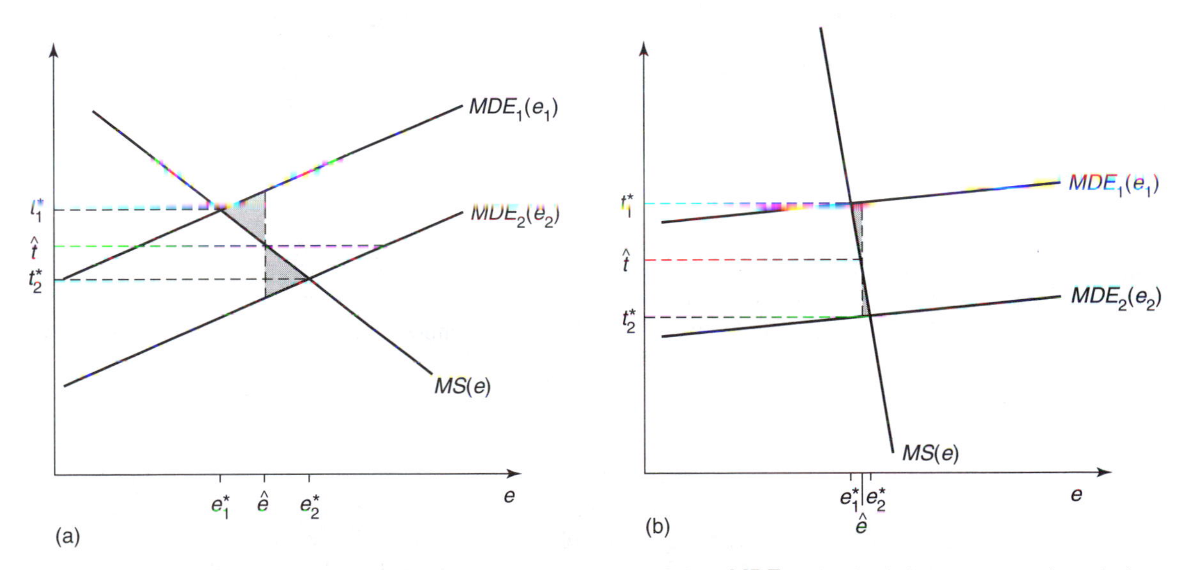

Figure 9.2 (a,b) Illustration of inefficiencies of uniform emission fee. *MDE*$_1$, Marginal damage per unit emissions, firm 1; *MDE*$_2$, marginal damage per unit emissions, firm 2; *MS*, marginal savings from emitting, firm 1, firm 2; t_1^*, efficient Pigouvian fee, firm 1; t_2^*, efficient Pigouvian fee, firm 2; efficient emissions firm 1; e_2^*, efficient emissions, firm 2; $\hat{t}$, uniform emission fee; $\hat{e}$, emissions from firm 1 or firm 2, uniform emission fee.

only one tax. If a uniform tax ($\hat{t}$) is used, both firms will emit the same ($\hat{e}$). The deadweight loss associated with this uniform emission fee is the shaded area in Figure 9.2. The "optimal" uniform tax is one that minimizes the total area of the two triangles.

The loss from a uniform fee depends on the nature of the marginal cost and damage functions. Figure 9.2b shows the same thing as 9.2a except with different slopes for marginal costs and benefits. Clearly, the deadweight loss is much lower. In fact, the steeper the marginal cost functions and the flatter the marginal damage, the smaller the deadweight loss.[3]

D. Marketable Ambient Permits

Marketable emission permits that take into account the effect of ambient pollution concentrations are somewhat more complicated, but in theory can work just as well as ambient-differentiated emission fees.

First we need to define what we mean by an ambient pollution permit.

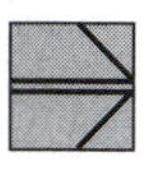

Definition *An ambient pollution permit for a receptor* j *gives the holder the right to emit at any location, provided the incremental pollution at receptor* j *does not exceed the permitted amount.*

An ambient pollution permit system is a set of permits, distributed to sources in a region, a well-defined way of computing the effects of emissions on ambient pollution at receptors, along with a right to buy and sell these permits.

1. Two Firms. First consider the case of two firms and one receptor (this is easiest). Suppose the EPA issues L_1 ambient permits to firm 1 and L_2 ambient permits to firm 2 for a total of $L = L_1 + L_2$ permits. Buying and selling of permits can then take place. Let ℓ_1 and ℓ_2 be the number of permits eventually held (after trade) by the two firms. Obviously,

$$L_1 + L_2 = \ell_1 + \ell_2 \tag{9.12}$$

We have several questions: (1) What will the price of permits be? (2) How much will each firm emit? Answering the second question first, a firm can emit whatever is allowed by its permit holdings. Remember permits are a right to degrade ambient pollution levels. Emissions and ambient pollution levels are connected by the transfer coefficient. Thus

$$a_1 e_1 = \ell_1 \tag{9.13a}$$

$$a_2 e_2 = \ell_2 \tag{9.13b}$$

provided all the permits are used. Now suppose the price of permits is π (an unknown). How much will be emitted? However many permits a firm may hold, if the price of permits is greater than the marginal savings from emitting, the firm will want to sell some permits and emit less. In contrast, if the permit price is lower than the firm's marginal savings from polluting, buying permits is easier than controlling emissions: the firm will buy permits and increase emissions. We seek a price for which the desired emission levels for each of the two firms corresponds to the number of permits issued.

Total costs (TC) for each firm are

$$TC_1(e_1) = C_1(e_1) + \pi(\ell_1 - L_1)$$
$$= C_1(e_1) + \pi(a_1 e_1 - L_1) \tag{9.14a}$$

and

$$TC_2(e_2) = C_2(e_2) + \pi(\ell_2 - L_2)$$
$$= C_2(e_2) + \pi(a_2 e_2 - L_2) \tag{9.14b}$$

where $C_i(e_i)$ is the direct cost to firm i, excluding permit costs. To minimize total costs, each firm sets the marginal total costs (MTC) to zero:

$$MTC_1(e_1) = MC_1(e_1) + a_1\pi = 0 \tag{9.15a}$$
$$MTC_2(e_2) = MC_2(e_2) + a_2\pi = 0 \tag{9.15b}$$

which implies

$$\frac{MC_1\ (e_1)}{a_1} = \frac{MC_2\ (e_2)}{a_2} = -\pi$$

or

$$\frac{MS_1\ (e_1)}{a_1} = \frac{MS_2\ (e_2)}{a_2} = \pi \tag{9.16}$$

Equation (9.16) should look familiar. It says that marginal savings, normalized by the transfer coefficient, should equal the permit price. This is analogous to how an ambient emission fee works—marginal savings, normalized by the transfer coefficient, equals the same number for all firms.

There is one additional equation that gives us information. Combining Eqs. (9.12) and (9.13), we know

$$a_1 e_1 + a_2 e_2 = L \tag{9.17}$$

Equations (9.16) and (9.17) constitute three separate equations in three unknowns (e_1, e_2, and π) and thus can be solved.

Example: Consider the previous example of the municipality and the river. We concluded that the right amount of pollution was 3.5. Suppose instead that permits for 5 tons of pollution were issued. What would be the resulting emissions? We know from Eqs. (9.16) and (9.17) that

$$\frac{MC_1\ (e_1)}{a_1} = \frac{(-14 + 7e_1)}{2} = -\pi \tag{9.18a}$$
$$\frac{MC_2\ (e_2)}{a_2} = \frac{(-14 + 7e_2)}{3} = -\pi \tag{9.18b}$$
$$a_1 e_1 + a_2 e_2 = 2e_1 + 3e_2 = 5 \tag{9.18c}$$

Equations (9.18a) and (9.18b) indicate that whatever the price of permits might be, the two firms will emit so that marginal costs, adjusted by the transfer coefficient, are equal to the permit price. A high price will result in fewer emissions than a low price. Equation (9.18c) indicates that not just any emissions will do; additional pollution concentrations must be exactly 5 units. The price of permits, as well as emission levels, are unknowns in Eq. (9.18). These three equations can be solved for $(e_1, e_2, \pi) = (1.23, 0.85, 2.69)$.

2. Multiple Sources and Receptors. Suppose we have $e = 1, \ldots, I$ polluters and $j = 1, \ldots, J$ receptors. Suppose the government determines that the efficient pollution level is $\bar{s}$ at each receptor and thus distributes $L_j = \bar{s}$ ambient pollution permits for each receptor, j. How will firms buy and sell these permits? Let ℓ_i^j be the number of permits held by source i for polluting receptor j. Because of buying and selling, ℓ_i^j need not equal the number of permits it initially received. But because permits cannot be created (legally), we know that the "law of conservation of permits" applies:

$$\bar{s} = L_j \geq \Sigma_i \, \ell_i^j \tag{9.19}$$

If all permits are used and none is lost or destroyed, we would have equality in Eq. (9.19). Equation (9.19) simply states that after trading has occurred, the number of permits held by various firms for polluting each receptor j must be less than or equal to the number of permits initially issued for that receptor.

How much then can source i emit, assuming it has a portfolio of permits? Basically, the set of permits the firm has allows certain pollution levels at the various receptors. If the firm emits e_i, its pollution at receptor j will be $a_{ij}\, e_i$. The permit holdings require that

$$a_{ij}\, e_i \leq \ell_i^j \qquad \text{for all } j \tag{9.20a}$$

or

$$e_i \leq \ell_i^j / a_{ij} \qquad \text{for all } j \tag{9.20b}$$

This means that allowed emissions for source i are given by

$$e_i{}^* = \min_j \{\ell_i^j \,/\, a_{ij}\} \tag{9.21}$$

This probably seems more complicated than it really is. Consider Figure 9.1 again but suppose there is a swimming area between the city and polluter B. Call the swimming area receptor S and the city receptor C. Each polluter has permits to degrade river quality at C and permits to degrade quality at S. Emissions are governed by whichever receptor is most sensitive, taking into account permits held. This is Eq. (9.21).

With some receptors being more sensitive than others, it is entirely possible that some receptors will have less pollution than is allowed. For instance, if we are regulating a large region that contains an urban area, some receptors will be in the urban area and some in the surrounding countryside. It is entirely possible that pollution will be at the limit at the urban receptors and below the limit at the rural receptors. If the pollution is below the limit, it means not all of the initially issued permits are being used and thus the price of those permits must be zero.

This brings us to the question of what prices will result from trading. The permits are rights to trade the ambient environment. There will be a different price for this at each receptor, π_j. To a firm holding such a permit, the price per unit of ambient pollution is π_j and per unit of emissions is $a_{ij}\ \pi_j$.

To see this, assume a firm emits e_i. This results in ambient pollution at j of $a_{ij}\ e_i$ for which permit costs are $a_{ij}\ e_i\ \pi_j$. Change emissions by one unit and change permit costs for receptor j by $a_{ij}\ \pi_j$. Total permit costs (over all receptors) change by $\Sigma_j\ a_{ij}\ \pi_j$. The firm will emit so that

$$MC_i(e_i) = \Sigma_j\ a_{ij}\ \pi_j \tag{9.22}$$

There are three results that we will state but not prove:[4]

1. A market equilibrium exists in buying and selling ambient pollution permits for any initial issuance of permits.
2. Emissions from each source in a permit market equilibrium are efficient [i.e., are the least-cost way of attaining $\bar{s}$ in Eq. (9.19)], no matter how the permits are initially distributed.
3. If the price of permits in equilibrium equals the marginal damage from pollution, efficiency has been obtained.

These are important results. First, they indicate that a marketable ambient pollution permit system will work. Trade will occur and will eventually reach equilibrium. Second, trade in permits will yield efficiency in the sense of being the least cost way of attaining $\bar{s}$ at each receptor. For full efficiency, $\bar{s}$ must have been chosen initially so that marginal damage and marginal costs are equalized, in the sense of the discussion in the previous section on emission fees. If there is only one receptor, $\bar{s}$ must have been chosen so that

$$MC_i\ (e_i^*)/a_i = -MD(\bar{s}), \qquad \text{for all sources } i \tag{9.23}$$

This condition is analogous to Eq. (9.6).

Although this system may seem complicated, a close analog has been used in the United States and Germany. In both countries, there are caps on pollution levels in most urban areas. For new economic activity to start up in a city, existing sources must be "convinced" (i.e., paid) to reduce emissions so that there is no net increase in pollution at any receptor in the urban area. This is the offset system and closely resembles the ambient marketable permit system described above.[5]

E. Zonal Instruments

We should mention that there is an intermediate territory between completely undifferentiated emission fees or permits (where space is ignored) and ambient fees or permits. Those are called zonal fees or permits. Consider first the case of a fee.

In the case of an undifferentiated emission fee, the same fee would apply to all polluters in a region, no matter what their effect on ambient pollution levels. An ambient fee is set at a different level at different locations. With a zonal system, a region or river basin is divided into zones—perhaps a few or perhaps many. Within each zone, the same emission fee applies. However, different zones have different fees. The advantage of a zonal system is that there is more flexibility with resulting gains in efficiency over the undifferentiated system, which is really a zonal system with one zone. The disadvantages are qualitatively the same as for ambient fees. In fact, as the zones become numerous and small, a zonal system becomes equivalent to an ambient system.

Zonal marketable permits are somewhat more complicated. Within a zone, permits are traded on a one-for-one basis, with none of the complexities of ambient permits discussed in the previous section. However if trades are made between firms that are in different zones, trade is as in an ambient permit system with different transfer coefficients for the different zones. Once again, there are efficiency gains from a zonal system as well as disadvantages due to complexity not encountered in an emission permit system.

We will not consider zonal systems further since they are a straightforward adaptation of the ambient permit or fee systems we have already discussed.

II. TIME

A. Stock Pollutants

So far, we have been concerned about emission fees and marketable permits in a purely static setting. Time has been of no concern. However, time does make a difference for many types of pollutants, particularly those that accumulate. When pollutants accumulate, the damage from emissions today is not just the damage today, but the damage in future time periods when today's emissions are still resident in the environment. Eventually virtually all pollutants are cleansed by the environment. But eventually may be a very long time.

Pollutants that accumulate over time are called *stock* pollutants. We are concerned with the stock of pollutants rather than today's flow of pollutants.

Definition *Assume a pollutant accumulates in the environment according to the process*

$$s(t) = \delta s(t-1) + e(t) \tag{9.24}$$

where s(t) *is the stock of the pollutant at any given time* t, e(t) *is emissions of the pollutant, and* δ *is the persistence rate of the pollutant. Further assume that pollution damage depends only on* s(t). *Then if* $\delta > 0$, *the pollutant is a stock pollutant (with respect to the time period). If* $\delta = 0$, *the pollutant is a flow (or fund) pollutant.*

The idea is that when we put a pollutant into the air or water, the environment will dispose of the pollutant, perhaps quickly or perhaps slowly. For instance, when organic matter is discharged into a river, microbes work on the organic matter, decomposing it into harmless substances. Or when sulfur dioxide is emitted into the atmosphere, atmospheric processes eventually cleanse the air, removing the sulfur. The measure of the rate at which the environment is cleansed is δ. More specifically, δ indicates what fraction of the total stock of pollutants is still left in the air after one time period. Alternatively $(1 - \delta)$ is the fraction of the stock that is cleansed out of the environment in one time period. Thus Eq. (9.24) describes the dynamic process of pollutant accumulation.

Basically, pollutants that accumulate over time are stock pollutants whereas pollutants that quickly fade away are flow pollutants. Greenhouse gases that have residence time in the atmosphere of decades are classic examples of stock pollutants, as are cholorofluorocarbons (CFCs) that deplete the stratospheric ozone layer. Particulates (e.g., soot) emitted into the atmosphere are flow pollutants since they settle out within a matter of hours or days.

The definition is not really quite as precise as it might seem. All pollutants accumulate over some period of time. If the time period is seconds, then even large particulates could be considered a stock pollutant. If the time period is centuries then greenhouse gases might look more like flow pollutants. The point is that there is no sharp line between flow and stock pollutants. Despite the potential ambiguity, the sense of the definition is clear. If pollutant accumulation is a dominant characteristic of the pollutant, it is not possible to have much effect on pollutant damage by adjusting emissions, at least in the short term.

Efficiency for a stock pollutant is somewhat more complex than for a flow pollutant. The marginal cost of emissions must be traded off with the marginal damage of the pollutant, where the marginal damage will occur over a period of time. This is more complex than the simple static analysis in which we trade off marginal savings from emitting with marginal damage. At a conceptual level it is the same, however; marginal damages must include damage over the time period when the pollutant is resident in the environment.

Suppose we are sitting in the present at time $t = 1$. The net costs of pollution emissions e_t, can be expressed as

$$\text{NetCosts} = NC = \Sigma_{t=1}^{\infty} \beta^{t-1}\{C_t(e_t) + D_t(s_t)\} \tag{9.25}$$

where β is used to discount the future [equal to $1/(1 + r)$ where r is the discount rate], indicating how we should trade off costs today with costs tomorrow. To determine the amount of emissions at any point in time, e_t, that minimizes net costs, we need to look at the marginal of Eq. (9.25) with respect to e_t. We can start by trying to determine the best current level of emissions, e_1:

$$\begin{aligned}\frac{\Delta NC}{\Delta e_1} &= \Sigma_{t=1}^{\infty} \beta^{t-1}\{\Delta C_t(e_t)/\Delta e_1 + \Delta D_t(s_t)/\Delta e_1\} = 0 \\ &= \Sigma_{t=1}^{\infty} \beta^{t-1}\{\Delta C_t(e_t)/\Delta e_1 + [\Delta D_t(s_t)/\Delta s_t](\Delta s_t/\Delta e_1)\} = 0 \end{aligned} \tag{9.26}$$

Focusing on the first term in braces in Eq. (9.26), note that we need to worry about marginal costs only when $t = 1$. All costs for $t \neq 1$ are constants with respect to e_1. Thus $\Delta C_t(e_t)/\Delta e_1 = 0$ for $t \neq 1$. Furthermore, inspection of Eq. (9.24) indicates that the stock of pollution at time t is simply the sum of previously emitted pollution, appropriately decayed:

$$s_t = e_t + \delta e_{t-1} + \delta^2 e_{t-2} + \cdots + \delta^i e_{t-i} + \cdots + \delta^{t-1} e_1 + \delta^t s_0 \tag{9.27}$$

Thus

$$\Delta s_t/\Delta e_1 = \delta^{t-1} \tag{9.28}$$

Putting all of this together allows us to rewrite Eq. (9.26) as

$$\frac{\Delta NC}{\Delta e_1} = MC_1(e_1) + \Sigma_{t=1}^{\infty} \beta^{t-1} \delta^{t-1} MD_t(s_t) = 0$$

or

$$MS_1(e_1) = \Sigma_{t=1}^{\infty} \beta^{t-1} \delta^{t-1} MD_t(s_t) \tag{9.29}$$

What does Eq. (9.29) tell us? It indicates that for efficiency, we should set the marginal savings from emitting a unit of pollution today equal to the sum of all marginal damages that may occur in the future, with those marginal damages discounted by two factors: the discount factor (β) and the persistence rate of the pollutant (δ). Both act in a similar way. The discount factor diminishes future marginal damages. The persistence rate also diminishes future marginal damages because some of the pollution will have been cleansed from the environment. If the pollutant is a flow pollutant, $\delta = 0$, in which case Eq. (9.29) defaults to the familiar efficiency condition that marginal savings equals marginal damage.

B. Temporal Variability

Most people intuitively recognize that pollution levels vary over space. In an urban area, we expect pollution to be worse in some areas and better in others. But pollution also often varies through time, both because of temporally varying emission rates (the dead of night vs. mid-day), as well as temporally varying assimilative capacity (wind vs. still air). It should also be pointed out that this variation could be over the course of a day, a week, a season, or a year.

Because of temporal variability, emissions will cause different damages, depending on when they are emitted. Air emissions at night that quickly dissipate are less worrisome than emissions at mid-day. Air emissions in the winter may be less worrying than summer emissions in smog-prone areas.[6] This suggests that regulation could be time-varying to promote efficiency.

The analysis of time-varying regulation is very similar to the analysis of spatially varying emissions except that the transfer coefficient relates emissions at one point in time to pollution concentrations at another point in time. In essence, the previous section of this chapter can be viewed as a treatment of temporal variability by substituting the word "time" for "space." For this reason, we will not examine temporal variability in any more detail.[7]

III. MULTIPLE POLLUTANTS

Often several pollutants are responsible for an environmental problem. For instance, the generation of urban ozone requires both nitrogen oxides and volatile organic compounds to be present with sunshine. Acid deposition typically involves deposition of both nitric acid and sulfuric acid, derived from nitrogen and sulfur oxides. In either case, controlling the problem involves controlling both pollutants. Furthermore, each pollutant should be controlled taking into account the cost of controlling it as well as the contribution the pollutant makes to damage.

Multiple pollutants can be handled in much the same way as space or time. Think of there being $i = 1, \ldots, I$ types of compounds emitted into the environment (e_i) and $j = 1, \ldots, J$ types of pollutants that injure the ambient environment (p_j). The different compounds might be nitrogen oxides and sulfur oxides. The different pollutants might be acid deposition at different locations or different types of deposition at the same location.

So damage depends on pollution levels: $D(p_1, \ldots, p_J)$ and control costs depend on emissions of various compounds: $C_i(e_i)$, $i = 1, \ldots, I$. A transfer coefficient, a_{ij}, indicates the ratio of increased pollution of type j from increased emissions of compound i. Although a linear transfer coefficient worked well in discussing spatial issues, different pollutants often interact in a nonlinear way, making the concept of a single constant transfer coefficient inappropriate.

Efficient levels of pollution require the same two conditions as set forth in Eq. (9.5) and (9.6). Marginal savings from emitting the different compounds must be equated, taking into account the transfer coefficients. And these marginal costs must be equal to marginal damage, adjusted for the transfer coefficient.

An ambient permit system can function well in this environment. Permits would be issued to degrade the ambient environment. For instance, a permit might be issued for a certain amount of acid deposition. That permit could then be converted into emission rights based on different transfer coefficients. However, because of the similarity with the spatially differentiated permit system, we will not discuss this case further.

IV. IMPLEMENTING MARKETABLE PERMITS

Emission fees are significantly simpler to implement than marketable permits. The major problem with emission fees is that they require a more active regulatory authority to set and adjust their levels. This is over and above the problem of instituting a fee to begin with, given the confiscatory nature of the fee. For these reasons, marketable permits seem to be more popular, at least in the United States.[8] For this reason, we will examine in a little more detail the problems of implementing a marketable permit system.

A. Initial Permit Issuance

We saw in the previous chapter that for the purposes of efficiency it does not matter which firms initially receive marketable permits. This is one of the primary advantages of maketable permits. In the previous section, this result was extended to ambient permits. But permits do have to be distributed. If they are auctioned off, the price at which they are sold will presumably be equal to the price that will prevail in the market, which will be equal to what an efficient emission fee would be. This raises a major political problem of permit systems—permits initially may involve a significant resource transfer from the polluters to the government. This will undoubtedly be vigorously resisted in the political process and may be successfully blocked.[9] For this reason permits are often given to polluters rather than auctioned. This is not really a problem for economic efficiency. A problem arises, however, regarding new entrants. Do they also receive free permits? Probably not. Incumbents could potentially use permits to retard entry.

One way of allocating permits without the problem of a wealth transfer to the government is a zero-revenue auction, originally proposed by Hahn and Noll (1982). The idea is simple. In advance of the permit auction, the political process decides what fraction of the auction revenue should accrue to each existing polluter. This might be based on current emissions, historic emissions, or ability to pay. The permits are then auctioned but the revenues are distributed to existing polluters in an agreed-on manner. This system can also be used to ensure new entrants adequate opportunity to acquire permits. Anyone can bid for permits. The revenue from the auction is distributed to those parties who might resist (politically) a permit auction without revenue recycling. Such an auction is used in the United States annually for a small portion of the acid rain emission permits. It appears to have the added advantage of providing a clear signal about the current market price of the emission permits, information that facilitates market operation.

B. Dominant Firms

One potential problem with the initial distribution of permits may arise if there is market power in the permit market. In other words, if there is a single large firm (or potentially, several large firms), it can make a difference how permits are initially allocated.[10]

To see this, suppose there is one large firm in an area with many small firms which together make up a price-taking competitive fringe. Suppose the large firm has costs associated with emitting of $C(E)$ where E is emissions by the large firm. Assume that the competitive fringe has inverse demand for permits of $P(e)$. In other words, at a permit price of $P(e)$, the fringe will demand e permits. It is through manipulation of this inverse demand curve that the dominant firm can exercise market power. Suppose that L emission permits are issued initially, with L_B going to the big firm. There are no spatial considerations here; we are considering emission permits only.

The total cost of the big firm is given by

$$TC(E) = C(E) - P(L - E)\,[L_B - E] \tag{9.30}$$

The last term is the revenue to the large firm from selling excess permits, $(L_B - E)$. The price at which the large firm sells permits is determined by the inverse demand function and how many permits the fringe is buying, $e = L - E$. The large firm wishes to minimize costs by choosing the appropriate E. The problem is not unlike that of the monopolist choosing the right amount of output, driving price up by restricting output.

To solve this problem, we will see what happens to total costs as we change E by a small amount, ΔE. Costs will be minimized when a small change in E causes no change in TC. First increase E by $\Delta E = -\Delta e$:

$$\begin{aligned} TC(E + \Delta E) &= C(E + \Delta E) - P(L - E - \Delta E)\,[L_B - E - \Delta E] \\ &= C(E + \Delta E) - P(e + \Delta e)\,[L_B - E] + P(e + \Delta e)\,\Delta E \end{aligned} \tag{9.31}$$

Now if we subtract Eq. (9.30) from this and divide by ΔE, we obtain

$$\Delta TC/\Delta E = \Delta C/\Delta E + \Delta P/\Delta e\, [L_B - E] + P(e + \Delta e)$$
$$\approx MC(E) + MP(e)\, [L_B - E] + P(e) \quad (9.32)$$

The right-hand-side of Eq. (9.32) is the sum of marginal costs (MC), a marginal price term (MP), and permit price (P). Cost minimization calls for Eq. (9.32) to be equated to zero:

$$MS(E) = MP(e)[L_B - E] + P(e) \quad (9.33)$$

Denote the emission level at which price equals marginal savings $[MC(E) + P(e) = 0]$ as E^*. Denote the solution to Eq. (9.33) as E^{**}. Referring to Eq. (9.33), efficiency basically calls for marginal savings from emissions ($-MC$) to be equated to permit price. Unfortunately, the first term on the right-hand side of Eq. (9.33) drives a wedge between marginal savings and price. We know that marginal price is negative since demand is downward sloping (quantity increases when price declines).

So, what is the relationship between E^* and E^{**}? Inspection of Eq. (9.33) indicates that when $L_B = E^*$, we have $E^{**} = E^*$. So we will consider the two other cases, one in which $L_B < E^*$ and one in which $L_B > E^*$. In the first case, $L_B < E^*$, we ask if $E^{**} > E^*$ or $E^{**} < E^*$? Note that in this case of $L_B < E^*$,

$$E^{**} < E^* \Rightarrow MS(E^{**}) > P(e^{**}) \Rightarrow MP(e^{**})\ (L_B - E^{**}) > 0 \Rightarrow L_B < E^{**} \quad (9.34a)$$

$$E^{**} > E^* \Rightarrow MS(E^{**}) < P(e^{**}) \Rightarrow MP(e^{**})\ (L_B - E^{**}) < 0 \Rightarrow L_B > E^{**} \quad (9.34b)$$

But Eq. (9.34b) implies $L_B > E^*$, which contradicts our original assumption. Thus it must be that $L_B < E^{**} < E^*$. A similar argument can be used to show that if $L_B > E^*$, then $E^* < E^{**} < L_B$. To sum up,

$$L_B > E^* \Rightarrow E^* < E^{**} < L_B \quad (9.35a)$$

$$L_B < E^* \Rightarrow L_B < E^{**} < E^* \quad (9.35b)$$

$$L_B = E^* \Rightarrow L_B = E^{**} = E^* \quad (9.35c)$$

This is illustrated in Figure 9.3, which shows both marginal savings ($-MC$) and permit price (P), as functions of emissions from the large firm. These are the two solid lines and they intersect at E^*. Note that the more the large firm emits, the fewer permits available to the fringe, which drives up the price. Also shown is the first term on the right-hand side of Eq. (9.33), for an initial permit issuance to the big firm of $L_B < E^*$ and the sum of this term and the price, $P(L - E)$—line A. Equation (9.33) holds where lines A and C cross, yielding emissions E^{**}—lower emissions than the efficient E^*.

The intuition behind this result is that the large firm gains market power by selling or buying permits. So if the firm is initially allocated too few permits, it will buy, but not quite enough, in an effort to keep price down. Thus its emissions will be lower than is efficient. Similarly, if the large firm is given too many permits to start with, it will sell some, but not quite enough, in an effort to keep price high. The only way the large firm can be induced to emit at the proper level is to give it exactly the right amount of permits to start with.

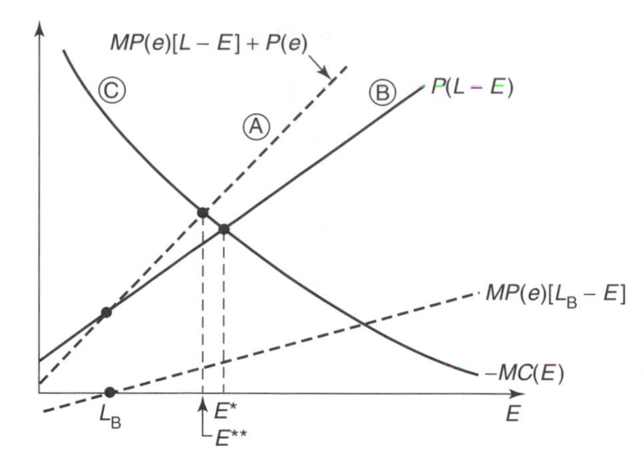

Figure 9.3 Illustration of how an inefficient initial distribution of permits can lead to a distortion when one firm has market power. $-MC$, Marginal savings from emitting for large firm; $P(L - E)$, permit price; L_B, initial permit issuance to the large firm; E^*, efficient level of emissions; E^{**}, manipulated level of emissions.

It is important to realize that the total amount of pollution is unaffected by how the permits are initially allocated. The total amount of pollution will be L. But if the large firm is emitting too little or too much, the total costs of the large firm and the fringe firms will be too high. Thus the inefficiency is on the cost side.

Unfortunately, this is more than an exercise in marginals. Often in controlling pollution in a region, there will be dominant firms. This suggests that in such a case, some of the gain from using marketable permits may be lost. How much is lost is an empirical question.[11]

C. Market Thinness

A related problem with marketable permits arises if the market generally is thin. If there are only a few firms in the market, transactions may be infrequent or nonexistent. Thus the main rationale for marketable permits—equalizing marginal control costs among firms—will not be accomplished. Of course, the correct amount of pollution will result (equal to the permit issuance), but costs will be excessively high.

Often the firms in a permits market will compete with one another in a goods market. Withholding permits can then be used as an entry deterrent. Or permits can be used in other anticompetitive ways.

D. Transaction Costs

A virtue of tradable permits is that any initial misallocation of the right to emit can be remedied by trading of permits; furthermore, the ability to trade makes the opportunity cost of emitting equal to the permit price. But if there are impediments to trade, these advantages disappear or are significantly diminished. Impediments to trade can be viewed as transaction costs associated with trade. These can be quite significant for tradable permit markets.

Robert Stavins (1995) identified three sources of transaction costs that might arise in permit markets, (1) search and information costs, (2) bargaining and decision-making costs, and (3) monitoring and enforcement costs. For a trade to take place, a potential buyer and a potential seller must identify each other (search and information costs), they must negotiate with each other and finally agree on a price for a trade (bargaining and decision-making costs), and the trade must be monitored and enforced. This last category of cost may be borne by the regulatory agency, but there are also costs that may be borne by firms. If a trade must be approved by a regulatory agency, that can add time and costs to consummating a trade.

Search and information costs become large the more heterogeneous a permit market may be. If emission permits involve a single pollutant with no spatial or temporal considerations, the market is homogeneous—there is only one "good," the emission permit. On the other hand, if we are dealing with an ambient permit market in which every permit must be viewed differently, in terms of its impact on ambient pollution, identifying likely buyers and sellers may be very difficult. Furthermore, if there are complex rules governing who may trade with whom, the problems of identifying trading partners becomes even more significant. Robert Hahn (1989) surveyed several attempts at establishing permit markets in the United States and Europe and identified transaction costs as a major reason for failure. In an examination of pollution trading in Los Angeles, Foster and Hahn (1995) examine trading in small and large quantities of pollution. They argue that transaction costs are likely to be higher per unit of pollution for the small trades. They then show empirical evidence that trading is less likely for small quantities of pollution, in support of their hypothesis on transaction costs.

The main conclusion is that a nicely designed complex system of ambient pollution permits may look efficient on paper but transaction costs may make it fail. A simpler, though perhaps less efficient, permit system may perform better. Alternatively, when permit markets are designed, there should be a conscious effort to minimize transaction costs.

V. COMPARATIVE REGULATORY ANALYSIS

One of the themes that permeates environmental economics is that command and control is inferior to market-based incentives. But the real question is, how inferior? There have been a number of authors who have tried to answer this question by conducting studies of actual environmental regulations, attempting to simulate the economic and environmental consequences of various regulatory regimes.[12] This sort of regulatory analysis can be very useful to regulators, legislatures, and other bodies contemplating environmental regulations or contemplating changes in a regulatory regime.

Most analyses focus on the spatial aspects of pollution generation, transport, and damage. Furthermore, most of the analyses share a common structure and solution technique—mathematical programming. Mathematical programming is a method of solving large maximization problems and has been implemented in many popular spreadsheet programs. Thus it is not necessary to be a master of complex algorithms to solve a typical mathematical program.

Perhaps the best way to explore how typical regulatory analyses are conducted is to consider an example. Scott Atkinson and Thomas Tietenberg conducted an analysis of the St. Louis metropolitan area in the United States.[13] Their goal was to examine current stationary sources of particulate pollution (soot and dust) as well as alternatives to the current regulatory regime. The alternatives the authors considered were marketable permits, both ambient permits and emission permits (undifferentiated with regard to space), as well as zonal permits.

One of the shortcomings of the analysis from a pedagogical point of view is that they do not consider pollution damage. Rather, they consider the U.S. standards for ambient levels of particulates to be the goal of regulation. The U.S. standard is a maximum highest concentration of particulates during the course of the year.[14] Thus theirs is a cost–effectiveness analysis.

The pollution problem can best be appreciated by considering a map of the St. Louis area with its sources and receptors. Such a map is shown in Figure 9.4. Note that there

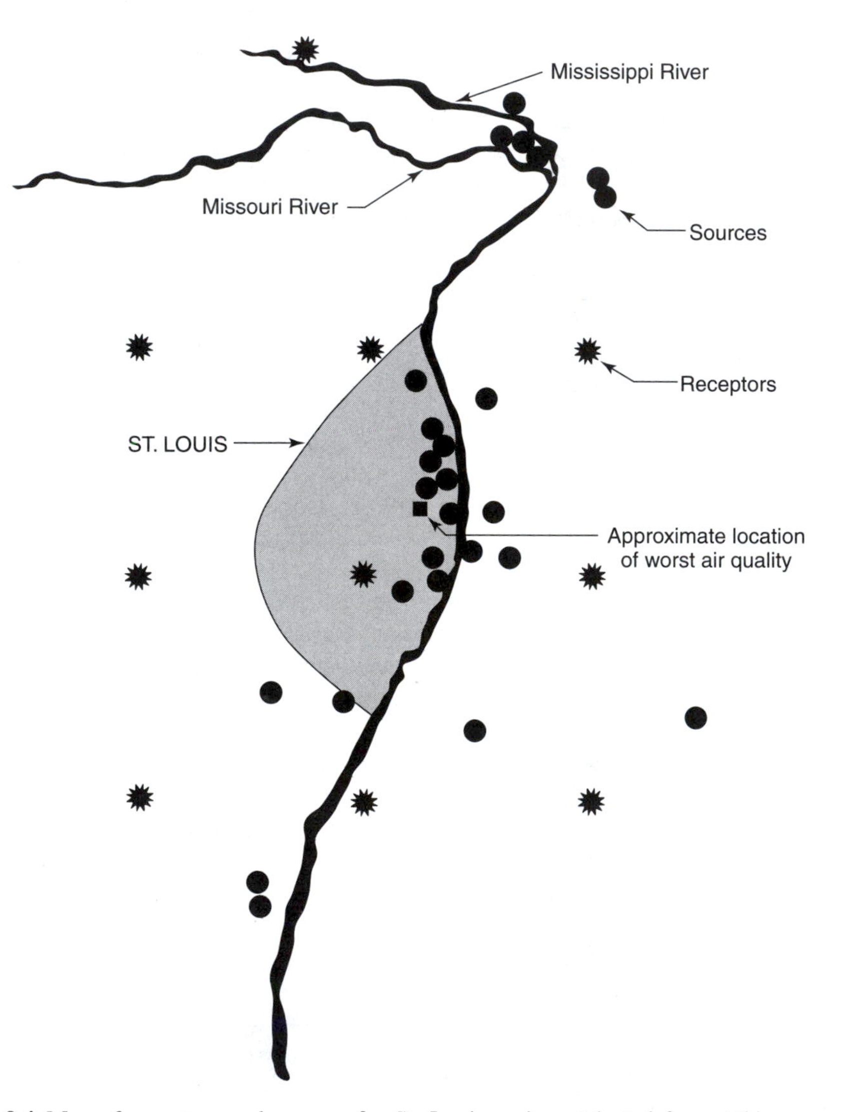

Figure 9.4 Map of receptors and sources for St. Louis region. Adapted from Atkinson and Tietenberg (1982), p. 111.

are 27 major stationary sources of particulate matter and nine receptors scattered around the metropolitan area. The authors' goal is to measure the short-run pollution control effects of different regulatory regimes. Thus they do not need to examine exit or entry. Further, they do not consider effects on the product markets of different emission control regulations, which of course could raise or lower product costs and thus sales and profits.

The authors obtain information on emission control costs, which is the basic driving force of the analysis. For each source, i, the control costs are quadratic in the amount pollution that is reduced:

$$C_i\,(e_i) = c_i\,(\bar{e}_i - e_i) + d_i\,(\bar{e}_i - e_i)^2 \tag{9.36}$$

where $\bar{e}_i$ is the amount of pollution the ith source will emit with no regulations and e_i is the amount of pollution the ith source will emit with regulation. The c_i and d_i terms are simply constants.

To find the least-cost way of controlling pollution, we must define the relationship between emissions and ambient concentrations. The following is a representation of the ambient standard:

$$\Sigma_i\, a_{ij}\, e_i + b_j \leq \bar{s} \tag{9.37}$$

where a_{ij} is the transfer coefficient between source i and source j, b_j is the background level of pollution, from other man-made sources and nature, and $\bar{s}$ is the standard—the maximum allowable concentration of particulates. Thus Eq. (9.37) must hold at all receptors.

The complete model can then be specified. We seek a set of e_i that satisfies the following:

$$\min_{e_i} \quad \Sigma_i\, \{c_i\,(\bar{e}_i - e_i) + d_i\,(\bar{e}_i - e_i)^2\} \tag{9.38a}$$

such that

$$\Sigma_i\, a_{ij}\, e_i + b_j \leq \bar{s}, \qquad \text{for all receptors } j \tag{9.38b}$$

$$0 \leq e_i \leq \bar{e}_i\,, \qquad \text{for all sources } i \tag{9.38c}$$

This is what is known as a quadratic program (a special type of optimization problem) for whose solution there are numerous computer packages available. We will not concern ourself with how this is solved and move directly to results.

First, note that the solution of Eq. (9.38) defines an efficient level of emissions. That means that the solution to Eq. (9.38) corresponds to what happens with an ambient permit system. This is our lowest attainable cost of control; no other regulatory regime can do better.

To simulate an emission permit system, replace Eq. (9.38b) with

$$\Sigma_i\, e_i \leq E \tag{9.39}$$

where E is the cap on emissions in the region. To meet an ambient standard with this type of regulation, slowly reduce E from a large number (with consequent high levels of ambient pollution), thus slowly reducing ambient pollution levels at all the receptors. When the ambient standard is just met at all the receptors, that is the correct amount of emission permits, E^*.

Finally, the current regulatory regime is simulated by replacing Eq. (9.38b) by

$$e_i \leq E_i, \qquad \text{for all sources } i \tag{9.40}$$

where E_i is the source-by-source emission limit specified in the command-and-control regulations.

Figure 9.5 shows the results of this analysis for different values of the maximum particulate concentration at the nine receptors ($\bar{s}$). There is no question that the command-and-control regulations are very costly at accomplishing their objectives—as much as a factor of 10 in excess of what is necessary to achieve the objective. As expected, the undifferentiated emission permit does not do as well as the ambient-differentiated permit but is much better than command and control.

Note also from Figure 9.5 that as the desired maximum level of pollution becomes increasingly lower (the environmental regulations become increasingly tighter), the costs of control increase at an increasing rate. Thus in lowering the maximum concentration ($\bar{s}$) from 3 to 2 $\mu g/m^3$, the cost increase is much greater (for all three regulatory approaches) than for lowering the maximum concentration from 9 to 8 $\mu g/m^3$. Costs really accelerate as one tries to tighten the standard.

It is this type of information that can guide a regulator in deciding what type of regulatory structure to govern emissions. Perhaps the regulator will decide that an emission permit system is relatively easy to implement and can save industry a considerable amount of money.

The point of going through this case is some detail is that the approach is very similar to what might be done in analyzing an urban area anywhere in the world. This type of analysis can be vitally important to decision makers contemplating changes in emission regulations.

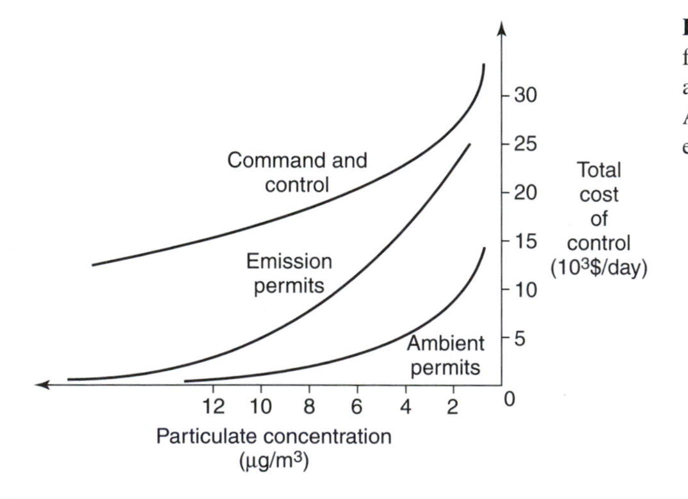

Figure 9.5 Total control costs for three regulatory systems, as a function of air quality ($\mu g/m^3$). Adapted from Atkinson and Tietenberg (1982), p. 114.

SUMMARY

1. Sources of pollution emit; receptors are points at which we care about pollution concentrations. The transfer coefficient is the ratio of the increase in pollution at a receptor to the increase in emissions at a source.

2. Efficient emissions in a spatial environment with one receptor require that the ratio of marginal savings from emissions to the transfer coefficient be equated for all sources. That ratio would in turn be equal to the marginal damage from pollution.

3. Spatially differentiated emission fees for each source should be set equal to marginal damage at the efficient amount of pollution, multiplied by the transfer coefficient.

4. Ambient pollution permits (with one receptor) will be priced, in equilibrium, at any source's marginal savings from emissions, divided by the transfer coefficient.

5. Stock pollutants accumulate over time; thus simply controlling emissions does not directly control the damage from the stock. Flow pollutants do not accumulate significantly.

6. In a competitive permit market, the initial allocation of permits is not important. With market power on the part of any polluter, the initial distribution matters.

7. Empirical analyses of efficiency of different approaches to regulating pollution have shown very significant savings from moving to ambient pollution permits from command and control.

PROBLEMS

1. Write the analog for Eq. (9.16) for the case of multiple receptors. Interpret your result.

***2.** An ambient differentiated emission permit gives the holder the right to emit anywhere provided emissions have no more impact on receptor j than one unit of emissions at source i. Let ℓ_i^j be the quantity of such permits a firm holds.

a. Suppose there is only one receptor. Write an equation showing allowed emissions for source k if it holds a portfolio of ambient differentiated emission permits $\{\ell_1^1, \ell_2^1, \ldots, \ell_I^1\}$.

b. Generalize your answer to part (a) for the case of multiple receptors.

3. Acid rain is a problem in Scandinavia. Electricity-generating plants burn coal and emit sulfur dioxide. That sulfur dioxide is converted to sulfates, which travel long distances and are washed out of the air as acidic rain, particularly affecting Swedish lakes. Assume that power plants in England and Denmark are the primary culprits. Also assume that in a year one tonne of sulfur emitted in England generates deposition in Sweden of 1 gram of sulfate deposition and one tonne of sulfur from Denmark generates 3 grams of deposition in Sweden. Damage from acid rain is approximately 1 pence per gram of sulfates deposited. If q is the reduction in annual sulfur emissions (in tonnes) from uncontrolled levels, the marginal cost of controlling sulfur emissions is £q for Denmark and £$2q$ for England [1£ = 100p].

a. Write the damage function, giving damages as a function of total sulfates deposited.

b. Write the transfer coefficients for England and Denmark.

c. First ignore the effects of pollution on Sweden and suppose we want simply to reduce emissions by 12 tonnes. How much of that reduction should come from Denmark, on the basis of cost efficiency? (*Hint*: Think of the marginal cost of emission control just like you think of the marginal savings for emissions.)

d. Now suppose we want to reduce emissions by 12 tonnes but we want to do it in a way that takes into account effects on pollution in Sweden. Although this might not be the right amount of pollution reduction, balancing the marginal cost of controlling pollution in Sweden from the two source countries would require how much of an emission reduction in Denmark?

***4.** Consider an abstraction of the Santiago, Chile air pollution problem. Assume all pollution damage is in downtown Santiago and is equal to $p^2/2$ where p is the pollution concentration in parts per million (ppm). Pollution comes from automobile emissions and a steel mill on the east side of town. Each unit of automobile emissions results in 2 units of pollution in downtown Santiago. Each unit of steel mill pollution results in one unit of pollution in downtown Santiago. Uncontrolled, there are 10 units of automobile emissions and 4 units of steel mill emissions. The cost of reducing emissions by q units is q^2 for the automobiles and also q^2 for the steel mill.

a. What are the transfer coefficients for the automobiles and for the steel mill?

b. What is the total control cost in terms of emissions for each source?

c. What is the total control cost per unit of pollution (not emissions) contributed by each source?

d. What is the marginal damage from pollution, in dollars per unit of pollution (\$1 ppm)?

e. What is the marginal damage in dollars per unit of emissions from each of the two sources?

f. What are the efficiency conditions for the optimal amount of pollution?

g. What efficient level of emissions would come from each of the two sources?

5. Consider paper manufacture. The raw material in paper manufacture is pulp which can be made from recycled paper and/or trees (virgin wood). Why does the use of command-and-control environmental regulations instead of emission fees for making pulp from virgin wood (a polluting process) result in lower prices for virgin pulp? Why does this result in less use of recycled paper in paper manufacture?

6. Two identical firms save money from polluting. A firm's marginal savings from emitting an amount e are given by $10 - 2e$. The two firms differ in their impact on ambient pollution concentrations. Two units of emissions from firm 1 result in one unit of ambient pollution. Firm 2 has twice the impact on the ambient environment from the same amount of emissions.

a. What are the transfer coefficients for each of the two firms?

b. If firm 1 is given two emission permits and firm 2 is given four emission permits and they are allowed to trade, how many permits will each firm end up with and what will be the price?

c. If instead each firm is given two ambient pollution permits and trading takes place, how much will each firm end up emitting and what will be the permit price?

7. Consider the case of carbon dioxide being emitted into the atmosphere. Assume that a ton of CO_2 emitted decays at a very slow rate; assume only 1% of the stock in the atmosphere decays in any given year. Also assume that the marginal damage of a ton of CO_2 in the atmosphere is \$1 per year regardless of the amount of CO_2 in the atmosphere. Using a discount rate of 3% per year, calculate the marginal damage caused by emitting 1 ton of CO_2 into the atmosphere. What CO_2 emission fee would you recommend to control this greenhouse gas problem?

NOTES

1. For a detailed analysis (and one of the first) of river pollution, cleansing, and regulation, see Kneese and Bower (1968).
2. Often, linearity will apply up to some carrying capacity of the environment. For instance, in the case of the river, linearity is a good approximation up to the point at which the river becomes so overloaded with organic material that oxygen (needed for aerobic bacteriological decomposition) is depleted. At that point the river's capacity to clean itself is greatly diminished. Many urban air pollutants follow linear chemistry and physics, but one major pollutant (urban ozone) does not. See Braden and Proost (1996) for a discussion of ozone and its control in the United States and in the European Union.
3. This problem of optimally designed uniform fees and their inefficiencies is explored further in Kolstad (1987).
4. These results are fully developed in Montgomery (1972).
5. Actually, the offset system as described is what is termed an ambient differentiated emission permit system. It turns out that an ambient differentiated permits system cannot be guaranteed to work. This was shown in a simple example in McGartland and Oates (1985). The distinction between the two systems is subtle and need not concern us here. See Montgomery (1972) for a discussion of ambient emission permit systems and ambient differentiated emission permit systems. See Foster and Hahn (1995) for a discussion of the U.S. offset system in practice.
6. In smog-prone areas, sunshine and heat facilitate ozone formation. Thus the summer is the worst time of year for this pollutant.
7. A fairly technical treatment of this issue can be found in Chapter 11 of Baumol and Oates (1988).
8. Europe has more of a history of emission charges. These, however, are usually set at very low levels, too low to be considered much of an incentive for pollution control (Howe, 1994).
9. See Buchanan and Tullock (1975).
10. Refer to Hahn (1984).
11. Hahn (1984) concludes that this may not be a serious issue based on his calculation that a dominant firm in the Los Angeles area would not be able to manipulate the market much unless it received over half the permits issued in the region.
12. For example, see Kneese and Bower (1968), Atkinson and Lewis (1974), Atkinson and Tietenberg (1982), Atkinson (1983), McGartland and Oates (1985), Kolstad (1986, 1990), Oates et al. (1989), Seskin et al (1983), Klaassen (1995), and O'Ryan (1996).
13. These authors have several papers on this topic. The paper discussed here is Atkinson and Tietenberg (1982).
14. The standard is somewhat more complicated than this. There are both primary standards and secondary standards for many pollutants. Some of these standards excuse one violation per year. Further, there are different standards for different averaging times. For example, for some pollutants there is a 3-hour standard, a 24-hour standard, and an annual standard, where the time period refers to the length of time over which concentrations are averaged. We will ignore these details here.

10 REGULATION WITH UNKNOWN CONTROL COSTS

We now turn to a very real problem in regulation not confined solely to environmental regulation. That problem arises from differences in the amount of information possessed by the polluter and the regulator. Most frequently, this difference is that the polluter has private information that the regulator needs.[1]

One of the earliest cases of this type in the environmental literature had to do with uncertainty (on the part of the regulator) in control costs and damages and whether that uncertainty tended to induce a social preference for emission permits or emission fees. A second issue that arose somewhat later concerns control costs. If polluters know their own control costs and regulators need to know those costs, is there a type of regulation that can induce polluters to truthfully reveal that cost? These are the questions we will examine in this chapter.

I. A SIMPLE MODEL OF INCENTIVES IN ENVIRONMENTAL REGULATION

One of the key features of the schematic presented in Figure 8.1 is the disconnect between the organization seeking pollution control and the entity actually doing the pollution control. This is one of the central problems in regulation generally and environmental regulation in particular. The problem arises because of different objectives. The legislature may want least-cost pollution control. But the firm wants profits to be highest. Thus it may object to a regulation, saying that the regulation will drive the firm out of business because its control costs are so high. This clash of objectives between the polluter and the regulator would be less of a problem if there were full information. If the regulator knew the control costs of the firm, the firm could not falsely threaten to go out of business. This clash of objectives with incomplete information is the source of the regulatory problem we will examine.

To address this question, we will simplify Figure 8.1, reducing it to a two-agent model consisting of an environmental regulator (the "EPA") and a firm (the "polluter").

The EPA is not privy to the detailed information on the polluter's operations, particularly its precise pollution control costs. For some of this information, the EPA must rely on the firm's statements and reports. But the EPA will not know whether the firm is telling the truth. Our goal then is to design a regulation that makes the most of this imperfect state of affairs.[2]

A. Unknown Polluter Characteristics

The problem we consider is one in which the EPA is uncertain about particular characteristics of the polluter, most frequently pollution control costs. The polluter may be able to control pollution relatively cheaply or it may find pollution very costly to control. This is a classic issue in environmental regulation. Typically the EPA will claim pollution can be controlled at reasonable cost while the polluter claims that the environmental regulation will force it out of business. It is a credible threat because the polluter knows its own costs much better than the EPA. To simplify things, we will assume that the polluter is one of two types: either a high-cost polluter or a low-cost polluter. Suppose, to the EPA, there is a 50:50 chance of a firm being one type or the other.

The EPA's goal is to design a regulation that induces the firm to do the right thing, from the EPA's perspective. Although regulations can take many forms, let us suppose that the regulation takes the form of an emission fee (r), based on the amount of emission generated, e. Thus the emission fees collected will be re. The polluter's problem is to minimize total costs where costs include the emission fee:

$$\min_{e} \quad TC(e) = C(e) + re \tag{10.1}$$

Of course the polluter will choose to emit where the marginal savings (negative of marginal costs) equal the emission fee:

$$-MC(e) = MS(e) = r \tag{10.2}$$

But the two types of polluters have different costs and thus will pollute at different levels. If the polluter is high cost, it will decide to generate e_{H}, whereas if it is low cost, it will generate e_{L}. We should think of these pollution levels as being dependent on the level of the fee. If r changes, both e_{H} and e_{L} will change.

The EPA's goal is to find a fee level that minimizes expected societal costs, assumed to be expected pollution control costs plus pollution damage. The problem is the EPA does not know whether the firm is high cost or low cost. This problem is illustrated in Figure 10.1, which shows marginal cost for the two types of polluters as well as marginal pollution damage. If the EPA chooses some emission fee level, say $\hat{r}$, emissions will be at the point at which marginal costs equal that fee: either $\hat{e}_{\mathrm{L}}$ or $\hat{e}_{\mathrm{H}}$. For the high-cost firm, we end up with too much pollution, with the shaded triangle A being the social loss. For the low-cost firm, we end up with too little pollution, with the social loss being the triangle B. Since there is a 50:50 chance of the firm being of one type or the other, the expected social cost of the fee $\hat{r}$ is the average area of the two triangles. The best level of the fee is the level that results in the smallest average loss (or area). One can see by inspecting Figure 10.1 that if $\hat{r}$ is raised, triangle A gets smaller while triangle B gets larger. There is clearly some best r that yields the smallest average of triangles A and B.

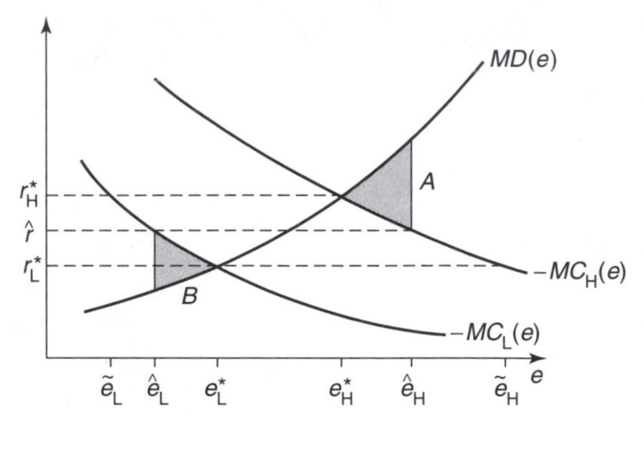

Figure 10.1 An illustration of regulating a firm with unknown pollution control costs, using an emission fee. $MC_H(e)$, Marginal costs of emitting, high-cost firm; $MC_L(e)$, marginal costs of emitting, low-cost firm; $MD(e)$, marginal damage from pollution emissions; e_L^*, e_H^*, efficient level of emissions, low-and high-cost firms; r_L^*, r_H^*, efficient emission fee, low- and high-cost firms; $\hat{r}$, best single emission fee; $\hat{e}_L, \hat{e}_H$, low and high-cost firms' response to $\hat{r}$; $\tilde{e}_H$, emissions from high cost firm when it claims to be low cost; $\tilde{e}_L$, emissions from low cost firm when it claims to be high cost.

The problem is that no single emission fee can eliminate the triangles. So why don't we just ask the polluter whether it is high or low cost? This is the obvious solution. However, it is easy to see that whatever the costs of the firm, it is in its best interests to claim to be a low-cost firm. This can be seen by looking at what a high-cost firm gains by claiming to be low cost. The emission fee will be set at r_L^* in Figure 10.1, which will allow the polluter to emit much more than it otherwise would ($\tilde{e}_H$), thus lowering its overall costs. A low-cost firm would never claim to be high cost since that would result in a high emission fee, higher total costs, and overcontrol of pollution.[3]

The ideal regulation in this case is one in which it is in the polluter's best interest to tell the truth. This is the incentive problem. Without knowledge of the important information the firm has, regulations will fail. To obtain the private information, the regulation must be designed so that it is in the polluter's best interest to help the regulator. The solution here is to somehow compensate a high-cost firm for admitting that it is high cost, so that telling the truth is preferred by both firms. We need a reward R for admitting to be a high-cost firm. The reward needs to be as low as possible (for the EPA's benefit) but still make telling the truth advantageous. In other words, for the high-cost firm

$$C_H(e_H^*) + r_H^* e_H^* - R < C_H(\tilde{e}_H) + r_L^* \tilde{e}_H \tag{10.3}$$

The left-hand side of the inequality is net costs when admitting to be high cost. The right-hand side is net costs from falsely claiming to be low cost.

In addition, the reward should not be so high as to encourage the low-cost firm to lie:

$$C_L(e_L^*) + r_L^* e_L^* < C_L(\tilde{e}_L) + r_H^* \tilde{e}_L - R \tag{10.4}$$

The left-hand side is net costs from truthfully admitting to be low cost; the right-hand side is net costs (including the reward) from falsely claiming to be high cost.

So the optimal regulation consists of two parts, an emission fee and a payment, both dependent on information provided by the firm. If the firm says it is high cost, it receives a payment R but must pay an emission fee r_H^*. If the firm says it is low cost, it receives no reward and must pay an emission fee r_L^*. This is a specific example of a class of regulations that involves a payment by the firm (positive or negative) that depends on emissions and stated type (costs). With the structure of costs and damages assumed here, it is always possible to find a regulation that induces firms to admit their true costs and results in a socially desirable amount of pollution.[4]

This problem illustrates the potential difficulties in designing regulations. This class of problem is known as the case of unknown characteristics or unknown "types" (in this case, the characteristic or "type" is whether the polluter is low cost or high cost). It is also known as the adverse selection problem.[5]

The basic point is that it is often not easy to simply direct a polluter to generate the right amount of pollution. It is necessary to recognize the different levels of information that may exist between the regulator and the polluter. Futhermore, if the regulations are not structured with the correct incentives, the polluter's response may be socially detrimental.

B. An Example of Water Pollution Regulation

The example of the previous section is somewhat stylized but not far from real-world applications. In fact, Alban Thomas (1995) studied water pollution regulation in a part of France, noting that the regulator (the Water Agency) was lacking one key piece of information necessary to efficiently control polluters: pollution control costs. The Water Agency would obtain that information by offering a pollution control subsidy that varied depending on control costs. The subsidy is the reward of the previous section. By observing the operation of the regulator and the polluters, Thomas was able to infer information about the objectives of the regulator, the efficient Pigouvian fee, and the costs of the polluters.

As with much of the "real world," the case examined by Thomas differs slightly from the example of the previous section. In the case study, there is an emission fee, but it is set by political factors outside the control of the Water Agency. Typically, it is lower than the Pigouvian fee. Thus the Water Agency supplements the emission fee by directly regulating the amount of pollution control equipment a polluter uses. However, to determine what the correct amount of pollution control equipment should be for a particular polluter, the Water Agency needs to know that polluter's marginal savings from polluting, information that is private to the polluter. To coax this information out of the polluter, the regulator offers to subsidize the purchase of pollution control equipment, with the level of the subsidy depending on the reported pollution control costs. If pollution control costs are reported to be high, the regulator is likely to require less abatement capital than when pollution control costs are reported to be low. Thus a straight question about costs would likely be met with all firms claiming to be high cost. A subsidy must be paid to those firms admitting to being low cost.

As in the model of the previous section, Thomas (1995) computes the smallest subsidy necessary to induce firms to truthfully reveal costs. Firms that admit their costs are directed to utilize the optimal amount of abatement capital and in return they receive a lump-sum subsidy for their cooperation.

Suppose the variable control costs for a firm are given by $C(Q,K,\theta)$ where Q is pollutant emissions, K is abatement capital, and θ is a parameter that shifts costs—a higher

θ means higher variable costs and higher marginal variable costs. But θ is private information to the firm and is not know by the Water Agency. The Water Agency's objective is to maximize weighted (by β) social surplus:

$$\max_{T,K} \quad \beta \{S(Q) + t\,Q - T\} - (1 - \beta)\{C(Q,K,\theta) + p_K\,K + t\,Q - T\} \qquad (10.5)$$

where $S(Q)$ is the consumer surplus associated with pollution levels Q and p_K is the price of capital services. Note in Eq. (10.5) that the objective of the Water Agency consists of two parts. The first term in braces is the total surplus accruing to consumers, consisting of consumer surplus [$S(Q)$] and pollution tax receipts ($t\,Q$), less subsidies to the polluter (T). The second term in braces is the costs of the firm, consisting of direct costs [$C(Q,K,\theta) + p_K K$], pollution tax payments ($t\,Q$) less subsidies (T). The Water Agency may weight consumer benefits more heavily than polluter costs, which is the reason for the β, which may be any number between 0 and 1. As written, Eq. (10.5) assumes the firm reveals θ and calls for the Water Agency to choose T and K so that surplus is maximized. The pollution fee, t, is not a choice variable but set exogenously.

Note in Eq. (10.5) that the Water Agency is not choosing Q. That is chosen by the polluter. In fact there are two requirements that must also hold while maximizing Eq. (10.5). One is that the firm chooses Q to minimize costs (direct costs plus emission fees):

$$C_Q(Q,K,\theta) = -t \qquad (10.6a)$$

where C_Q is the increase in variable costs associated with a unit increase in emissions. Variable costs would of course be expected to decrease with increases in Q. If there were no regulation of levels of K, the firm would choose K so that the price of capital equals the negative of marginal variable costs with respect to K.

$$C_K(Q,K,\theta) = -p_K \qquad (10.6b)$$

Variable costs would be expected to decrease with increases in K. Let $[Q_0(\theta),K_0(\theta)]$ be the levels of pollution and abatement capital we would expect to see with the emission fee but with no directives on K. Let $C_0(\theta)$ be the total costs, including emission fee payments, associated with $[Q_0(\theta),K_0(\theta)]$.

Returning to Eq. (10.5), we need the value of Q. Q is determined by the firm according to Eq. (10.6a). But there are two other requirements. For the firm to be willing to divulge its costs, θ, it must be assured that its costs will be lower than if the firm remained silent, incurring costs $C_0(\theta)$:

$$C_0(\theta) > C(Q,K,\theta) + p_K\,K + t\,Q - T \qquad (10.6c)$$

A second requirement is that telling the truth about θ must be more profitable than lying about θ. Let $H(\theta,\hat{\theta})$ be the firm's costs when its true type is θ but it tells the regulator it is really type $\hat{\theta}$. Telling the truth then requires that

$$H(\theta,\hat{\theta}) \geq H(\theta,\theta), \qquad \text{for all } \hat{\theta} \qquad (10.6d)$$

In other words, telling the truth results in lower costs than lying.

To sum up, the Water Agency tries to choose K and T to maximize social surplus [Eq. (10.5)]; given the constraints that the firm chooses Q [Eq. (10.6a)], the firm must be better-off reporting θ than not [Eq. (10.6c)], and lying must not be attractive [Eq. (10.6d)].

Of course it is entirely possible that the Water Agency will be better off remaining ignorant of the costs of some firms: the gain from learning θ may not offset the cost of obtaining the information (T). It turns out that there is some θ^* that separates the firms that admit their costs from the firms that refuse to participate. Firms with $\theta < \theta^*$ will tell the Water Agency their costs (θ) because the subsidy is sufficiently high. Firms with $\theta > \theta^*$ will choose to remain silent, leaving pollution and abatement capital at $[Q_0(\theta), K_0(\theta)]$.[6]

On the assumption that the Water Agency is acting optimally (which may be a big assumption), it is possible to infer both the variable costs as well as the weight β in Eq. (10.5) from a statistical analysis of actual pollution regulation. Thomas (1995) used a data set with information on subsidies, required abatement capital, emission fees, and emission levels for the Adour-Garonne Water Agency in southwest France. The data set contained 185 observations. Adopting a number of assumptions,[7] he was able to infer that the β in Eq. (10.5) was approximately 0.74 and that the existing emission fee was approximately half the Pigouvian fee. The β greater than 0.5 means that the Water Agency weighted the consumer's welfare more than the producer's welfare. The results also show that the chemical industry has the lowest pollution control costs, whereas the iron and steel industry has the highest. Recall that subsidies tend to flow to those industries truthfully admitting to being low cost.

C. Implications

Although the examples considered above may seem abstract, the lesson is very practical in the context of environmental regulation. The basic point is that it is often not easy simply to direct a polluter to generate the right amount of pollution. It is necessary to recognize the different levels of information that may exist between the regulator (the EPA) and the polluter. Furthermore, if regulations are not structured with the correct incentives, the polluter's response may be socially detrimental.

II. UNOBSERVED CONTROL COSTS: PERMITS OR FEES?

A. Emission Fees or Quantity Regulation?

One of the puzzles in the environmental economics literature of the past three or four decades is why economic incentives have not been used more when economists generally prefer them to direct regulation. A related issue concerns marketable permits and emission fees. Marketable permits are beginning to find acceptance in parts of the world. Emission fees, in contrast, are rarely used as real incentives for pollution control. Most applications are in Europe and the former Soviet Union, where fees are usually used to finance regulatory activities rather than provide incentives for pollution control. Economic theory suggests that a marketable permit system and an emission fee system should work equally well in controlling pollution. It is from this debate that Abba Lerner and Martin Weitzman[8] suggested that if there were uncertainty and it was necessary to set a regulation and live with it, perhaps marketable permits and emission fees would not do an equally good job of controlling pollution. This in fact turns out to be the case.

The situation we consider involves a regulator and a firm. The regulator has two choices of regulation for controlling emissions: an emission fee and a quantity regulation. The emission fee means that the polluter pays the fee on every unit of pollution, and thus sets marginal savings from polluting equal to the emission fee. The quantity regulation means that the firm emits exactly the quantity of pollution specified by the regulation. Since there is only one firm, there is no issue of trading permits among firms.

To be a little more concrete, suppose we have a polluter with a marginal savings from emissions function that remains uncertain to the regulator. The marginal savings function is not, however, unknown to the polluter. We also have a marginal damage function that for the time being we assume is known with certainty. Our question is, which is better, an emission fee or a quantity control? Figure 10.2 illustrates this. Shown is the marginal damage function (*MD*) as a function of emissions. Also shown is the marginal savings function (*MS*), also a function of emissions. However, to capture the regulator's uncertainty about marginal savings, low and high marginal savings curves have been plotted (MS_L and MS_H). The regulator knows the true marginal savings function must be one of these two but is unsure which. These low and high values average out at $\overline{MS}(e)$.

Note first that if there is no uncertainty or variability in marginal savings, then $MD(e)$ and $\overline{MS}(e)$ are the marginal damage and marginal savings functions. And in this case, there is no difference in terms of efficiency if the firm is told to produce e^* or told to pay p^* per unit of emissions (since in this latter case, the firm will emit where the fee equals marginal savings—e^*).

But with uncertainty, things change. If the regulator directs the firm to emit e^*, that is how much will be emitted, regardless of which marginal savings function is correct. If the regulator charges an emissions fee of p^*, the firm will emit e_H if the marginal savings function is high, and e_L otherwise. Although these *may* average out to e^*, we clearly do not have the perfect symmetry between price and quantity regulation that we had in the case of certainty.

So which is better, prices or quantities? To answer this, we must determine the inefficiency of these two types of regulations. Figure 10.3 shows a portion of Figure 10.2: the marginal damage function and the high marginal savings function. Also shown are the price and quantity levels that would be optimal without full knowledge of firm costs, p^* and e^*. Note that if it turns out that the true marginal savings is $MS_H(e)$, the best level of emissions would be e^{**}, where marginal savings and marginal damage functions in-

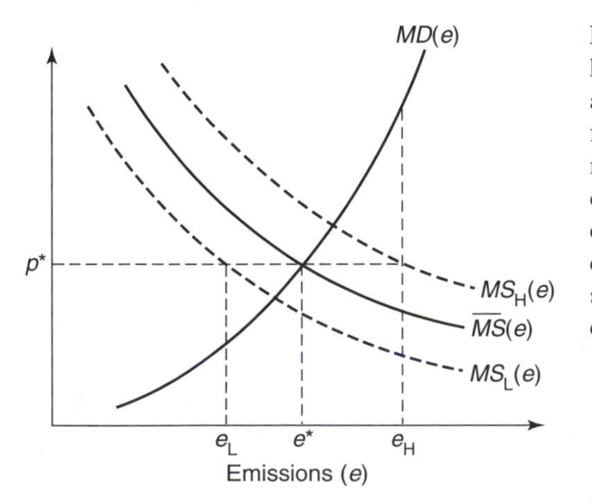

Figure 10.2 Fees vs. quantity regulations, unknown control costs. $MD(e)$, Marginal damage from emitting; $MS_H(e)$, marginal savings from emitting if firm is high cost; $MS_L(e)$, marginal savings from emitting if firm is low cost; $\overline{MS}(e)$, average marginal savings from emitting; p^*, optimal emission fee; e^*, optimal quantity regulation; e_L, emissions from emission fee if firm is low cost; e_H, emissions from emission fee if firm is high cost.

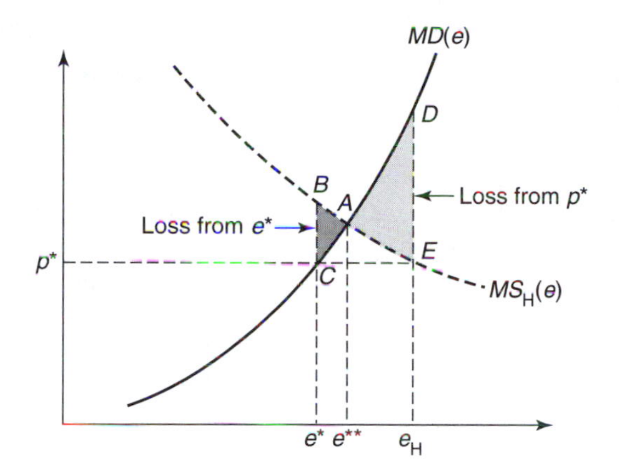

Figure 10.3 Losses if firm is high cost. $MD(e)$, Marginal damage from emissions; $MS_H(e)$, marginal savings from emitting if firm is high cost; e^*,p^*, optimal quantity and price regulations; e^{**}, optimal emissions if firm is known to be high cost; e_H, emissions from fee p^*.

tersect. Clearly, both e^* and e_H are less that perfect. The inefficiency associated with e^* is the dark shaded triangle, *ABC*, and the inefficiency associated with e_H is the lighter area, *ADE*. In this particular example, it looks as if the price regulation results in more inefficiency than the quantity regulation (since *ADE* appears to be larger than *ABC*).

Of course, Figure 10.3 tells only half of the story since we do not know whether the true marginal savings function is MS_L or MS_H. Thus we must examine the area of the welfare loss triangles if marginal savings turns out to be MS_L. We then add up the areas of the two loss triangles for an emission fee (one for MS_L and one for MS_H), weighting the two areas in our sum by the likelihood that marginal savings are low vs. high. This gives the *expected* welfare loss from an emission fee. We then do a similar thing for the case of a quantity regulation. Comparing the two *expected* losses tells us which is better, the emission fee or the quantity regulation (smaller losses are of course better).[9] Figure 10.4 puts this all together. The light shaded areas are the losses associated with an emis-

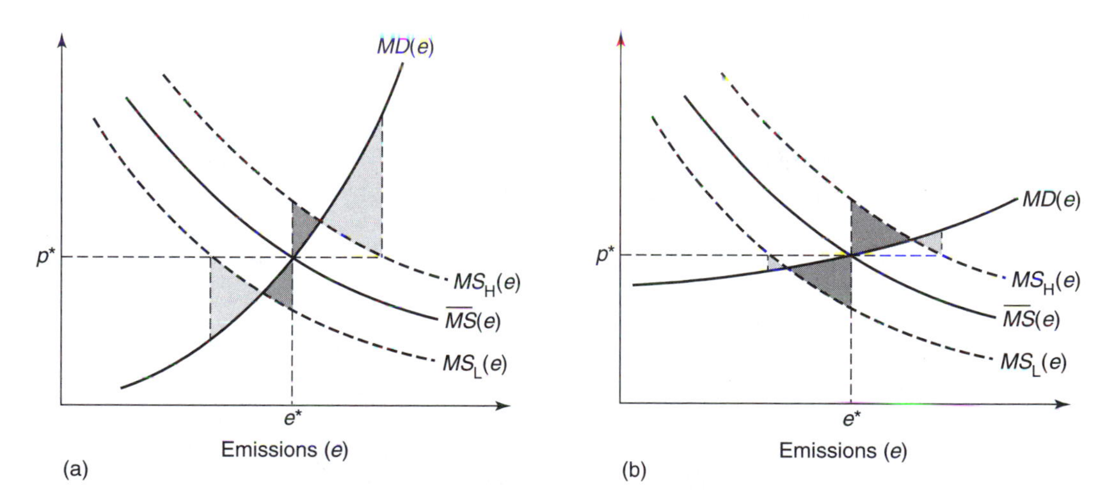

Figure 10.4 (a,b) Welfare losses from price and quantity control. $MD(e)$, Marginal damage from emissions; $MS_H(e)$,$MS_L(e)$, marginal savings from emitting for high (H)- and low (L)-cost firms; e^*,p^*, optimal quantity and price regulations; dark shaded area, inefficiency from emission fee; light shaded area, inefficiency from quantity control.

sion fee, whereas the darker areas are the losses associated with a quantity regulation. Whatever the likelihood of high vs. low marginal savings, it is clear that in Figure 10.4a a quantity regulation results in smaller welfare losses and is thus to be preferred; in Figure 10.4b the regulatory preferences are reversed: the losses associated with an emission fee (light shaded area) are considerably smaller than the losses from a quantity regulation (dark shaded area). The only difference between Figure 10.4a and 10.4b is the slopes of the marginal damage and savings functions.

So we see that there are situations in which an emission fee might be best and other situations in which a quantity regulation might be best. The relative slope of the marginal damage and marginal savings functions are key to this. Focusing on Figure 10.4a, mentally rotate the marginal damage function clockwise around the (e^*,p^*) point. Visualize the changing size of the loss triangles, noting that the loss from a fee is reduced and the loss from a quantity regulation is increased. As the marginal damage function becomes more horizontal, the loss from a quantity regulation becomes larger than that of an emission fee (something like Figure 10.4b). So a small slope for marginal damage favors the emission fee; a large slope favors a quantity regulation. Analogously, try mentally rotating the marginal savings curves clockwise in Figure 10.4a. The same thing happens—as the marginal savings function becomes steeper, the loss from the emission fee is reduced and the loss from the quantity regulation is increased. So a shallow slope for marginal savings favors a quantity regulation; a steep slope favors an emission fee. Keep in mind that marginal savings have a negative slope. By more steeply sloped, we mean the absolute value of the slope is greater. These two results can be put together into the following[10]:

Proposition (Weitzman). With uncertainty over marginal costs of emissions, quantity regulations are preferred if marginal damages are more steeply sloped than marginal savings from emissions; emission fees are preferred if marginal savings are more steeply sloped than marginal damages.[11]

This is an important result because it explains a good deal of the preponderance of quantity regulation in much of the world and also because of the policy implications for currently unregulated environmental problems, such as greenhouse gas accumulation.

For instance, much of health-based environmental regulations are based on the concept of a threshold of damage—no damage below a threshold and large damage above a threshold. In the vicinity of the threshold, the damage increases dramatically. This is roughly equivalent to very steep marginal damage in the vicinity of the threshold. This is illustrated in Figure 10.5, which shows three possible damage functions and the associated marginal damage functions. Linear damage (D_1) results in horizontal marginal damage. Quadratic damage (D_2) results in linear marginal damage. Damage that is very low until it reaches the vicinity of a threshold, T^*, and then rises rapidly (D_3) yields a marginal damage function that is very similar (though not necessarily identical)—low below the threshold and steeply sloped in the vicinity of the threshold.

Unless marginal savings are even more steeply sloped, the above proposition suggests that a quantity regulation will be preferred in the situation in which damages ex-

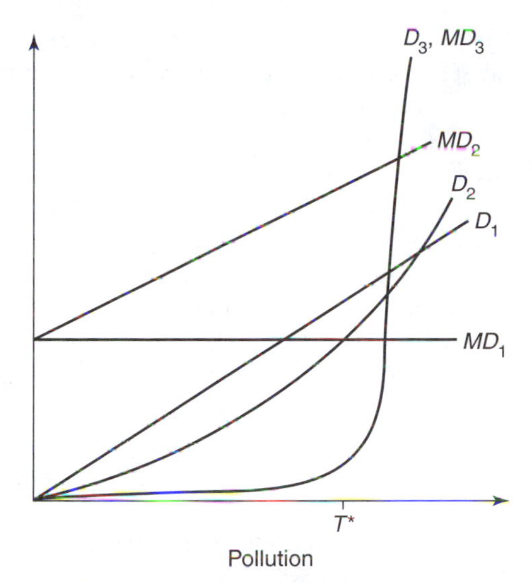

Figure 10.5 Three damage functions and their marginals. T^*, Threshold; D_1, MD_1, linear damage, constant marginal damage; D_2, MD_2, quadratic damage, linear marginal damage; D_3, MD_3, highly curved damage (and thus marginal damage) in vicinity of threshold, T^*.

hibit a threshold effect. And in fact that is what we see around the world—limits on the quantity of pollution emitted are much more common than emission fees.

Another way of grasping this is to view a quantity regulation as yielding a sure quantity of pollution with an unsure marginal cost of control. An emission fee yields a sure marginal cost of control but an unsure quantity of pollution. With steeply sloped marginal damage (e.g., thresholds in damage), a small error in the quantity of pollution can have very large effects on damage. (As an extreme example, a level of poison just over the lethal dose has much greater effects than an amount just under the lethal dose.) Erring on the control cost side has a relatively modest effect.[12]

On the other hand, if a pollutant is such that every unit has more or less the same marginal effect on the environment, marginal damage is constant and the proposition suggests that an emission fee is best. Over a broad range (though not necessarily a complete range), pollutants such as CO_2 may exhibit such constant marginal damage. This suggests that on efficiency grounds, an emission fee may be preferable to quantity regulation. There are of course many other issues that bear on the political desirability of one regulatory instrument over another, particularly in the context of global climate change.

Example: In the early 1980s, the U.S. Congress was considering ways to improve on the Clean Air Act, the primary law governing air pollution in the United States. As part of this effort, Congress established the National Commission on Air Quality to examine specific air pollution problems and recommend improved methods of regulation. One of the concerns of the Commission was air quality protection in the Southwest United States, an area with many national parks but also significant energy resources, the combustion of which would threaten regional air quality. There were two air quality goals in the region: protecting ambient air quality in national parks from combustion sources located nearby and preventing the buildup of regional haze due to region-wide emissions. The ambient air quality goal is specified in terms of a maximum concentration of pollutants within a park. The regional haze goal is simply to have as low a level of regional emissions as possible—there is no threshold. The uncertainty at work in this market relates to energy demand. If energy demand is high, marginal cost of emission control will be high; if demand is low, marginal control costs will be low.

Consistent with the theoretical analysis above, the first air pollution goal, the maximum ambient concentration, would appear to be best served with a quantity control on emissions. The second goal would appear to be best served by an emission fee, since there is no threshold effect. The issue is, what type of regulatory instrument might best suit these dual and conflicting goals? To answer this, a computer model of the relevant region of the United States was constructed and used to simulate the effects of different types of regulations, specifically a marketable emission permit system and a regional emission fee system for control of SO_2 emissions.[13] Further, the dual goals were assumed to imply a particular shape for marginal damage from emissions. The analysis concluded that a marketable permit system was approximately 20% lower in cost than an emission fee system except when marginal damages were close to linear. In this case, the fee systems resulted in 5–10% lower costs. Because damage was not thought to be linear, these results suggest that a marketable permits system is the best choice.

B. Hybrid Price/Quantity Regulations

The problem with a pure emission fee or a pure quantity regulation is that if it turns out to be cheap to control pollution, we would like the firm to undertake more pollution control. On the other hand, if it turns out to be very expensive, perhaps we can be a little more lenient. And we would like that flexibility built into a regulation. Unfortunately, pure price or quantity regulations do not have that flexibility. It is a relatively straightforward step, however, to propose a quantity regulation coupled with a subsidy for emitting less than the firm is required and a penalty for emitting more.[14]

This situation is shown in Figure 10.6. As before, marginal savings can be high or low and the regulator does not know which (the regulator knows the mean value of the marginal savings—$\overline{MS}$). The regulatory scheme is that the firm is told to emit e^* but if it

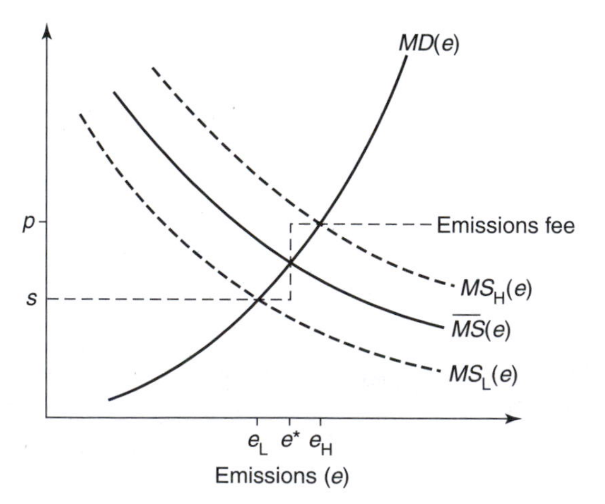

Figure 10.6 Permit system with penalty for overemitting and subsidy for underemitting. $MD(e)$, Marginal damage from emissions; $MS_H(e)$, $\overline{MS}(e)$, $MS_L(e)$, marginal savings from emitting: high-cost, average, low-cost assumptions; e^*, level of permit issuance; s, subsidy rate for underemitting; e_L, emission rate if firm is low cost; p, penalty rate for overemitting; e_H, emission rate if firm is high cost.

emits less, it will receive s for each unit of pollution under the limit, for a total payment of $s(e^* - e)$. This of course is the same as the firm having to *pay* $s(e - e^*) = se - se^*$, which is a negative quantity if $e < e^*$. Similarly, if the firm emits more than e^*, it will pay a penalty—a total of $p(e - e^*) = pe - pe^*$. In the first case the marginal cost of emissions is s; in the second case, the marginal opportunity cost of emissions is p. This is the reason for drawing the emission fee in Figure 10.6, which has the jag at e^*. In Figure 10.6 $s < p$.

In Figure 10.6, if the marginal savings turn out to be low (MS_L), the firm will choose the subsidy. Why wouldn't the firm just emit e^* as told? At e^*, the marginal saving from emitting pollution is quite low—lower than s. Thus it is advantageous to receive a subsidy to cut back emissions, since that subsidy exceeds the marginal cost of cutting back emissions. Emissions are cut back to the point at which marginal savings equal the subsidy (e_L). Analagously, if marginal savings turn out to be high, at e^* the marginal savings from emitting a unit of pollution are high—higher than the penalty from overemitting. Thus it is cost effective to emit more than allowed and just pay the penalty. Emissions will occur when marginal savings equal the penalty (e_H).

What is the inefficiency of such a scheme? Remarkably, in the simple example of Figure 10.6, there is no inefficiency. No matter what the true marginal savings, the firm will emit exactly the right amount. Of course if marginal savings can take on more than two values, which is probably more common, we cannot achieve complete efficiency with this approach. Intuitively, since we have more flexibility with this approach than with pure fees or quantities, we must be able to do better than a pure emission fee or a pure quantity regulation. This is in fact the case.

One way to understand this result is to look at the complete solution to our problem. That is to present the marginal damage schedule to the firm and tell the firm to emit wherever the marginal damage schedule crosses the true marginal savings function (which is unknown to the regulator but known to the firm). A pure quantity regulation can be viewed as approximating this marginal damage function with a vertical line through e^*. A pure emission fee can be viewed as approximating this with a horizontal line through p^*. The hybrid system can be viewed as approximating the marginal damage with a step function (shown as a dashed line in Figure 10.6). Clearly we can get a better approximation with a step function.

III. OBTAINING PRIVATE CONTROL COST INFORMATION

In the last section, we assumed the regulator simply could not know the marginal cost or marginal savings to the firm from emitting. We now take the next logical step and assume the regulator can ask the firm what its marginal costs of emitting are. Now we have to be concerned about truthfulness.

The regulatory setup now consists of two steps. First the polluter reports to the regulator the marginal savings from emitting—$MS_R(e)$. Then the regulator issues a regulation, and the firm responds by choosing an amount to emit.

Let us look at the price and quantity instruments in this framework. Figure 10.7 shows the true marginal savings function (*MS*—unknown to the regulator) and the true

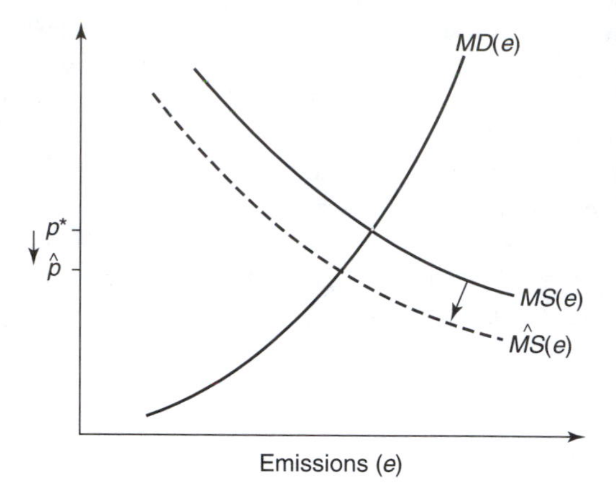

Figure 10.7 Emission fee encourages underreporting of marginal savings. $MD(e)$, Marginal damage of emission; $MS(e)$, marginal savings from emissions; $\hat{MS}(e)$, reported marginal savings from emissions; p^*, efficient emission fee; $\hat{p}$, emission fee from misreported marginal savings.

marginal damage function (*MD*—known to the regulator). An emission fee will be set at the point at which marginal damage and marginal savings are equal. Clearly, if the polluter reports a low marginal savings function ($\hat{MS}$), the emission fee will be reduced, which is in the polluter's best interest. Thus with an emission fee, there is an incentive for the polluter to understate the marginal savings from emitting. This means that the cost of reducing emissions is understated—the firm says it can cheaply reduce emissions.

Figure 10.8 shows the other case, that of a quantity regulation. In this case, it is clear that it is in the polluter's best interest to exaggerate the marginal savings from emissions to increase the amount of pollution the firm is allowed to emit.[15]

So clearly neither a fee nor a quantity instrument will induce truthful revelation of costs. It turns out that a hybrid of the two works quite well, at least in this simple context.[16] To examine this hybrid, we must add more than one firm, so that we may have a market for emission permits with a known market price. The simplest way to do this is to assume we have many identical firms. After the regulator receives information on mar-

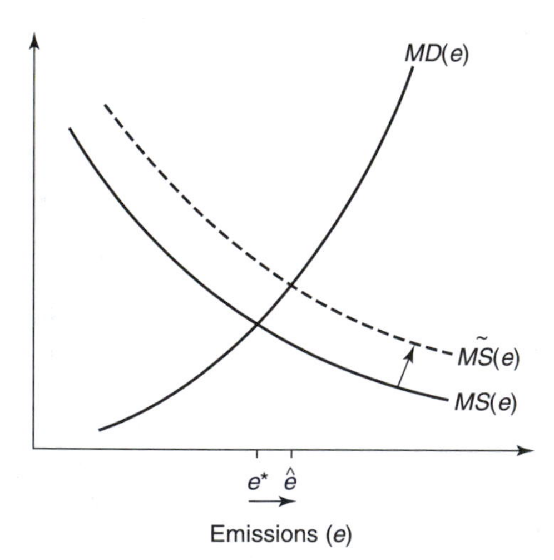

Figure 10.8 Emission permit encourages overreporting marginal savings. $MD(e)$, Marginal damage from emissions; $MS(e)$, marginal savings from emissions; $\tilde{MS}(e)$, reported marginal savings from emissions; e^*, efficient level of emissions; $\tilde{e}$, emission from misreported marginal savings.

ginal savings functions, the regulator auctions off a certain number of marketable emission permits and announces a subsidy rate for firms emitting less than allowed by the permits they hold. Thus this is similar to the hybrid system in the previous section except that overemitting is not allowed.

Figure 10.9 illustrates what the regulator will do. The regulator receives reports of marginal savings functions from each firm and then aggregates these to a marginal savings function for the entire industry. This is shown in Figure 10.9 as $\overline{MS}(e)$—the *reported* aggregate marginal savings function. This intersects the marginal damage function at (L^*,s^*). The regulator then auctions off L^* marketable permits to pollute and announces the subsidy rate s^*.

Suppose the polluters lied when telling the regulator their costs. Suppose they understated their marginal savings—their true marginal savings function is MS_H. With only L^* permits available, the various firms will compete for those permits, driving the market price up to p_H, the value of $MS_H(L^*)$. Since the subsidy is lower than this, none of the firms will choose to receive a subsidy to emit less than the number of permits held. If they had told the truth about the marginal savings functions, the subsidy rate would be higher (s_H**), the number of permits would be larger, and thus the market price lower.

Now consider the other case, that the polluters' true marginal savings functions were lower—$MS_L(e)$. In this case, the market price might end up being p_L, using the same logic as before. But this is lower than s^*. Consequently, firms will choose to emit less than the number of permits held. Every permit held costs p_L but yields s^*. As long as $s^* > p_L$, there will be excess demand for permits. This will drive the price of permits to s^*. But if the firms had told the truth, the subsidy rate would be even lower, at s_L**, which would result in a lower permit price. Thus by lying, the polluters have increased the price of permits in the market.[17]

In summary, not telling the truth about the marginal savings function has the effect of increasing the price of emission permits. The lowest permit price is obtained when the firms truthfully reveal their marginal cost functions.

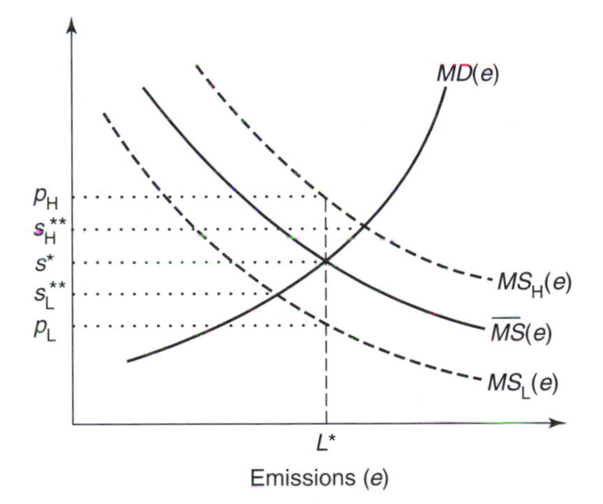

Figure 10.9 Hybrid permit-subsidy system. $\overline{MS}(e)$, Reported aggregate marginal savings from emitting; $MS_H(e)$, true aggregate marginal savings if underreporting; $MS_L(e)$, true aggregate marginal savings if overreporting; L^*, number of permits issued; s^*, subsidy rate with reported $\overline{MS}(e)$; s_L**, subsidy rate with reported $\overline{MS_L}(e)$; s_H**, subsidy rate with reported $MS_H(e)$; p_H, permit price when $\overline{MS}(e)$ is reported but $MS_H(e)$ is true; p_L, permit price with no subsidy when $MS(e)$ is reported but $MS_L(e)$ is true.

SUMMARY

1. Regulation problems involving asymmetric information typically are associated with the regulator having less information than the polluter.

2. Adverse selection problems involve polluters having hidden characteristics (hidden from the regulator) such as their cost of pollution control. This makes efficient regulation difficult.

3. When a regulator must make a decision about the level of an emission fee or marketable permit system with uncertainty on the marginal cost of pollution control, the relative slopes for marginal savings and marginal damage functions determine which instrument is more desirable. If marginal damage is more steeply sloped than the marginal savings from pollution, marketable permits are preferred; otherwise, emission fees are preferred.

4. A marketable permit system with a subsidy for emitting less than permitted and a penalty for emitting more than permitted can perform better than either an emission fee alone or a marketable permit system alone, when marginal savings are uncertain to the regulator.

5. If a polluter has private costs about marginal savings from pollution, an emission fee will cause the polluter to understate marginal savings. A marketable permit system will cause polluters to overstate marginal savings. A marketable permit system coupled with a subsidy for emitting less than permitted will induce telling the truth about pollution control costs.

PROBLEMS

1. In the hybrid fee-subsidy scheme described in Figure 10.6, suppose there are two additional marginal savings functions, one even higher than MS_H and one lower than MS_L. In this case, graphically show the inefficiencies associated with a fee-subsidy system.

2. Consider the case of a regulator and a single polluter. Suppose the regulator knows the marginal damage from pollution but is unsure about the firm's marginal savings from emitting. The regulator asks the firm to reveal its marginal savings from emitting schedule. Regulators know that if they use an emission fee, the firm has an incentive to lie about its marginal savings. Regulators also know that if they use a permit system, lying may still result. So the regulator announces that after being told the firm's marginal savings, the regulator will flip a coin to determine whether an emission fee or an emission permit will be used to control pollution. In this case, will the firm have an incentive to truthfully reveal its marginal savings from emissions? Why or why not? [*Hint*: Assume a firm is low cost; look at the possible costs or cost savings from telling the regulator it is high cost.]

3. Suppose the total cost of controlling the pollution in Bangkok is given by $TC = (3 + r)q^2$ where q is the amount of emissions controlled. Uncontrolled, there would be 2 units of emissions. Thus $q = 2 - e$, where e is emissions. The variable r is unknown to the pollution control board. All they know is that it could take the value of either $r = 0$ or $r = 4$, and with equal likelihood. Marginal damage from emissions is given by $MD(e) = 4e$.

   a. Write the total cost of pollution control in terms of e. Graph this total cost as a function of e.

b. Graph the marginal damages from emissions and the expected value (i.e., average over the two possible rs) of the marginal savings from emissions. Be as accurate as you can.

c. What level of emission fee or emission permit should be chosen, not knowing what value r will take? Show your answer on the graph.

d. Suppose after you have set the fee or permit in part (b). it turns out that $r = 4$. Show the deadweight loss from the permit and fee, assuming these instruments cannot be changed. Which instrument appears to be better?

***4.** Suppose we have two polluters that have a hidden characteristic, θ. The θ does not have to be the same for the two firms. Assume that θ can take on one of two values: 1 or 2. These two firms emit pollution, with marginal savings functions equal to $MS(e,\theta) = 1 - \theta e$. Total savings from polluting for each firm are $S(e,\theta) = 1 - \frac{(1 - \theta e)^2}{2\theta}$. Damage from pollution is $D(e_1+e_2) = (e_1 + e_2)^2/2$ with a marginal damage of $(e_1 + e_2)$.

a. Suppose the regulator knew the value of θ for each of the firms, θ_1 and θ_2. For all possible combinations of θ_1 and θ_2, what would be the optimal amount of pollution for each firm: $e_1^*(\theta_1,\theta_2)$, $e_2^*(\theta_1,\theta_2)$?

b. Now assume the regulator does not know θ but asks each firm its true θ. After receiving those reports from each firm, each firm i will be charged an amount, $T_i(e_i,\theta_i)$, based on the reported θ_i, the report by the other firm, θ_j, and actual emissions, e_i:

$$T_i(e_i,\theta_i) = D[e_i+e_j^*(\theta_1,\theta_2)] - S_j[e_j^*(\theta_1,\theta_2),\theta_j]$$

where i and j are the two firms. The firms know this before they report their values of θ. Show that it is in the best interest of each firm to tell the truth about θ and also to emit the right amount of emissions, e^*. [*Hint*: Prove in general or enumerate for possible θ_i,θ_j].

NOTES

1. This issue is considered generally in an advanced review paper by Baron (1989). See also Lewis (1996) and Sappington and Stiglitz (1987).
2. This is generally known as the problem of "mechanism design." Our goal is to design a way (a mechanism) to regulate polluters so that they behave in a socially desirable manner, given that we have imperfect control over the polluters.
3. It turns out the incentives are different when a marketable permit is used for regulation.
4. Dasgupta et al. (1980) show this. They also point out that if costs are interrelated among firms or if damages are more complex, it may not always be possible to find such a regulatory setup (called a "mechanism").
5. One of the classic examples of adverse selection (and that illustrates the origin of the term) is the used car market. Typically, some used cars have hidden defects (we call these cars "lemons") and some do not ("peaches"), and both are offered for sale in the same market. Rational consumers would assume that the car they are considering buying may have hidden problems and thus shade the price they would pay downward. Thus the price of cars will be lower. This lower price will discourage owners of good quality cars from offering theirs for sale. This tends to result in the used car market having a high number of cars with hidden defects relative to the overall population of used cars. This is the "lemons problem" highlighted by Akerlof (1970). The problem of lemons can be so severe as to cause the market to fail to form to begin with,

as often happens with insurance markets in which adverse selection is significant (e.g., private unemployment insurance).

6. For these firms that do not reveal θ, the Water Agency must calculate expected social surplus, based on the agency's assessment of the probability of different types occurring.
7. One of the major assumptions in his paper was that consumer surplus was linear in Q, which implied that the marginal damage from pollution is constant.
8. The definitive treatment of this issue is by Weitzman (1974), apparently motivated by a suggestion of Lerner (1971). Weitzman's work was criticized, and extended by Laffont (1977), among others. Weitzman generalized his results in a later paper (Weitzman, 1978).
9. We have glossed over the point that the optimal price or quanitity regulation in the case of uncertainty may not be the same as for the case of certainty. In other words, if p^* and q^* are best with certainty, they may not be best in the case of uncertainty, even if average costs and benefits under uncertainty are the same as under certainty. The best price regulation would be the emission fee that makes the expected welfare loss (the weighted average of the two triangles) smallest. For this graphic exposition, this issue can be ignored.
10. This result is due to Weitzman (1974) and has application beyond environmental regulation (e.g., to import tariffs vs. quotas). For a specifically environmental version, apparently developed independently, see Adar and Griffin (1976).
11. Some qualifiers are necessary here, particularly the assumption that marginal savings and marginal damage are linear and that uncertainty is in the intercept of these linear functions.
12. Even when damages lack such a threshold, regulators that are unfamiliar with economics often prefer the sure quantity of pollution to the sure marginal cost of control (and unsure quantity of pollution).
13. The model and the results are described in several reports and papers, including Kolstad (1986, 1990).
14. Roberts and Spence (1976) first proposed this scheme.
15. Bulckaen (1997) points out that the incentive to distort reported marginal savings functions is greatly diminished if the regulator requries the firm to behave according to the marginal savings function reported. When the firm reports a marginal savings function, the regulator determines the optimal emission fee (t^*) as well as the emissions the firm should generate if the reported marginal savings are true (e^*). The regulator then not only charges the emission fee t^* but also requires the firm to emit e^*. This of course is more complex than the arrangement in which the reported costs are used only to set the fee level or the number of permits issued.
16. The result and model presented here are due to Kwerel (1977). It is one example of a much larger set of methods for achieving a similar end. Dasgupta et al. (1980) discuss a more general approach to the same problem as does Spulbur (1988).
17. It should be pointed out that if other firms are not being truthful about their costs, then telling the truth is not necessarily the best strategy. This is a weaker notion of best: telling the truth is optimal for a firm if other firms are telling the truth. Telling the truth is a Nash equilibrium.

11 AUDITS, ENFORCEMENT, AND MORAL HAZARD

In the last chapter we were concerned with a very specific type of asymmetric information: regulating polluters when characteristics of the polluter are unknown to the regulator. We now turn to a slightly different case of asymmetric information, one in which the polluter takes some action that is hidden from the regulator.

One example of this type of problem involves firms in which emissions are not directly observable by the regulator. Thus the unobserved action is emission levels. A case would be area sources of pollution (automobiles, agriculture), in which it is too costly to monitor individual polluters. Is there a form of regulation that can work when such an important variable is unobserved? Related to this, suppose firms self-report emissions. If it is costly to verify such reports, what kind of enforcement mechanisms, including spot audits, can most efficiently ensure truthful reporting of emissions? These are all examples of unobserved polluter actions. Our concern in this chapter is in developing regulations that can effectively deal with this sort of asymmetric information.

A related question that we will take up at the end of the chapter concerns dynamics. There are specific problems of asymmetric information that arise when dynamic regulation is concerned. For instance, if the reward to a firm from investing in cost-reducing research and development is simply a tightening of the regulation, the incentive to undertake such research is diluted.

I. A SIMPLE MODEL OF INCENTIVES IN ENVIRONMENTAL REGULATION

As we did in the previous chapter, we will simplify Figure 8.1 reducing it to a two-agent model consisting of an environmental regulator (the "EPA") and a firm (the "polluter"). The EPA is not privy to detailed information on the polluter's operations, such as how much pollution the firm is generating. For some of this information, the EPA must rely on the firm's statements and reports. But the EPA will not know whether the firm is telling the truth.

In the last chapter we considered the case of adverse selection, in which the EPA is uncertain whether the polluter can control pollution cheaply. We now turn to the problem in which the polluter takes steps to control pollution but these steps are unobservable. For instance, the polluter must make efforts to maintain pollution control equipment, yet the EPA cannot know whether this is being done properly (without constant monitoring, which may be expensive). The problem we consider involves no uncertainty about costs or other characteristics of the firm but rather uncertainty about some action the firm is taking. The typical example is *effort* in controlling pollution. Effort by the polluter is costly and unobserved, and often significantly affects pollution levels. For instance, if a regulation requires the installation of certain pollution control equipment it is very difficult to also require that it be maintained diligently. Maintenance, of course, is important to obtaining the intended amount of pollutant emissions, but good maintenance is hard to monitor, observe, and enforce.

The set-up will be similar to the adverse selection case except that this time pollution abatement (a) will be monitored by the EPA. The firm will combine abatement activities with maintenance and operation effort (f) to generate pollution levels. Higher levels of effort will be more costly to the polluter but will also reduce pollution levels. We will write the pollutant emissions as $e(a,f)$. The EPA does not observe e directly; rather it can see what abatement equipment is in place—a. But it is pollution (e) that causes damage.

The EPA will choose a regulation to control the polluter. In the last section, we considered an emissions fee. To diversify a bit, here we will assume the EPA uses an emissions standard. (We could carry out the analysis with any type of regulation.) Specifically, assume the EPA specifies a required level of abatement, $\bar{a}$.

The polluter's problem then is to choose a level of effort, f, to minimize costs:

$$\min_{f} C(\bar{a},f), \qquad f_{\mathrm{L}} \leq f \leq f_{\mathrm{U}} \tag{11.1}$$

Note that abatement, a, is specified by the EPA, so the polluter has no choice. The polluter chooses effort and it is obvious that the polluter will choose the lowest level of effort, f_{L}. This is because the costs of effort are lowest at this level and it will not affect the relationship with the EPA.

The EPA's problem is to choose a to minimize social costs:

$$\min_{a} C(a,f) + D[e(a,f)] \tag{11.2}$$

The problem is that f affects e, which affects D, but the EPA has little control over f. Suppose the EPA had full knowledge of f. In such a case there are optimal levels of abatement and effort, call them a^* and f^*. Fixing effort at f^*, Figure 11.1 shows how the marginal cost of abatement curve and the negative marginal damage from abatement curve might look. Note that the abscissa is abatement and that the more abatement there is, the higher the marginal costs of abatement are. Furthermore, increasing abatement decreases pollution and thus damage. Marginal damage is negative and approaches zero as abatement becomes large. Shown in Figure 11.1 is the "first-best" outcome of a^*, assuming the polluter undertakes the desirable amount of effort, f^*. However, this will generally not be the case. The firm will take the minimum amount of effort, f_{L}, that will result in lower

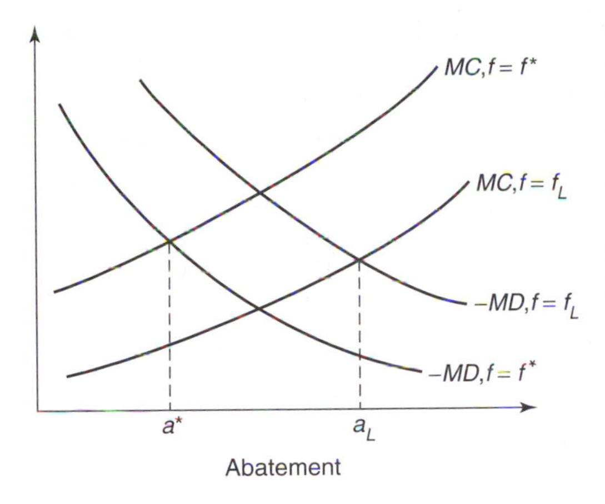

Figure 11.1 Illustration of unobserved effort. *MD*, $f = f_L$, Marginal damage, effort at lowest level; *MD*, $f = f^*$, marginal damage, effort at optimal level; *MC*, $f = f_L$, marginal cost of abatement, effort at lowest level; *MC*, $f = f^*$, marginal cost of abatement, effort at optimal level; a^*, efficient level of abatement; a_L, second-best level of abatement, with effort at lowest level; f^*, efficient level of effort.

marginal costs but also higher levels of pollution and higher marginal damages. This is shown in Figure 11.1. The optimal "second-best" level of abatement with minimal effort being exerted is a_L, which is clearly larger than a^*. Thus to compensate for low effort, the EPA must require extra abatement. This of course will not fully compensate. Control costs and damages will likely be higher.

How can we circumvent this problem? There are a number of solutions, a few of which we will mention here. One is to identify a variable that is related to effort, perhaps imperfectly. This might be full measurement of pollution levels (which would solve the problem). Or it might be imperfect measurements of pollution, perhaps obtained by spot checks on emisson levels. Alternatively, observable staffing levels in maintenance might have some correlation with actual maintenance effort.[1] Or the firm could be made to pay a fee related to pollution damage.

The bottom line is that if a polluter can take secret actions that impact pollution levels, it may be difficult to solve the pollution problem completely. Some way must be found to induce the firm to take the socially desirable level of effort. This is once again an incentives problem. Regulations must be designed to align with the incentives of the polluters. This particular problem of hidden actions is known more generally as the problem of moral hazard.[2]

II. MONITORING EMISSIONS

As the world makes progress in solving some environmental problems, the problems that remain are typically the more difficult ones. Thus controlling large stationary sources of pollution is the first target of regulators. One of the most difficult types of sources to control is dispersed small polluters that individually contribute modestly to pollution but collectively have serious environmental repercussions. This would include individual farms that contribute to water pollution through agricultural runoff (pesticides, fertilizers, soil, animal wastes), automobile pollution in urban areas, and small generators of waste—households generating solid waste or small firms generating hazardous waste. In all of these cases, the problem is fundamentally one of monitoring emissions. It is either too costly or too difficult for the regulator to learn what individual emissions are. We will consider two variants of this problem. In the first, polluters have different impacts on ambient concentrations and we examine a method of regulating individual polluters based only on observed ambient concentration. In the second, we are interested in encouraging socially desirable behavior, such as disposing of waste in a safe repository. It is necessary to reward firms for safe disposal rather than penalizing them for unsafe disposal.

A. Regulation with Unobserved Emissions

The first case we consider involves a number of firms generating emissions that lead to ambient concentrations of pollution. We observe ambient pollution concentrations but do not observe emission from individual polluters. The point is that it is not impossible to monitor individual polluters, just too expensive. In other words, for practical purposes, the polluting actions of individual polluters are hidden to the regulator. This is basically a moral hazard problem in that firm actions are hidden. The question is whether there are regulatory mechanisms that can work in this situation.

There is a regulatory design (in fact, there are quite a few) that works in this case. We consider a simple one.[3] Since the problem can quickly become complex, we will consider the case of two polluters and a single environmental monitoring point. To add a little realism, suppose that we have two farmers, both bordering a river (Figure 11.2). One farmer (Bush) raises chickens and the other (Dukakis) grows vegtables such as tomatoes and Belgian endive. Both contribute pollution through runoff to the river, although in different ways. We have one downstream monitoring point. Let p be the pollution level downstream and let e_i and a_i (for i = D, B) be the emission levels and transfer coefficients for Dukakis (D) and Bush (B).

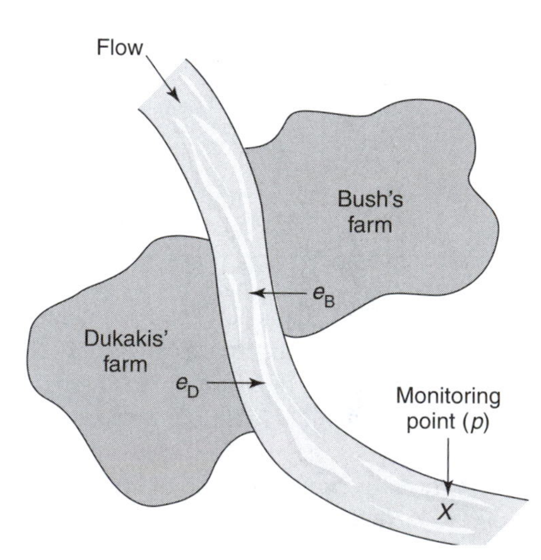

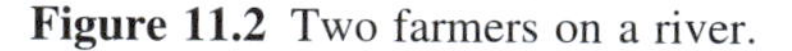
Figure 11.2 Two farmers on a river.

This is quite similar to the problem considered in Chapter 9. As in that chapter, the desirable amount of pollution is associated with emissions from each farm that equate marginal savings from emitting to marginal damage in emission terms:

$$-MC_i\ (e_i) = MDE_i\ (e_i) = a_i\ MD(p), \qquad \text{for } i = D,B \tag{11.3}$$

The compounding factor here is that the regulator cannot observe e_i, only p. Let p^* be the optimal amount of pollution, i.e., the amount of pollution that balances costs and damages as in Eq. (11.3). Consider the following tax scheme for i = D or B:

$$\text{Tax}_i = t(p - p^*) \tag{11.4}$$

Note that if the ambient pollution levels exceed p^*, each farmer pays a tax in proportion to the size of the exceedence. If the pollution levels are less than p^*, the farmers pay a negative tax—they are paid a subsidy. What will each farmer do?

The costs of each farmer i are given by

$$\begin{aligned}\text{Costs}_i\ (e_i) &= C_i\ (e_i) + t\ (p - p^*)\\ &= C_i\ (e_i) + t\ (\text{a}_\text{D}e_\text{D} + a_\text{P}e_\text{P} - p^*)\\ &= C_i\ (e_i) + t\ a_ie_i + t\ (a_je_j - p^*)\end{aligned} \tag{11.5}$$

In Eq. (11.5), the j refers to the other farmer (i.e., if i = D, then j = B). Notice that the last term in Eq. (11.5) is independent of farmer i's emissions. It is just a constant. Thus farmer i does not take it into account in setting emission levels. In fact, from Eq. (11.5) we see that farmer i will minimize costs by choosing emissions such that

$$-MC_i\ (e_i) = a_i\ t \tag{11.6}$$

Now we need to compare Eq. (11.3) and Eq. (11.6). Equation (11.3) is what is socially desirable. Eq. (11.6) is what the farmers will do. We can bring those together if we set t such that

$$t = MD(p^*) \tag{11.7}$$

So we see that each farmer is charged the same amount per unit of pollution, an amount equal to marginal damage.

You might say that this looks the same as the situation we considered in Chapter 9. The environmental problem is the same. The difference (a subtle one) is that instead of an ambient-differentiated emission fee, we have a fee based purely on observed levels of ambient pollution. And without observing emissions, the two farmers will cut back emissions to the socially desirable level.

This may seem a little counterintuitive. How can a charge paid by a motorist in London, based only on observed air pollution levels in London, encourage that motorist to reduce emissions? The answer lies in Eq. (11.5). Clearly if we charged motorist i the amount $t\ a_i\ e_i$, we would expect pollution control as in Eq. (11.6). Under the system here we are charging that plus a fixed amount, $t(a_je_j - p^*)$, which should not distort behavior. The total just happens to be $t(p - p^*)$.

The credibility of such a system breaks down when there are very many sources. In such a case, polluters may have to pay (or receive) huge sums of money, corresponding to the total incremental damages from all sources of pollution. Politically, this is unlikely to occur. It would seem that this system is most likely to work when there are only a small number of sources.

B. Midnight Dumping and Deposit–Refunds

The previous case concerned moral hazard—emissions were unobservable. We now consider a different though closely related problem of moral hazard. For simplicity, suppose a polluter can do two things with emissions—one socially desirable and easy to monitor and the other socially undesirable and difficult to monitor. The best example is the generation of solid waste. A generator has two options (more or less)—to dispose of the wastes "safely" in a regulated facility or to surreptitiously dispose of the wastes, e.g., dump the wastes illegally in the middle of the night. The first option is more costly to the firm, thus incentives are to "midnight dump." The second option, "midnight dumping," is more costly to society and also more difficult to monitor.[4]

A standard approach to discourage the generation of hazardous wastes would be to tax waste disposal at the marginal social cost of that disposal. In this case, that would have the unfortunate side effect of encouraging illegal dumping.

One solution is to reward proper disposal of wastes. Instead of taxing disposal, proper disposal (which is verifiable) is subsidized. Thus proper disposal becomes more attractive to the polluter than illegal disposal (which is not verifiable).

The problem we are discussing is a problem of moral hazard. Taking care to dispose of wastes properly involves a hidden action. We cannot observe when wastes are improperly discarded. If we are lucky, we can observe proper disposal, though this is not always the case. For instance, with the generation of smoke, proper disposal involves conscientious operation of the pollution control equipment. This effort is not verifiable. On the other hand, disposal of hazardous wastes in a properly constructed repository (also know as a dump) is verifiable—other, less desirable, ways of disposing of hazardous wastes are not verifiable.

In a world of perfect information, we would choose to tax the illicit disposal at a higher rate than the "safe" disposal, though both would be taxed at a positive level since both involve costs to society. The positive taxes on both would discourage the production of wastes. However, since we cannot tax illicit disposal, the best we can do is subsidize safe disposal, assuming it is possible to verify safe disposal. Safe disposal must be seen as more profitable to the polluter than illicit disposal.

Let us start by supposing that a firm generates wastes (w), and disposes of waste. Disposal is unobservable except in the case in which it is done safely (as in a landfill). The costs to the firm are $C_s(w)$ if the firm disposes of waste safely and $C_i(w)$ if the firm dumps waste illicitly. Clearly for any w, $C_i(w) < C_s(w)$; otherwise, there is no issue. The question is, what should we charge the firm for disposing of its wastes safely? Our primary goal is to prevent illicit dumping. Suppose we provide a subsidy for safely disposing of wastes. What should the subsidy be? Call the subsidy s. Total costs to the polluter are given by

$$TC(w) = \begin{cases} C_s(w) - sw & \text{if wastes are safely discarded} \\ C_i(w) & \text{otherwise} \end{cases} \quad (11.8)$$

Let w_i be the amount of wastes generated if illicit disposal is the choice of the firm and w_s be the amount of wastes generated if safe disposal is chosen. Clearly it is necessary to choose s so that

$$C_s(w_s) - sw_s < C_i(w_i) \tag{11.9}$$

Basically, we want safe disposal to be cheap. Obviously, if s is zero, Eq. (11.9) is not satisfied and illicit dumping takes place. If the subsidy is negative and equal to marginal damage (a pollution fee), the problem is exacerbated since this makes safe disposal even more expensive. The only solution is $s > 0$: the polluter must be subsidized to pursue safe disposal.

A subsidy for safe disposal solves part of our problem; set at the proper rate, the subsidy should discourage illicit dumping, which has much higher social costs than safe disposal. However, making waste disposal cheap encourages the generation of wastes. It is thus desirable (though not always possible) to identify some observable action on the part of the firm that is correlated with waste generation but unaffected by the choice of disposal methods. This action can then be taxed as a way of reducing waste production.

Suppose the generation of wastes (w) is observable but dumping is unobservable except when it is done safely. Damage from illicit dumping is $D_i(w)$ and damage from safe dumping is $D_s(w)$. It must be that $D_i(w) > D_s(w)$, otherwise there is no issue. The solution to this problem is to set a tax on waste generation, t, and a unit subsidy on waste disposal, s, so that polluters safely dispose of wastes and generate the right amount of waste. Total costs for the polluters are given by

$$TC(w) = \begin{cases} C_s(w) + tw - sw & \text{if wastes are safely discarded} \\ C_i(w) + tw & \text{otherwise} \end{cases} \tag{11.10}$$

We must next determine how much waste the polluter will generate. Let w_s and w_i be, respectively, the amount of waste generated if safe disposal and illicit dumping are pursued in the presence of a tax on waste generation. From Eq. (11.10), we can see that these two levels of waste are determined by

$$\begin{array}{lll} w_s(t-s): & MS_s(w_s) = t - s & \text{if safe disposal} \\ w_i(t): & MS_i(w_i) = t & \text{if illicit disposal} \end{array} \tag{11.11}$$

The last question we ask is what s and t should be? Basically, s should be set so that costs are lower with safe disposal than with illicit disposal (a non-marginal condition):

$$C_s[w_s(t-s)] + (t-s)\, w_s(t-s) \leq C_i[w_i(t)] + t\, w_i(t) \tag{11.12}$$

Assuming we succeed in encouraging safe disposal, it is important to induce the socially desirable amount of waste generation. In this case total costs to society are the sum of private costs (C) and damage (D):

$$SC[w_s(t-s)] = C_s[w_s(t-s] + D_s[w_s(t-s)] \tag{11.13}$$

Clearly we want to set marginal savings to the firm equal to marginal damages:

$$(t-s) = MS_s[w_s(t-s)] = MD_s[w_s(t-s)] \tag{11.14}$$

We are now at the end of our journey. Equation (11.11) defines how the firms respond to t and s. Equation (11.12) ensures that safe disposal is pursued. Equation (11.14) ensures that the right amount of wastes is generated. There will be a number of subsidy levels that will satisfy Eq. (11.12). (If a small subsidy will encourage safe disposal, then a large one will too.) If we choose the smallest subsidy that works, then Eq. (11.12) can be rewritten as an equality. Let t^* and s^* solve Eq. (11.11), Eq. (11.12) as an equality, and Eq. (11.14) with w_s^* and w_i^* the corresponding waste levels. Rewrite Eq. (11.12) as

$$s^* = \frac{\{C_s(w_s^*) - C_i(w_i^*)\} + t^* [w_s^* - w_i^*]}{w_i^*} \tag{11.15}$$

This is intuitive. Basically, the polluter saves money by disposing of wastes illicitly. The terms in braces in Eq. (11.15) indicate the direct cost savings from illicit dumping. The second term in the numerator indicates the waste generation tax savings from illegal dumping (associated with differences between w_s^* and w_i^*). These two terms give the advantage to illicit dumping. The subsidy must be just large enough to overcome this advantage.

Three examples of this type of regulation come immediately to mind. Consider disposal of organic solvents used in metal fabrication. We may observe the amount of solvent used, since it is generally purchased from a manufacturer and leaves an "audit trail." But we do not necessarily observe what happens to the solvent after it is used—some evaporates, some is reused, some is safely disposed of and (perhaps) some is illegally dumped. Fortunately, we can monitor safe disposal. The results here suggest we should tax the use of solvents and subsidize the safe disposal of solvents.[5]

Another example is household solid waste. A good portion of household waste in the developed world is product packaging. Although it is difficult to monitor the disposal of this waste, it is relatively easy to monitor the quantity of waste a household is generating, simply by monitoring the nature and quantity of product packaging purchased. The result suggests we should tax the use of packaging and subsidize proper disposal, either in a landfill or through recycling.

The clearest example of this type of regulation is the deposit–refund system for beverage containers. Simply taxing the use of these containers is insufficient to ensure their proper disposal. A payment is made when a beverage is purchased in a container; a refund is paid when the empty container is returned to the point of purchase or other suitable repository.[6]

III. ENFORCEMENT

We turn now to the problem of enforcing environmental regulations.[7] Enforcement is really a special case of incomplete information. It primarily involves identifying those polluters that are not doing as they should. Thus enforcement involves identifying private information on compliance with regulation, i.e., monitoring whether polluters are obeying the law.

In all of our discussion of regulation, we have either assumed that firms do as they are told and/or voluntarily provide private information to the regulator. We will relax that now and add the option that the regulator can audit a polluter, but *at a cost*. In other words, the polluter may have private information but the regulator may obtain it, at some cost to the regulator. Obviously, if the cost is very low, the regulator will usually obtain the information and we are back to the situation of perfect information. We are interested in the case in which the information is moderately costly so that the regulator will not seek perfect information from all polluters. The situation we have just outlined is the case of costly *audits* of polluters.[8]

This is very similar to the problem faced by a government taxation and revenue department. It is not cost effective to audit all taxpayers nor is it cost effective to audit no taxpayers (then many would cheat). The optimal amount of auditing is somewhere between these two extremes. Also, there is an interplay between the frequency of auditing and the penalities if caught cheating. If the penalties are low for cheating then no amount of auditing can induce taxpayers to behave appropriately. If the penalties are very high, quite infrequent auditing may be sufficient.

Thus enforcement of pollution control regulations has two components: finding out who is violating the regulations (auditing) and applying penalties to cheaters. The dual objectives of an efficient enforcement program are to induce polluters to pollute at the right level while keeping enforcement costs low.

Suppose we have a group of identical firms, each emitting pollution (e) at cost $C(e)$. As usual, C declines as e increases; thus polluting reduces firm costs and marginal costs with respect to e are negative. We are interested in what the typical firm will do when faced with an enforcement program as descibed above: the firm reports emissions to the regulator and has some probability, π, of being audited by the regulator. If cheating, the firm will pay a fine related to the size of the violation.[9]

A. Auditing an Emissions Standard

Consider the case of an emission standard. Suppose the regulator has decreed that each firm shall emit no more than $\bar{s}$ units of pollution. If a firm is audited and is found to be emitting more than allowed, the firm must pay a fine, f, per unit of emissions in violation plus a fixed fine, D, unrelated to the size of the violation. Thus the total costs to the firm, before it knows whether it will be audited, are

$$TC(e) = C(e) + F(e) \tag{11.16a}$$

where

$$F(e) = \begin{cases} \pi[f(e - \bar{s}) + D], & \text{if } e > \bar{s} \\ 0, & \text{if } e \leq \bar{s} \end{cases} \tag{11.16b}$$

The last term in Eq. (11.16a) is the *expected* fine from emitting at level e. By expected fine, we mean the amount of the fine times the probability (π) of having to pay it (i.e., the probability of being audited). And of course, it is in part proportional to the amount of the violation, $e - \bar{s}$. If there is no violation, there is no fine. A fixed fine (D) is also included in Eq. (11.16b). If the firm is found to be in violation, part of the fine is proportional to the violation and part is not.

This is shown graphically in Figure 11.3. In both Figure 11.3a and 11.3b, direct costs (C), expected fines (F), and total costs (TC) are shown as a function of emissions.

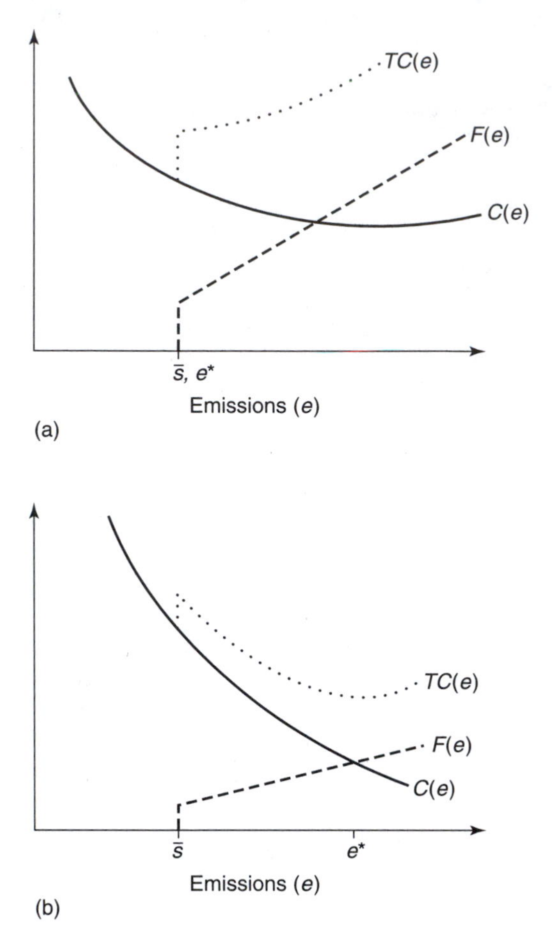

Figure 11.3 (a,b) Illustration of a fine with fixed and variable components. $C(e)$, Direct costs of emissions; $F(e)$, expected fine for emission level e; $TC(e)$, total cost: direct costs plus expected fine; $\bar{s}$, environmental standard; e^*, cost-minimizing level of emissions.

Focus first on Figure 11.3a. Note that the rather peculiarly shaped total cost function (TC) has a minimum at $\bar{s}$. Thus under this system, the firm will choose to emit at the regulatory standard and not violate the regulation. It is a different situation in Figure 11.3b; here the optimal behavior for the firm is to emit at $e^* > \bar{s}$. The fine (f or D) and audit frequency are simply too low to induce compliance.

In the case of Figure 11.3b, total costs can be written as

$$\begin{aligned} TC(e) &= C(e) + \pi f(e - \bar{s}) + \pi D \\ &= C(e) + \pi f e - \pi f \bar{s} + \pi D \end{aligned} \tag{11.17}$$

The firm will choose to operate at the minimum of total costs that occurs when marginal total costs are zero ($MTC = 0$). Note that costs are lower here than at $\bar{s}$. The marginal total costs can be separated into the marginal direct costs (MC) and the marginal fine (MF):

$$\begin{aligned} 0 = MTC(e) &= MC(e) + MF(e) \\ &= MC(e) + \pi f \end{aligned} \tag{11.18a}$$

or, alternatively,

$$-MC(e) = MS(e) = \pi f \tag{11.18b}$$

In other words, the firm will choose to pollute at the point at which the marginal savings from polluting are equal to the marginal expected fine. This is as expected. As long as the savings from emitting one more unit are less than the costs (in expected fines) from emitting one more unit, the firm will expand polluting. When the marginal savings are less than the marginal expected fines, the firm will cut back on pollution. When the two are the same, no further adjustments are necessary. In deciding whether to operate as in Figure 11.3a or 11.3b, the firm must compare total costs and choose the option that is least costly. Choosing between Figure 11.3a and 11.3b is not a marginal decision.

Note that what is important to the firm is the expected marginal fine, which is the product of π and f. Thus the specific values of π and f are not important, only their product. A small fine along with a large probability of audit will look exactly the same to the firm as a larger fine and a smaller probability of audit, as long as the product of the two remains the same. Alternatively, if D were large enough, we would return to the case in Figure 11.3a, in which marginal conditions are unimportant.

Now that we have characterized the firm's behavior, we turn to the regulator, which must choose the level of the fine and the audit probability. Note first that because it is costly to audit and only the product of π and f matter to the firm, it is clearly best to audit infrequently and levy large fines. This makes the auditing costs very small. The only limit to the size of the fine would be the assets of the firm. A firm cannot be made to pay a fine that exceeds its assets or else it will simply declare bankruptcy. This would destroy the incentive properties of the fine. The next question is what should πf be set to? Clearly if the standard is $\bar{s}$, then πf should be set so that Eq. (11.18) is satisfied at $\bar{s}$:

$$MS(\bar{s}) = \pi f \tag{11.19}$$

If the firm chooses not to violate the standard all is well; if the firm considers violating the standard, it will see that it is not in its best interest to do so and will instead choose emissions equal to the standard, according to Eq. (11.19).

There are two main messages to carry away from the model developed here. One is that it is not necessary to closely monitor polluters. Provided the sanctions for violating the regulations are high enough, the same outcome can be achieved with infrequent auditing as with frequent auditing. The second message to carry away from this is that it is not sufficient to simply establish a regulation and some sort of enforcement program. An inappropriately set enforcement regime will not induce polluters to behave. The costs of cheating must be seen to be higher than the costs of adhering to the regulations.

B. Extensions

The model of enforcement presented in the previous section is very simple. It is possible to add significantly to the model's realism and at the same time to its complexity. The simplest extension is to consider the case of an emission fee in which the firm reports to the regulator the emission fees due. The firm might be inclined to underreport emissions to reduce fee payments. An appropriately designed auditing system can avoid this.

Another extension would be to make the auditing probability and the unit fine depend on the size of the violation. This is somewhat more realistic than purely random audits, since there would probably be some information that would direct the auditors to firms that are more likely to be in violation.[10]

Another extension can involve a more complex auditing program, involving dynamics and reputation. Firms can be assumed to belong to one of two (or more) groups. Firms in group A have never been found to be in violation—either they have never been audited or, if audited, were found to be in compliance. Firms in group B have been audited and found to be in violation. Firms in group B are audited more frequently. Such a two-state auditing program can significantly reduce auditing costs.[11] Another interesting characteristic of such a multicategory auditing system is that fines can be set quite low, at least for firms in group A.[12]

IV. DYNAMICS AND COMMITMENT

Most of what we have been discussing in this and the previous chapter ignores the dynamics of regulations. In other words, our previous analyses have concerned one-time interactions between regulators and polluters. We have been concerned about structuring regulations so that the problems of asymmetric information are minimized. Reality is of course that regulators and polluters continue interacting. Furthermore, what each party does in one period may influence what happens in subsequent periods. There are two basic cases that we treat here, one concerning the cost of flexibility and the other concerning informational asymmetries in a dynamic setting.

First consider flexibility. Suppose the regulator promulgates a regulation today based on an assessment of costs and benefits of various levels of pollution control. There is a great deal of uncertainty about both costs and benefits. Regulators are uncertain about the right amount of control but promulgate a regulation anyway. The polluter sees a regulation but also perceives that the regulation could change, either becoming tighter or weaker in the future. Because of this, the polluter is reluctant to invest in expensive pollution control equipment, even though such an investment may be the cheapest way of controlling pollution in the long run. Instead, the polluter pursues a costly short-term strategy of pollution control and may lobby regulators to relax regulations in the future. Regulators in turn may observe costs and notice that they are higher than expected. Basically, it is more costly to be flexible. When regulators see that costs are higher, they may weaken the regulations. The desire on the part of the regulator to maintain flexibility has resulted in weaker pollution control. The regulator has legitimate reasons for retaining flexibility (in order to revise regulations when more becomes known). The down side is that when the regulator retains flexibility to revise the regulation, the polluter will in turn retain flexibility (at a cost) in anticipation of the regulation changing.

Why is flexibility always costly? Conceptually, flexibility is an additional constraint on production so it cannot decrease costs. Practically, the reason this makes any difference is that the cost of pollution control usually consists of a combination of short-run costs (such as using cleaner fuels) and longer-run costs (such as investing in pollution control equipment). It may be that the cheapest way to control pollution is to invest in pollution control equipment, perhaps with some use of clean fuel. However, if polluters think that the regulation is only temporary, they are unlikely to do any investing, instead relying exclusively on cleaner fuels. Although this is rational for the polluter, it makes pollution control more expensive than necessary. And if regulations are set to balance costs and damages, higher control costs will lead to weaker regulations and thus more pollution.

The other case draws from the adverse selection and moral hazard problems discussed earlier. Suppose the regulator is concerned with regulating a single firm. The pollution control costs are known to the firm but only imperfectly to the regulator. The firm reports its costs to the regulator. This is the adverse selection problem. We saw in the previous chapter that one solution to this problem is to reward low-cost firms for admitting that they are low cost, thus inducing them to tell the truth. But suppose we are dealing with multiple periods of time. If in the first period the firm admits to being a low-cost firm, in the second period, the regulator can change the rules and stop rewarding firms for admitting they are low cost (since the information is acquired in the first period). Because firms know this, how can we get them to cooperate in the first period?

As a concrete example of this, suppose the regulator is setting automobile emission standards for automobile manufacturers to follow. The regulator sets standards at some level in an initial period of time. Unfortunately, the regulator does not know how difficult it is for the automobile manufacturers to meet these standards. From the point of view of the automobile manufacturers, if they meet the standards easily, they are inviting even tighter regulations in the next period of time. If they complain about meeting the standard, they are more likely to get more lenient treatment in the next period.[13]

In both of these cases, the dynamic nature of the interaction between regulator and polluter leads to problems. In the first case, the polluter defers making long-run commitments to the cheapest form of pollution control because of the possibility that the regulation will change in the next period. This is the case of there being a cost of maintaining flexibility. In the second case, the firm realizes that the better it does in complying with regulations today, the more likely it will be that regulations will be tightened next period. This is called the "ratchet effect," from the tendency of the regulator to ratchet up the regulation if the firm behaves properly.[14]

The key factor generating these problems is the lack of commitment on the part of the regulator. Suppose the regulator could convince the polluter that the regulation will not change. In our first example, polluters would then be willing to invest in the pollution control equipment and achieve the lowest possible pollution control costs. In our second example, polluters would not worry about revealing their true costs. This is what is known as the problem of commitment. If the regulator cannot commit to a particular regulatory strategy, then polluters will not respond in the desired way. If on the other hand the regulator is able to commit, the polluter will have a different response. The downside for the regulator is that commitment is just that and precludes loosening controls later if the situation warrants.

The significant point is that commitment can be important to reduce costs and, as a result, pollution. Thus if a regulator does not need to maintain flexibility, it is in the regulator's best interest to pursue regulatory policies that involve commitment. Commitment is not easy, however. Even if a legislature were to promulgate a regulation and state that it will not be changed, in most countries it is not possible to prohibit the legislature from revisiting the issue and changing its mind at a later date.

What are examples of commitment or lack thereof in environmental regulation? One example is the case of marketable emission permits in the United States. In the sulfur allowance system governing emissions of sulfur dioxide in the United States, the legislation is quite specific that the tradable allowances being distributed to polluters are not property rights; in other words, they may be confiscated without compensation if the EPA changes its mind about relying on tradable permits. This sounds good for the EPA, but what is being sacrificed? The EPA is failing to commit to the allowance system and thus is discouraging (somewhat) long-term investments in pollution control, possibly raising the costs of pollution control.[15]

Interestingly, command-and-control regulation can have real advantages in the area of commitment. Often command and control specifies the control capital that must be used. This capital often becomes a sunk cost. By specifying that firms must make long-run investments, command and control is able to require firms to commit to a long-run strategy, an outcome that is difficult if the possibility is real that the emission fee can change next year.

A. Flexibility

We will examine an example that is overly simplified, but that illustrates the problem of retaining regulatory flexibility. The pollutant emissions of a firm are subject to regulation. Think of the regulations as being tight (T) or loose (L). The firm has the choice of buying abatement equipment to totally eliminate the pollution emissions. But once the abatement equipment has been installed, it is a sunk cost; abatement equipment cannot be uninstalled. Once installed, abatement equipment costs \$50 per year (this is the total cost). Alternatively, the firm may buy cleaner fuel to reduce emissions. With tight regulations, clean fuel will cost \$60 per year; with loose regulations, clean fuel costs only \$30 per year. So if the firm knows regulations will always be tight, it is clearly cheapest to buy the abatement equipment. If the firm knows it will be subject to loose regulations, it pays not to buy the abatement equipment and buy cleaner fuel instead.

Suppose the regulator promulgates tight regulations but leaves open the possibility of loosening the regulations in subsequent years. What would we expect the firm to do? To answer this question, we will divide our example into two time periods: the present and the future. In the present, regulations are tight. But the EPA is conducting a study to decide whether regulations should be tight or loose in the future. We expect the results of that study before the second period begins. But our current assessment of what may emerge from that report can be summarized by saying that in the future, the regulations will be tight with probability π and loose with probability $(1 - \pi)$. Figure 11.4 illustrates these possibilities. Two periods are shown in Figure 11.4, time period 1 and time period 2. The branches of the figure represent the state of regulations, T (tight) or L (loose). In the far right of Figure 11.4 is total costs under two conditions: (1) abatement equipment is purchased in period 1 and (2) abatement equipment is not purchased in period 1. With abatement purchased in period 1, there are no additional pollution control expenses, no matter what happens to regulations in the future. Thus the total cost in this situation is simply the cost of the abatement equpment, \$100 (\$50 per period). On the other hand, if no abatement equipment is purchased in the first period, the firm pays \$60 for clean fuel and revisits the decision again in the second period. In the second period, the firm sees

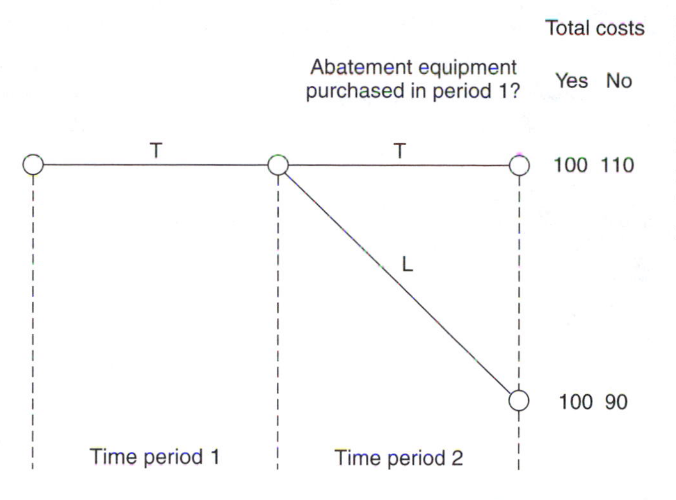

Figure 11.4 The decision to invest in abatement equipment. T, Tight regulation; L, loose regulation. *Note*: Abatement equipment costs $50 per period. Clean fuel costs $60 per period with tight regulation and $30 per period with loose regulation.

whether regulations remain tight (T) or become loose (L). If regulations turn out to be tight, the firm buys the abatement equipment, committing to $50 per year; if regulations turn out to be loose, it is cheapest to continue to buy clean fuel for $30. Adding these figures onto the $60 spent for clean fuel in the first period yields the total costs for the first 2 years shown in the far right-hand column (a similar figure could be shown for costs for more than two periods).

We are still faced with the question about whether the firm should buy the abatement equipment. The firm's total costs if it buys abatement equipment in the first period is $100, no matter what happens to regulations. On the other hand, consider the firm's expected costs if it does not buy abatement equipment immediately. Since the firm assigns a probability of π to regulations remaining tight, its expected costs are given by

$$EC_{NA} = \pi\ 110 + (1 - \pi)\ 90 \tag{11.20a}$$

$$= 20\ \pi + 90 \tag{11.20b}$$

Under what conditions would EC_{NA} (expected costs, no abatement equipment) be less than $100? Clearly, for $\pi < 1/2$, it is cheaper to defer the decision to buy abatement equipment until the second period and pay for the cleaner fuel in the meantime. In other words, if there is a 50% or greater chance that regulations will be loosened, it is worth paying higher emission control costs until the regulatory uncertainty has been resolved. Only when the probability of the regulations being loosened becomes sufficiently low, will the firm commit to lower cost abatement equipment. Suppose there is a 60% chance of regulations being loosened. Putting $\pi = 0.4$ into Eq. (11.20b) yields $EC_{NA} = \$98$, which is cheaper than committing to abatement equipment ($100).

The interesting thing is that an analysis of the cost of regulations without uncertainty indicates that tight regulations cost $50 per year whereas loose regulations cost $30 per year. However, introducing sufficient uncertainty over the two regulations has the effect of raising the cost to $60 per year as long as regulations remain tight and uncertain, not a figure somewhere between $50 and $30 as one might expect from uncertainty over whether the regulation should be loose or tight.

The bottom line is that flexibility costs money. If the regulator wants to retain flexability and is consequently unwilling to commit to tight regulations, firms respond by adopting more flexible control plans, though at higher cost. Although there may be a good reason for the regulator to want to retain some flexibility and ability to adjust regulations in the future (which is in fact the case in our example), it is important for the regulator to be aware that there is a cost associated with this. In particular, a lack of commitment on the part of a regulator may cause an underinvestment in pollution abatement equipment and thus unnecessarily high pollution control costs.

B. The Ratchet Effect

The ratchet effect occurs when the regulator has incomplete knowledge of the firm's costs and the two must repeatedly interact. If the firm is capable of controlling pollution at low cost, and the regulator knows it, the firm will be subject to tight regulations. What makes the problem interesting is that the firm and the regulator interact repeatedly. Thus it is in the firm's best interest to disguise its true nature, or the regulator will ratchet up the intensity of the regulation in subsequent periods.

The mathematics of the ratchet effect is fairly complex, so we will content ourselves with a description of how a model of the effect might operate. We will draw on the model of automobile pollution regulation mentioned earlier and examined in detail by Yao (1988).

We start with an automobile industry subject to regulation by the EPA. The goal is to set an emission limit for the industry that is tight but realistic. The industry will have to invest in R&D to meet the regulation. The EPA's job is to make sure the emission limit is attainable at acceptable cost (whatever that is).

To make the problem even simpler, suppose there are two time periods, today and tomorrow (which of course could mean this decade and next decade). So the EPA sets an emission limit in each period equal to L_1 and L_2. How does the industry meet this limit? In contrast to the other models we have examined, the industry knows it can meet the limit; the only question is the cost. Furthermore, the industry can invest r dollars in R&D to lower the cost of meeting the regulation. To make things a little bit more interesting and realistic, the productivity of industry R&D is not known to the EPA. Call this productivity θ. So the cost of meeting the regulation (including R&D costs), C, depends on the size of the regulation, L, the R&D productivity of the industry, θ, and the industry's investment in R&D, r: $C(L,r,\theta)$. C includes R&D costs as well as abatement costs. Clearly C is decreasing in L, decreasing in θ, and first decreasing, then increasing in r. The EPA can observe C but cannot observe θ or r.

This is both a moral hazard problem (r is an unobserved action) and an adverse selection problem (θ is an unobserved characteristic of the industry). What would happen if the EPA could perfectly observe θ and r? In that case the EPA would balance the costs of pollution control, including R&D, with the benefits of pollution control.

But what happens when the EPA is unaware of θ or r? As we saw in the previous chapter when we were considering adverse selection, the EPA must reward the industry for admitting that it is productive and adjust the regulation, L, depending on the admitted productivity of the industry. But if the industry admits its R&D productivity, in the next period the EPA will dispense with the reward for providing truthful information and impose as tight a regulation as possible. In other words, if the industry admits it is low cost, it will pay dearly in the next time period.

So what should the industry do? First, it should be less willing to admit to being a productive industry. Second, when the EPA observes costs after the first period, it will have a better idea whether the industry is productive or not. If the costs of meeting the regulation are low, the EPA will be more likely to think the industry is productive. To avoid sending that signal to the EPA, the industry is likely to underinvest in R&D (r), artificially keeping the cost of meeting the regulation high.

This behavior becomes more meaningful if we return to the automobile industry example that motivated us. The EPA imposes regulations on new cars in terms of their allowed emissions. The automobile industry complains that it is difficult to meet the regulation. Furthermore, some years later, it turns out that the automobile industry was right, it was very costly to meet the regulation. Thus the industry gets gentler treatment in subsequent regulatory rounds.

What is missing is that the industry may have intentionally underinvested in R&D to send a signal (through high costs of meeting the regulation) that the industry has a difficult time meeting automobile emission regulations. That signal can pay handsomely over years to come.

The solution to this problem is commitment. If the regulator can commit to a multiyear regulatory strategy, the gamesmanship the industry may play will not pay off. Such commitment is difficult, however, particularly for government regulators and legislators.

C. Why Is Commitment Rare?

We have seen that for both of the examples we considered, commitment on the part of the regulator is generally desirable. In the case of flexibility, if the regulator can commit, the cost of emission control will be lower. We saw in the case of the ratchet effect that multiyear commitment can eliminate perverse incentives and thus decrease the costs of emission control.

The fact is that environmental regulators are rarely willing to commit. Why is this? One simple explanation is that they are unaware of the advantages of commitment; however, there may be other reasons (see Freixas et al., 1985). Commitment may not be possible. Even if the EPA were to say it will not change the regulation, we know that regulators change as do legislatures. Commitment typically requires a constitutional guarantee, which is rarely possible.

Another reason for a lack of commitment is that it is very difficult and costly to design regulations to apply many years into the future. One must carefully weigh different outcomes and contingencies. This is not easy and may be undesirable.

A third reason for a lack of commitment is learning and information acquisition. If the EPA expects to acquire new information, perhaps it should keep its options open. In the arena of environmental regulation, acquiring new information is a very real possibility. More is being learned every day, particularly about the damages associated with pollution.

In conclusion, commitment is generally (though not always) desirable and should be pursued if possible. However, we must recognize that commitment will not always accompany environmental regulations and, for that reason, such regulations are likely to be second best, with higher costs and fewer environmental benefits than would otherwise be the case.

SUMMARY

1. Moral hazard problems involve polluters taking important actions that cannot be observed by the regulator, such as diligently maintaining pollution control equipment. Consequently, moral hazard involves "hidden actions."

2. One problem with many area sources of pollution (e.g., automobile pollution or agricultural runoff) is that emissions are not directly observed. A properly constructed levy on observed ambient levels of pollution can induce the optimal amount of emissions, even when those emissions are unobserved.

3. For some pollutants, safe disposal is important while illegal disposal is cheaper for the firm. Subsidies to safe disposal accompanied by a charge on waste generation can solve this problem. Such fee-subsidy systems are also known as deposit–refund systems.

4. In constructing an effective enforcement program, it is important to adjust both the fine for noncompliance as well as the probability of a firm being audited.

5. When a regulator retains flexibility by refraining from committing to a long-run regulatory strategy, polluters may fail to adequately invest in pollution control equipment, thus raising costs. Higher costs may result in pressures to relax the regulation in subsequent periods.

6. The ratchet effect involves repeated interaction between a regulator and a firm. If the firm is capable of controlling pollution at low cost and the regulator knows it, then the firm will be subject to tight regulation. Thus it is in the firm's best interest to disguise its true nature, or else the regulator will ratchet up the intensity of the regulation in subsequent periods.

7. Commitment can solve dynamic regulatory problems. However, it is difficult for governments and regulators to commit themselves not to take certain actions in the future.

PROBLEMS

1. Some policymakers argue that it would be desirable to reward generators of hazardous wastes for appropriately disposing of their wastes. Describe a possible reward scheme. Why is it a good idea? What are its shortcomings?
2. Describe how a tax scheme as detailed in Section II could be used to regulate automobile pollution in an urban area.
3. Identify whether each of the following is an example of adverse selection or moral hazard and why:
   a. A sawmill threatens to close down because of tight new environmental regulations.

b. Smog checks are required for automobiles in addition to emission requirements imposed on manufacturers.

c. There is difficulty in operating a program to buy the most polluting cars from consumers and retire the cars.

4. Contrast emission fees (fee per unit emissions) and marketable permits (permit to emit one unit in perpetuity) from the point of view of commitment in a society with well-defined private property rights (i.e., property cannot be confiscated without compensation).

***5.** Consider the problem of automobile pollution in a large city in Central Asia. There are a million cars in the city, each emitting 10 grams of carbon monoxide (CO) per kilometer (km) driven. On average, each car is driven 10,000 km per year. The average cost to each driver is approximately 1 Ruble/km whereas the marginal cost is lower and independent of km travelled, at about 0.3 Rubles/km (because of the large fixed cost of owning an automobile). A recent study of driver behavior determined that individual demand for kilometers driven was roughly linear; if a gas tax were instituted that raised the marginal cost of driving by 0.1 Rubles/km, the average car owner would reduce driving by about 1000 km per year. Each kiloton (1000 tons) of CO emitted in a year raises the average concentration of CO by 1 $\mu g/m^3$. A team of economists, biologists, and physicists determined that the marginal damage per unit of ambient concentration of CO is roughly constant, 10,000,000 Rubles per year per 1 $\mu g/m^3$.

a. For an individual, what are the annual emissions when the "price" of emissions is zero? What about when the price rises to R10 per kg? R20 per kg? Draw an individual's supply curve for emissions—the relationship between price (R per kg) on the vertical axis and quantity of emissions (kg) on the horizontal axis.

b. Draw the aggregate supply for emissions over all individuals, showing the price (R per kg) on the vertical axis and the quantity of emissions (kilotons—millions of kg) on the horizontal axis. In another graph, draw the equivalent curve for the supply of pollution, showing the quantity of pollution on the horizontal axis ($\mu g/m^3$) and the price in millions of Rubles per $\mu g/m^3$ on the vertical axis. This is the marginal savings from polluting. Draw the marginal damage from pollution on the same graph. What is the efficient ambient concentration of CO? Call this value S.

c. What are the average emissions per car associated with this efficient concentration of CO? If you could observe emissions from each automobile, what emission charge would you levy per kg of emissions per car to obtain the efficient concentration?

d. You are advising the Local Environmental Protection Committee and point out that emissions from individual cars are very difficult to measure. You recommend an ambient fee to control this pollution. In other words, each driver would be assessed a charge, based on observed ambient conditions only (such a fee was described in this chapter). How much would each driver have to pay if ambient pollution levels exceeded S by 1 $\mu g/m^3$?

e. Discuss the answers to parts (c) and (d).

6. Imagine that you are a manufacturer of inflatable rubber ducks. Unfortunately, the manufacture of ducks involves damaging water pollution. You have computed that your marginal savings in millions of dollars from emitting pollution is linear: $MS(e) = 10 - e$, where e is the amount of emissions in tons. The Ministry of the environment has decreed that your emissions should not exceed 2 tons.

a. How much money does it cost you to obey the regulation rather than emit as much as you want?

b. The Ministry has just instituted a program whereby it will audit you with a probability of 0.01. If you are found in violation, you must pay a fine of $10 million per ton of emissions in excess of the limit. How much is optimal for you to emit (optimal from your point of view)?

***7.** You are a polluter who is faced with the question of whether to comply with the regulations on emissions or to violate them. If you comply with the regulations, your costs are k. If you do not comply with the regulations, your costs are zero. The EPA has a new system whereby it audits your compliance with the regulation and gives you a black mark or gold star. If you have a gold star and are audited and found out of compliance, you pay a $50 fine and get a black mark. If you have a black mark and are audited and found in compliance, the EPA flips a coin and gives you a gold star with probability 0.5. If you have a black mark, are audited, and found out of compliance, you are fined $100. The probability of audit depends on whether you have a gold star (probability 0.1) or a black mark (probability 0.5). These probabilities and costs are annual.

Based on your costs, four strategies are available to you: never comply, always comply, comply only if you have a gold star, comply only if you have a black star. Call these strategies, nn, cc, cn, and nc. One is always best for you, and it depends on your costs of compliance, k. Suppose you repeatedly interact with the regulator year after year. Your discount factor is β (you are indifferent between $1 tomorrow and $\$\beta$ today) and is equal to 0.9. Define your expected costs from all future interactions as $C_{ij}(m)$ where ij is nn, cc, cn, or nc and m is gold star or black mark.

Calculate the value of $C_{ij}(m)$ for both ms and all four ijs. [*Hint*: write an equation for a particular $C_{ij}(m)$ in terms of this year's costs and next year's $C_{ij}(m)$, which must be the same as this year's.]

What is the highest value of k for which you always comply with the regulations? What is the lowest value of k for which you never comply with the regulations?

NOTES

1. Schmutzler and Goulder (1997) consider the problem of pollution generation in which taxing the output of emissions is difficult, generally because of monitoring costs. They consider the alternative approach of taxing output of the good associated with the pollution or inputs into production, both of which are imperfectly correlated with the pollution asociated with producing the good. They show that such a tax works particularly well when there are few ways of reducing pollution other than reducing the output of the good associated with the pollution. Eskeland (1994) examines the practicality of a tax on gasoline as a substitute for a tax on emissions associated with gasoline.
2. This is a rather peculiar term with its genesis in the insurance literature. If you are a smoker with your own house, you will be careful about smoking in bed, because of the risk of fire. But if your house is fully insured against fire, you lose some or all of the incentive to be morally upright, taking care to prevent fire. Furthermore, the insurance company cannot know how careful you are. Your care is the analog to effort in our example. The insurance has created a "moral hazard."
3. The model presented here is due to Segerson (1988). Cabe and Herriges (1992) extended Segerson's work. The problem we are considering is more generally known as the problem of moral hazard in teams. A team is a group (such as a football team) in which the actions of individuals are unobservable but clearly contribute to the sucess of the team. How do you reward individuals based solely on the performance of the team? Holmström (1982) provided the first

definitive treatment of this question. Xepapadeas (1991) considers a slightly different system of random fines, which can be tied to partial information on emissions.

4. Two empirical analyses of illegal disposal are worthy of note. Hilary Sigman (1998) examines the illegal disposal of waste motor oil in the United States. She shows that the higher the cost of legal disposal, the greater the rate of illegal dumping. Fullerton and Kinnaman (1996) examine the extent to which illegal dumping of municipal trash is affected by the cost of legal disposal. They conclude that higher legal disposal prices increase illegal dumping, though their data source for this conclusion is questionable.
5. Hilary Sigman (1995) examined several policies for reducing the quantity of lead that is discarded into the environment in the United States. Approximately 80% of lead used in the United States in 1990 was for batteries. She examines four policies to reduce lead discharges: a tax on virgin lead used in manufacturing, a deposit–refund system for batteries, recycled content standards for batteries (minimum percentage recycled lead), and subsidies to recycling lead. She concluded that a deposit–refund system or a virgin materials tax would be similar in operation and the best choices for reducing lead discharges. The worst policy is subsidies to recycling, primarily because it makes lead use cheaper and thus encourages the use of lead, and indirectly the disposal of a fraction of the lead used.
6. Porter (1983) conducts a detailed cost–benefit analysis of a beverage container deposit–refund program.
7. Russell et al. (1986) provide a good overview of the literature through the mid-1980s.
8. The problem of when to conduct costly audits goes beyond the environmental economics literature. See, for example, Baron and Besanko (1984).
9. Harford (1978) provided one of the earliest analyses of this problem in the context of environmental regulation. See also Downing and Watson (1974).
10. Malik (1990) considers this for a permit market.
11. See Harrington (1988), Harford (1987), and Harford and Harrington (1991) for more on models of this type. Also refer to problem 7 at the end of this chapter.
12. The apparent motivation for Harrington's (1988) development of the multicategory auditing system was his observation for the United States that (1) there was a low frequency of monitoring, (2) fines are relatively low and infrequent, and (3) most firms were not in violation of the law. He showed that the multicategory auditing system can have just such characteristics. The threat of being audited more frequently leverages relatively low fine levels. Heyes (1996) offers a different explanation for low fines, suggesting that high fines would discourage cleaning up environmental accidents.
13. This is precisely the problem considered by Yao (1988).
14. See Freixas et al. (1985) for a discussion of the ratchet effect in the context of central planning. See Laffont and Tirole (1993) for a broader though advanced treatment of this issue.
15. Hahn (1989) cites the lack of permanence as a major factor in the failure of some experiments with marketable permits.

12 RISK AND UNCERTAINTY

As a government makes its first moves to provide environmental protection, it initially focuses on environmental problems for which the relationship between pollution sources and damage is clear and certain. Attention is first directed to the large industrial plants that spew smoke or effluent that causes people to sicken, crops to die or wilt, or buildings to decay. These sources, with a clear connection between pollution and damage, are usually the easiest to bring under control.

As time goes by, and these large sources of pollution are cleaned up, attention inevitably turns to the more difficult sources of pollution. Often these more difficult sources of environmental damage are in the realm of risk, often characterized by accidental discharges of pollution. A landfill is built to contain the material it is storing; accidents do happen and occasionally the groundwater under a landfill is contaminated. Oil tankers are designed to carry, not to spill, oil. Yet large amounts oil are, on occasion, unintentionally spilled. These are both examples of risk. Because accidents do happen, these activities for which environmental degradation is not an intentional by-product are risky and need to be regulated as much as large smokestack industries spewing smoke over the city and countryside. The goal is not to eliminate risk; that would be futile. The goal is to manage risk.

The environmental risks described above are harmful. A conceptually similar uncertainty relates to the benefits that the environment may provide; these benefits may be threatened by pollution and development. Certain ecosystems or natural areas may not appear to be particularly valuable, yet there is a possibility that significant benefits (such as new drugs or improved crops) may emerge from such ecosystems. Pollution and land development may threaten these environmental resources that have uncertain benefits. How can economic activities that threaten these uncertain benefits be managed?

In this chapter we will seek to understand risk from a conceptual perspective as well as understand the various means for reaching socially acceptable levels of risk. This may involve the private market, as with insurance; it may involve decentralized regulation, such as liability; it may involve public provision of risk-reducing technologies; or it may involve conventional direct regulation of activities that are risky or lead to exposure to risk.

I. WHAT IS RISK?

Risk is more complicated to manage than "simple" deterministic pollution problems. The cause and effect links are fuzzy and exposure to pollution does not ensure an adverse effect on people, property, or ecosystems, only an elevated *probability* of an effect. In addition, the effects may be so subtle that science is unable to evaluate the seriousness of the risk. It is quite common to see a chemical cause adverse effects in Petri dishes or in laboratory animals; it is quite another thing to demonstrate the effect in a human population.[1] Of course this uncertainty about whether a risk exists is just part of the risk itself—we are unsure about whether an action may lead to any adverse effects in any given individual.

A. Environmental Risks Are Complex

Think of the chain of causality associated with environmental risks. Consider two examples, first the case of disposing of benzene in a well-constructed landfill. If all goes well, the benzene stays in the landfill (or evaporates). But there is a chance that the landfill will leak. Even if it leaks, the benzene may not leave (the leak may be in a portion of the landfill that contains no benzene). But the benzene may leak from the landfill and if it does, it may or may not make its way into the groundwater. If it makes it into the groundwater, it may find its way into a drinking water well. But it is possible that it will be detected before it makes it into the drinking water supply. Finally, if drinking water is contaminated, it may or may not cause an adverse effect in the individual drinking the contaminated water.

Pesticide residue poisoning starts with the application of pesticides to an agricultural field. The pesticides may be unevenly distributed, perhaps resulting in parts of the field having particularly high concentrations. The pesticides may or may not be present in the food that is harvested and may or may not be present after the food is washed by the processor or consumer. The pesticide may then be ingested by the consumer. The likelihood that an adverse reaction occurs (such as cancer) depends on a variety of factors, including other synergistic pesticides that may have been ingested and the age and other characteristics of the consumer.

In both of these examples, we see a long and tortuous chain of events between introduction of the hazard into the environment (spraying the pesticide or depositing the benzene in the landfill) and the possible deleterious health effects. Parts of the process are uncertain but potentially knowable, such as the amount of pesticide residue in the food supply. Other parts may involve a much greater element of randomness, such as whether an individual may develop cancer from ingesting a specific amount of pesticide residue.

B. Causation

Given that an adverse health effect occurs, it is not easy to work backward, to ascribe the effect to a particular cause. There is uncertainty about the cause of the effect. For instance, if a large quantity of oil washes up on a beach, it is likely that someone has spilled the oil; what is uncertain is who spilled it. Or if a solvent shows up in groundwater, it is clear that the solvent has leaked from a containment area or been deliberately dumped; what is uncertain is who is responsible for the contamination. More complicated is the case where damage has occurred (such as groundwater being contaminated or someone contracting cancer) and several parties are potentially responsible. Perhaps several firms accidently dumped solvents on the ground. Which accident caused the adverse effect? Perhaps an individual was exposed to contaminated drinking water and also air pollution. Which caused the cancer?

It is particularly difficult to establish causation in the case of environmental hazards, in part because of the long time delays that are often involved. In may take years or even decades for a hazard to turn into a risk. Establish a landfill and it may take some time to leak. After leaking, it may take a long time for the contaminants to penetrate into the groundwater aquifer. After reaching the aquifer, it may take some time before it migrates to a drinking water well. After drinking contaminated water, cancer (or other effects) may take decades to develop.

C. Subjective versus Objective Risk

A question that has become central in public policy is whose assessment of risk is most appropriate? Nuclear power is the classic example. The risks of nuclear power have consistently been perceived to be greater by the general public than they have been by experts.[2] In setting public policy, whose assessment of risks do we use?

Although there can be some debate over the probability of a nuclear accident (because they happen so infrequently), there can be little debate over the frequency of death from tornadoes, botulism, or heart disease. These happen frequently enough that we have objective statistics that indicate the probability that a randomly selected individual will succumb to one of these. Table 12.1 shows a number of risks, all of which are estimated to involve a risk of loss of life of one in a million in a particular year. Most of us are surprised at the magnitude of many of these risks; although the activities described are relatively familiar, we have a difficult time assessing the risks. Psychologists have shown that individuals have a difficult time assessing probabilities or risks. Individuals consistently overestimate the likelihood of low probability events and underestimate the likelihood of high probability events.[3] This anomaly leads to the term *objective risk* to refer to the statistical relationship between exposure and adverse effects and the term *subjective risk* to refer to an individual's *perception* of his or her objective risk. Figure 12.1 is a classic example of the difference between objective risk (the horizontal axis) and subjective risk (vertical axis) as articulated by a large number of lay people.

In economics, there is a tradition of distinguishing between subjective and objective uncertainty. When the probability of occurrence of random events is objectively known, we say we are dealing with a risk. If the probabilities are subjective, we say we are dealing with uncertainty. For instance, when we place a bet on "red" based on the turn of a roulette wheel, it is easy to compute the objective probability of winning; this would be risk. On the other hand. if we are betting on who will win the next World Cup, there is no objective probability of a particular team winning; each of us has a subjective probability. Such subjective probabilities are termed uncertainties.[4]

TABLE 12.1 Equivalent Risks

Risks that increase the chance of death by one in a million per year	Causes of death
Smoking 1.4 cigarettes	Cancer, heart disease
Drinking 1/2 liter of wine	Cirrhosis of the liver
Spending 1 hour in a coal mine	Black lung disease
Spending 3 hours in a coal mine	Accident
Living 2 days in New York or London	Air pollution
Traveling 6 minutes by canoe	Accident
Traveling 10 miles by bicycle	Accident
Traveling 500 km by car	Accident
Flying 1000 miles by jet	Accident
Flying 10,000 km by jet	Cancer caused by cosmic radiation
Living 2 months in an average stone or brick building	Cancer caused by natural radioactivity
One chest x-ray taken in a good hospital	Cancer caused by radiation
Living 2 months with a cigarette smoker	Cancer, heart disease
Eating 40 tablespoons of peanut butter	Liver cancer caused by aflatoxin B
Drinking Paris drinking water for 1 year	Cancer caused by chloroform
Living 5 years at site boundary of a typical nuclear power plant in the open	Cancer caused by radiation
Living 20 years near a polyvinylchloride (PVC) plant	Cancer caused by vinyl chloride
Living 150 years within 30 km of a nuclear power plant	Cancer caused by radiation
Eating 100 charcoal broiled steaks	Cancer from benzopyrene
Risk of accident by living within 8 km of a nuclear reactor for 50 years	Cancer caused by radiation

Source: Adapted from Wilson (1979).

The risk analysis literature makes a somewhat different though consistent distinction between risk and uncertainty. Lichtenberg and Zilberman (1988) define risk as the probability that a randomly selected individual will contract an adverse health effect from some environmental hazard, such as the storage of wastes in a landfill. However, this probability may not be known with certainty. There may be some uncertainty over exactly what the magnitude of the risk may be. This is termed uncertainty. For example, we may say that the risk of death in a given year from smoking 1.4 cigarettes is one in a million. We may also say that there is uncertainty in this risk, making the confidence interval from one in 500,000 to one in 1,500,000. The connection to the economic definition is that the one in 1,000,000 figure may come from scientific measurements. However, we may know that these measurements are subject to errors, so that we posit a range of possible risk figures.

In any event, we should not be too concerned with the semantic difference between the terms risk and uncertainty. The main point is that risk and uncertainty can be evaluated subjectively and objectively. When objective and subjective assessments of the same risk are different, it is not easy to decide which estimate should be used for public pol-

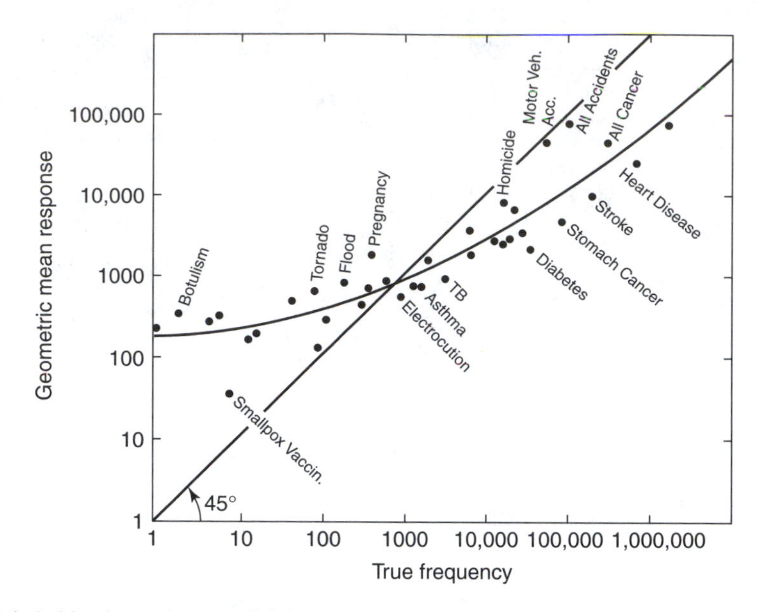

Figure 12.1 Subjective estimates of risk versus true risk for group of 40 subjects. *Note*: Data points are geometric means of estimates of risk from a group of 40 subjects told that there are 50,000 motor vehicle accidents per year in the United States. Frequencies are deaths per year in the United States. From Slovic et al. (1979).

icy purposes. For example, a policymaker may know the public overstates the risks from nuclear power. In making decisions regarding nuclear power, on behalf of the public, should the decision maker use subjective or objective risks?

D. Voluntary versus Involuntary Risk

Leaving aside disagreements over the probability of an unpleasant event happening, it appears that our determination of how important a risk is depends on whether our exposure is voluntary or involuntary.[5] Studies of the wage differential necessary to induce people to undertake risky jobs suggest that the extra compensation is much less than what people demand for a similar involuntarily imposed risk, such as the risk from a nearby hazardous waste landfill. We are quite willing to take significant risks for often very marginal benefit when we drive in a car. If we are exposed to risks involuntarily, we demand a much higher level of safety.

II. MAKING CHOICES ABOUT RISK

Making choices that involve risks is not easy. Should we buy the inexpensive airline ticket on the carrier that has a spotty safety record? How frequently should we test our well water for contamination? Should we drive home from a party after one drink or take a taxi?

Should we install a dam and flood a pristine canyon and lose uncertain future environmental services from the canyon but gain hydroelectric power? Should we install a safe municipal water system to replace wells in an area with many hazardous waste dumps that may leak into the groundwater? Should we protect a biologically diverse ecosystem from commercial development on the off-chance that some obscure species in the ecosystem may be the source of a wonder drug?

These are all choices that involve risk and uncertainty. In some cases, the risk is that something nasty may occur, such as a hazardous waste landfill contaminating the drinking water supply. In other cases, the risk is that something good will occur, such as the discovery of a useful species in the environment. There are several questions that arise in either case. How do individuals value risky outcomes? Can we extend utility theory to cover risks? Another issue concerns the value of investments that reduce or eliminate risk. This would include a public water supply to ensure drinking water is safe, even if a landfill leaks. (Such a water supply would have positive value as a reducer of risk.) Or it might include development that destroys an ecosystem, eliminating the possibility of discovering a wonder drug from the members of that ecosystem. (Such development would have negative value in terms of risk reduction.)

We consider these issues in turn.

A. Expected Utility

We start with a simple risk. Suppose we have a single good, x, but are unsure how much we will have to consume. That is the risk. Two things can happen and we do not know which. With probability π_A, event A will occur. With probability π_B, event B will occur. If event A occurs, the quantity of x we receive is x_A and if event B occurs, we receive x_B. We call A and B "states of the world" and we call x_A and x_B payoffs under these different states of the world. Furthermore, these states of the world will occur with known probabilities. We might think that these states of the world are exhaustive and mutually exclusive (one or the other but not both will occur). In that case, $\pi_B = 1 - \pi_A$.

If we start with a conventionally defined utility function, $u(x)$, which is defined for payoffs that are sure (known with certainty), is there an easy way of extending this utility function to risks? The answer is yes. If we make some assumptions about individual behavior under uncertainty, we can define the utility of a risk as the expected utility[6] from the individual payoffs in different states of the world:

$$U(x_A, x_B, \pi_A, \pi_B) = \pi_A\, u(x_A) + \pi_B\, u(x_B) \tag{12.1}$$

When utility of a risk can be evaluated in this fashion, we say that this utility function defined over risks (U) is a von Neumann–Morgenstern utility function, named after John von Neumann and Oscar Morgenstern, who first developed the concept. What we have done is start with an ordinary utility function, u, and developed another utility function, U. U is identical to u when goods are not risky (π_A and π_B are 0 or 1). However, u is not defined over risks, whereas U is.

To make the concept a little more understandable, let us substitute numbers for x_A, x_B, π_A, and π_B, but assume events A and B are mutually exclusive and exhaustive. This means $\pi_B = 1 - \pi_A$. And let us measure x in monetary terms. Let x_A be $-\$5{,}000$ (a loss),

x_B be \$1,000, (a gain) and π_A be 0.6. The first question to ask is what would we expect this risk to produce on average (if it were repeated over and over again) in terms of money? The expected number of dollars would be $0.6 \times (-\$5000) + 0.4 \times \$1000 = -\$2600$. So a risk in which you might lose \$5000 or might win \$1000, on average will yield a loss of \$2600.

How much this risk is worth can be seen in Figure 12.2. Shown is a utility function, which depends on only one commodity, measured in monetary terms. The utility function increases at a slower rate as the amount of the good increases. In other words, when you are poor, you receive more utility from an extra dollar than when you are rich. Payoffs from events A and B are shown on the horizontal axis. Each event yields utility as indicated on the graph: event A yields x_A with utility $u(x_A)$ and event B yields x_B with utility $u(x_B)$. The expected quantity of the good that results from this risk is shown as EP (the expected payoff), which is $\pi_A x_A + (1 - \pi_A)x_B$. In our numerical example, $EP = -\$2600$. EP is a weighted sum of x_A and x_B, with the weights being π_A and $1 - \pi_A$.

The *utility* of this risk is similarly, $\pi_A u(x_A) + (1 - \pi_A)u(x_B)$, which is shown as *EU* (expected utility) in Figure 12.2. Basically, the point *Z* in Figure 12.2 is located between points *W* and *V* in proportion to π_A and $(1 - \pi_A)$. *Z* is a weighted average of *W* and *V* with the weights being π_A and $(1 - \pi_A)$. The horizontal coordinate of *Z* is *EP* whereas the vertical component of *Z* is *EU*.

Note that because of the curvature of the utility function, there is some certain amount of money (less that *EP*) that would yield the same amount of utility as the risk. Calling that amount of money the *certainty equivalent of the risk*, *CE*, it is defined by $U(CE) = EU$. In Figure 12.2 $CE < EP$. The person is indifferent between having *CE* for sure as opposed to *EP* on average as a risk. The difference between these two quantities, $EP - CE$, can be called the risk premium (*RP*). This is the extra amount of money needed to induce our consumer to take a risk as opposed to a sure thing. In Figure 12.2 *EP* is $-\$2600$ while *CE* is $-\$3700$. Our consumer is indifferent between losing \$3700 for sure versus taking this risk, even though on average only \$2600 will be lost. This is because

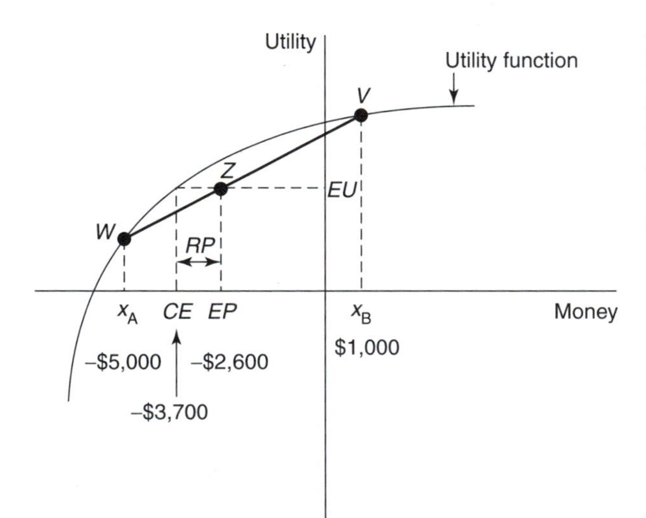

Figure 12.2 Expected utility. x_A, Payoff with probability π_A; x_B, payoff with probability $1 - \pi_A$; *EP*, expected payoff $= \pi_A x_A + (1 - \pi_A) x_B$; *EU*, expected utility from risk; *CE*, certainty equivalent of risk; *RP*, risk premium $= EP - CE$.

of the risk involved, the risk of possibly losing $5000. The difference between the two figures, $1100, is the risk premium, the amount of money our consumer would be willing to pay to convert the risk to a sure average value of *EP*.

The utility function, as drawn in Figure 12.2, is for someone who is risk averse. It is easy to visualize the utility function drawn in Figure 12.2 being more linear. As the utility function becomes closer to a straight line, the risk premium would decline. If the utility function were a straight line, there would be no risk premium—the consumer would be risk neutral. It is also possible to draw utility functions for consumers who are risk loving. This is not the common assumption however; risk aversion is most commonly observed and most consistent with intuition.

B. The Value of Risk Reduction

The next question concerns actions that reduce risk. If, at a cost, we can reduce a risk, how much is it worth to reduce that risk? In the previous section, we saw that the certainty equivalent was one way of converting a risk into a sure thing. We now look at that question a little more carefully and from several different perspectives.

To be concrete, think of two examples of risk reduction. In the first, we have a small community with a number of hazardous waste landfills in the surrounding area. The people in the community have private water wells that are risky because they may become contaminated if a landfill leaks. The risk reduction policy could take many forms, but let us assume that a municipal water supply, with frequent testing for water quality, is being proposed.[7] It will cost money but it will reduce risk. How much is it worth to the community to reduce (perhaps eliminate) the risk?

Our second example is a small but unique ecosystem located in a highly desirable coastal area, near a large and growing city. We know from past experience that many of our most valuable drugs have come from obscure plant and animal species. Although we do not know how valuable our coastal ecosystem is, we know there is some chance that one of the species may become very valuable. The risk, then, is that we may find something valuable in the ecosystem or we may not.[8] We usually think of environmental risks as being harmful, but this is a risk that is beneficial, in contrast to the previous example of the hazardous waste landfill. The risk reduction policy that is being considered is housing development, which will reduce the risk by destroying the ecosystem. The housing development has value (for housing). How does that compare to the lost value from reducing the risk?

Focusing on the example of the hazardous waste landfill, there are two possible states of the world that may materialize: the landfill leaks (L) or it remains sound and secure (S). The probability of L is π_L, which means that the probability of S is $1 - \pi_L$. We are interested in how an individual views risk reduction, though ultimately we will need to aggregate the views of individuals to determine if the risk-reducing municipal water supply (*W*) is a good idea. For our individual, let utility depend on three things: income, the state of the world (L or S), and whether the municipal water supply is available ($W = 0$ or $W = 1$): $U_L(Y, W = 0 \text{ or } 1)$ if L occurs, $U_S(Y, W = 0 \text{ or } 1)$ if S occurs. We will assume our individual has income *Y* available. This is slightly different from the previous section where the payoff was state-dependent. Now we make the utility itself state-dependent. This is basically the same thing: if the landfill leaks, the consumer will have less utility holding everything else constant than if the landfill does not leak.

We now turn to valuing the water supply in terms of risk reduction. We are interested in a measure of the risk-reducing benefit of the water supply so that we may compare that to its cost. Conceptually, there are two ways of thinking about the benefit. One would be to look at how beneficial the water supply is if the landfill leaks (state L) and how beneficial it is if the landfill remains sound (state S). The expected value of these benefits (taking into account the probability of the two states occurring) would be one measure of benefit. We call this the expected surplus.

Another way to look at the benefits would be to ask what quantity of money would our consumer be willing to pay up-front in order to assure the presence of the water supply, no matter which state of the world turns up. We call this the option price. The intuition is that this would be the price of having the option of the water supply, no matter which state turns up (S or L).

Are these two measures the same? It turns out that risk aversion (as well as income effects) makes the two measures different. Which is a better measure of benefits? That is difficult to say although if money needs to be generated in advance to pay for the water supply, the option price may be the best measure. In any event, it is useful to examine these two measures in more detail.

First consider the expected surplus measure. To compute this, we need to know the value of the municipal water supply in each state of the world and then look at the expected value of this based on the probability of the different states of the world occurring. We would compute the economic surplus of the water supply if the landfill were to leak, V_L, and the economic surplus from the water supply if the landfill does not leak, V_S. The *expected surplus* would be defined as

$$ES = \pi_L V_L + (1 - \pi_L) V_S \tag{12.2}$$

This is one measure of the value of the municipal water supply and, in fact, is very commonly used to value public projects under uncertainty. Where can we obtain V_L and V_S? They can be derived from the underlying utility functions:

$$U_L(Y - V_L, 1) = U_L(Y, 0) \tag{12.3a}$$

$$U_S(Y - V_S, 1) = U_S(Y, 0) \tag{12.3b}$$

Equation (12.3) states that we should compare utility with the water supply ($W = 1$) with utility without the water supply ($W = 0$). The value of the water supply under state of the world L is the amount income must be reduced to make utility with the water supply (and with lower income) equal to utility without the water supply (but with no income reduction). Equation (12.3) can in theory be solved for the unknowns, V_L and V_S.

The second measure of the value of the municipal water supply is how much money our individual would be willing to pay for the water supply, up front, before uncertainty is resolved, the *option price*. Option price can also be defined using our state-dependent utility functions:

$$\pi_L U_L(Y - OP, 1) + (1 - \pi_L) U_S(Y - OP, 1) = \pi_L U_L(Y, 0) + (1 - \pi_L) U_S(Y, 0) \tag{12.4}$$

What Eq. (12.4) indicates is that option price is the amount of money taken from income regardless of which state of the world is realized, so that the expected utility with the water supply but with the reduced income is equal to the expected utility without the water supply but with full income. Our consumer would be indifferent between the two sides of Eq. (12.4). Equation (12.4) can be solved for the unknown OP.

An obvious question is, are there any differences between option price and expected surplus? To answer that question, we define yet a third way of measuring how the consumer feels about the municipal water supply. But before doing so, we need to introduce the idea of a pair of *state-dependent payments* for the water supply (P_L,P_S). The idea is that you sign an agreement, in advance, to pay P_L if in fact the L state develops and P_S if in fact the S state develops. We ask, what are all the possible pairs of state-dependent payments (P_L,P_S) that will equate expected utility with the water supply with expected utility without the water supply (and without the payments):

$$\pi_L U_L(Y - P_L,1) + (1 - \pi_L)\, U_S(Y - P_S,1) = \pi_L U_L(Y,0) + (1 - \pi_L) U_S(Y,0) \qquad (12.5)$$

You will note that Eq. (12.5) is a slight variation on Eq. (12.4). In Eq. (12.4), the same value (OP) is used for both states of the world: $P_L = P_S = OP$. In Eq. (12.5), we allow the payments to be different in the two states of the world, but in such a way that expected utility with the water supply and with the payments is the same as expected utility without the water supply and without the payments.

All of the possible (P_L,P_S) pairs that satisfy Eq. (12.5) have been plotted in Figure 12.3. The locus of points is concave downward primarily because of risk aversion on the part of our consumer. If the consumer were risk neutral and there were no income-effects, the locus would be a straight line. One point we know is on the locus: (OP,OP). Since Eqs. (12.4) and (12.5) are so similar it is easy to see that (OP,OP) satisfies Eq. (12.5) and is thus the point on the locus in Figure 12.2 that intersects the 45° line.

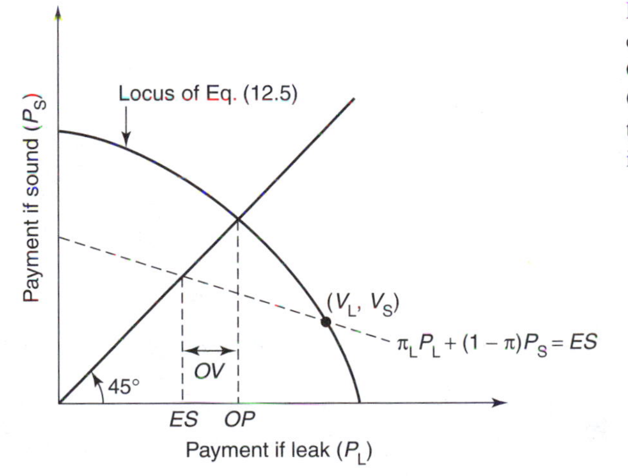

Figure 12.3 Willingness-to-pay locus for risk reducing investment. OP, Option price; ES, expected surplus; OV, option value; (V_L,V_S), state-contingent surplus from risk reducing investment.

Another fact to notice is that V_L and V_S as defined by Eq. (12.3) satisfy[9] Eq. (12.5); thus this pair is on the locus in Figure 12.3. But what is *ES* on the graph? If we plot a variant of Eq. (12.2), namely the line

$$\pi_L P_L + (1 - \pi_L) P_S = ES \tag{12.6}$$

on the graph, it must pass through the point (V_L,V_S), since that point satisfies Eq. (12.6). This line has a slope of $-\pi_L/(1 - \pi_L)$ and is shown in Figure 12.3, passing through (V_L,V_S). But note that (*ES*,*ES*) also satisfies Eq.(12.6). This means that the point at which Eq. (12.6) intersects the 45° line defines expected surplus. This is shown on Figure 12.3.

Note that these two values (*ES* and *OP*) are different, at least they appear different in Figure 12.3. This is because of risk aversion. If our consumer were risk neutral, the locus shown in Figure 12.3 would be essentially a straight line. And it is easy to see that if the locus is a straight line, $ES = OP$. But with risk aversion, the two quantities are different. The difference is called the *option value*: $OV = OP - ES$. Recall that option price is the amount of money that can be paid up front to guarantee the option of having the water supply, no matter which state of the world materializes. The option value is the extra amount of money, over and above expected surplus, that the consumer would be willing to pay to be sure of having the municipal water supply in place under either state of the world.

Option value is analogous to the risk premium we developed in the previous section. There is a difference however. Option value depends on attitudes toward risk as well as income in different states of the world and is thus a little more complex than the risk premium. In fact, option value can be negative (see Schmalensee, 1972), as is alluded to in one of the problems at the end of this chapter.

Which measure, option price or expected surplus, is the most appropriate measure of the value of the municipal water supply? Since money is needed ahead of time to build the municipal water supply, it is logical that the quantity of money our individual is willing to pay in advance is the best measure—the option price.[10]

We return briefly to our other example of the unique ecosystem that will be destroyed by development. The two states of the world in this case might be discovery of a useful species or nondiscovery of a useful species. The expected surplus would involve the probability of these two states of the world occurring as well as the economic surpluses associated with discovering a useful species and not discovering a useful species. The option price would be the amount of money an individual would be willing to pay, up front, to preserve the option of finding a valuable species. The option value would be the amount that would need to be added to expected surplus to account for keeping the option of discovery open.

C. The Value of Information

In the previous section we implicitly assumed that the municipal water supply had to be installed before the hazardous waste landfill leaked. This was because it takes time to build a water supply and it also may be difficult to know when the landfill leaks. But suppose we had better information on whether the landfill was intact or not. If we know that the landfill has leaked, we will have plenty of time to build a municipal water supply to counteract the damage to the water supply. What would be the value of such advance information on the condition of the landfill? This is not idle speculation because it may be possible, at a cost, to reduce or eliminate the uncertainty about whether the landfill has leaked (for instance, using monitoring wells). This may save resources because we may need to take action, such as building a water supply, only *if the leak occurs*. If the leak does not occur, we need do nothing.

The important question here is, how much would we be willing to pay to eliminate (or reduce) uncertainty? This does not mean the elimination of the possibility of the landfill leaking. It just means we will know about the leak in time to build the water supply to ameliorate the adverse effects. It turns out that we do not need risk aversion for the information about the leaky landfill to have value. To see this, we will assume our consumer is risk neutral and that there are no income effects between the two states of the world (thus option price equals expected surplus).

The situation is that the municipal water supply costs C_W. If we do not know whether the landfill will leak, we will assume municipal water is a good idea (i.e., $ES = OP > C_W$). But if we could wait and see if the landfill leaks before constructing the water supply, we would build it only if a leak occurred, and not otherwise. In other words, $C_W > V_S$. Recall that V_S is the value of the water supply if the landfill remains sound.

We consider two cases, one in which the water supply has to be built before uncertainty is resolved (U) and the other case in which the water supply will be built only if it turns out that the landfill leaks; the case of perfect information (P). If we build monitoring wells, we have the case of perfect information, P. If we choose not to build monitoring wells, we have case U where we must build the water supply before knowing the state of the world. We are interested in the net benefits (expected surplus less expected cost) for both uncertainty (B_U) and perfect information (B_P):

$$B_U = ES - C_W = \pi_L (V_L - C_W) + (1 - \pi_L) (V_S - C_W) \tag{12.7a}$$

$$B_P = \pi_L (V_L - C_W) \tag{12.7b}$$

In Eq. (12.7a), we have used the definition of expected surplus found in Eq. (12.2). In Eq. (12.7b), the water supply is built only if a leak occurs. If a leak does not occur (event S), the water supply is not built so the net benefits are zero; thus the form of Eq. (12.7b). The expected value of perfect information (*EVPI*) is simply $B_P - B_U$:

$$EVPI = B_P - B_U = (1 - \pi_L) (C_W - V_S) \tag{12.8}$$

Note that Eq. (12.8) is positive because, by assumption, the water supply is a bad idea if the landfill does not leak (which is the same as $C_W > V_S$). As might be expected, having more information is valuable because it gives more flexibility. The water supply need be installed only under certain conditions. If a set of monitoring wells around the landfills will provide perfect information, those monitoring wells will be a good idea provided they cost less than *EVPI*.

D. Irreversibilities

A variant on the discussion of the previous section applies when information arrives over time. Our example of the unique ecosystem threatened by development is particularly appropriate here. Consider two time periods, today and tomorrow. We can do two things. We can build houses, destroying the ecosystem, or we can defer building, protecting the ecosystem. If we defer building houses today, we can always change our mind tomorrow. If we build houses today, destroying the ecosystem, we cannot reverse our action tomorrow. The ecosystem is gone. This is an example of an *irreversibility.* Development is irreversible, at least as far as the environment is concerned. There is a certain asymmetry here. If we preserve a piece of land and tomorrow decide to build houses on the land, preservation has not hindered our ability to use the land for development. On the other hand, if we destroy an ecosystem and then realize next year that we really need that ecosystem, we cannot reconstruct it. Often (but not always) environmental damage is irreversible, whereas the economic activity associated with that damage is deferrable.

In the first of our two time periods, we are uncertain about the value of our ecosystem (and perhaps also uncertain about the value of development, but that is less important). Between the first and the second time period, we acquire information. In the extreme, we could assume we acquire perfect information. Thus in the second period we know exactly how valuable the ecosystem is and how valuable development is. In the second period we can take action under conditions of certainty.

What might a rational policymaker do? One alternative would be to measure the expected surplus in the first period and decide if development is a good idea, ignoring the fact that information will be acquired between the two periods. If development looks good, then the decision maker will develop. Now information arrives. If it turns out that the ecosystem is quite valuable, decision makers may want to undo their previous decision to develop. But they cannot; destruction of the ecosystem is irreversible. If decision makers hold off in the first period to see how valuable the ecosystem really is, all that is lost is the value of the development in the first period. Development can always be pursued in the second period.

If we hold off on development in the first period, the information that arrives will have value because it allows us to condition our actions on the information. We call the value of that information, provided development does not occur first, the *quasioption value.*[11]

Example: Fisher and Hanemann (1986) provide a nice numerical example of quasioption value,[12] though the example is based only partially on real data. They are concerned with availability of genetic material for maize (corn), which they indicate is the Western Hemisphere's most important food crop. They point out that current high-yield varieties are very homogeneous and thus highly susceptible to disease and other stresses. In the 1970s, a perennial maize plant that grows wild in a very small part of Mexico was discovered. Such a plant has great potential for strengthening the varieties used in commercial food production. Yet agricultural and other development could have destroyed this promising addition to our genetic toolbox. They use this discovery to motivate protecting ecosystems for the potential benefit of future agricultural crops.

The setup[13] is a two-period model with an area of land that contains an ecosystem that has potential benefits to commercial maize production. The land has potential development benefits of $3 billion. The value of preserving the land is harder to quantify. They

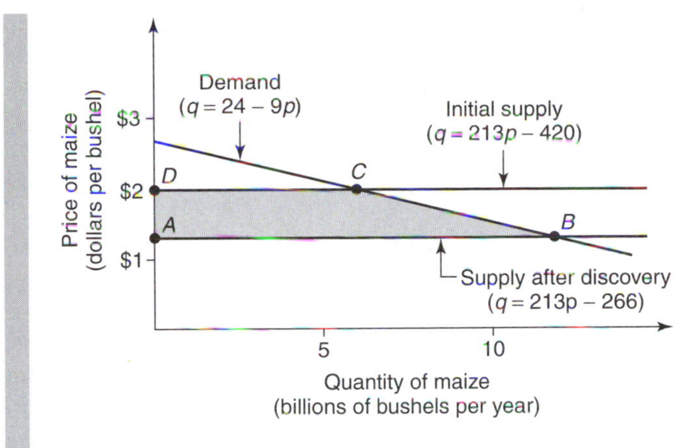

Figure 12.4 Demand and supply for maize, with surplus from genetic discovery. *ABCD*, Surplus gain from lowered supply curve.

assume that there is a 25% probability that new genetic material will be discovered that will benefit commercial maize production. Figure 12.4 shows demand and supply curves for maize in the United States, based on empirical estimates of the market in the United States (ignore the fact that the United States is only a portion of the world maize market). The upper supply curve is the current supply curve for maize, resulting in a price of approximately $2 per bushel. If new genetic material is discovered in the threatened ecosystem, a more productive maize plant will result, costs will be lowered and the supply curve will drop as shown, resulting in a price of approximately $1.30 per bushel. The benefit (surplus) of this discovery is the shaded portion of Figure 12.4. The shaded portion has a value of approximately $3 billion (similar to the development benefits), but because there is only a 25% chance of it occurring, the expected benefit is only about $0.75 billion.

So what can we conclude from this? The expected surplus from avoiding development is less than a billion dollars, whereas the benefits of development are $3 billion. So development looks like a good idea. But if we wait, we get to see whether the ecosystem has any genetic material that is of benefit to agriculture. It turns out that the quasioption value is $0.3 billion, nearly half the expected value of the ecosystem benefits. This means that if the cost of waiting, the cost of deferring development until next period, is less that $0.3 billion, it is worth deferring development.

We see that what looks like a case for development may in fact be a case for deferring such development until the future. This is why the quasioption value is such an important concept when actions result in irreversible damage to the environment. In fairness, the quasioption value is not always significant; this example could have been constructed to result in a very small quasioption value.

III. REGULATING RISK WITH LIABILITY

Risk is of course not new to our society. Driving a car, or in earlier days riding a horse, can lead to accidents. Despite this, we clearly do not want to eliminate risky activities from society, only reduce risks to acceptable levels. How can we be assured that people will take appropriate care when driving, riding, or otherwise engaging in risky activities? We can develop laws that indicate how people should behave in undertaking risky activities—for instance the vast body of law in most countries regarding driving motor vehicles. However, there is an element of moral hazard (hidden actions) that cannot be proscribed (in the case of driving, one must be alert and drive cautiously, behavior that is not easily observed). It is also difficult, or at least burdensome, to develop a set of laws for all possible risky activities.

Around the world, a body of law has developed over the centuries that deals with risky activities—liability. Although such laws may differ somewhat from country to country, the English common law approach of liability is illustrative.[14] The basic idea is that anyone engaging in an activity that involves risks to other people may have to compensate any victim of an accident. When a victim is injured in an accident, the victim may sue the injurer to recover damages. Contrary to popular belief, the primary purpose of liability law is not to compensate the injured but rather to provide incentives to potential injurers to behave responsibly and take precautions when engaging in risky activities. Liability effectively internalizes the accident damage into the cost–benefit calculation of the person undertaking the risky activity. This should ensure the "right amount" of risk or the "right amount" of precautionary behavior.

A. A Simple Model of Liability

The action of liability is shown in Figure 12.5 for a simple example of risky behavior. We will call the person or firm undertaking the risky activity the "potential injurer" and the individual who is damaged should an accident occur, the "victim." The potential injurer can take steps to either reduce the severity of the accident or reduce the probability of an accident occurring or both. We will bundle all of these steps together and call them "precaution." So the potential injurer can take precaution to reduce the effects on the victim.

With no precaution, the effects on the victim may be large. If we multiply the probability of an accident by the damage to the victim should an accident occur, we obtain the expected damage. For instance, if there is a 10% probability of an accident that causes

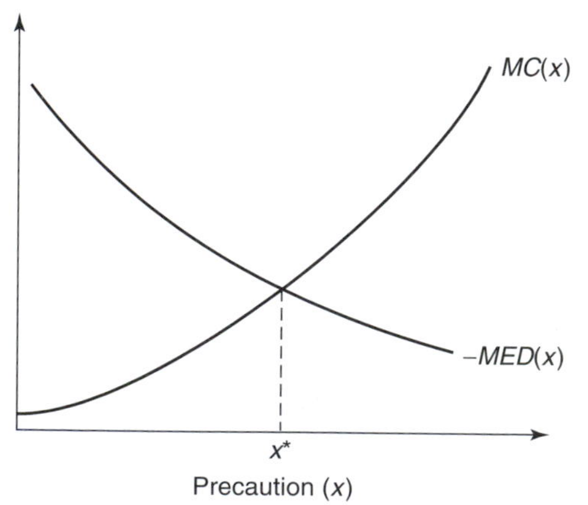

Figure 12.5 The socially optimal amount of precaution. $MED(x)$, Marginal expected damage; $MC(x)$, marginal cost of precaution; x^*, efficient amount of precaution.

£10,000 of damage, we say the expected damage is £1000. This allows us to reduce the uncertain event to something certain. So with little precaution, the expected damage will be quite high. The more precaution taken, the lower the expected damage. We can speak of the marginal expected damage as a function of the level of precaution. Since expected damage decreases as precaution increases, the marginal expected damage is negative. Furthermore, we would expect that at low levels of precaution, the marginal expected damage would be quite large (in absolute terms)—a little precaution goes a long way in reducing expected damage. As levels of precaution increase, the marginal expected damage will be smaller (again, in absolute terms). This is shown in Figure 12.5. The negative of the marginal expected damage function ($-MED$) is shown. It exhibits the behavior just described.

Also shown in Figure 12.5 are the marginal costs of taking precaution. These costs of course fall on the potential injurer. Intuitively, when levels of precaution are low, the marginal cost of precaution should also be low. The first precautionary steps will be the cheapest. Similarly, as precaution levels rise, and fewer and fewer opportunities for precaution remain, the marginal cost of providing it should rise. This is shown in Figure 12.5.

Now what will the potential injurer do? Without liability or any other form of intervention, no precaution will be taken. Precaution is costly and the benefits do not accrue to the potential injurer. But if the law says that the costs of any accident must be borne by the potential injurer, then total costs are the sum of the costs of precaution and expected damages. The potential injurer will now seek to minimize this total, which occurs at the point at which the marginal costs of precaution equal the (negative) marginal expected damages, x^*. Note that for any level of precaution less than x^*, the potential injurer can increase precaution at a cost that is less than the money saved in lowered expected damages. Thus it is in the potential injurer's best interest to increase precaution. In an analogous manner, if precaution is in excess of x^*, it is in the potential injurer's best interest to reduce precaution. The optimal level of precaution will be x^*. This is optimal from the perspective of both the potential injurer and society.

B. Strict Liability versus Negligence

Two basic types of liability exist in most countries—fault-based liability (negligence) and no-fault (strict) liability. Strict liability is what was described in the previous section. If an accident occurs, the potential injurer compensates the victim. No matter how careful the potential injurer is, he or she must pay compensation in the event of an accident.[15]

Negligence, on the other hand, is somewhat more lenient toward the potential injurer. Only if the potential injurer has taken less than the right amount of precaution will he or she be liable for damages to the victim from an accident.[16] The idea is that risky activities are an integral part of society. Particularly in the past, risk was all around. The primary goal of negligence is to ensure that all potential injurers are being good citizens and taking the right amount of precaution without being told directly to do so. We all have a *duty* to exercise a certain amount of precaution. How does negligence work? We present a simplified economic perspective on liability, not a legal perspective. If an accident occurs, the victim takes the injurer to court. The court then determines the two curves in Figure 12.5 and "computes" x^*. This x^*, the socially desirable level of precaution, is termed the *legal standard of care*. If the potential injurer is found to have been taking precaution at least equal to the legal standard of care, no damages need be paid. If, on the other hand, the injurer is taking an insufficient amount of precaution, the injurer is liable for all damages.

This is shown in Figure 12.6. In contrast to Figure 12.5, here we see the total costs and damages, not the marginals, though still as a function of the level of precaution. $C(x)$ is the direct cost of precaution to the potential injurer and $ED(x)$ is the expected damage to the victim. The broken line is the sum of these two costs and it reaches a minimum at x^*, the legal standard of care. The total costs to the potential injurer are shown by the heavy line. At levels of precaution below x^*, the potential injurer's costs include the damage to the victim. For levels of precaution in excess of x^*, the potential injurer's costs are simply direct precautionary costs, since he or she will not be held liable for damages. Note that costs for the potential injurer are still a minimum at x^*. So if potential injurers are acting in their best interest, they will choose x^* as their level of precaution. This is a socially efficient outcome.

So are these two approaches basically the same, since they both achieve the socially desirable level of precaution? From the point of view of ensuring the right amount of risk in society, both rules give the same efficient result. However, distributionally they are different. In the case of negligence, your neighbor may be engaging in dangerous activity that causes you harm. But provided the neighbor was taking the right amount of precaution, you have to bear the costs of your ill fortune, at least under negligence. So although you do not benefit from your neighbor's risky activities, you may have to bear the costs. Some would argue that is not fair.

One striking advantage of negligence, however, is that the courts are much less burdened. And establishing the legal standard of care can be a costly process. Only when an injurer may have been negligent, do the courts need be involved. And, arguably, no rational injurer who is aware of the costs and benefits of his or her actions should take precaution that is less than the legal standard. On the other hand, with strict liability, the courts become involved every time an accident occurs, to assess damages.

When does negligence apply and when does strict liability apply? Negligence is the traditional form of liability. In recent years in the United States, more and more types of

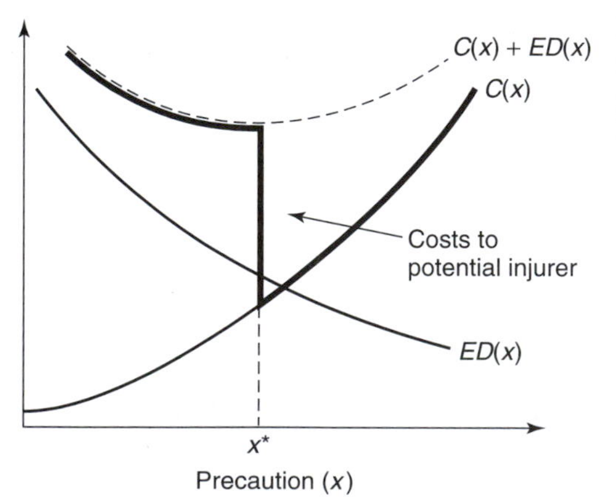

Figure 12.6 Total costs under a negligence rule. $C(x)$, Costs of precaution; $ED(x)$, expected damage to victim; x^*, socially optimal level of precaution.

harm have been legislatively moved to the realm of strict liability. The largest category of harms that falls under strict liability is product liability—when a product injures a customer or user.

C. Joint and Several Liability

Over the past several decades in the United States and Europe, another form of liability has emerged that is particularly important for environmental damage—joint and several liability. Refer back to the example of the tortuous chain of causation connecting hazardous waste in a landfill and contaminated groundwater that is used for drinking. This can present a problem for establishing causation, usually necessary to collect damages under negligence or strict liability. The groundwater becomes contaminated and an individual develops cancer. Whom should the victim sue? Whose pollution caused the cancer? Suppose a landfill is owned by a small company that is paid to dispose of hazardous wastes from several large local chemical companies. The landfill leaks and the landfill company goes bankrupt. Whom do you sue?

These problems are common to environmental hazards and risks. Long chains of causation make it easier to establish that someone caused damage than to establish who caused the damage. This is compounded by the very long time lags involved in environmental damage. During these long lags, ownership of companies change, some firms go bankrupt, records are lost, and memories fade. The end result is that liability is more difficult to establish and thus damages are more difficult to assess and collect. The important role of liability is to encourage potential injurers to take precautions when they engage in risky activities. If one is unlikely to be sued, even if behaving negligently, then the incentive effects of liability are lost and there will be an insufficient amount of precautionary behavior.

One solution to this problem is the concept of *joint and several liability*. If several parties are potentially responsible for causing an accident, then suit may be brought against any one for all of the damages. Thus in the case of the bankrupt landfill, injured parties may choose to sue the richest of the various chemical companies that contributed waste to the landfill, and sue for all damages. The defendant may then choose to try to recover damages from other potentially responsible parties. The idea is that someone has to absorb the costs of the accident and it might as well be one of the parties that contributed to the accident, rather than the victim. If the victim has to recover from each individual that might have contributed to the accident, it is much less likely that suit will be brought and thus it is less likely that the incentive properties of liability will work in the proper way.[17]

IV. LIABILITY VERSUS DIRECT REGULATION

Though liability requires the involvement of government, to promulgate laws and administer courts, liability is much more decentralized than direct regulation of safety or environmental hazards. In direct regulation, the government indicates both the level of safety to be provided and the appropriate precautionary behavior. This requires a great deal of *ex ante* (in advance) information collection and rule construction. Similar information must be collected in the case of liability, but *ex post* (after an accident). If risk is low and accidents infrequent, liability has significant informational and thus efficiency advantages.

A. Disadvantages of Liability

So why isn't liability always preferred to direct regulation? One answer is that for hazards that are common, and risks significant, it may be more efficient for a regulator to define appropriate precautionary behavior than to burden the courts with that task for countless lawsuits. This is a cost argument: regulation in some cases can involve lower costs for information acquisition.[18]

Another problem with liability is that suit is not always brought, even when it should be. This may be for a variety of reasons, not the least of which is the cost (including inconvenience) to the plaintiff of bringing suit. If an injurer is not held responsible for damage when it should be, the incentive properties of liability will be diluted.

Another problem with liability is bankruptcy. A firm cannot be successfully sued for damages that exceed the net assets of the firm; the firm will simply go bankrupt if it must pay damages that exceed its value. This means that a firm will not take the proper precautions to avert an accident with very large damages; at least liability does not provide the proper incentives.[19]

Another problem with liability is that the firm, *ex ante*, is uncertain about how the courts will actually determine the legal standard of care, should an accident occur. If a firm is risk averse, it may provide excessive care because of this uncertainty (hedging its bets). Furthermore, there may be a systematic bias in how a firm perceives the legal standard of care as determined by a court. This may be due to deep pockets of the firm, the reputation of the firm, or many other reasons.

B. Combined Use of Liability and Regulation

Frequently we see the combined use of liability and direct regulation. For instance, with prescription drug safety in many parts of the world, new drugs must pass a rigorous approval process before being marketed to the public. However, drug manufacturers still face liability for any problems the drugs might have. Laws govern the proper transportation and disposal of hazardous wastes, whereas other laws govern liability for hazardous waste spills or accidents.[20]

So why do we have both types of regulation? Is use of both redundant? Or are the two types of regulations somehow complementary? There are two views of this question. Both involve elaborating on the simple view of liability and accidents that we have described above. Both approaches involve defects in liability and direct regulation; thus the use of both approaches achieves a better result than either alone.[21]

Suppose we are concerned with regulating an activity that leads to accidents of various sizes. The activities are not all identical and potential injurers know what the accident size is associated with their activities. For any accident size, D, define the optimal level of precaution as $x^*(D)$. Figure 12.7 illustrates how x^* rises with D, up to the maximum accident size of D_{max}.

Now consider what direct regulation might achieve. Here we will assume that the regulator does not have enough information to distinguish between activities that lead to large accidents vs. activities that lead to small accidents. Thus the regulator has to choose one single best level of precaution. That is shown in Figure 12.7 as x_R and it is the same, whatever D is. Note that for small accidents, the regulatory level of precaution is too great; for large accidents it is too small.

Finally, we introduce liability. With liability, the courts are able to determine, after the fact, the level of damage and the amount of precaution that should have been taken. In a perfect world, the amount of precaution associated with liability will be $x^*(D)$, a quantity that varies with D. But suppose that not all suits that should be filed actually are filed. This may happen for a variety of reasons, discussed above. This will tend to lower the level of precaution, since it lowers the expected liability costs for the potential injurer. The level of precaution that will be associated with liability is shown as $x_L(D)$ in Figure 12.7. Note that for any accident size, it is strictly less than x^*. Note also that x_L flattens out for damage in excess of A. This reflects that fact that the firm will not take precautions to avoid accidents that exceed the net assets of the firm, A. If an accident exceeds A, the firm will simply declare bankruptcy. This effectively places an upper limit on precaution.

Which is best? It depends on the accident size. For any given accident size, we prefer the approach (direct regulation or liability) that gets us closest to x^*.[22] Figure 12.7 shows three regions, with the center region favoring direct regulation and the other two regions favoring liability. Which is best? It really depends on the distribution of damages. If most accidents occur with size that favors direct regulation, direct regulation may be best.

What about using both together? If both are used together, direct regulation effectively provides a lower bound on precaution. For large accident sizes, liability will be the driving force behind precaution; for smaller accidents, direct regulation will apply. How-

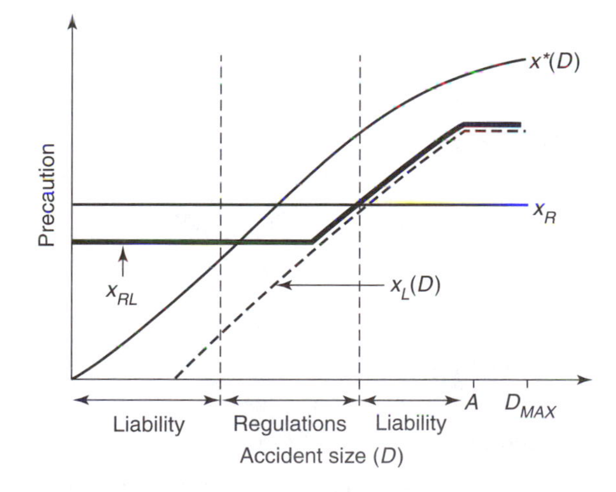

Figure 12.7 Illustration of joint use of liability and direct regulation. $x^*(D)$, Optimal level of precaution; x_R, optimal level of precaution with direct regulation; $x_L(D)$, precaution provided by liability; $x_{RL}(D)$, precaution with joint use of liability and direct regulation.

ever, since regulation applies just to smaller accidents, it should be set at a lower level than when it applied to all accidents. In Figure 12.7, the optimal combination of direction regulation and liability is shown as $x_{RL}(D)$.

Note that the optimal regulation with liability is lower than without liability. This result has significant implications. Suppose, for instance, that hazardous waste storage is being regulated. This means that the regulation governing how wastes are stored should not be designed for infrequent large accidents but rather for smaller accidents; the threat of liability will take care of providing precaution for large accidents.

V. INSURANCE

Environmental risks often involve the slight possibility of very large accidents and damages. Certainly this would characterize the risks associated with nuclear power and hazardous waste disposal. We have seen that both liability and regulation can provide some protection against these risks, though not perfect protection. Many have advocated the use of environmental hazard insurance in conjunction with liability. Insurance has a number of advantages. An obvious one is that there would be some assurance that if an accident were to occur, victims would be compensated; the prospects of a bankrupt injurer would be less onerous. Another advantage would be that insurance companies could provide a level of expert advice to potential injurers intermediate between the centralized regulator and the potential injurer. Insurers that help an insured reduce risks would reward the insured with lower rates. There is an incentive for the two parties to work together. A third advantage to insurance is the standard one: risk exposure is reduced and spread among many. Thus if a firm is risk averse regarding large environmental accidents (and any but the largest firms probably are), insurance provides a way of converting a lottery into a sure payment.

Probably the major question to ask in the environmental context is, will insurance spontaneously emerge from the private sector or are there impediments to the private provision of insurance? If so, can governmental action remove these impediments?

A. Conditions for Insurability

For an insurance market to spontaneously emerge to insure a particular environmental risk, it is necessary that the risk be attractive to insurers and that the premiums be attractive to potential injurers. Generally, for a potential loss to be insurable six basic criteria must be met:[23]

1. The loss must be amenable to risk pooling.
2. There must be a clear loss.
3. The loss must be in a well-defined period of time.
4. The frequency of loss must allow a premium calculation.

5. Moral hazard must not be too severe.
6. Adverse selection must not be significant.

Each of these conditions needs some explanation.

1. Risk Pooling. Because most of us are risk averse, we dislike being exposed to risks that have the potential to create large losses—large with respect to our individual assets. This is why we seek fire insurance on our house or collision insurance on our new car. On the other hand, we bear the risk of small losses without much problem—we typically do not insure small appliances from the risk of breaking. Although it might appear that insurance works by contracting with a less risk-averse large firm to bear a risk, that is not really the case. Insurance works through risk pooling.

Fire insurance is a good example. Suppose the risk of fire causing $10,000 damage to a house is 0.001 in any particular year (one in a thousand). The expected value of this risk is $\$10{,}000 \times 0.001 = \$10$. If an individual does not insure the house, in most years there will be no damage; but occasionally, a big loss will occur. Suppose 10 people get together and annually contribute $10 each to a pool to cover their fire losses. The pool will increase by $100 annually and roughly every 100 years an accident will occur. It is still possible that several accidents may occur in a row and bankrupt the group. Now suppose 100,000 people get together and each contributes $10 annually to a fund. The fund can expect approximately 100 fires per year. There may be a few more one year and a few less another year. What happens as we expand the number of houses in the pool is that the variability of the annual loss declines. With the 100,000 houses in the pool, it is almost certain that there will be approximately 100 houses that are damaged each year. We have moved from a situation of no losses most years with a large loss once in a thousand years (on average) to a case in which we have 100 losses every year with much smaller variability. This is risk pooling.

For risk pooling to work, the multiple risks must be uncorrelated as in the above example. Whether a fire happens in one house is unrelated to whether a fire happens in another house. Some risks have a systematic component. Earthquake losses are an example of this. Suppose a company insures 100,000 houses against earthquakes. What will happen is that in most years there will be no claims, but when an earthquake hits many houses will file claims, particularly if there are only a few areas susceptible to severe earthquakes. Thus insuring the 100,000 houses has not smoothed out the losses as in the fire insurance case. An insurer may be subject to wild fluctuations in claims. Nothing may have been solved. The risk has been shifted to a company with more assets, but the company is covering more risks. In situations such as this, effective insurance is unlikely to be provided by the market.

2. Clear Loss. For an insurance market to function, it must be clear when a loss has occurred. For instance, it would be difficult to insure against mental anguish from an accident because it is impossible to determine the severity of mental anguish when a loss has occurred—it is private unverifiable information possessed by the injured party.[24] And most losses that are difficult to verify are equally easy to counterfeit. There are of course losses that are easy to verify and difficult to fake—death for instance.

The loss from an environmental hazard may be difficult to establish. For instance, suppose it is proved that company X has contaminated the groundwater with a known carcinogen, and a city's water supply has been contaminated exposing the citizenry to a risk. Has a loss occurred? Is exposure to the risk a loss or must someone contract cancer? Has anyone who contacts cancer suffered a loss? Keep in mind that cancers are not uncommon and it may be difficult, if not impossible, to establish that a particular cancer is due to the contaminated water.[25]

Another feature associated with environmental losses is defining exactly what is covered. Is it feasible to insure a company against all possible environmental obligations, including those as yet undiscovered? In covering a risk, it is important that the potential losses be clearly delineated as well as verifiable.

3. Loss in Well-Defined Period. Why is it difficult to insure a risk for which the damage may occur at any time in the future, without time limit? The simplest explanation is that eventually nearly everything happens. Any house will eventually burn down (if it is not torn down). Every hazardous waste landfill will eventually leak. Furthermore, as time goes on, information changes. It is difficult to appreciate fully the potential damage from a landfill over the coming year, let alone in the coming 10 or 20 years. In 10 years, the institutional environment, as well as the understanding of the physical environment, may have changed substantially. This is particularly the case with environmental risks. So for an insurer to offer coverage over a very long time period, the premiums will probably be so great as to be unattractive to potential injurers.

Unfortunately, many environmental risks have very long gestation periods. If a chemical company "accidentally" dumps solvents on the ground behind the factory, it may take many years for those solvents to leak into the water table. Furthermore, it may take many years of consumption of contaminated water for people to receive a substantial dose of the carcinogen. Finally, cancers typically have long gestation periods, with decadal delays between exposure to the carcinogen and diagnosis of the cancer. Thus several decades may transpire between the risky activity (dumping the solvent) and the damage (the cancer).

4. Frequency of Loss Allows Premium Calculation. Obviously it is necessary to compute the premium to charge for the insurance. And the premium that is charged is not a minor detail; an incorrect premium can cause the market to fail. If the premium is too large, no insurance will be sold; if it is too small, insurers will lose money or possibly become bankrupt. If we have experience with a loss, it is possible to calculate a premium based on the historic frequency of loss. For instance, we have a great deal of information on house fires and can thus compute the probability that a randomly chosen house will burn down in a year—exactly the information needed to compute a premium.

If we do not have much experience with a loss, it is necessary to speculate on the loss frequency and magnitude. If there is a great deal of uncertainty about the risk (as there often is with environmental risks), it is likely that the insurer will err on the high side in estimating damage or loss frequency. After all, experience tells us that with environmental risks, time has the tendency to generate bad news more frequently than good news.

5. No Moral Hazard. The opportunity for the insured to influence the occurrence of the accident must be minimal. This does not preclude observable actions by the insured, such as using a double instead of a single hull on a tanker for transporting crude oil. The problem is that when there is a risk of loss, people are careful. If the insurer removes the risk of loss, the incentive to be careful is gone. If the insurer can monitor the amount of care, there is no problem. If there are significant activities that are hidden to the insurer, the insurer must assume the worst. This raises premiums, often to the point at which they become unattractive to the injurer.

An example of this would be insuring bicycles on a college campus, where bicycle theft is a popular means of raising revenue. If an insurer noted that approximately 1% of all bicycles on campus are stolen every year, and the average value of those bicycles is $200 (which gives an expected loss of $2 per year), and then offers a theft policy for $5 per year, covering $200 of loss, the insurance company could expect to make money. But what happens is that those with insurance will be much less likely to lock their bike, more likely to leave it unattended overnight, and generally to take less care to prevent it from being stolen. Thus the insurer may be bewildered to find that the loss rate is 5%, for an average loss per policy of $10, twice the premium. The insurer raises its rates to $15 and finds that her loss rate goes up because people are buying insurance instead of locks. Eventually the premiums rise so high that the market collapses. This is an example of how moral hazard can prevent an insurance market for a particular risk from emerging.

Fortunately, many environmental risks are generated by large companies in which it is easier to monitor how much care is being taken to prevent hazards. With a large company it becomes cost effective for the insurer to monitor the activities of the injurer, using audits and inspections as well as involvement in safety procedures. This tends to reduce the significance of moral hazard.

6. No Adverse Selection. Adverse selection is a problem for regulation, as we saw in the previous chapter. It is also a problem for insurance. If there are some injurers that are more likely to have large losses, the insurance company must be aware of that in calculating premiums, since premiums are related to expected loses. If there are characteristics of the injurers that (1) influence the likelihood and magnitude of a loss and (2) are unobservable to the insurer, the insurance company must assume the worst and charge a comparatively high premium. This premium will be unattractive to injurers who are comparatively good risks; thus they will not buy insurance. This means that the only injurers that will be attracted to insurance will be the bad risks; this may further increase rates. The outcome will be either no market at all, or a much thinner market than desired.

An example of adverse selection is private unemployment insurance. Suppose a firm offered a policy that would pay you your wage should you be terminated in your job during the term of the policy. The insurance company has no economical way of determining how likely a person is to lose his or her job. Furthermore, the people most likely to take out such insurance are the people who feel their job is in real danger. Because of this, the insurance company must set rates high, which makes insurance unattractive to all except those in the greatest danger of job loss. In the end, only those people who have a very large chance of losing their job will be interested in insurance; premiums for this will be so high as to be unattractive to nearly everyone. The market collapses.

Adverse selection is probably not a severe problem for environmental risks because the insurance company can determine whether a company is more likely to have an accident based on past performance or facility characteristics (in the case of a landfill).

B. Insurability of Environmental Risks

The next logical question is, how well do particular environmental risks satisfy these conditions of insurability? Take for instance the insurability of a hazardous waste dump. Wastes can leak from the dump into the groundwater, leading to contaminated drinking water and resulting in illnesses and deaths. There are a large number of such dumps in existence now and any leakages should be uncorrelated from one dump to another. So condition (1) would appear to be satisfied.

It is less clear when a loss has occurred. The damage that is being insured against is health effects from contaminated drinking water. It may be very difficult to determine what the effects actually are. There may be elevated cancer rates, but there may be other processes that could have led to these cancers. Mental trauma from consuming contaminated drinking water will be harder to verify than cancer. Thus condition (2) is not well satisfied for our example of a waste dump.

As is well known, a leaking waste site can take many years to manifest itself in adverse health effects. Clearly an insurer can limit the length of time of coverage, but that may leave the risk underinsured. To be fully covered, the length of time must be nearly indefinite. This violates condition (3).

Leaking hazardous waste dumps, at least those built in recent years, are quite reliable and leak very infrequently. Although this is good from an environmental point of view, it is not good from the point of view of being able to calculate insurance premiums. Consequently, condition (4) is not met.

Does moral hazard exist? Here, chemical risks would appear to pass the test. Although a dumpsite can take action to influence the probability of an accident, most of these are verifiable and observable. An insurance company can inspect an operation from time to time and can observe physical precautions that have been installed to reduce the likelihood of an accident. Thus condition (5) is met.

Adverse selection can probably be handled by observing the claims record of an insured. A new policyholder will be charged premiums assuming they are a bad risk. Over time, with a good record, those premiums could drop. Thus adverse selection would not appear to be a serious problem for chemical risks.

In summary, insurance may emerge to cover some environmental risks but for others, the conditions of insurability are not met. This is unfortunate. However, it may be possible to design insurance instruments or adjust liability law so that more risks become insurable.

SUMMARY

1. Objective risk involves direct observation of the probability of risky events occurring (based on historical experience), such as the probability of a randomly chosen airplane flight ending in a crash. A subjective risk is an individual's perception of the probability of the same event. Subjective risk is often termed uncertainty, whereas objective risk is simply called risk.

2. Risks to which exposure is voluntary are usually valued less than risks that are involuntary.

3. Expected utility is a common way of evaluating the utility of a risk. The certainty equivalence of a risk is the sure quantity of a good, consumption of which gives the same utility as the risk.

4. In considering the value of a risk-reducing investment, the expected surplus is the expected value of the state-dependent surpluses from the risk-reducing investment. State-dependent values are the values after the uncertainty has been resolved and the risky event has occurred. The option price is the maximum sure payment that would be made for the risk-reducing investment prior to uncertainty being resolved. The option value is the difference between option price and expected surplus. It is positive if option price is greater than expected surplus.

5. The expected value of perfect information is the amount that total expected value of a risk will increase if perfect knowledge of the outcome of uncertain events is known in advance.

6. Often environmental damage is irreversible. In such a case, there is an additional value associated with being cautious about causing environmental damage. That is called the quasioption value and is the value of perfect information, conditional on initially preserving the environment.

7. Strict liability involves the injurer always paying damages to the injured in the case of an accident. Negligence involves payment of damage only if the injurer has been negligent, defined as taking less precaution than is socially optimal (the "legal standard of care").

8. When there are several potential parties that could have caused an accident, joint and several liability holds each potentially responsible for all damages.

9. The primary purpose of liability is to induce injurers to take the proper amount of precaution. A secondary purpose is to seek a fair share of the burden of damages in the case of an accident.

10. A major advantage of liability is that the government need acquire detailed information on damages and control costs only when an accident happens that ends up in court. This reduces regulatory costs.

11. Major disadvantages of liability are bankruptcy and factors that discourage valid cases from being brought to court (including the cost of bringing suit). Both tend to dilute the incentives for potential injurers to take the right amount of precaution.

12. Direct regulation (*ex ante* regulation) can be combined with liability (*ex post* regulation) to correct some of the problems of liability. In this case, the level of direct regulation should be set below the appropriate level when liability is not involved.

13. Six basic conditions are needed for insurance markets to spontaneously emerge: (1) risk pooling, (2) clear loss, (3) limited loss period, (4) frequent loss, allowing premium calculation, (5) no moral hazard, and (6) no adverse selection.

PROBLEMS

1. You are in charge of designing a new hazardous waste landfill for Kiev. You are trying to decide whether it is worth spending R1,000,000 to install a special liner that reduces the probability of a leak from 0.001 to 0.0001. If a leak occurs, the expected number of fatal cancers in Kiev can be expected to rise a total of 100. What value of a fatal cancer (in R) is necessary to justify your installation of the special liner? Assume risk neutrality.

2. If a potential victim can take precautions to avoid an accident (such as looking both ways before crossing a street), which liability system yields more efficient levels of precaution, strict liability or negligence? Why?

3. One of the arguments for punitive damages (damages paid to the plaintiff in excess of actual damages) is that such damage payments compensate for the fact that lawsuits are not always filed, even if justified. Explain.

4. Assume you are the President of British Petroleum (BP). You note that the Exxon Valdez spill cost Exxon Corp. approximately $8 billion. You are trying to decide on the thickness of the hull of your new supertanker. Your engineers tell you that the probability of an accident ever occurring (p) is inversely proportional to hull thickness: $pt = 0.002$ where t is the hull thickness in centimeters. But the costs of a thicker hull rise rapidly as the thickness rises. Costs (in million $) are given by $C(t) = t^2$ where t is the thickness in centimeters. Assume an accident's damage is the same as with Exxon. What is the socially desirable hull thickness? If there is a probability of 10% that BP will not be found liable in case of an accident, what will BP choose as a hull thickness (ignoring the fact that BP may be socially responsible and take the socially desirable route)? (*Hint*: Use calculus or graphs and pay attention to units.)

5. The local dry cleaner is concerned about the risks from its use of chemicals, which evaporate into the air and occasionally are spilled onto the ground. The cleaner is concerned about liability for environmental damage. How likely is it that insurance will be available to cover the liability exposure of the dry cleaner? Why?

6. In Figure 12.3, option value is shown as positive (option price is greater than expected surplus). Using the same locus and the same probabilities of each state occurring, change the state-contingent value of risk reducing investment to show how option value could be negative. Why is this counterintuitive? Discuss.

*7. In the context of Figure 12.3, suppose that the risk involved with the hazardous waste landfill is objective. This may be because leaks occur often enough that the probability of a leak can be objectively established. Suppose a bank is willing to loan the municipality money for the water supply with the loan paid off by a payment whose size depends on whether the landfill leaks. The expected value of the payment is equal to the amount loaned. Show on a copy of Figure 12.3, the maximum amount that could be borrowed from the bank to pay for the water supply. Explain.

NOTES

1. Problems arise when the effect is very slight. In such cases, it is very difficult to detect adverse effects using epidemiology, i.e., statistics on the general population. An alternate approach involves laboratory studies. However, in these cases the problem of detecting slight effects also exists. It is compounded by the problem of extrapolating from one species (e.g., mice) to another (humans). In the United States, two major food additive issues of this sort were the artificial sweetener sodium saccharin and the apple growth enhancer Alar.
2. See Slovic et al. (1979).
3. See Slovic et al. (1979) for more discussion of this point.
4. This distinction between risk and uncertainty is due to the English economist Frank Knight. Whether uncertainty is subjective or objective can make a significant difference to the theory of choice under uncertainty. See Mas-Colell et al. (1995) or Kreps (1990) for advanced but general discussions of this issue.
5. For the most part, evidence about how we perceive risks comes from the psychological literature. See Slovic et al. (1979) for a discussion of voluntary vs. involuntary risks as well as other attributes of risk that shade our perceptions of their seriousness.
6. The key assumption necessary to use expected utility theory to value risks is called the independence axiom. Basically, it states that if we compare two risks, our relative ranking of those risks should be independent of other risks we may be taking. Varian (1996) discusses this at an elementary level. Although expected utility theory is the most common way of evaluating risks, there are other approaches that use different assumptions about how people evaluate risks. For an advanced treatment of this issue, see Mas-Colell et al. (1995).
7. This example is suggested by Smith and Desvousges (1988).
8. This is a stylized example, since there may be many other reasons for preserving an ecosystem.
9. Premultiply Eq. (12.3a) by π_L and premultiply Eq. (12.3b) by $1 - \pi_L$. Then add the two equations together and you have Eq. (12.5).
10. Actually, if there is a risk-neutral agent that will fairly exchange lotteries on the future into sure payments today, then there is a quantity even larger than the option price, called the fair bet point, that is the true maximum value of the water supply (see Graham, 1981).
11. The term quasioption value originated with Arrow and Fisher (1974), though its precise meaning has been debated for many years. Hanemann (1989) appears to have had the last word on this, defining the quasioption value in a two-period model such as we have described as the expected value of perfect information conditional on no development occurring initially. Others have defined it as the loss in overall value due to the fact that the environmental destruction is irreversible.
12. See Krutilla and Fisher (1985) for a number of other more in-depth analyses of the importance of quasioption value. For an application to cumulative pollutants such as greenhouse gases, see Kolstad (1996a,b).
13. The model presented here has been simplified somewhat from the model presented in Fisher and Hanemann (1986).
14. The common law tradition is found in England and former colonies, such as the United States. Most of the rest of the world, including most of Western Europe and Asia (and one state of the United States—Louisiana), follow a civil law tradition. Civil law is made by legislatures and courts enforce the written law. Common law is drawn from more diverse sources, particularly past judicial decisions. Refer to Cooter and Ulen (1997) for further discussion of these differences and the apparent convergence of the two approaches in modern times.
15. This assumes that one can establish that what the potential injurer did caused the accident. This may be nontrivial; for instance, consider the difficulty of establishing that a particular chemical "caused" a cancer that the victim developed.
16. We are assuming here, and throughout this section, that the victim is passive and can take no action to avoid an accident. This is obviously not always the case. When the victim can take

some action to avoid or reduce the size or probability of an accident (such as looking both ways before crossing a road), we have negligence with a defense of contributory negligence. An analysis of contributory negligence would take us too far afield. See Cooter and Ulen (1997) for a complete treatment of this and other issues associated with liability.

17. Tietenberg (1989) provides a discussion of joint and several liability as well as an economic interpretation of the concept.
18. One could also argue that having multiple firms carry out detailed cost–benefit analyses of the same risky behavior is more costly than doing it centrally, by the regulator.
19. Shipping oil internationally illustrates this point. In recent years, one-ship companies have emerged to transport oil, thus limiting liability to the value of one oil tanker (MacMinn and Brockett, 1995). Beard (1990) investigates the problems of bankruptcy.
20. In the United States, the primary governing statute is the Resource Conservation and Recovery Act (RCRA). In the United States, the primary law is the Comprehensive Environmental Response, Compensation and Liability Act (CERCLA), which includes provision for the "Superfund," a fund financed by chemical producers to clean up waste sites without establishing fault in advance.
21. The two primary results on the simultaneous use of regulations and liability are given by Shavell (1984) and Kolstad et al. (1990). The example here is based on Shavell (1984).
22. The goal is actually to minimize the welfare losses from deviating from x^*. For the discussion here, a reasonable approximation is to try to get as close to x^* as possible.
23. For a discussion of insurability in general, see Mehr (1986). Insurability in the case of catastrophic environmental risks is covered by Katzman (1988). Cook and Graham (1977) cover the case of insuring irreplaceable goods.
24. A court may award a victim of an accident, which insurance may then cover. The court's finding is observable.
25. This issue is explored by Harr (1995) in a very readable account of a court case in which an attempt was made to connect leukemias with hazardous waste spills.

13 INTERNATIONAL AND INTERREGIONAL COMPETITION

In this chapter, we are concerned with competition and collusion between jurisdictions in the setting of pollution control policy. Jurisdictions may be states or localities within a country or jurisdictions may be sovereign nations. The basic determining factor is that what goes on in one state influences actions in other states. To this end, we first consider how incomes in a jurisdiction affect pollution control. As per capita income increases, can we expect pollution to decrease or does it follow a more complicated path, first increasing and then decreasing (or vice versa)? The next question we ask concerns competition among jurisdictions for capital and the role that environmental regulation may play in that competition. Would we expect states to weaken their environmental regulations to attract capital and thus jobs? A closely related question is the effect differences in environmental regulations among countries have on trade and the location of industry. Will countries with lax environmental regulations end up as "black holes," specializing in dirty industries? The last question we ask concerns the problem of transboundary pollution—pollution generated in one state that causes damage in other states. These transboundary pollutants are becoming more and more important in the world—carbon dioxide leading to climate change and chlorofluorocarbons leading to ozone depletion are only two examples. Control of these pollutants requires agreement among sovereign states. It is not easy to forge a treaty that is effective, enforceable, and to which countries will voluntarily agree.

I. THE INCOME EFFECT AND THE DEMAND FOR ENVIRONMENTAL QUALITY

Why are oil refineries located in poor rather than rich sections of cities? Why do developing countries tend to have more lax environmental regulations than rich countries? Do we expect environmental protection to become more or less important globally as the decades pass? The answers to these questions hinge on how income influences demand for environmental protection. As people become wealthier, they seem to demand more and more enviornmental protection. Conversely, at any point in time, the poor seem to demand less protection (in terms of being willing to pay less for protection) than the rich.

Consider an individual's demand for any good. As the individual's income increases, we would expect the quantity of the good consumed to change, keeping other factors (such as prices) constant. If the quantity consumed increases, we call the good a *normal good*; if the quantity decreases with increases in income, we call it an *inferior good*. If the quantity demanded of a normal good goes up by a greater percentage than the percentage income increases, we call the good a *luxury good*; otherwise it is simply a *necessary good*.[1] These terms are not terribly imaginative, but they do offer an important classification scheme that helps explain differences in concern for the environment among countries and people of different income levels.

It is generally accepted that environmental quality is a normal good; indeed, most goods are normal. Whether it is a luxury good is less clear. As we all know, environmental quality is not a homogeneous good; it encompasses diverse concerns such as drinking water quality, urban air quality, global climate, wilderness access, and species diversity. The question is not whether each of these is desirable, but how demand for these goods increases when income increases. This is a very important empirical question.[2] As the world becomes wealthier, the extent to which environmental protection becomes more or less important depends on the income elasticity of demand for environmental quality. Differences in pollution levels in different countries can be explained by differences in income.

Making the connection between individual demand for environmental quality and societal demand involves a number of steps and assumptions. How income affects individual demand must be translated into aggregate demand. Aggregate demand from the populace must then be translated into demand on the part of government. The government translates consumer demand into binding regulations that apply to polluters. If a government is democratically elected, the actions of legislators will at least approximate the will of the people. If a government is less representative, general demand for environmental quality may fail to be translated into environmental quality.

What makes measuring the effect of income changes difficult is that observed levels of pollution are the result of the interaction of supply of environmental quality and demand for environmental quality. Figure 13.1 illustrates this. Shown in Figure 13.1 is a demand curve for environmental quality for one income level (AB), and two demand curves associated with higher income, one ($A'B'$) corresponding to a modest income effect on demand and one corresponding to a stronger effect ($A''B''$). Also shown in Figure 13.1 is a supply/marginal cost curve for environmental quality (CD), along with a possible alternate supply curve, $C'D'$. The forces that are causing income to increase for the consumer will most likely also change the supply conditions for the polluting firms and in Figure 13.1 we show marginal costs rising from an income increase. This is consistent with the notion that a richer economy will have a larger industrial base, which will increase the marginal cost of providing a given level of environmental quality. (Of course the supply curve could also drop. When incomes rise, institutions may develop that make it easier and cheaper to institute pollution controls.) Note that the original efficient level of environmental quality is point E. With a small upward shift in the demand for environmental quality, the efficient level decreases to F; with a larger shift in the demand curve, the efficient level becomes G, an improvement relative to E.[3]

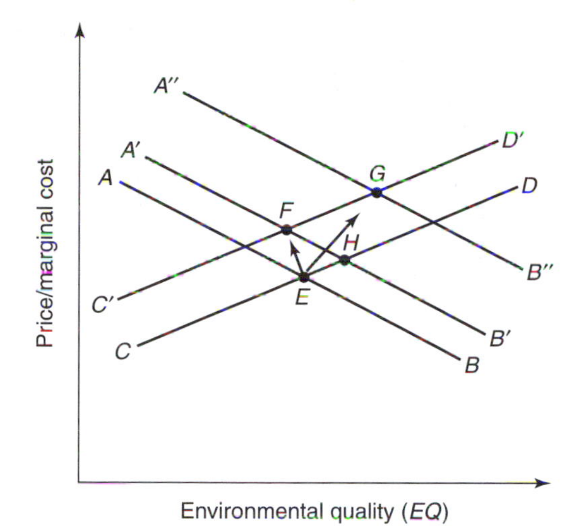

Figure 13.1 Effect of a shift in demand and supply/marginal cost for environmental quality. *AB*, Base demand for environmental quality; *A′B′*, demand for environmental quality with increased income, modest income effect; *A″B″*, demand for environmental quality with increased income, large income effect; *CD*, base supply of environmental quality; *C′D′*, supply of environmental quality, higher income; *E*, base provision of environmental quality; *F*, environmental quality, modest income effect; *G*, environmental quality, high income effect; *H*, environmental quality, increased income, no supply effect.

An income increase, keeping everything else constant has a clear impact on the quantity of environmental quality demanded—increased income increases demand, resulting in a shift from *E* to *H*. But if factors that increase income also shift the supply curve for environmental quality, the overall effect is ambiguous. We could have either deterioration or improvement in environmental quality.

Because differences in environmental quality are most pronounced in comparing different countries rather than regions of a single country, a number of researchers have examined different countries around the world, trying to infer the relationship between national income and environmental quality. Figure 13.2 shows the relationship between income and the most basic of environmental goods—access to clean drinking water. There are two graphs in Figure 13.2, one for 1975 and one for 1986. These graphs were estimated by Shafik (1994) using 86 observations from different countries around the world. The data were more scattered than is obvious in Figure 13.2; only 46% of the variability of the variable of interest was explained by the analysis. Nevertheless, Figure 13.2 clearly shows that as incomes increase, this environmental variable improves. As our intuition would suggest, these graphs show that richer countries have higher levels of at least this measure of environmental quality. Figure 13.3 shows a similar curve for smoke (sus-

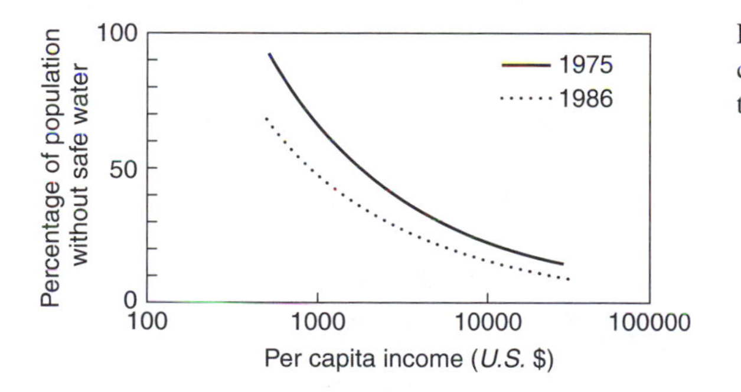

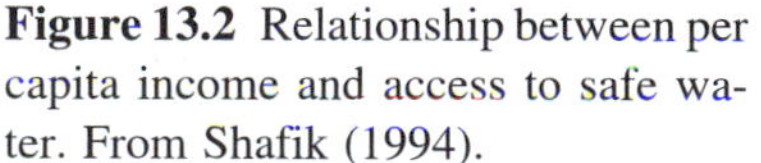
Figure 13.2 Relationship between per capita income and access to safe water. From Shafik (1994).

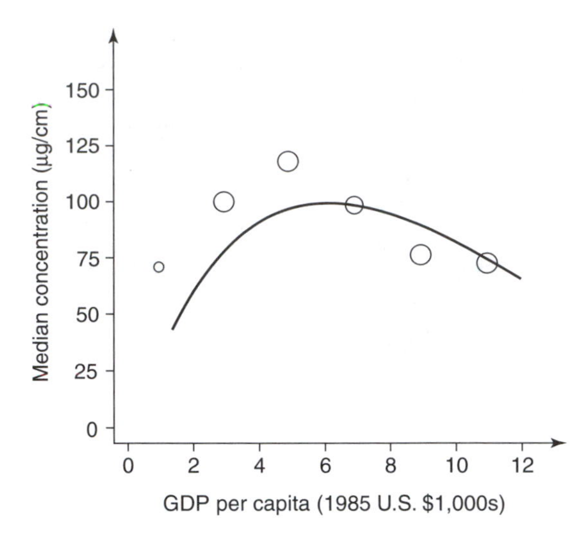

Figure 13.3 Relationship between per capita GDP and smoke concentrations in cities. Circles represent the mean value of concentration median over an interval with the diameter proportional to number of data points. Line is the best statistical fit to data. From Grossman and Krueger (1995), p. 363.

pended particulates). This is somewhat more complicated to present since the smoke concentration varies widely within a country. What is shown is a modified scatter plot of the median concentration of suspended particulate matter (basically smoke) for assorted cities of the world,[4] as reported by the World Bank. Also shown is a statistical fit to nearly 500 data points, done by researchers at Princeton University (Grossman and Krueger, 1995). The data clearly show that for low incomes, as income levels increase, smoke levels increase. But this effect peaks for incomes in the neighborhood of U.S. $6000 per capita, from which point increases in income reduce the concentration of smoke. Some decades ago, Simon Kuznets developed curves showing how income inequality changed as per capita income in a country increases. He found that income inequality initially increased and then decreased. Because of the similarity, curves showing how environmental quality or pollution change with changes in income in a country are known as *environmental Kuznets curves.* Such curves are shown in Figures 13.2 and 13.3.

It is important to be somewhat cautious in interpreting curves such as the ones presented in Figures 13.2 and 13.3. These curves were generated using data for different countries for the same year. However, most people interpret them as showing what will happen to environmental quality as income rises in a specific country over time. This is not the same thing as comparing countries at one point in time. Unfortunately, we do not have a long enough time series of data to examine what happens to individual countries over time. Even if we did, the confounding effects of innovation and technical progress would make interpreting the data difficult. If all countries represented in Figures 13.2 and 13.3 were exactly the same except for income and environmental quality, we might be justified in assuming a country would follow such a curve over time as it develops. However, individual countries are very different from one another and there may be factors that sort countries by income that we are not taking into account. For instance, poor countries may tend to have very high population densities and rich countries may have low population densities and these differences may explain part of the patterns in Figures 13.2 and 13.3. If this were the case, we would not expect a country with high population density to follow the same curve as its income rises. Population density may also play a role in the cost of pollution control. Low population density correlates with a high assimilative capacity of the environment, which is the environment's ability to naturally cleanse itself. At the simplest level, the more pollution is spread out, the easier it is for the environment to deal with it. Thus countries with high population density would have higher costs of providing environmental quality.

The empirical literature on environmental Kuznets curves divides pollutants into three categories, depending on what happens to pollution levels as incomes rise. Figure 13.2 represents the case of environmental quality rising as incomes rises. Figure 13.3 represents the case of environmental quality first falling and then rising. A third category involves continual deterioration as income rises. Evidence suggests that pollutants falling into this category would include generation of municipal solid waste and emissions of greenhouse gases, leading to global warming.[5]

Why are environmental Kuznets curves interesting and/or useful? The policy question is how to improve the level of environmental quality in poor countries.[6] One approach would be for international organizations to invest heavily in cleaning up the environment and protecting natural resources, while at the same time directing economic growth away from pollution-intensive sectors. However, if one believes a relationship such as is illustrated in Figures 13.2 and 13.3, then a viable policy would be to focus energy on increasing incomes. Once the peak has been reached, environmental quality will start to improve without any extra intervention.

II. JURISDICTIONAL COMPETITION

Firms compete against one another in market economies to produce a variety of goods of differing quality and price. Governments also compete against one another for economic gain. However, instead of competing in terms of the products they offer, governments compete to attract firms, for the purposes of providing jobs, tax revenue, income, and advantage to existing domestic firms. How do governments compete? The primary tools available to governments are tax policy, including tariffs, and regulatory policy, including environmental regulation. A government may offer low taxes to entice a company to locate in its jurisdiction. A government may subsidize production by a company to enhance the company's competitiveness overseas. A government may erect tariff barriers to protect domestic industries. One worry shared by many environmentalists and nonenvironmentalists is that governments will use environmental regulations to attract business, to the detriment of the environment. They worry that to bring in that new automobile manufacturing facility, the local community may offer to relax environmental regulations.

This governmental competition may go on at various levels. Within the United States, individual states or localities may compete with one another to attract businesses and jobs. Within the European Union, members states may compete with one another for economic activity. And internationally, where income differentials are the greatest, countries may compete with one another to attract industry. From our point of view, the critical issue is the extent to which environmental regulations are used for such competition. We might expect to see jurisdictions loosening environmental regulations to attract business. We might see jurisdictions specializing in waste assimilation, whether generating pollution from their own factories or by storing wastes generated elsewhere. We might see jurisdictions subsidizing pollution control to give their own industries a competitive advantage without sacrificing environmental quality.

Within federations, such as the United States or the European Union, the federal government may restrict competition among jurisdictions, if it is determined that such competition leads to undesirable consequences. Most Federal environmental regulation in the United States is specifically written to minimize the extent to which localities can use environmental regulations as a tool to attract business, despite the fact that marginal damage from a unit of pollution may vary considerably from one location to another as a result of variations in population density and physical characteristics of the locale. An issue is the proper division of responsibility for environmental protection between the Federal government and member states.

At the international level, environmental regulations vary substantially from one country to another. A simple reason for this is that environmental quality is a normal good (as discussed in the previous section): the wealthier people are, the more environmental quality they demand. Thus we would expect the citizens of a rich country to demand more environmental quality than the citizens of a poor country. But the real question is, do these differences among countries make much difference to potential polluters? If they do, we might expect to see the emergence of "pollution havens," countries that specialize in polluting industries. At a less extreme level, we might expect countries with weaker environmental regulations to have a relatively high concentration of "dirty" industries.

A related question concerns the extent to which a country might *tighten* environmental regulations to protect domestic industries.[7] Denmark introduced a requirement that beverage containers be reusable. Denmark was accused of discriminating against firms in other countries exporting to Denmark, because of the extra costs they would incur in first retrieving and then refilling their containers.

A. A Simple Model of Jurisdictional Competition

Would we expect a jurisdiction to weaken environmental regulations to attract capital? The answer to this question is mixed. If the jurisdiction is operating in the best of all possible worlds, it should neither subsidize nor tax capital movements into a jurisdiction. Furthermore, it is not in a jurisdiction's best interest to relax environmental regulations to attract business. However, if we are in a second-best world in which capital has been excessively taxed, it may be efficient to offset that taxation with a reduction in the stringency of environmental regulations. We can see this with a simple model.[8]

Start with a large number of jurisdictions with fixed populations in each. Jurisdictions compete with one another for capital (K), imposing tax rates on capital (which may be negative—subsidies) and regulations on pollution emissions within the jurisdiction (E). Localities need capital to provide jobs which provide income. Local residents also desire environmental quality. Additionally, capital taxation may be necessary to raise money for running the local government and providing local public goods. Industrial activity uses labor and capital and produces goods plus pollution. The production function can be written as $Y(K,L,E)$, where K is inputs of capital, L is inputs of labor, E is allowed emissions of pollution, and Y is goods output. We can assume the units for output are such that its price is one. Figure 13.4 shows how output changes with capital. Also, higher emissions allow higher levels of output. One assumption we will make is that industrial activity exhibits constant returns to scale in labor and capital inputs. In other words, if labor and capital inputs are doubled, both goods and pollution output are doubled.

Finally we need to characterize the preferences of the residents of each jurisdiction. We will assume residents are identical within a community with a utility function that depends on per capita consumption of goods (C) and pollution: $U(C,E)$. Environmental quality (the pollution level) is chosen by the locality. Consumption comes from the wage income associated with the firms that are attracted to the jurisdiction plus any tax revenue raised by taxing the firms in the jurisdiction plus any profit generated by local industry after labor and capital have been paid.

We now focus on a single, arbitrarily chosen jurisdiction and ask what level of capital taxation and environmental quality it should choose, considering that it is competing against all of the other jurisdictions. Thus labor is now fixed at $\bar{L}$, since we assumed no mobility among communities.

It takes a bit of development to reach our final result, so a preview may be in order. First we will look at how industry may respond to a capital tax, t, and pollution limit E. How much capital will industry choose to locate in a municipality with a tax of t and pollution limit E? Having established how the firm responds to jurisdictional tax and environmental policies, we then determine what levels of the capital tax and environmental quality will maximize utility for our representative consumer. We do this in two steps: (1) For any level of E, what is the optimal capital tax? (2) What is the optimal E? We are then in a position to develop our results that (1) the optimal capital tax is zero, but (2) a positive capital tax induces a lowering of environmental quality standards.

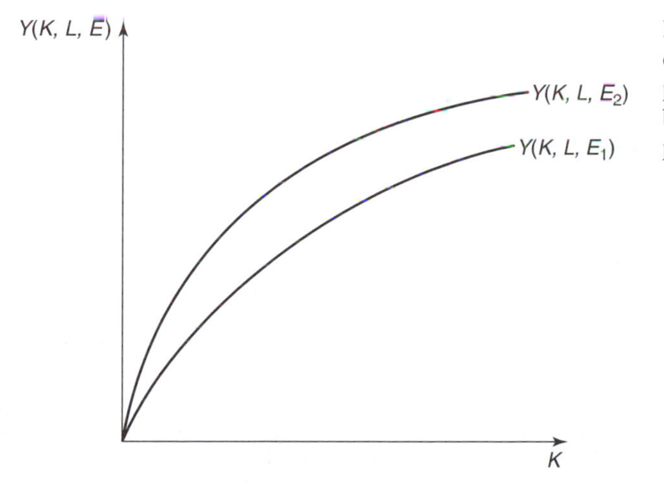

Figure 13.4 Output with two emisson levels, $E_2 > E_1$. K, inputs of capital; L, inputs of labor; E, allowed emissions of pollution; Y, goods output.

B. The Optimal Tax on Capital Is Zero

Keep in mind that the capital market is competitive; in fact the competitive world price of capital is r. Thus if the market provides the amount of capital K to the community, the owners of capital are paid rK. This leaves $Y(K,L,E) - rK$ for the residents of the community. If capital is taxed at the rate t, the wages and profits drop to $Y(KL,E) - (r + t)$ K but the total income of the population is supplemented by the tax revenues, tK, to give total consumption (equal to income) of $Y(K,\bar{L},E) - (r + t)K + tK = Y(K,\bar{L},E) - rK$. It might seem that the tax rate has no effect on income; however, the tax rate influences the amount of capital that is attracted and this in turn affects income.

If firms are faced with a capital tax of t and pollution limit E, in a locality with a wage w, the profits are

$$\Pi(K,\bar{L},E) = Y(K,\bar{L},E) - wL - (r + t)K \tag{13.1}$$

As we know, profits are maximized by setting the value of marginal product for each factor (labor and capital) equal to its price:

$$MP_K\ (K,\bar{L},E) = r + t \tag{13.2a}$$

$$MP_L\ (K,\bar{L},E) = w \tag{13.2b}$$

Since the labor supply in the municipality is fixed, the wage will adjust to clear the market. Figure 13.5 shows how the value of the marginal product of capital changes with the amount of capital. Also shown in Figure 13.5a is the choice of capital in response to a tax of t and no tax, when the emission level is E_2. Remember that L is fixed. Note that as the tax level increases, the amount of capital decreases. Figure 13.5b shows how the

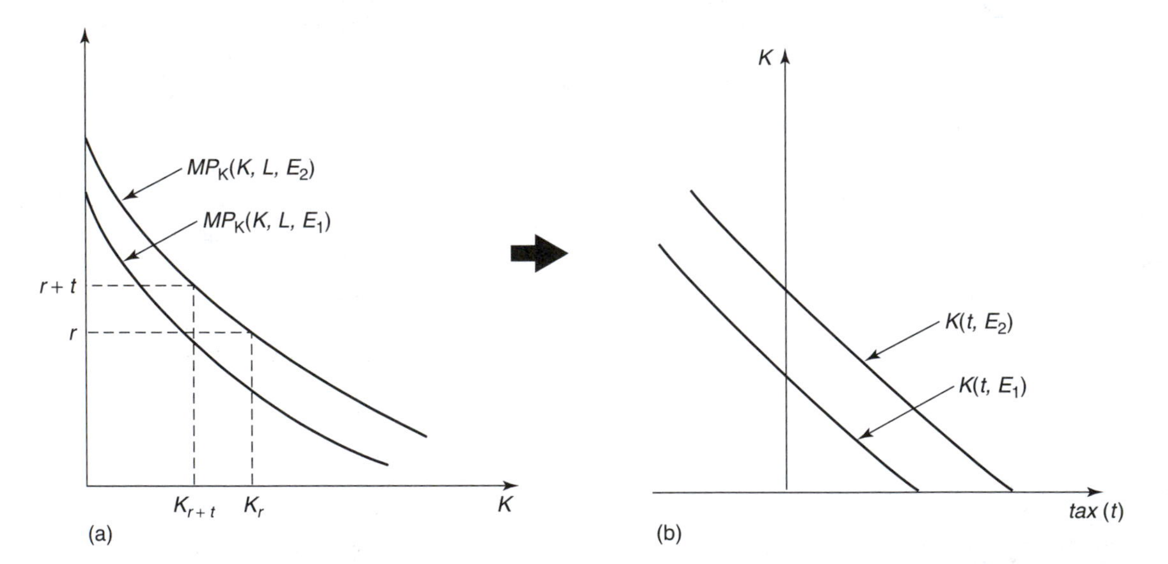

Figure 13.5 (a,b) Value marginal product of capital (MP_K), two emisson levels ($E_2 > E_1$) and capital (K) as a function of the capital tax rate (t) and emission levels. K_r, Amount of capital demanded with no tax on capital, emissions at E_2; K_{r+t}, amount of capital demanded with tax t on capital, emissions at E_2.

level of capital the firm chooses varies with the tax rate. Note also that just about any level of capital can be attracted to the jurisdiction, simply by choosing the right t.

What is the optimal tax on capital? To answer this we first determine the right amount of capital for the community. The consumption level for a tax of t, capital K, and emissions E is simply

$$C\bar{L} = Y(K,\bar{L},E) - (r + t)\,K + tK \tag{13.3a}$$

$$= Y(K,\bar{L},E) - r\,K \tag{13.3b}$$

In Eq. (13.3b), note that the level of the tax does not appear explicitly, only implicitly in that K is determined by t. A higher t gives a lower K and a lower Y. Now for any particular level of E, what should the community want K to be? That is simple—whatever maximizes per capita consumption, C. That will give maximum utility. From Eq. (13.3b), consumption is maximized when K is chosen so that the value of the marginal product of capital is equal to the price of capital services, r:

$$MP_K\,(K,E) = r \tag{13.4}$$

But this happens only when $t = 0$ [as can be seen from Eq. (13.2a)]. If t is high, there is too little capital and output suffers; with a large subsidy to capital, the difference between output and capital payments (i.e., consumption) is low. So our first result is that it is always desirable to have no tax on capital (neither a tax nor a subsidy).

C. The Optimal Level of Emissions

Now that we have determined the optimal capital tax rate, what should the optimal emission limit be? The representative consumer's utility depends on consumption and emissions: $U(C,E)$. The consumer will maximize this utility subject to the constraint that C and E must be attainable. E is under the control of the municipality but C depends on how much capital is attracted. Since K depends on t and E, we can write it as $K(t,E)$. We can also write the identity for consumption:

$$C = \{Y[K(t,E),\,\bar{L},E] - r\,K(t,E)\}/\bar{L} \tag{13.5}$$

Equation (13.5) defines the feasible consumption levels, given t and E. If we plot the (C,E) combinations that satisfy Eq. (13.5) for $t = 0$ we obtain a positively sloped line as shown in Figure 13.6 (line BD). As allowed emissions rise, output increases (as shown in Figure 13.4), and as a consequence consumption also increases. But in addition, the capital stock rises (see Figure 13.5b), which increases output and surplus available for consumption.

Also shown in Figure 13.6 are indifference curves for the consumer (lines RS, VW). The optimal choice for $t = 0$ is the combination (E_0,C_0). The slope of the indifference curve is the marginal rate of substitution between consumption and pollution for our representative consumer. Furthermore, the slope of the production feasibility line is the marginal rate of transformation between emissions and consumption. They are equal at the optimum. This is the first-best amount of pollution. So our second result, analogous to our first result about the capital tax rate, is that emission regulations should not be relaxed (or tightened) to attract capital.

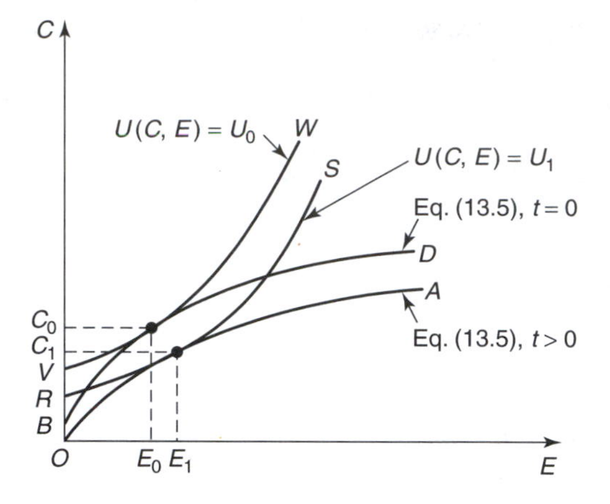

Figure 13.6 Optimal levels of emissions (E) and per capita consumption (C) for different capital taxation rates (t). Equation (13.5) is feasible combinations of C and E. BD, Feasible combinations of C,E with $t = 0$; OA, feasible combinations of C,E with $t > 0$; VW,RS, indifference curves for typical consumer; (E_0,L_0), efficient levels of E and C, $t = 0$; (E_1,C_1), efficient levels of E and C, $t > 0$.

D. Emissions Levels When the Tax on Capital Is Positive

Now suppose that for some reason the tax on capital is positive ($t > 0$). One understandable reason for this would be that the community has a need for tax revenue and finds it necessary to obtain this revenue, at least in part, from capital taxation. This is very common. A property tax is a tax on capital. Most business income taxes are basically taxes on capital income.

Our question now is, what happens to the limit on pollution when t becomes positive? Figure 13.6 illustrates the result. Shown in Figure 13.6 is the production possibility frontier for a capital tax $t > 0$ (line OA). For any emission level, attainable consumption is less with a positive tax. Note that in Figure 13.6 allowed emissions increase to E_1, largely to compensate for the overly high tax on capital.[9] Even though the indifference curve is tangent to the production function at this point, the tax on capital results in a difference between the marginal value of capital in production ($r + t$) and its price (r). Thus we do not have efficiency.

This is our third result. If for some reason there is a positive tax rate on capital, a jurisdiction may in fact loosen its environmental regulations relative to what we would find in a first-best world.

To summarize: if jurisdictions/countries offer foreign firms tax-free status, it would not be in the best interest of these jurisdictions to loosen environmental regulations to attract capital. If, on the other hand, conditions are such that the jurisdiction must tax capital, we might see the complementary use of weak environmental regulations to make up for the capital taxation. In reality, we might expect positive capital taxation to exist in countries that are not excessively worried about attracting capital. It seems logical that such countries would not feel strongly enough about attracting capital that they would resort to substantial weakening of environmental regulations. This is just speculation, however; what happens in reality is an empirical question.

III. INTERNATIONAL TRADE

In the previous section we examined competition among jurisdictions. We focused on what distortions jurisdictions ought to introduce into capital taxation and environmental regulation. Industry was assumed homogeneous. We did not examine the extent to which incomes and thus environmental quality differed among jurisdictions. We did not examine the effects of such differences on industries with differing propensities to pollute. As we move to the international arena, the perspective changes. In fact, a dominant characteristic of different countries competing in international trade is differences in income. Poor countries typically can offer low wage labor, attracting labor-intensive manufacturing. Poor countries can also specialize in pollution-intensive industries. In fact, there are several ways in which lower incomes may influence trade.

Because environmental quality is a normal good, we would expect citizens of poorer countries to demand lower levels of environmental quality. This could result in specialization in pollution-intensive manufacture. Alternatively, it could result in the importation of wastes for long-term disposal, including hazardous, nuclear, and municipal solid waste. Although the driving factor in both of these cases is the income effect in demand for environmental quality, the issues are different than we encountered in the case of the environmental Kuznets curve literature. Here we are more concerned with how income differences among countries influence the relative specialization in waste generation or disposal. This issue is the concern of the next section.

Another issue, totally unrelated to income differences, is the extent to which environmental policy can be used as a nontariff trade barrier. Examples were given earlier in this chapter of environmental regulations that were judged by some to be trade barriers in disguise. This issue will also be taken up here.

A. Pollution Havens

Environmental regulations cost polluters money. Firms subject to tighter environmental regulations will incur higher costs than firms subject to weaker or nonexistent environmental regulations. If we have two countries identical in all respects except for the tightness of environmental regulations, economic theory would indicate that the country with weak environmental regulations would offer a cost advantage to polluting industries and would thus tend to specialize in these industries.[10] The country with the tighter regulations would tend to specialize in cleaner industries and import the output of dirty industries. This is a natural conclusion from standard international trade theory: countries will have a comparative advantage in goods produced with factors that are in relative abundance. In this case the environment as an allowable dumping ground for pollution would be the factor that is scarce or abundant.

Although this theoretical result is not in much dispute, its empirical significance is of great policy interest. If we were to expect a significant amount of such specialization, the establishment of free trade areas may be of concern not only to traditional constituents, such as labor unions, but also to environmental groups. In fact, the environment was a major issue in the debate in the United States over the North American Free Trade Agreement. The issue was whether a free trade zone involving Canada, the United States, and Mexico would result in a significant number of Canadian and U.S. industries moving to Mexico, seeking out weaker environmental regulations. Similarly, should Eastern Europe join the European Union, will firms migrate to the East where environmental regulations may be laxer? Furthermore, if tight environmental regulations tend to drive firms elsewhere, this will be another factor militating against tighter environmental regulations at home.

It turns out to be not very easy to detect an effect of different environmental regulations on international competitiveness and trade, probably because the effect is a weak one at best. There are two basic approaches to examining this question. One is to determine whether countries with lax regulations actually do specialize in exporting goods associated with pollution in their manufacture. A second approach is to examine capital flows in polluting industries and try to determine whether countries with lax regulations tend to attract relatively more of that capital compared to countries with tighter regulations.[11]

One of the primary models explaining international specialization is the Heckscher–Ohlin model of international trade.[12] A basic result of the model is that countries will have a comparative advantage in goods produced with endowed factors that are in relative abundance. A country with plentiful hydroelectric resources will specialize in goods with significant electricity content (such as aluminum); a country with plentiful labor will specialize in labor-intensive goods (such as clothing); a country with a large capital stock will specialize in capital-intensive goods (such as automobiles). The natural extension to the environment is that a country with a large assimilative capacity for pollution should specialize in pollution-intensive goods. However, it is not just assimilative capacity that determines the receptiveness of a country to polluting firms; income is just as important. The citizens of countries with high incomes will demand high levels of environmental quality. The degree of strictness of environmental regulations will thus result from two factors: the endowment of assimilative capacity in a country and the average or median level of per capita income in a country (or some other proxy related to citizen demand for environmental protection).[13]

One way of determining if lax environmental regulations tend to attract dirty industries is to use the Heckscher–Ohlin model to explain trade in goods that are associated with pollution and try to determine if environmental variables are significant. For instance, let X_{ij} be the net exports of industry i from country j. Let E_{jk} be the endowments in country j of basic factor k, such as capital, skilled labor, land, and natural resources (assume there are K different basic factors). Finally, let R_j be the strictness of environmental regulations in country j (a higher R is more strict). The Heckscher–Ohlin model would have X_{ij} explained by E_{jk} and R_j:

$$X_{ij} = \alpha_i + \beta_{i1} E_{j1} + \beta_{i2} E_{j2} + \ldots + \beta_{iK} E_{jK} + \delta_i R_j + u_{ij} \tag{13.6}$$

or some other functional form involving the same basic variables. In Eq. (13.6), u_{ij} is an error term, representing variables that have been omitted, measurement errors, or other factors than may lead to errors in characterizing the relationship between right-hand-side variables in Eq. (13.6) and net exports. If Eq. (13.6) is estimated over a large cross section of countries, an indication of whether the laxity of environmental regulations influences trade is the statistical significance and sign of δ_i, the coefficient on R_j, the strictness of environmental regulation.

James Tobey (1990) has done just such a statistical analysis. He uses observations on 23 countries, about half industrialized and half developing, examining net exports of five particularly dirty industries: mining, paper, chemicals, steel, and other metals. Using 11 different categories of resource endowments (such as capital, different kinds of labor, different kinds of land, and different natural resources), he fails to find any statistical significance in δ_i, the coefficient on the strictness of environmental regulations.[14]

One problem with finding evidence that differential environmental regulations influence international trade is that the effect is probably small. Because of this, some authors have focused on the margin most likely to be sensitive to the effect: international flows of capital. The logic is that the nature of the capital stock in any country and thus the country's specialization in one industry or another is the result of investments made over many decades. Looking for effects in the capital stock of environmental regulations that may have existed for a decade or two at most is likely to be futile. Any effects are far more likely to show up in marginal changes in the capital stock: current investment. Movements of capital from one country to invest in industry in another are called foreign direct investment (*FDI*). If lax environmental regulations tend to attract dirty industries, then we should expect to see *FDI* for dirty industries concentrated in countries with lax environmental regulations, all else being equal. As with explaining trade in the Heckscher–Ohlin model, we start with an equation that attempts to explain *FDI* in industry i into county j:

$$FDI_{ij} = \alpha_i + \beta_{i1} F_{1j} + \beta_{i2} F_{2j} + \ldots + \beta_{iK} F_{Kj} + \delta_i R_j + u_{ij} \tag{13.7}$$

where F_{kj} is the level of variable k influencing *FDI* in country j (for instance, tax policy), R_j is the strictness of environmental regulations in country j, and u_{ij} represents variables and noise that are unaccounted for. Although Eq. (13.7) is linear, it could take many forms, involving the same basic variables. Unfortunately, there has not been much empirical work in trying to measure δ_i in Eq. (13.7). A number of authors have statistically estimated variants of Eq. (13.7), although few with an explicit variable for environmental regulation.[15] Xing and Kolstad (1997) have examined *FDI* from the United States for several sectors. Although their sample is small, they find δ_i to be significantly negative for the chemical industry, a highly polluting industry, but statistically indistinguishable from zero for the electronics industry, an industry with relatively low levels of pollution.[16] This suggests that lax environmental regulations do attract *FDI* for polluting industries.

The general consensus of the empirical literature on pollution havens and the effect of environmental regulations on trade is that the effect is very weak at best. Some authors find no effect; other authors find an effect, though only a weak one.

B. Strategic Trade

The next issue we confront is the use of environmental regulations as tools to increase the competitive advantage of domestic firms engaged in international trade.[17] The standard view of international trade is that any barriers to trade that a country may impose, either barriers to protect domestic industries or subsidies to exports, can make the country only worse-off as a whole (though parts of the country may obviously be made better-off). If environmental regulations are weakened, domestic firms benefit via lower costs; however, residents who must put up with the pollution are harmed even more. Overall, it is not welfare improving to weaken environmental regulations to enhance trade. This is essentially the same result we saw earlier in this chapter in the context of interjurisdictional competition.

This result breaks down somewhat if industries within a country have market power in the global market. With market power, the overall welfare of a country can be improved by subsidizing exports, although this result applies to very specific circumstances and thus is not universally true. Since weakening environmental regulations is one way of providing a subsidy, it might be the case that there is a real incentive to relax environmental regulations. Several authors have examined this question in some detail.[18] Their models are quite complex, so we will sketch only a rudimentary analysis of this issue here.

The basic setup of our analysis is a specific industrial sector located in two different countries. Each of these two countries has a firm in this sector that is large in the global market and thus enjoys market power. For simplicity, we will assume these firms sell to a third country. This way, we do not have to worry about the welfare of consumers in the two countries, only the welfare of the firms. We will first show that a subsidy to production can be in the best interest of the country offering the subsidy.[19] The firms are assumed to engage in Cournot competition with each other, meaning that each will maximize profits taking the level of output of the other as fixed. We will then examine the subsidy implications of different environmental regulations.

As is the common approach in examining a problem of this nature, we first fix the subsidies at arbitrary levels and see how the firms will respond. Knowing how firms respond to arbitrary subsidies, we can determine what the optimal subsidies are. Or, in our case, the two countries will be competing against each other so we are interested in the equilibrium subsidy levels that result from the competition.

Let the outputs of the large firm in each of the two countries be denoted by D and F, for domestic and foreign. Assume the inverse demand function for their output is $P(Q) = a - b\,Q$ where $Q = D + F$. Inverse demand is thus linear, with a and b the two positive parameters of the inverse demand function. To make things even more simple, we assume fixed costs are C_D and C_F while variable costs are zero. This is the simplest possible arrangement we can define. Assume that the domestic firm receives a subsidy s per unit production. Profits for the domestic firm are then given by

$$\Pi_D = [a - b(D + F)]\,D + sD - C_D \tag{13.8a}$$

$$= (a + s - bF)\,D - bD^2 - C_D \tag{13.8b}$$

For any particular level of output, F, chosen by the foreign firm, how much should the domestic firm produce? Clearly, the domestic firm should produce so that marginal profits are zero; this will maximize profits. To do so, we use Eq. (13.8b) to generate marginal profits, which we then set to zero:

$$M\Pi_D = (a + s - bF) - 2bD = 0 \tag{13.9}$$

which implies that

$$D = (a + s - bF)/\,(2b) \tag{13.10}$$

This is shown graphically and generally in Figure 13.7. The figure shows a series of concentric isoprofit lines. These are lines of constant profit, defined by Eq. (13.8). Each of these lines represents (F,D) combinations that yield the same level of profit for the domestic firm. These isoprofit lines indicate what profit-maximizing level of D should be chosen for any particular level of F. Fix F; the D associated with the highest profit is the D of choice. These Ds are the "best responses" to any given level of

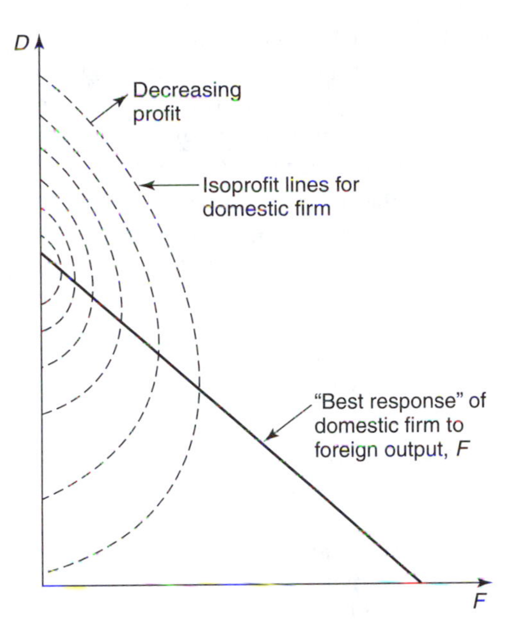

Figure 13.7 Best response of domestic firm to foreign output levels.

F. The solid line in Figure 13.7 shows the best response line—the optimal Ds. This is also Eq. (13.10).

Now we can do the same thing for the foreign firm, deriving its best response to different output levels by the domestic firm. The best response lines for D and F are shown in Figure 13.8. The point at which the two lines intersect is the equilibrium, (D^*,F^*), the D and F levels at which each firm is doing the best it can, given what the other firm is producing. This is what is known as the Cournot equilibrium.[20]

So far, we have ignored the subsidy level implicit in Eqs. (13.8) and (13.10). If we assume the best response lines in Eq. (13.8) are associated with a zero subsidy level, $s = 0$, then what happens if the domestic country increases s to a positive number? From Eq. (13.10) we see that if we increase s, the best response D will increase as well, for any

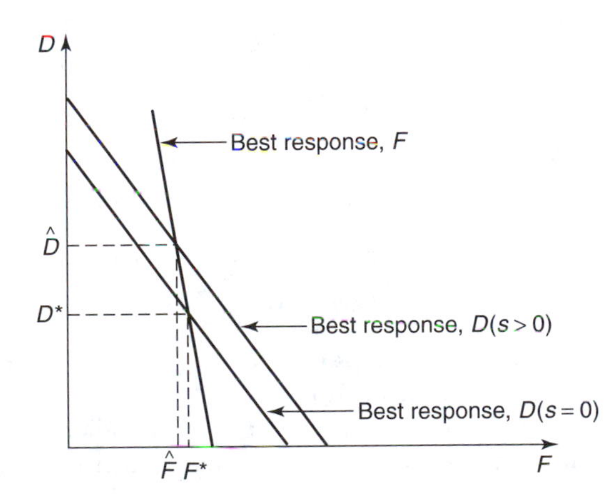

Figure 13.8 Effect of a subsidy on Cournot equilibrium output levels. F^*, D^*, Foreign and domestic output, no subsidy to D; $\hat{F}$, $\hat{D}$, foreign and domestic output, subsidy to D.

given F. Thus the best response line for D is shifted up as shown in Figure 13.8 for $s > 0$. Note that this results in a new equilibrium, one with higher D and lower F. As would be expected, a subsidy to firm D results in lower costs for this firm and thus expanded output. It comes at the expense of output from the other firm.

Profits clearly expand for firm D, but to a certain extent these increased profits are due to the subsidy that has to be financed by the general tax revenues of country D. It turns out, however, that the net profits for country D (i.e., taking into account the cost of the subsidy) are actually higher for a positive subsidy than for no subsidy. Thus it is in the best interest of country D to subsidize firm D.[21] In effect the subsidy allows D to take output and profits from F.

Now suppose both countries decide to subsidize exports. This turns out to be a little more complicated. In this case, both countries will try to impose a positive subsidy, but the result will not increase net profits for the countries. In fact, with both countries subsidizing, it turns out that a negative subsidy (an export tax) maximizes net profits if both countries were able to coordinate policies. The reason for this is that the two firms competing against each other produce more output than if they were coordinating output and acting as a cartel. A positive export tax will restrict output and drive the two firms closer to this monopoly cartel level of output and profit. The important point, however, is that two countries competing against each other will subsidize production, and not obtain the point of maximum net profit.

So far, these results are not directly related to environmental matters. However, suppose we have exactly the same situation except that pollution is generated in conjunction with output in each country. Pollution causes damage at home (forget about transboundary pollution). Now assume that it is not possible to directly subsidize a domestic industry involved in an export market of the type described above. An obvious reason for being unable to use direct subsidies is that such subsidies may violate free trade agreements. A natural way of subsidizing the industry is to relax environmental regulations. Instead of directly paying a subsidy to the polluter, the relaxed regulations reduce polluter costs at the expense of the citizens who must endure greater levels of pollution.

Consider a firm that generates pollution in direct proportion to output. We will focus on only one firm, but the setup is as above: there are two firms/countries with market power selling into a third country. If output of our firm is D, then emissions are αD. Suppose the efficient Pigovian fee is t^*. If the firm is subject to the Pigovian fee, its profits are

$$\Pi_t = [a - b(D + F)]\, D - C_D - t^* \alpha D \tag{13.11}$$

If the country would like to offer a subsidy s^* per unit output but is unable to do so, it can instead lower the Pigovian fee to $(t^* - s^*/\alpha)$ resulting in profits

$$\begin{aligned}\Pi_s &= [a - b(D + F)]\, D - C_D - (t^* - s^*/\alpha)\, \alpha D \\ &= [a - b(D + F)]\, D - C_D - t^*\alpha D + s^* D\end{aligned} \tag{13.12}$$

Lowering the appropriate pollution tax is a way of granting a subsidy of s^*D. The down side is that there will be more output than desired and more pollution.[22]

What does all of this mean? The basic result is that subsidies can be in the self-interest of a country when its industries enjoy market power internationally (but not always, as we showed). Since direct subsidies may be difficult to pursue, subsidies in the form of lax environmental regulations may be attractive. This is somewhat worrisome in that it may be difficult to determine whether a country is being too lax in protecting the environment.

IV. TRANSBOUNDARY POLLUTION

The last issue we will consider is that of transboundary pollution—pollution that migrates beyond the jurisdiction with the power to control that pollution. Greenhouse gas emissions are a classic example. Greenhouse gases emitted in Norway contribute to climate change worldwide. Or consider acid rain pollution from emissions from power plants in the United Kingdom that may be blown across the North Sea to Sweden and Norway. How do you control such pollutants?

There are three basic situations that we will consider. The simplest one is two adjoining countries with pollution and goods produced in one country; the pollution migrates across the international boundary and is consumed in the other country. Goods are also shipped to the victim country. One way of solving this is for the victim to levy a tariff on imports that were produced with pollution. The question is, how well will such a countervailing tariff work if the tariff is related to pollution damage but applied to goods imports?

The second situation is an extension of this two-country model with pollution migrating across borders. But here these two countries are involved in a plethora of international transactions, including but not limited to trade in pollution-intensive goods. Is it possible that sanctions in unrelated markets can be used to drive the polluting country to generate pollution levels that are Pareto optimal?

The third situation involves a group of countries with a common pollution problem. The pollution problem is common in that each country pollutes and each country suffers damage from the aggregate levels of pollution. Carbon dioxide, leading to global warming, is a good example of such a pollutant. Effective control of the externality must involve international agreements. Can such agreements be effective and binding, considering the fact that there is no international supergovernment that can enforce such environmental agreements?

A. Countervailing Tariffs

We first consider a very simple situation involving two countries. One country, the polluter, has an industry that produces a good that is exported to the other country. Unfortunately, the industry also generates pollution that is entirely exported to the other country, which we will call the victim country (for example, the factory is on the border and all the pollution blows into the victim country). To correct the pollution problem, the victim country levies a tariff on imports of the good, with the tariff equal to the total pollution damage divided by the quantity of goods being imported. Thus the tariff revenue the victim country collects will be exactly equal to the damage inflicted on it by the polluter country. Does this correct the problem in an efficient manner? The answer is "no," except in very particular circumstances. Only when all of the goods produced are exported to the victim country and when marginal damage is equal to average damage from pollution could the countervailing tariff be efficient. In such a case, the victim country has complete control over the producer of the good. However, because of this, the victim country is a monopsonist and will likely levy an import tariff, in the absence of pollution. Consequently, efficient pollution control will be difficult to establish.

Consider a simplification of this example. Suppose the marginal and average damages from pollution are the same, so that we do not have that complication. Pollution is also produced in direct proportion to the amount of goods produced. This is simpler than the case of allowing abatement technology—in order to reduce pollution, output must be reduced. Also suppose that the good is consumed in the polluter country as well as the victim country. Finally, assume that the victim country does not levy a tariff to exploit its monopsony power but only to correct for the pollution externality. Figure 13.9 shows this situation. Shown are the two markets: the polluter market (13.9a) and the victim market (13.9b). Shown is the marginal private cost of producing the good as well as the demand for the good in each of the two countries. The first-best pollution policy would be a Pigovian fee of t^* on all production of the good, equal to marginal damage from the pollution. This raises the marginal cost of production in both markets, as shown in Figure 13.9. Only those consumers for whom the marginal value of the good exceeds the private marginal costs (MC_{private}) plus the marginal pollution damage (t^*) will consume the good. Sales of the goods will be q_P^* and q_V^* in the two countries and this will be efficient. Now suppose that such a first-best correction of the externality does not take place. Instead, there is no emission fee in the polluter country, leading to output of q_P^-. The only

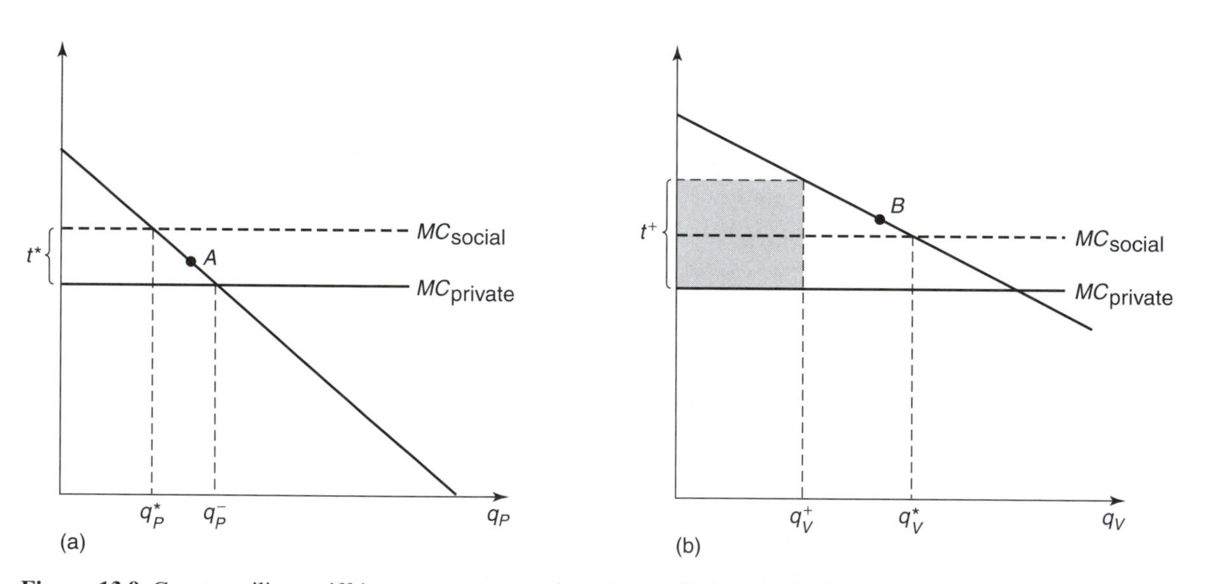

Figure 13.9 Countervailing tariff in response to transboundary pollution. (a) Polluter market; (b) victim market. t^*, First-best Pigovian fee; t^+, countervailing tariff levied by victims; q_P^*, q_V^*, first-best consumption levels, polluter and victim markets; q_P^-, consumption in polluter market without emission fee; q_V^+, consumption in victim market with countervailing tariff; shaded area, pollution damage with countervailing tariff t^+.

recourse the victim country has is to levy a tax per unit of the good imported (t^+) set so that total tax revenue ($t^+ q_V^+$) is equal to total damage [$t^*(q_P^- + q_V^+)$]. This is shown in Figure 13.9.

Why is this not an efficient correction of the externality? In other words, why is this not Pareto efficient? One reason is that there are consumers in the victim market who are precluded from consuming the good but for whom the value of the good exceeds the value of consumers in the polluter market. For instance, the consumer corresponding to point *A* in the polluter market consumes the good whereas the consumer labeled *B* in the victim market does not consume the good. Yet, the marginal value of the good for *B* exceeds the marginal value for *A*. In fact, with an efficient Pigovian fee, consumer *A* would not be consuming the good—the consumer's value is less than the marginal social cost of producing the good. So we see that there is room for a Pareto improvement among consumers. What about the quantity of pollution? Is that efficient? Note from Figure 13.9 that the efficient amount of output (which is directly proportional to the amount of pollution) is $q_P^* + q_V^*$ but the quantity that will be produced with a countervailing tariff is $q_P^- + q_V^+$. These two quantities are unlikely to be the same.

Efficiency in pollution generation really requires that all consumers of the dirty good see the total social cost of their consumption. That is the basic problem with the countervailing tariff. And the problem gets worse if more than two countries are involved. If there are multiple consuming countries each country will levy a tariff related to damage in that country only, ignoring damage in other countries. Such an arrangement is unlikely to lead to efficiency.

In summary, a countervailing tariff can help correct a transboundary pollution problem but is unlikely to totally correct the problem.

B. Sanctions

The previous section assumed that the victim country would levy countervailing tariffs equal to the pollution damage. Although that is one approach to dealing with transboundary pollution, the fact is that countries deal with each other on a multitude of issues, not only economic but political and cultural as well. It may be that the victim country has little ability to coerce the polluter to reduce emissions; however, the victim country may have a great deal of power on a totally unrelated issue, such as involvement in a joint defense treaty.[23] By withholding agreement on the joint defense treaty, the victim country may be able to force the polluter to reduce pollution levels. In essence, the victim country can threaten sanctions of one sort or another against the polluter country if the polluter country does not adopt pollution regulations to achieve the efficient amount of pollution. This is more of a political than economic question so will not be addressed further.

C. International Environmental Agreements

We now turn to the most general of international pollution problems, one in which there are many countries contributing pollution to a common pool that damages all countries. Only an international agreement among countries can hope to solve the problem. An example is the emission of greenhouse gases such as carbon dioxide. The emission of greenhouse gases may lead to global climate change, such as warming and sea level rise. It makes no difference where the gases are emitted; the effect on the global climate is the same. However, due to climate change some countries may incur greater damage than others. Some arctic countries may benefit from warming whereas some low-lying coastal countries may suffer greatly from sea level rise. The emissions of chlorofluorocarbons (CFCs) to the atmosphere, leading to ozone depletion, is another example of a global pollutant. In this case, a treaty (the Montreal Protocol) has been forged to deal with the problem.[24]

The basic problem in designing an effective international environmental agreement is that there is no supernational organization to enforce such an agreement. In effect, the agreement must not only be self-enforcing but it must be sufficiently appealing to all parties involved for them to agree to it in the first place. Thus there are three desirable characteristics of an international environmental agreement: (1) the agreement should be self-enforcing; (2) each country should be better off as part of the agreement than not; and (3) pollution levels obtained by the agreement should be a Pareto improvement over the status quo and, ideally, a Pareto optimum. It is easy to contemplate an international agreement that meets two of these three criteria; it is quite another thing to design an agreement that meets all three conditions.[25]

The difficulties in forging an effective international environmental agreement are not unlike the problems faced in constructing effective cartels. A cartel is a group of producers that coordinates its production to increase its market power and thus increase its profits. Such cartels typically cannot rely on enforcement of agreements by third parties. If the cartel consists of a group of domestic firms within a single country, the cartel is usually illegal and thus any agreement is nonenforceable in court. If the cartel is a group of countries (such as OPEC), just as with environmental agreements, there is no supernational enforcement body. Successful cartels must rely on self-enforcement, voluntary participation, and effective restrictions on output. The work that has been done in understanding what makes cartels work can be applied to understand what makes international environmental agreements work (or not work).

To simplify our analysis somewhat, assume that we have two countries, $i = 1,2$, each emitting pollution e_i. The damage to each country, $D_i(E)$, depends on the total pollution $E = e_1 + e_2$. The benefits to each country from being able to emit (e.g., lower production costs) are given by $B_i(e_i)$. For measurability reasons, we assume that benefits and damages are denominated in the same monetary units. Prior to examining how a treaty might work, we will specify two noncooperative emission levels. One involves a total absence of pollution regulation in each country. Call this the unregulated amount of pollution (U). Another level of emissions involves regulation, but only insofar as damages within the same country are concerned. Country 1 sees cost-reducing benefits of polluting equal to $B_1(e_1)$ and pollution damage of $D_1(E)$. Regulation will be set so as to equate marginal benefits (MB) and marginal damages (MD): $MB_1(e_1) = MD_1(E)$. Damage to country 2 is specifically excluded from country 1's calculations. Emissions for country 2 are determined similarly. Call this the noncooperative level of pollution, N.[26] There will be somewhat more pollution control with N than with U. Neither N nor U will be Pareto optimal.

This situation is illustrated in Figure 13.10. Shown in Figure 13.10 is net benefits for each of the two countries, $B_i - D_i$, resulting from specific emission levels, (e_1,e_2).

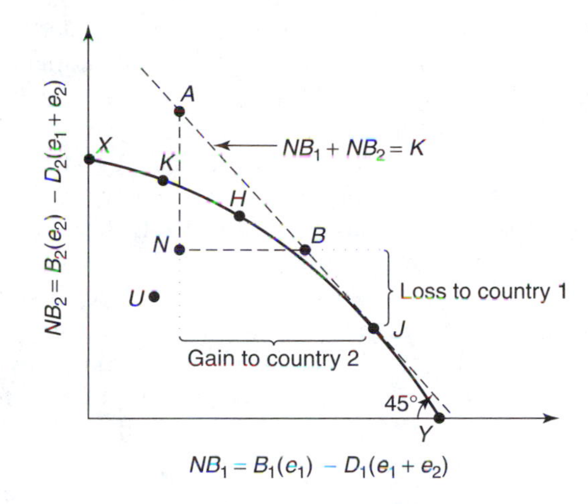

Figure 13.10 Possible benefits to two countries from an international environmental agreement. *U*, Net benefit in the absence of regulation; *N*, net benefits with noncooperative regulation; *XY*, Pareto frontier without compensation; *H*, possible efficient cooperative outcome without compensation; *J*, joint net benefits maximum; *K*, efficient point that is not a Pareto improvement on *N*; *AB*, Pareto efficient outcomes with compensation.

Each point in Figure 13.10 can be associated with a particular (e_1, e_2). Point U is shown in the figure as well as point N. As can be seen, each country does better with N than with U, as would be expected. But also shown in Figure 13.10 are the maximum attainable levels of net benefits, the curve XY. This is the Pareto frontier. Each point on this frontier is the solution to the following problem:

$$\max_{e_1, e_2} \quad B_2(e_2) - D_2(e_1 + e_2) \tag{13.13a}$$

$$\text{such that} \quad B_1(e_1) - D_1(e_1 + e_2) \geq \tilde{N} \tag{13.13b}$$

where $\tilde{N}$ is some arbitrary constant. Based on Eq. (13.13), if we guarantee country 1 the level of net benefits $\tilde{N}$, how large can we make the net benefits for country 2? We can then vary $\tilde{N}$ over a wide range; the results of the solution to Eq. (13.13) as we vary $\tilde{N}$ will trace out the Pareto frontier for the two countries. Note that we have explicitly excluded monetary transfers from one country to another—large gains to one country do not offset small losses to the other.

It would be desirable for an international agreement to result in emissions that are on the Pareto frontier. Points K and H are both on the Pareto frontier but H will be much easier to attain since it involves a welfare improvement for both countries. Point K will be difficult to attain without transfers of resources from country 2 to country 1. This is because country 1 is better-off at N than K. Thus point K violates the condition that every country must be better-off with the environmental agreement than without the agreement. However, we note that the gains to country 2 of moving from N to K are much greater than the losses to country 1 of moving from N to K. (Both axes are denominated in monetary terms and we can assume the same units are used for both axes.) Thus country 2 has enough resources to transfer to country 1 to induce it to move to K.

This suggests that what we really should be seeking is the largest possible total gain for the two countries as a whole. The maximum of the total or joint benefits to the two countries occurs at point J. This is the joint net benefits maximum. Note that the gains to country 2 are more than sufficient to offset the losses to country 1 in moving to point J from point N. If compensation is possible (resource transfers among countries), any division of net benefits along the broken line in Figure 13.10 is possible. In particular, we would expect to see the division on the line segment *AB* since all profits on this segment are Pareto preferred to point *N*. Emissions will be at point *J*. But without the possibility of compensation, country 2 will veto any attempt to move to *J* from *N*.

The importance of being able to compensate one country for losses it incurs in the process of generating benefits to the other country cannot be overemphasized. Typically, compensation is politically difficult. We thus distinguish between two possible environmental agreements, a perfect environmental agreement in which compensation is possible and an imperfect environmental agreement in which compensation is not possible.[27] We see from Figure 13.10, that the possible outcomes can be very different in these two cases.

Cartel theory suggests that there are five problems in maintaining an effective international cartel.[28] Extending this to environmental agreements, we infer the following basic problems in implementing environmental agreements are: (1) to locate the Pareto frontier, (2) to choose a point on the frontier, (3) to detect cheating, (4) to deter cheating, and (5) to discourage countries outside the agreement from exploiting and weakening the agreement.

To illustrate each of these five conditions, consider the example of a successful agreement to control emissions of carbon dioxide and greenhouse gases leading to global warming and climate change. Such an agreement has been difficult to forge and has not yet been reached. Condition (1) requires that countries know the costs and benefits of pollution reduction. In the case of climate change, neither is known well, particularly the benefits of pollution reduction. Condition (2) involves choosing a point on the frontier. This is the sharing rule. How much of the gains from the agreement will go to which countries? One of the problems in the climate change context is that many proposed initiatives are not Pareto improvements on the status quo. Many proposed agreements involve the OECD (developed world) paying much of the costs of pollution reduction with most of the benefits going to the developing world. Although this is understandable on the grounds of fairness, it also means that the proposed burden sharing is not a Pareto improvement, which will cause problems in convincing countries to join the agreement. Another problem is in deterring and detecting cheating. Although detecting cheating may not be too difficult in this age, effective sanctions on countries that violate their obligations may be difficult. One popular proposal is to use countervailing tariffs against any country that violates the agreement.[29] A country that emits more than the treaty allows would have the carbon content of all exports taxed by countries of destination at such a rate as to discourage noncompliance. The final problem is that of nonsignatory countries exploiting the agreement. In the case of a climate change agreement among developed countries, the concern is that developing countries would specialize in carbon-intensive activities and export the goods associated with those activities to the signatory countries. This "carbon leakage" would erode the effectiveness of the agreement. Once again, countervailing tariffs can be used to deter this activity.

SUMMARY

1. Income is a significant determinant of the demand for environmental quality. Environmental quality is most likely a normal good; thus we can expect demand for environmental protection to increase as income increases around the world.

2. An environmental Kuznets curve describes the level of environmental quality in a country as a function of per capita income in that country. There are indications that as income increases from very low levels, environmental quality first deteriorates but then, as the country begins to become more prosperous, begins to improve and continues to improve as income increases. However, this only applies to some pollutants, not all.

3. Different regions of a country (e.g., states of the United States), different countries of a federation (e.g., countries of the European Union), or independent countries of the world are interested in attracting capital and businesses to provide jobs and income. Countries may use preferential tax rates or concessions on environmental regulations to attract business. In a competitive environment the first-best strategy for a country is to neither subsidize nor tax capital and to grant no concessions on environmental protection. If, however, it is necessary to tax capital, it may be in a country's interest to weaken environmental regulations to make up for the tax on capital.

4. Because of the income effect on demand for environmental quality, we would expect environmental regulations to be more lax in poorer countries. This might result in poor countries attracting and then accumulating the particularly dirty industries of the world. The empirical evidence on this effect is quite weak. Most studies are unable to find evidence of this "pollution haven" effect. A few studies do find some evidence that lax environmental regulations attract dirty industry.

5. When a country has an industry that has market power internationally (for instance, because of its size), there are ways in which the country can help the industry, increasing income for the industry and the country. A subsidy to the industry can be in the best interest of the country. If a subsidy is not possible (for instance, because of free-trade agreements), weakening environmental regulations can be a more subtle way of offering the industry a subsidy.

6. Transboundary pollution is difficult to control because the damage from the pollution occurs in a locality that has no authority over the generator of the pollution. There are three ways of controlling this problem. One involves the victim country levying tariffs on imports from the country generating the pollution. Another approach involves linking international agreements between the two countries on totally unrelated issues to controlling the pollution problem. A third and most common approach is to forge international agreements on pollution control, such as the Montreal protocol on control of ozone-depleting substances (e.g., chlorofluorocarbons). The problem with international agreements is that participation is of course voluntary, but at the same time sanctions for violating the terms of the agreement must be biting. It is not always possible to forge an effective environmental agreement to control transboundary pollution.

PROBLEMS

***1.** In the example (Section III,B) of the two countries with firms with market power, competing internationally, determine the optimal subsidy for country *D*. Assume only country *D* uses a subsidy. Do this in two parts. First examine how the firms in each country will produce, given a subsidy level, s, in country *D*. Then solve these reaction functions for the equilibrium level of output in each of the two countries. These should depend only on a, b, and s. Then, for the second part, write an expression for net profit for country *D*. The expression should be in terms of a, b, and s only. Next determine the s that maximizes this net profit. Use calculus or show the result graphically for several different values of a and b.

2. How does Figure 13.2 support the following statement: "Looking at how countries of different income at any point in time supply environmental quality (the environmental Kuznets curve) does not necessarily indicate how an individual country's environmental quality will change as that country's income rises."

3. In Figure 13.6, the optimum (E_0,C_0) is shown as the point at which the marginal rate of transformation of E for C equals the consumer's marginal rate of substitution of E for C. Yet E is a public bad. Why does this appear to violate the Samuelson conditions for optimal provision of a public bad/good?

4. In the example (Section III,B) of the two countries with market power, assume that inverse demand is $P(Q) = 10 - Q$ where $Q = D + F$ is measured in tons and P in dollars per ton. Assume fixed costs are equal to one for both firms.
   a. First assume no subsidy ($s = 0$). In a plot of D (vertical axis) vs. F (horizontal axis), plot several isoprofit lines for the domestic firm. Plot the best response for the domestic firm, showing the domestic firm's profit maximizing choice of D as a function of F. (This is similar to Figure 13.7.)
   b. Since the foreign firm has identical costs to the domestic firm, add the foreign firm's best response line to your graph. What is the resulting level of output for the two firms, the resulting profit for each of the firms, and the profits for each of the two countries (where a country's profit is equal to its firm's profit less the cost of the subsidy)?
   c. What happens to output for each firm if the domestic country only institutes a subsidy of \$0.50 per ton? Show your answer graphically and calculate it as well.
   d. Calculate the profit for each of the two firms and profits for each of the two countries. Compare your answer with that of part b. Which country has done better with the subsidy?
   e. What happens to output if both countries institute subsidies of \$0.50 per ton? Show this on your graph and calculate the result as well.
   f. For this case of two subsidies, calculate the profit for each of the two firms and the profit for each of the two countries. Compare your answer with that of part b. How have the two countries fared in comparison to the case of no subsidies?

5. In the early 1990s, the chief economist of the World Bank came under a great deal of criticism for suggesting that developing countries take advantage of opportunities to import hazardous and nonhazardous wastes for storage and disposal. Why was he making this argument? What are the arguments against his position?

NOTES

1. The ratio of the percentage change in quantity demanded to the percentage change in income is of course the income elasticity of demand. If the income elasticity is negative, we have an inferior good; otherwise we have a normal good. If the income elasticity is greater than 1, we have a luxury good; if it is between 0 and 1, we have a necessary good.
2. Kriström and Riera (1996) review a number of studies, primarily European, and conclude that the evidence suggests a positive income elasticity of demand less than one, i.e., environmental

quality is a normal good but not a luxury good. In an interesting study of voter actions on environmental issues in California, Kahn and Matsusaka (1997) conclude that the environment is a normal good for mean income individuals, but that some environmental goods are inferior for higher income individuals. One explanation for this is that some environmental goods, such as urban parks, are provided privately by higher income people. Thus this result does not necessarily mean that park-like open spaces are inferior, only that *publicly provided* open spaces may be inferior.

3. Kelly (1997) presents a theoretical model of what determines this shifting of supply and demand for environmental quality.
4. The data are found in the Global Environmental Monitoring System (GEMS), a joint effort of the United Nations Environment Program and the World Health Organization.
5. The interested reader is referred to Thompson and Strohm (1996) for a review of this literature and Grossman and Krueger (1995) for a quantitative estimation of a number of Kuznets curves for enviornmental quality. More detail is included in Shafik and Bandyopadhyay (1992) and Shafik (1994). Bohn and Deacon (1997) focus on poorly developed property rights in explaining how environmental quality changes as income changes; Kelly (1997) adapts a simple optimal growth model to show how the environmental Kuznets curve emerges from standard economic theory.
6. The World Bank's 1992 World Development Report (World Bank, 1992) is devoted to the environment and development; there is considerable discussion in that document of the policy significance of environmental Kuznets curves.
7. Farber and Hudec (1996) review some of these issues in the context of the General Agreement on Tariffs and Trade (GATT). Bhagwati and Srinivasan (1996) discuss some of the economic issues related to differences in environmental regulation among countries.
8. The model presented here is based loosely on Oates and Schwab (1988). Markusen et al (1995) provide a different treatment of the use of environmental regulations to attract capital.
9. The graph does not unequivocally show this since perversely shaped indifference curves could conceivably yield the opposite result, much as one can show how quantity demanded of a Giffen good decreases when its price goes up. However, Oates and Schwab (1988) give a mathematical proof of the result stated here.
10. Professor Michael Porter of Harvard University has advanced the proposition that tight environmental regulations can actually enhance international competitiveness over the long run; tight regulations induce firms to innovate and ultimately lower costs ahead of their competitors (Porter and van der Linde, 1995). This is a controversial proposition not widely accepted within the economics profession (see Palmer et al., 1995).
11. Jaffe et al. (1995) provide a highly readable summary of the evidence on the effect of environmental regulation on international industrial "competitiveness."
12. Refer to any standard textbook on international trade, such as Husted and Melvin (1997).
13. Another factor, other than income and assimilative capacity, that may lead to relaxed environmental regulation is described by Chichilnisky (1994). She argues that poorer countries have less developed institutions for providing public goods and regulating public bads. Thus even if the citizenry demands a certain level of environmental protection, the country will be unable to provide it.
14. There are two obvious explanations for Tobey's results. One is that a sample of 23 countries is too small to find a significant effect unless it is very strong indeed. Alternatively, it may be that total exports are too coarse a measure to pick up subtle effects of environmental regulations. Low and Yeats (1992) look at a slightly more sensitive measure, the *revealed comparative advantage*. This is the ratio of the fraction an industry makes up of a country's exports relative to the world average for the industry. They find only slightly more evidence than Tobey that lax environmental regulations attract industry. Another explanation has to do with his measure of environmental regulatory strictness. It is based on a subjective assessment of regulations made by other researchers. It is not easy to summarize the degree of strictness of a large body of law and regulation. There may be many regulations, but they may be poorly enforced. There may be few regulations, but they may be very effective and stringently enforced. This

problem of measuring the strictness of environmental regulations in the context of finding evidence of pollution havens is considered by Xing and Kolstad (1997).

15. Wheeler and Mody (1992) estimate the determinants of *FDI* using a variety of explanatory variables; however, the closest they get to an environmental regulation variable is one associated with "bureaucracy and red tape" in obtaining approvals and permits generally.
16. The Xing and Kolstad (1997) analysis uses approximately the same size data set as Tobey (1990)—roughly 20 countries. The study also uses a second equation to explain the stringency of environmental regulations, treating the stringency as an unobserved variable. Although this is an innovation in the measurement of stringency, it is not without problems. Basically, it introduces another layer of uncertainty into the problem of determining if regulations affect *FDI*.
17. It should be pointed out that this is a different issue than the use of environmental regulations as disguised nontariff barriers to trade. For instance, the Danish requirement of a deposit on beverage bottles was alleged to be a trade barrier, hindering distant beverage suppliers (who must incur large costs to return their bottles for refilling). However, the courts found otherwise (Rehbinder, 1993).
18. Barrett (1994) provides one of the more complete analyses of strategic environmental policy at a theoretical level. See also Copeland (1992), Ulph (1992), and Barrett (1997).
19. This result that export subsidies are desirable is due to Brander and Spencer (1985).
20. The Cournot equilibrium involves several firms choosing quantities to produce, assuming the level of output by the other firms is fixed.
21. This result is relatively easy to show using calculus. Write the net profit for country D and find the s that maximizes that net profit. It will be positive.
22. A Pigovian fee provides an incentive to provide the correct amount of pollution and for the polluter to pay for damage from that pollution. In a more general framework, if the firm is directed to emit a certain amount of pollution and not required to pay damage from that pollution, that can be considered a subsidy without sacrificing quantities of pollution. In this example, a quantity regulation on pollution would be preferred to an emission fee.
23. Folmer (1993) couches this in terms of the two countries playing repeated games on a variety of issues. If a country has no power in one particular game, the country can threaten in another game in which it may have power. Folmer refers to this situation as interrelated games and suggests that transboundary pollution problems can be solved in this context.
24. See Murdoch and Sandler (1997).
25. Barrett (1994, 1997) develops a theory of self-enforcing environmental agreements that meets the first two conditions and concludes that the gain resulting from the agreement is very small. Hoel (1992) discusses another approach to international agreements.
26. This is the voluntary provision of the public good, emission reduction. Such voluntary provision was discussed in Chapter 5.
27. This terminology follows cartel theory: a perfect cartel involves pooling of joint revenue; an imperfect cartel does not.
28. Osborne (1976) discusses these problems of maintaining an effective cartel.
29. Barrett (1997) and Nordhaus (1997) propose such countervailing tariffs to strengthen environmental agreements.

14 ECONOMY-WIDE EFFECTS OF ENVIRONMENTAL REGULATIONS

For the most part we have been concerned with the behavior of firms subject to environmental regulations. How do we construct optimal regulations for such firms? How will firms react to specific regulations? We turn now to a broader consideration of the consequences of environmental regulations. There are two primary dimensions to this. One concerns properly accounting for environmental protection in our aggregate measures of economic performance, such as gross national product (GNP) and productivity. It is commonly held that environmental regulations retard growth and "hurt" the economy. But standard measures of growth and economic health omit the benefits of a clean environment.

A second issue concerns the effects of environmental regulation on preexisting inefficiencies in the economy, primarily due to taxes. This is a subtle issue. All modern economies raise a great deal of revenue through taxation, particularly labor taxation. Such taxes introduce distortions into the economy. It has been suggested that pollution taxes be substituted for distortionary taxes and that, as a result, everyone will be better-off, even before the benefits of reduced pollution are taken into account. This is the so-called "double dividend" of environmental taxation. The view turns out to be somewhat optimistic.

I. PRODUCTIVITY GROWTH

Why do personal incomes increase and standards of living improve over time? This is a fairly fundamental question in economics. The three primary reasons are usually thought to be expansion of the workforce, accumulation of capital (including human capital—educated and trained workers), and innovation/invention. Innovation and invention allow an economy to make more with the same sets of inputs. Productivity is the general term for the ability of an industry or an economy to produce goods and services with a given set of inputs. Productivity and the growth in productivity are of fundamental importance to an economy for the simple reason that they are major determinants of the level and growth of per capita income. We are concerned when productivity growth is slow; in such cases we look for reasons for the slow growth. In the past, fingers have been pointed at environmental regulations as a reason for slow productivity growth.

Innovation and invention can have two effects on economic activity. They can quite simply reduce the costs of production, allowing an increase in output over time, keeping inputs of labor and capital constant. Alternatively, they may improve the quality of a product. The electric light bulbs manufactured a century ago burned out more quickly than today's electric lights and were thus inferior.

Productivity growth is the standard measure used to gauge progress of this sort due to innovation and invention. Suppose we are comparing two time periods, this year and last year. Technology in a particular industry has improved over that period. This technological improvement has lowered the cost of production. This will decrease the cost of the output of the industry, and thus output price. Output will in turn expand. So how do we measure the growth in productivity? The productivity growth over a year is defined[1] as the rate of growth in quality-adjusted output less the rate of growth of quality-adjusted inputs.[2] The rate of growth would typically be measured in percentage terms per year. As an example, suppose we note that over a year the output of electricity has grown by 2% while the inputs into electricity manufacture (fuel, capital, labor) increased only by 0.5%. We would label this difference as productivity growth: 1.5% per year (2%–0.5%). As a more realistic example, from 1948 to 1979 output in the United States grew by 3.4% per year, on average. But inputs of capital and labor grew by 2.6% per year, leaving 0.8% as a residual, attributable to productivity growth (Jorgenson, 1988). Because the unobservable "productivity growth" is the difference between two measurable quantities (output growth less input growth), we speak of productivity growth as a residual. Table 14.1 presents estimated productivity growth rates economy-wide for a number of industrialized countries over the past four decades. Note that the 1960s were a period of rapid productivity growth in many of the countries listed, particularly those that needed the most rebuilding following World War II. Starting in the 1970s there was an across-the-board slowdown in productivity growth. There has been a great deal of speculation and research into the causes of this somewhat mysterious "productivity slowdown" of the 1970s. One possible culprit is the rise in environmental regulation in the 1970s.[3]

TABLE 14.1 Total Factor Productivity Growth in Selected Countries (Percentage per Year)

Country	1960–1973	1973–1979	1979–1996
United States	1.9	0.1	0.6
Japan	5.7	1.1	1.1
Germany	2.6	1.8	0.6
France	3.7	1.6	1.3
Italy	4.4	2.0	1.2
United Kingdom	2.7	0.7	1.5
Sweden	1.9	0.0	1.1
Australia	2.1	1.1	0.8
Switzerland	1.3	−0.9	−0.1

Source: OECD (1997).

A. Measuring Total Factor Productivity Growth

To understand how environmental regulation could have contributed to the productivity slowdown, it is important to first understand how we measure productivity growth. Start with a standard production function of the form

$$Y = A\, f(L,K) \tag{14.1}$$

where Y is output, L is input of labor, K is input of capital, and A is simply a parameter that changes over time as technology changes. The parameter A allows output to expand without additional inputs. Thus it represents the level of technology.[4] Equation (14.1) typically is for an individual firm, but may represent the aggregate output–input relationship for an industry or even the entire economy. If we move forward one step in time and let all of the variables change a small amount, we obtain

$$Y + \Delta Y = (A + \Delta A)\, f(L + \Delta L,\, K + \Delta K) \tag{14.2}$$

Equation (14.2) represents the equilibrium input and output choices of a firm or industry in the second time period. We can subtract Eq. (14.1) from Eq. (14.2), and manipulate the resulting equation a bit to obtain

$$\Delta Y = \Delta A\, f(L,K) + MP_L\, \Delta L + MP_K\, \Delta K \tag{14.3}$$

where MP_L and MP_K are the marginal products of labor and capital, respectively.[5] If we divide Eq. (14.3) by Eq. (14.1) we obtain

$$\Delta Y/Y = \Delta A/A + (L\, MP_L/Y)\, \Delta L/L + (K\, MP_K/Y)\, \Delta K/K \tag{14.4}$$

Note that the terms in parentheses in Eq. (14.4) are the shares of each input (in value terms) in the total value of output. We would expect the price of an input to be set to the value of the marginal product of that input and for total cost to equal the value of output (zero profits). Thus ($p_Y\, MP_L$) would be the price of labor (with p_Y the price of output) and $p_Y\, Y$ would be the value of output. We can rewrite Eq. (14.4) as

$$\Delta A/A = \Delta Y/Y - s_L\, \Delta L/L - s_K\, \Delta K/K \tag{14.5a}$$

where

$$s_L = p_L\, L/(p_L L + p_K\, K) \tag{14.5b}$$

$$s_K = p_K\, K/(p_L\, L + p_K\, K) \tag{14.5c}$$

We can simplify Eq. (14.5a) further by dividing the equation by Δt and substituting $\dot{A}$ for $\Delta A/\Delta t$, $\dot{Y}$ for $\Delta Y/\Delta t$, etc.:

$$\frac{\dot{A}}{A} = \frac{\dot{Y}}{Y} - s_L \frac{\dot{L}}{L} - s_K \frac{\dot{K}}{K} \tag{14.6}$$

Note in Eq. (14.6) that s_i is the value share of factor i in cost.[6] Equation (14.6) states that the rate of change of A ($\dot{A}/A$) is equal to the rate of change in output less the rate of change of inputs, weighted by their relative importance (defined by value shares). A is the parameter that stands for the level of technology. Equation (14.6) defines what is known as multifactor or total factor productivity growth, because it recognizes changes in both output and all of the input factors. The productivity growth figures in Table 14.1 are for total factor productivity.

B. Single Factor Productivity

Sometimes it is easier to measure some of the inputs than others. For instance, it is far easier to observe labor inputs than capital inputs, simply because of the difficulty of measuring capital and the ease of counting hours worked. Consequently, growth in labor productivity is often used as a simple measure of productivity growth. Labor productivity growth is defined as the percentage change in output less the percentage change in labor inputs. As we can see in Eq. (14.6), if labor and capital stay in fixed proportions and grow at the same rate, there is no loss in accuracy. In other words, if we assume $\dot{L}/L = \dot{K}/K$, then Eq. (14.6) reduces to

$$\begin{aligned} \dot{A}/A &= \dot{Y}/Y - s_L\, \dot{L}/L - s_K\, \dot{K}/K \\ &= \dot{Y}/Y - (s_L + s_K)\, \dot{L}/L \\ &= \dot{Y}/Y - \dot{L}/L \end{aligned} \qquad (14.7)$$

Of course, labor and capital may not grow at the same rate. If they do not, labor productivity may be lower or higher than multifactor productivity growth. For instance, suppose from one period to the next, the price of capital increases relative to labor but technology does not change. This is illustrated in Figure 14.1. Shown in Figure 14.1 is a unit isoquant for a firm using labor and capital. The price of capital is initially p'_K, which results in labor use L' and capital use K'. Now raise the price of capital to p''_K. We see that capital use declines (to K'') and labor use increases (to L''). We see a substitution into labor, away from capital. There is no change in technology from one period to the next because the unit isoquant in Figure 14.1 remains fixed from one period to the next (instead of moving inward as would occur with gains in productivity). If we were observing only labor productivity, we would see a decline in productivity since more labor is

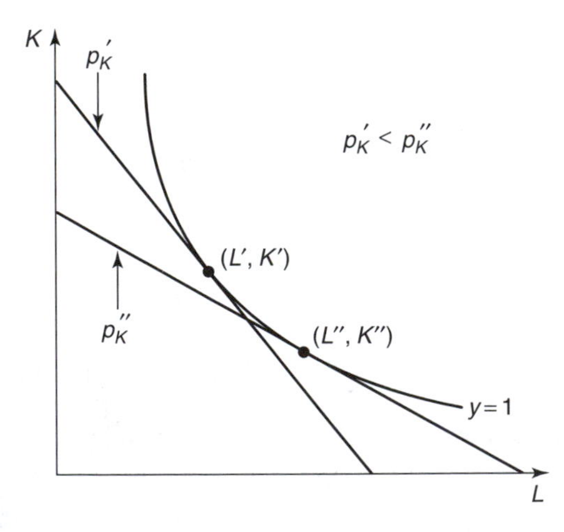

Figure 14.1 Change in factor use from a price increase for capital, no change in technology. L', K', Base labor and capital use (price of capital = p'_K); L'', K'', labor and capital use after price increase for capital to p''_K; $y = 1$, combination of labor and capital that gives one unit of output.

being used to produce the same output. Thus labor productivity growth would give a distorted view of multifactor productivity growth.[7]

To see this more generally, suppose we are comparing two periods in which productivity as well as the relative price of capital and labor (capital has become more expensive) have changed. Because of the price change, we would expect $\dot{K}/K = \dot{L}/L - B$ where B is some positive number. In other words, capital use will grow more slowly than labor use ($\dot{K}/K < \dot{L}/L$) due to the relative price change. Ignoring the small changes that may occur in factor shares, Eq. (14.6) can be written as

$$\begin{aligned} \dot{A}/A &= \dot{Y}/Y - s_L\,\dot{L}/L - s_K\,\dot{K}/K \\ &= \dot{Y}/Y - \dot{L}/L + s_K\,B \end{aligned} \tag{14.8}$$

which in words is

$$\text{Total Productivity Growth} = \text{Labor Productivity Growth} + s_K\,B$$

Because both s_K and B are positive, a measure of labor productivity growth will understate the rate of total factor productivity growth.

C. Bias from Omission of Environmental Factors

So how does this all relate to the environment? The environment is an input into the production of many goods. Steel cannot be manufactured without using the environment as a dumping ground for pollution. Let E denote the input of environment into production for a firm. If the firm uses a little E, that is equivalent to disposing of only a small amount of pollution; a large E would correspond to dumping a large amount of pollution. Production of output now involves K, L, and E: capital, labor, and environment. If the environment is cheap to use (no regulation), firms will tend to use a good deal of E. If the environment becomes more expensive to use (tight regulations or a high Pigouvian fee), firms would substitute other factors (such as capital), using less of E.

What has happened over the last three decades throughout the world is that the effective price of the environment has gone up dramatically. In the 1960s in many places, the environment could be freely used for pollution disposal. That is typically not the case today. If the environment is neglected as a factor in measures of productivity and the price of the environment increases, measures of productivity growth will understate the extent of real productivity growth. This is analogous to the case we just discussed in which labor productivity growth will understate multifactor productivity growth when there is a simultaneous increase in the price of nonlabor inputs (capital).

How serious is this potential understating of productivity growth? Equation (14.8) indicates the error that occurs when productivity growth ignores one of the inputs. In Eq. (14.8), the distortion was introduced by ignoring capital, focusing on labor productivity. But a version of Eq. (14.8) could be generated with the environment as the missing input. As can be seen from Eq. (14.8), the size of the value share of environmental inputs in total costs is an important determinant of the extent of the understatement of productivity gains. The larger the share, the greater the understatement.[8] Thus industries with significant pollution or pollution control would be expected to show the greatest divergence between true productivity growth and measures that omit the environment.

Several studies have tried to correct productivity measures for environmental regulations. Robert Repetto and colleagues at the World Resources Institute have calculated multifactor productivity growth for several industries, taking into account the use of the environment as an input.[9] Figure 14.2 shows the growth of productivity over the period 1970–1990 for two industries in the United States, electricity generation and agriculture. Electricity generation in particular has moved from a situation of free use of the environment to one of strict regulation of emissions. Shown in Figure 14.2 is the value of A in Eq. (14.1), normalized to a base year. The change in A over time would yield information needed to calculate the rate of productivity growth. Figure 14.2a shows productivity for electricity generation. Note that the conventional measures of productivity in electricity show a decline over the 20-year period. However, when the environment is included, the result is reversed—productivity (negative productivity growth) increases, though modestly. Figure 14.2b presents a similar picture for agriculture in the United States. Conventionally measured multifactor productivity increases significantly over the 20-year period. However, in this sector, correcting for the omission of the environment only modestly increases the estimate of productivity levels. Pollution is less significant to agriculture than to electric power.

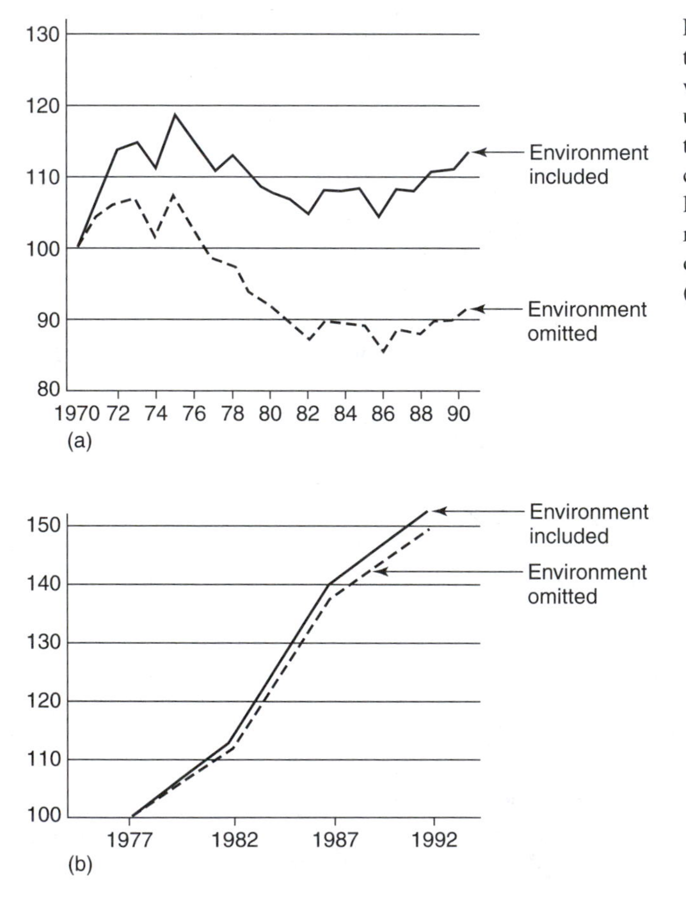

Figure 14.2 Sectoral multifactor productivity in United States with and without adjustment for use of environment. (a) Electricity (1970 = 100); (b) agriculture (1977 = 100). Solid line accounts for use of environment; broken line omits use of environment. From Repetto (1996), pp. 27, 41.

Conrad and Morrison (1989), in another interesting study, compare traditional measures of productivity growth with "corrected" measures, taking into account the environment, for the manufacturing sectors of the United States, Canada, and Germany. To compute the corrected measure, they make two assumptions: that pollution regulations are efficient and that only capital expenditures can reduce pollution emissions. The implication of these assumptions is that the marginal cost of pollution abatement capital to the firm must be equal to the marginal damage from the pollution avoided by one more unit of abatement capital. These are strong assumptions, but making them allows the authors to show that for the United States in the period 1960–1967 (a period of weak environmental regulation), conventional and adjusted productivity growth measures are about the same (2.9% per year), whereas in the period 1972–1980 period (a period of tightening environmental regulations), the adjusted measure is 2.4% per year compared to 2.2% for the conventional way of measuring productivity growth. The results for Germany and Canada were less pronounced.

What do we learn from this discussion? Simply that in judging productivity gains, if we give signals to firms to save on the use of one input (the environment), we should count efforts to innovate to reduce the firm's reliance on the environment, even if that comes at the expense of other inputs. Thus if we want a true measure of how much innovation has taken place, we should include all of the relevant factors. Excluding the environment from measures of multifactor productivity growth would be expected to understate productivity growth in a period in which environmental regulations are tightening.[10]

II. GREEN NATIONAL ACCOUNTING

We have all become accustomed to news reports on the state of our economy. Such reports rely on such basic statistics as the unemployment rate, the rate of growth of gross domestic product (GDP), or the imbalance in foreign trade accounts. Such information on the economy has not always been so readily available. In the early 1930s Simon Kuznets, an economist born in Russia but who had emigrated to the United States, developed the first detailed accounting of the U.S. economy. During the 1930s and 1940s this system of accounts developed, in part, to assist with planning the war effort in World War II. From that point on, the United States has maintained a set of National Income and Product Accounts (NIPA). Other countries have set up similar systems of national accounting. After World War II, the United Nations established a System of National Accounts (SNA), defining standards to use in national accounting that would permit comparisons across countries.

Other than providing news releases for the media, what purposes do the national accounts serve? Although private firms and investors can benefit from knowledge about the condition of the economy, the primary purpose of the national accounts is to assist policymakers in government.[11] Incentives to encourage saving may result from knowledge that the national savings rate is low compared to similar countries. Monetary policy is guided by statistics on the money supply, economic growth and inventory levels, among other things. Nearly all government policies that are designed to intervene in the economy rely substantially on economic data contained in the national accounts. Even ostensibly noneconomic policies such as environmental regulation and funding of scientific research are influenced by data in the national accounts. A major justification for investing in R&D is the demonstrated gains to economic growth that have been contributed by innovation and invention (as discussed in the previous section). Environmental policy can be affected in a negative way. National accounting data can indicate the extent to which environmental protection is retarding conventionally measured economic growth. Or, in many developing countries, significant harvesting of forests or mining of minerals is promoted on the basis of its effect on growth in GNP (it raises the growth rate, some say misleadingly).

National income accounting can be easily illustrated with an example. Suppose we have a very simple economy, consisting of two companies and consumers: the AJAX coal company (a mining company), the Edison Power Company, an electric power producer (who uses coal to generate electricity), and consumers who consume electricity.

AJAX has underground coal reserves and operates an open pit mine. A new law requires that following mining, mined land be returned to its original state, regraded, and vegetated. AJAX has complained bitterly that this new law is costing the company a great deal and, as a result, coal prices have had to be increased. The company also maintains an exploration division that works to add new coal reserves to the company's inventory of reserves. In 1997, AJAX produced 5000 tons of coal and sold it for £20 per ton. AJAX's exploration department succeeded in finding 2500 tons of additional reserves on AJAX lands. A summary of AJAX's operations is given in Table 14.2. All of AJAX's coal was sold to the Edison company for the generation of electricity.

Edison produced £300,000 worth of electricity in 1997, along with 100 tons of sulfur dioxide air pollution. These emissions were lower than they used to be, before environmental regulations were enacted. A summary of Edison's income and expenses is also given in Table 14.2.

TABLE 14.2 Sample Accounts

AJAX Mining Company, 1997		
Expenses		
Wages		£50,000
Mining labor	£30,000	
Exploration labor	£10,000	
Land reclamation labor	£10,000	
Depreciation of mining equipment		£10,000
Income		
Sales of 5000 tons @ £20/ton		£100,000
Edison Power Company, 1997		
Expenses		
Wages		£100,000
Pollution control labor	£10,000	
Production labor	£90,000	
Coal purchases		£100,000
Depreciation of generating equipment		£20,000
Income		
Sales of electricity		£300,000

How might we measure the total output of this little economy? One way would be to add together the value of everything produced from the two firms: £100,000 + £300,000 = £400,000. But this is not quite right because coal is counted twice, both as output of AJAX and as part of the value of Edison's electricity. A more appropriate way would be to look at the *value added* to purchased raw materials by each company (labor and capital are not considered purchased raw materials). AJAX does not buy anything except labor and capital so its value added is £100,000. Edison purchases £100,000 of coal so its value added is £200,000. This makes the total value added by these two firms £100,000 + £200,000 = £300,000.

The goal of National Income and Product Accounting is to summarize the productive activities of the economy. The overall output of a nation's economy is termed the gross domestic product. The term "gross" is used because depreciation is excluded. In the process of producing, equipment gets older, some equipment is retired, and some equipment becomes obsolete. This depreciation of equipment should really be subtracted from the measure of output to obtain a better idea of true output. If depreciation is subtracted from GDP, we obtain NDP—net domestic product. Depreciation is not so easy to calculate. (Think of some capital good you may own—perhaps a radio. How much did it depreciate last year?) Hence the common use of GDP rather than NDP. In our example, there is depreciation of mining equipment and electric power equipment. Deducting depreciation would bring the measure of output down by £30,000 to £270,000.

Although the methods of generating national accounts have evolved over the decades, they are not perfect. Illegal activities are generally excluded, for example. Housekeeping for pay is included whereas housekeeping within the family is excluded. The value of the environment to the economy is also excluded, and this has been the source of significant criticism over the past three decades. For instance, when oil is pumped from the ground, the oil is counted in GDP. But it could be argued that the inventory of oil in the ground is being depleted and that this inventory reduction should be a subtraction from GDP.[12] Using up our oil is making us poorer, not richer. Similarly, pollution is not included in the national accounts. Pollution is a product like any other with two exceptions: it is not marketed and it has a negative value. If a firm produces $100 of goods and $30 of pollution damage, it seems reasonable to value its total output at $70, rather than at $100 as is currently done in most national accounting systems.

Let us examine these resource and environmental issues one at a time. Consider first the case of nonrenewable resources (such as oil). The traditional way of dealing with depletion of nonrenewable resources is to ignore the depletion. The justification is that the value of additions to the reserve stock is ignored so depletion may also be ignored. This is somewhat like saying we ignore inventory additions so we can ignore inventory reductions.[13] But this would only be reasonable if additions balanced subtractions. Reserves are decreased by production and increased by exploration. In our little economy, perhaps we should deduct the value of the net amount of coal reserves that are depleted. The value of these reserves is difficult to determine, but one way of doing this is to estimate the cost of finding new reserves.[14] In the example, £10,000 of exploration effort resulted in 2500 tons of new coal—a finding cost of £4 per ton. AJAX found 2500 tons and used 5000 tons for a net loss of 2500 tons. Thus the 2500 tons lost could be viewed as reducing reserves by £4 × 2500 = £10,000. Taking this deduction reduces our measure of output to £270,000 − £10,000 = £260,000. Excluding depletion is somewhat like saying that your income has gone up when you decide to sell all the furniture in your house! Your cash flow may increase but you are no richer for liquidating your assets.

What about pollution? In some sense pollution is conceptually simple. If one of the outputs has negative value then it should reduce the measure of net output. One problem is that there is no market and thus no obvious value to impute to the pollution. However, as we have seen, the nonexistence of a market does not mean the pollution has no value. There are methods for imputing the value of pollution. Take our example of the Edison Power Company. If the damage from sulfur dioxide is £100 per ton, then the 100 tons of emissions are causing £10,000 of damage. This reduces net output to £260,000 − £10,000 = £250,000.

What about expenditures on pollution abatement (Edison) and land reclamation (AJAX)? These are necessary expenses of doing business. Land reclamation expenses are necessary to prevent the land from degrading. The pollution control expenses for Edison are necessary to prevent even more pollution from being emitted. In much the same way that janitorial services for keeping the workplace tidy are legitimate inputs into production, these pollution control expenses are valid expenditures. Thus no further adjustment is needed.

In summary, we have seen that a conventional measure of gross output, GDP, starts at £300,000 in our economy. Depreciation of man-made capital, depletion of natural resources, and damage from pollution reduce this to £250,000. Although our numbers here are fabricated, these adjustments can be significant in reality. In fact, Repetto et al. (1989) examined GDP numbers for several resource-exporting developing countries.[15] They conclude that on the basis of conventionally measured GDP figures these countries appear to be growing rapidly, but in fact they are growing much more slowly. A significant portion of these economies is based on mining and forestry. When the value of depleted natural resources (such as forests) is included, the rates of growth are significantly reduced. To the extent public policy encourages depletion of natural resources, correcting the accounts (which are using in formulating public policy) can have significant policy implications.

Although our example may make the inclusion of the environment in national accounts appear straightforward, that is not the case. In the case of natural resources, a large issue is determining the value of resources in the ground or standing in forests. For pollution, it is ambient concentrations, which cause damage, and those concentrations are often imperfectly observed. This is compounded by the difficulties in determining damage from ambient pollution.

Many governments are very interested in extending their national accounts to include the environment, and as a result, have thought through many of the details necessary for the inclusion. The United Nations has proposed a System of Environmental and Economic Accounting (SEEA) as a "satellite system" to the conventional System of National Accounts (SNA). Similarly, the United States has proposed a "satellite system" termed the integrated economic and environmental satellite accounts (IEESA).[16] Undoubtedly, these companion accounting systems will undergo a great deal of maturation in the coming years.

III. THE DOUBLE DIVIDEND

Governments must raise revenue. Unfortunately, most taxes tend to introduce inefficiencies and distortions into the economy. A labor tax (wage income tax) makes labor more expensive and thus discourages work. A capital tax makes capital more expensive and thus discourages investment. A number of authors have tried to measure the inefficiencies associated with taxation. Edgar Browning (1987) estimates that the last dollar collected through wage taxation in the United States has an additional welfare loss of approximately 40¢.[17] In other words, at the margin, the inefficiencies of such taxation run 40% of revenues collected—it costs $1.40 to raise $1 more of revenue.

In recent years, there have been a number of proposals to substitute pollution taxes for more traditional revenue raising taxes, such as income taxes. The idea is that such taxes would not only reduce pollution but would also reduce the distortions associated with existing taxes—a "double dividend." This policy suggestion has prompted an examination of the efficiencies of pollution taxes when there are preexisting distortions.[18] When distortionary taxes already exist (such as an income tax), pollution taxes generate an indirect cost in terms of exacerbating the preexisting inefficiencies. In fact, reducing pollution by any means has a hidden cost when there are existing tax distortions. A pollution tax can offset some of this indirect cost by raising revenues that can replace distortionary taxes, keeping total revenues collected constant. However, an environmental regulation that does not raise revenue (such as command and control) does not have this feature. In such cases, it is entirely possible that the indirect inefficiencies are greater than the gains in reduced pollution damage. These are classic results in what is known as the theory of second best. What is advisable in an economy without distortions is not necessarily a good idea when there are distortions, even in apparently unrelated parts of the economy.

To see this result, we start with a representative consumer who chooses how much labor, L, to provide (in exchange for a wage, w) and how much of a polluting good to consume, X. A polluting good means that pollution is generated when the good is produced. Suppose there is a tax on wages, t_L (which, for simplicity, assume the firm pays). With this tax on labor, a certain amount of labor is provided, L^*. With a lower tax, employers could pay more and thus more labor would be supplied; with a higher tax, less labor would be supplied. Figure 14.3a shows a supply curve for labor and a horizontal demand curve set at the value marginal product of labor (assumed constant). We see that the labor tax lowers the amount of labor supplied to L^*. We can also look at this in terms of demand for leisure, since workers divide their total time into labor and leisure. A tax on labor is the same as a subsidy to leisure. The higher the wage tax, the greater the subsidy to leisure and the more leisure will be provided (less labor). This is shown in Figure 14.3b, where H^* is the amount of leisure that will be consumed when the labor tax is t_L (ignore the broken line and the shaded area for the time being). Without the tax on labor, there would be more labor and less leisure. The deadweight loss from the labor tax is shown in Figure 14.3b as the triangle ABC. Thus to raise revenue $t_L L^*$, a deadweight loss of ABC is incurred.[19] Define V as the marginal deadweight loss, i.e., the deadweight loss that would be incurred to raise one more dollar of revenue when the tax rate is set at t_L.

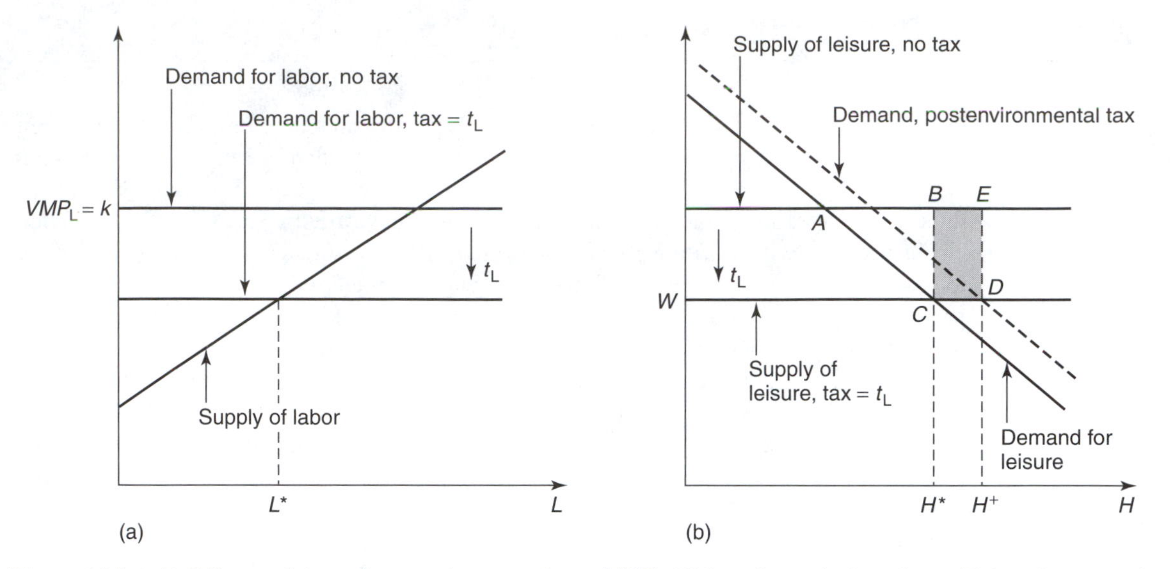

Figure 14.3 (a,b) Effect on leisure from environmental tax. VMP_L, Value of marginal product of labor; t_L, tax on labor; w, wage rate; H^*,L^*, supply of leisure and labor with $t_X = 0$; t_X, tax on X; H^+, quantity of leisure with $t_X > 0$; $BCDE$, wage tax revenue loss from $t_X > 0$.

Now consider demand for a polluting good, i.e., a good whose manufacture generates pollution. Figure 14.4 shows a typical demand curve for the good, as well as a marginal private cost curve and a marginal social cost curve that includes the damage from pollution. Without any intervention, the amount of the good produced is X^*. Now suppose we wish to reduce the amount of this good produced. This may be accomplished by taxing the production of the good (or the production of pollution) or simply by command and control, directing the firm to reduce pollution and thus incur additional production costs. In either case, assume the price increases by an amount t_X and the output is reduced from X^* to X^+. This is shown in Figure 14.4. If tax revenues result, then labor taxes will be reduced to keep the total government revenues constant.

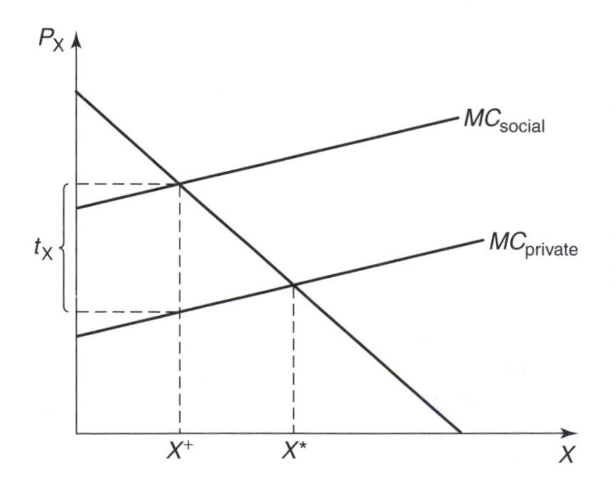

Figure 14.4 Supply and demand of polluting good. MC_{social}, Marginal costs of producing X, including pollution damage; $MC_{private}$, marginal costs of producing X, excluding pollution damage; X^*, production without internalizing externality; t_X, product tax necessary to internalize externality; X^+, production with tax t_X.

The increased price of X has an effect on the demand for leisure. In general, we would expect X and leisure to be substitutes.[20] Thus an increase in the price of X will shift the demand for leisure outward, as shown by the broken line in Figure 14.3b. This results in an increase in the quantity of leisure to H^+. The amount of this shift can be computed from the cross-price elasticity of demand for leisure with respect to the price of good X:

$$\eta_{HX} \equiv (\Delta H/H)/(\Delta p_X/p_X) = [(H^+ - H^*)\ /t_X]\ (p_X\ /\ H^*) \tag{14.9}$$

In Eq. (14.9), η_{HX} indicates how much the demand for leisure changes when the price of X changes. This is the cross-price elasticity of demand for leisure with respect to the price of X. For a substitute, this will be positive.

We see from Figure 14.3b that the reduction in X has increased the quantity of leisure and thus decreased tax revenues by the shaded area in Figure 14.3b. The area of this shaded area is $(H^+ - H^*)\ t_L$. This loss must be made up and every dollar that is raised through labor taxation actually costs society $1 + V$. Thus [using Eq. 14.9] the indirect loss from the environmental controls on X is

$$IE = (1 + V)\ (H^+ - H^*)\ t_L = (1 + V)\ t_L\ t_X\ \eta_{HX}\ H^*/p_X \tag{14.10}$$

This is termed the "tax interaction" effect (IE). Though IE in Eq. (14.10) is positive, remember that it is a loss. The tax interaction effect is the extra inefficiency from a pollution regulation due to the fact that labor is already taxed. Note that if the labor tax rate is zero ($t_L = 0$), the tax interaction effect is zero. Furthermore, the greater the value of η_{HX}, the larger the effect. Basically, what is happening is that the pollution tax is generating a deadweight loss by marginally changing the very significant preexisting deadweight loss associated with labor taxation.

Although IE is unequivocally a welfare loss (provided the good is a substitute for leisure), there are two welfare gains that are possible from pollution regulation. The most obvious one is the reduction in damage associated with the pollution. Less pollution means less damage. Term this the Pigouvian effect (PE). In the context of the double dividend debate, suppose any tax revenue raised by the pollution tax is offset by a reduction in labor tax. Call this the revenue recycling effect (RE). In the context of Figure 14.4, clearly

$$RE = V\ t_X\ X^+ \tag{14.11}$$

The revenue raised is $t_X X^+$, which is used to offset labor taxes that have a marginal deadweight loss to revenue ratio of V. It is clearly positive—a gain. Note, however, that if no revenues are raised through the environmental regulation, RE is zero. If the regulation is command and control, no revenue is raised. If the regulation is marketable permits for pollution, then revenue is raised if the permits are initially auctioned. If the permits are given away initially (as is often the case), then $RE = 0$.

To summarize, we conventionally think of the Pigouvian effect (PE) as the primary gain from a pollution control regulation. We have just seen that if pollution tax revenues can be used to reduce distortionary taxes, we have a bonus of the revenue recycling effect (RE). But we have also seen that reducing pollution will generate a loss through the indirect effect on preexisting distortionary taxes (IE). The big question is, how big a problem is this tax-interaction effect? That question is empirical.

Using a very simple model and typical parameter values for the United States, Parry (1995) calculates that the optimal pollution tax should be 63% of marginal damage (not equal to marginal damage as suggested by conventional wisdom). Goulder et al. (1997) use a more detailed computer model to examine the control of sulfur emissions from electric utilities in the United States. They reach similar though more detailed conclusions. The 1990 Clean Air Act calls for a 10 million ton per year reduction (from about 20) in emissions of sulfur dioxide from electric power plants. This is to be achieved using a marketable permit system with the permits distributed without charge based on historic levels of electricity generation. Thus this is a regulation without a revenue recycling effect. An alternative way of achieving the same effect but with revenue recycling would be to auction the permits. Goulder et al. (1997) calculate that the preexisting taxes on labor increase the total cost of the program by 71%. They further calculate that more than half of that increased cost is due to the fact that permits are given away rather than auctioned with the proceeds going to reduce taxes on labor. Another result is that the tax interaction effect (*IE*) is on the order of $100 per ton of pollution reduction (independent of how much pollution is reduced), about a third of their assumed private marginal cost of pollution abatement (on average $300 per ton of pollution). Thus if the damage from pollution does not exceed $100 per ton, no pollution control is justified unless there is revenue recycling.

The lesson to be learned is that pollution regulation can result in significant indirect costs due to preexisting tax distortions in the economy. Much of this distortion can be eliminated if pollution taxation raises revenue, which is used to offset the preexisting distortion. Without revenue recycling, only much weaker environmental regulations are justified on efficiency grounds.

It might seem that it is "unfair" to have the environment bear the cost of the inefficiency introduced by the wage tax. But this is not really the correct interpretation of the result. In fact, in this sort of second-best environment, it is difficult to interpret what is happening with the emission tax without looking at the distortions throughout the economy introduced by the income tax. For instance, if the polluting good and leisure are substitutes as assumed, the subsidy to leisure may already have reduced consumption of the polluting good in the economy. Thus we do not need to reduce it as much with the emission tax. In fact, it has been shown that when there are no commodity taxes and an optimal income tax, it can be welfare improving to subsidize the consumption of goods that are substitutes for leisure.[21] Although we have made no assumption that the income tax is optimal, if it were and there were no tax or subsidy on the polluting good, the optimal action would be to simultaneously tax the polluting good at the full marginal damage but then add a subsidy to account for the distortions of the income tax. The net effect is that the tax on the polluting good is less than marginal damage.

SUMMARY

1. Productivity growth is defined as the rate of growth of output less the rate of growth of inputs into production. Both inputs and outputs should be measured in constant quality units. The rate of growth of inputs as a whole is a weighted average of the rates of growth of all inputs, with weights being value shares in cost. This definition of productivity growth corresponds to multifactor or total productivity growth.

2. Single factor productivity growth is defined as the rate of growth in quality-adjusted output less the rate of growth of a single input. The most common measure of single factor productivity growth is labor productivity growth.

3. Single factor productivity growth may be higher or lower than multifactor productivity growth, depending on changes that may occur in relative prices of inputs, or relative intensity of input use, among other things.

4. Standard measures of productivity omit the uses of environmental resources. To the extent that environmental resources have become more costly to use over the past few decades, measures of productivity growth that omit these resources will understate true multifactor productivity growth. This is particularly true for heavily polluting industries.

5. Gross domestic product is a conventional measure of output of an economy and consists of the value added by all of the productive enterprises in an economy. When capital depreciation is subtracted, net domestic product is obtained. The depletion of natural resources and damage to the environment are usually excluded from this measure. "Green" national accounts aim to correct this. With green accounts, net depletion of natural resources would be a deduction from the overall measure of output as would damage to the environment. A major problem with moving to green accounts is valuing resource depletion and environmental damage.

6. The "double dividend" refers to the possibility that imposing a revenue-neutral tax on pollution will have two effects: (a) reduce pollution and thus pollution damage (the Pigouvian effect) and (b) reduce distortionary taxes on labor and thus the deadweight loss associated with those taxes (the revenue recycling effect). There is a third effect that acts in the opposite direction, the tax interaction effect.

7. When pollution is reduced, the demand curve for leisure is shifted outward (if the polluting good and leisure are substitutes). If labor is taxed, this shift will generate an additional deadweight loss that must be attributed to the reduction in pollution. This is the tax interaction effect.

8. A tax on pollution increases welfare through the Pigouvian effect and the revenue recycling effect but reduces welfare through the tax interaction effect. A marketable permit systems acts similarly, though if the marketable permits are initially given away rather than auctioned the revenue recycling effect is zero, which reduces the overall efficiency of the pollution control.

9. The size of the tax interaction effect is an empirical question and will vary from one polluting industry to another. Ian Parry calculates that a pollution tax should be set at 63% of marginal damage to account for the tax interaction effect.

PROBLEMS

1. The table below shows the use of labor and capital in the caviar industry of Beluga for two particular years. Also shown is the output of caviar and water pollution. Caviar output is shown in millions of eggs, pollution in tons of biological oxygen demand (BOD), labor in

person-years, and capital in Rubles. The price of labor is constant at R10,000 per person year, and capital services at R0.20 per Ruble of capital stock.

Calculate productivity growth for these 2 years (using value shares for 1997). In particular, calculate labor productivity growth and total factor productivity growth (excluding pollution). Assuming pollution damage is R10,000 per unit, calculate total factor productivity growth, including pollution (considering pollution as an input).

Year	Caviar output	Pollution	Labor	Capital stock
1997	100,000	20	50	1,000,000
1998	103,000	10	51	1,000,000

2. In the example of the AJAX coal mining company and the Edison Electricity company, suppose there was no law requiring reclamation of the land after coal mining. Thus the coal mining company had no expenditures for land reclamation. The damage to the land is just as serious, and approximately equal to what is now spent on land reclamation. What would the net domestic product (NDP) be before the reclamation law, taking into account natural resource depletion and environmental degradation? Explain.

***3.** It is possible to show that the marginal welfare loss from a labor tax is $V = t_L \epsilon / [w - t_L \epsilon]$ where ϵ is the labor supply elasticity evaluated at the net of tax wage received by the consumer, w. Using this, show that the absolute value of the ratio of the tax interaction effect to the revenue recycling effect is given by $IE/RE = (X^*/X^+)(\eta_{XH}/\epsilon)$ where η_{XH} is the price elasticity of demand for good X with respect to the wage rate. Assume elasticities are compensated elasticities. If on average we expect $\eta_{XH} = \epsilon$, what can you conclude about the size of IE relative to RE in this case? What is the significance of this result?

4. Assume the marginal savings from emissions for an industry are given by $MS(e) = 30 - e$ and that the marginal damage from emissions is given by $MD(e) = e$. Suppose that the tax interaction effect corresponds to a welfare loss of \$10 per ton of emissions reduced. Suppose further that the cost of collecting revenue from labor taxes is \$1.40 for every dollar collected (\$1 for revenue plus 40¢ deadweight loss).

a. Plot marginal damage and marginal savings from emissions for this industry. What is the efficient level of an emissions tax, ignoring the revenue recycling and tax interaction effects? Derive this result graphically.

b. Now take into account the tax interaction affect. What is the efficient number of marketable permits, if the permits are initially distributed for free? Derive this result graphically.

c. If the permits in part (b) were initially auctioned, what would be the total surplus and what would be the magnitude of the tax interaction effect, the revenue recycling effect, and the Pigouvian effect.

NOTES

1. Strictly speaking, productivity growth is the percentage increase in output, keeping inputs constant, thus excluding returns to scale in the measure of productivity growth. If the economy ex-

hibits constant returns to scale, which empirical evidence suggests is generally the case, these two definitions are identical.

2. By quality adjusted, we mean taking into account any changes in the quality of inputs and outputs that may have occurred. Alternatively, simply keep quality fixed in making the calculation.
3. Jorgenson (1988) provides a readable overview of the growth evidence in the United States. Repetto et al. (1996) present a very accessible discussion of the role of environmental regulation in the productivity slowdown. Jorgenson and Wilcoxen (1990) estimate that environmental regulations in the United States are responsible for reducing conventionally measured economic growth (not productivity growth) in 1974–1985 from 2.7% per year to 2.5% per year. This estimate does not include the benefits of environmental regulations, only the cost.
4. Technology could make individual factors more productive. The representation of Eq. (14.1) is the most straightforward, where technology affects all factors equally ("neutral" technical change).
5. The marginal product of labor is $A[f(L + \Delta L,K) - f(L,K)]/\Delta L$. In deriving Eq. (14.3), we have assumed that $f(L + \Delta L, K + \Delta K)$ is essentially the same as $f(L,K)$, which is true if ΔL and ΔK are very small. Calculus simplifies the derivation considerably.
6. We have condensed several steps into one in moving from Eq. (14.4) to Eq. (14.5). One assumption is that in equilibrium the price of a factor is equal to the value of marginal product. The second assumption is that with a constant cost industry, costs will be equal to the value of output.
7. A change in relative prices for inputs is not the only thing that can introduce a wedge between multifactor and single factor productivity growth. Technical change can be biased towards the use of a particular factor or towards saving the use of another factor.
8. The rate of change of use of environmental inputs relative to other inputs is also important in Eq. (14.8).
9. Refer to Repetto et al. (1996). In their study they impute a value to the emissions of the industry in question based on estimates of damage from pollution. Thus as pollution levels are reduced, the net value of output rises more rapidly than when pollution is neglected. This is not quite the same as changing the price of the environment over time, though conceptually very similar.
10. In a closely related analysis, Squires (1992) estimates productivity in a Pacific fishery, in which the environmental input of the stock of fish is usually excluded from measures of productivity. Although the stock of fish was sometimes growing and sometimes declining in his sample, his results clearly indicate the importance of including the fish stock in a measure of productivity growth.
11. National accounts are used by a very diverse set of people and institutions. For example, investors use national accounts to gauge profitability of different industries. Demand for output of specific industries (eg, automobiles) may be highly correlated with the health of the overall economy.
12. When a firm produces goods that go into inventory, the value of those inventory additions increases measures of output. When goods are drawn from inventory and sold, there is no addition to measures of output.
13. The United States Department of Commerce (1994a) points out that in the United States in the early 1940s, depletion was deducted from measures of output. This was dropped in 1947 because of the omission of explicit treatment of additions to reserves.
14. The marginal (not average) cost of finding new reserves is equal to the marginal value of new reserves only if the firm is efficiently exploring.
15. See Crowards (1996) for an example of adjusting for resource depletion in the national accounts of Zimbabwe.
16. See United States Department of Commerce (1994a).
17. There is considerable uncertainty over this figure for the marginal welfare cost of taxation. Browning (1987) suggests the figure is between 32¢ and 47¢ (compared to an *average* welfare cost of 16¢). Other authors (cited by Browning) suggest figures outside this range.
18. A readable survey of this literature can be found in Goulder (1995). Bovenberg and de Mooij

(1994) and Parry (1995) are two of the earlier analyses of the double dividend. The Parry (1995) paper provides a graphic analysis from which the presentation here is drawn. Goulder et al. (1997) provide an advanced though accessible treatment of the theoretical results on the double dividend as well as an application to sulfur emission regulations in the United States.

19. To the extent that other prices in the economy change as a result of the wage tax, there may be other deadweight losses from the wage tax.
20. In general, goods are substitutes for one another. If the price of one item in a consumption bundle increases, we consume less of it and more of other things. There are of course exceptions.
21. This result is due to Christiansen (1984). It is somewhat intuitive. We have too much leisure so we have too little of goods that are substitutes for leisure. The problem can be partially ameliorated by a subsidy to those substitutes.

15 ENVIRONMENTAL DEMAND THEORY

One of the cornerstones of economics is understanding consumer preferences for goods. The typical way those preferences are represented is through demand functions—relationships that give the amount of a good an individual at a given income level will desire, when facing a particular set of prices. A demand curve (another name for a demand function) is a way of summarizing how important a particular good is to an individual. Since every individual has a limited income, the demand curve tells us how much money the individual devotes to a specific good, out of the many choices available. Additional information is provided by indicating how much of the good the individual chooses to forgo if the price increases. If a good is absolutely indispensable with no clear substitutes, demand will probably be insensitive to price (inelastic). If, on the other hand, the good is important, but there are ample substitutes, an increase in price will drive the consumer to substitutes, with a correspondingly sharp decrease in consumption of the good whose price rose.

All of this information is useful in understanding markets and the effects of changes in those markets. If we summarize an individual's preferences for a good by a demand curve, we can summarize the market's overall consumption of that good by an *aggregate demand curve*, the sum of individual demand curves. We can then match that with a supply curve for that good to find out how much of the good is likely to be supplied in a competitive market and conduct analyses of the effects of government interventions, such as taxes and regulations, in those markets.

In much the same way that demand curves are useful summaries of consumer preferences for use in analyzing markets for private goods, a demand curve can also be useful in summarizing preferences for environmental goods. The problem with demand curves for environmental goods is that there are typically no markets, so we have no observations on how much of an environmental good is consumed at different prices. Yet the idea of a demand curve as a representation of underlying preferences is just as valid for environmental goods as for tomatoes, bread, or gasoline. It is just harder to measure.

This chapter examines demand theory for environmental goods and bads. Much of this theory is the same as for conventional goods and services. There are some unique extensions however.

I. WHAT IS SO SPECIAL ABOUT AN ENVIRONMENTAL GOOD?

There is a wide variety of goods that fall under the rubric of environmental goods. Included would be commodities such as air pollution and water pollution, or viewed as goods, air quality and water quality. Also included would be amenity values such as a beautiful vista over land or ocean, or negatively characterized, a smoggy vista over a city. Even more distant from the conventional notion of a good might be the existence of a species (which is important to many people) or the lives of particular members of a species (such as birds that might be killed in an oil spill). The challenge to economists is to be comprehensive, viewing all aspects of the environment, over which people have preferences, within the paradigm of consumer theory.

Why force the things we value in the environment into consumer theory? The most important reason is that all environmental problems really involve a trade-off between using resources (money) for conventional goods and services and using those same resources for environmental protection. We really are talking about trade-offs. And that is precisely what a demand curve represents, at least from the consumer's point of view: How much is the consumer willing to give up (pay) for particular levels of an environmental good? No matter how esoteric or intangible the environmental good might be, its protection usually requires money. And it is usually appropriate to enquire how much people are willing to contribute to protect that environmental good. Eliciting true values is not easy; but that does not diminish the validity of the principle of demand for environmental goods and services.

One of the clearest distinctions between environmental goods and ordinary goods is the existence of a market. With an ordinary good, such as milk, the market allows us to observe a variety of prices (usually) and different purchases at different price levels. Thus it is "straightforward" (there are even problems here) to deduce the relationship between prices and quantity demanded (a market demand curve). It is harder still to measure an individual's demand curve. Figure 15.1 shows a simple plot of the quantity of gasoline consumed in particular years in the United States vs. the average price in those years. Although this does not take into account all of the variables that might influence consumption from year to year, it illustrates the ease with which we can determine the relationship between prices and consumption for conventional goods. However, even for market goods, moving from price–quantity correlations as presented in Figure 15.1 to an actual demand curve is no easy task.

It is of course much more difficult to generate a demand curve for clean air in an urban area. Leaving aside how precisely we would define "clean air," we have no observations on different consumption levels and prices. Yet clearly, individuals who value clean air would be willing to pay for it. The more expensive clean air is to provide, the more people would be willing to tolerate some pollution; similarly, the cheaper clean air is, the more of it people would demand.

The absence of a market is the major factor that complicates finding the demand curve for environmental goods. But the absence of a market is the problem with nearly all public goods—including but not limited to environmental goods. For instance, schools are provided publicly in many parts of the world (education has elements of nonrivalry). Finding the demand for public schooling is no easy task, basically because of the lack of

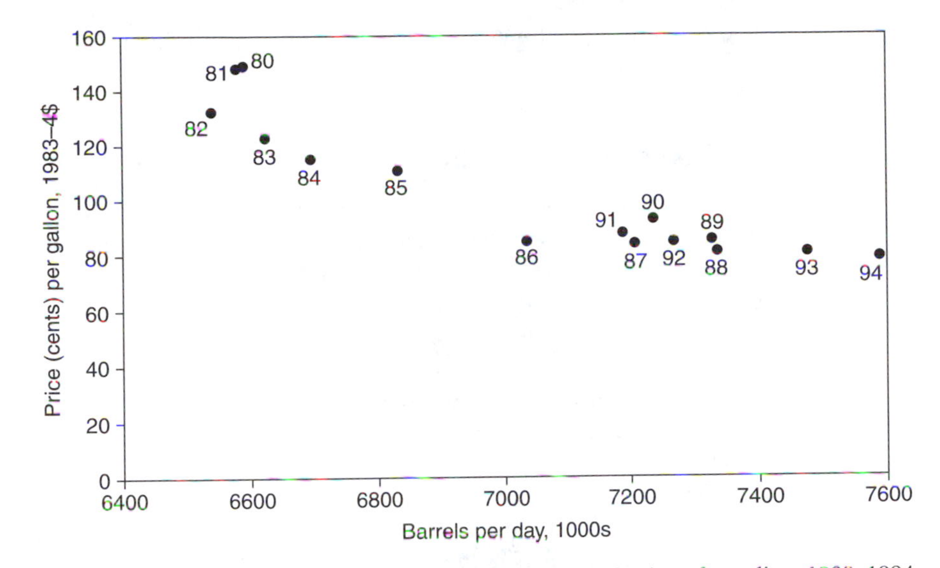

Figure 15.1 Gasoline consumption (United States) vs. real price of gasoline, 1980–1994.

markets.[1] But the big difference between environmental goods and ordinary public goods is the supply side. Ordinary public goods are produced at a cost. Thus citizens have a reference point of the cost of supply to take into account when determining how much of the good they want at particular prices. Although strictly speaking this should not impact the value to consumption, it can make the process of valuation easier. Furthermore, people may have experience with different points on their demand curve; perhaps they have lived in different communities with different amounts of the public good or are aware of the cost of private schools.

In contrast, environmental goods are often quite disconnected from the supply side. In determining how worthwhile the protection of a wilderness area might be, it is difficult to anchor that value, since the wilderness area is not produced. Air quality is in fact supplied, since there is a cost associated with cleaning up the air. However, this is a much more dispersed cost, much different than that associated with constructing a public school or park. This too makes the job of determining demand more difficult.

II. WILLINGNESS TO PAY

A conventional demand curve plots quantity demanded as a function of price. Without a market there is, of course, no price. This is not really a problem, though we need to understand what underlying information is contained in a demand curve. Similarly, the consumer surplus associated with consuming a certain amount of a good is defined in terms of a market transaction. In this section we see how similar concepts are equally applicable to environmental goods in the absence of markets. The key concept corresponding to price and surplus is willingness to pay. We first discuss how marginal willingness to pay is analogous to price and then show how total willingness to pay is analogous to consumer surplus.

A. Prices and Marginal Willingness to Pay (MWTP)

In the next few chapters we will be taking a closer look at how the demand for environmental goods is measured. We start with an example. Leaving aside how it was obtained, Figure 15.2 shows the demand (at the household level) for nitrogen oxides (NO_X) air pollution in Boston, based on a study done in the 1970s. Figure 15.2 shows three different demand curves for NO_X, corresponding to three different levels of household income. The demand curves are downward sloping but since NO_X is a bad, they are in the fourth quadrant (MWTP negative, NO_X positive). If you are observant, you will note that the vertical axis is labeled MWTP (marginal willingness to pay) instead of price, the variable used in a conventional demand curve. Obviously, without a market, there can be no price. But it is clear that these two terms are equivalent.

To see this, consider the case of gasoline. In a time period such as a month, what is the marginal willingness to pay for one more gallon of gasoline if the price of gasoline is $1 per liter? We can do a mental experiment, imagining a consumer trying to decide how much to consume at that price. The consumer starts at zero liters of gas and decides more is needed, because at that level of consumption a liter is worth more than $1. So the consumer expands the amount desired first to 10 liters, then 20 liters, all the while asking if the next liter of gasoline is worth more than the $1 price. This process stops when the value of one more liter is exactly the price, $1. The result of this experiment is for a consumer buying a good in a market, the value of one more unit of the good to the

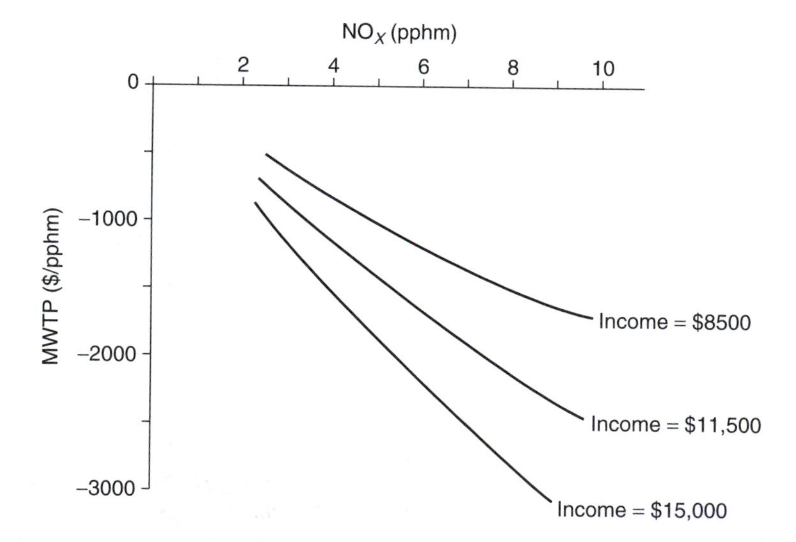

Figure 15.2 Marginal willingness to pay (MWTP) for nitrogen oxides (parts per hundred million, pphm) for Boston households at three different income levels. Figures for MWTP are annual, in 1970 US$ per pphm. Adapted from Harrison and Rubinfeld (1978).

consumer is exactly the price. If at a price of $1 per liter the consumer chooses to purchase 200 liters in a particular month, the consumer's marginal willingness to pay for one more liter has to be $1.

In fact, this is really what any sort of individual demand curve indicates. Each point on a demand curve indicates, for a particular quantity of the good, how the consumer values one more unit of the good—the marginal willingness to pay. In a market, it is fundamental that for every consumer the price is the marginal willingness to pay. There is a nuance that has to do with whether the consumer is acquiring one more unit of the good or giving up a unit. If the consumer is giving up a unit that is possessed (e.g., selling a liter of gas), the appropriate term is marginal willingness to accept compensation (MWTA). In most contexts, these two concepts are equivalent. We will have more to say about differences between these two terms later in this chapter.

There is one more issue of terminology that arises in the case of pollution. Often, we speak of the damage from pollution or the benefit of having less pollution. The marginal benefit of one less unit of pollution is the same as the marginal damage of one more unit of pollution. Furthermore, the marginal willingness to pay for one more unit of pollution is negative and equal to the negative of the marginal benefit of one less unit of pollution or the negative of the marginal damage of one more unit of pollution. Confusing? Just keep in mind that benefits of less pollution are the other side of the coin of damages from more pollution. When expressed in monetary terms, these benefits and damages are simply willingnesses to pay, making sure that the sign (positive or negative) is correct.

B. Willingness to Pay versus Marginal Willingness to Pay

With pollution, we are often concerned with the total damage from pollution or total benefits from reducing pollution. For instance, Table 15.1 shows an estimate of the total benefits (1970–1990) from the cleaner air provided by the 1970 U.S. Clean Air Act Amendments. The table is taken from a U.S. Government study of benefits over the period 1970–1990 and compares the situation of pollution continuing to grow without the legislation with the levels of pollution that in fact resulted over the 20-year period. Thus Table 15.1 does not compare existing levels of pollution with pristine conditions but rather existing levels of pollution with much dirtier conditions that might have prevailed had there been no regulation of air pollutants (a level that is very hard to forecast). The point is that Table 15.1 shows total benefits of the Clean Air Act, equivalent to the total willingness to pay for the attained levels of air quality, not marginal willingness to pay.[2]

It is important to distinguish between total damage/benefits and marginal damage/benefits, or to use the terms we used earlier, total willingness to pay and marginal willingness to pay. Suppose we are in an urban area with high levels of air pollution, such as Mexico City. We may think of the total willingness to pay for an individual citizen to eliminate all air pollution from the city. This would be the total willingness to pay to eliminate air pollution. It is the same as the damage (in monetary units) to the individual from all of the air pollution in Mexico City. We can also consider the willingness to pay to eliminate one unit of air pollution. This would be the marginal willingness to pay for air pollution reduction. These will be very different numbers.

TABLE 15.1 Present Value of 1970 to 1990 Monetized Benefits[a] of Clean Air Act in Continental United States, by Endpoint (Billions of 1990 \$, Discounted to 1990 at 5%)

Endpoint	Pollutant(s)	Present value (\$)		
		5th %ile	Mean	95th %ile
Mortality	PM-10	1,991	13,542	30,968
Mortality	Pb	125	1,550	4,096
Chronic bronchitis	PM-10	2,143	7,156	12,613
IQ (lost IQ pts. + children with IQ < 70)	Pb	299	466	656
Hypertension	Pb	77	99	120
Hospital admissions	PM-10, O_3, Pb, and CO	18	24	33
Respiratory-related symptoms, restricted activity and decreased productivity	PM-10, O_3, NO_2, SO_2	31	76	149
Soiling damage	PM-10	7	76	196
Visibility	Particulates	55	72	91
Agriculture (net surplus)	O_3	11	23	35
Total		10,500	23,000	40,600

[a]*Benefits are with respect to a hypothetical baseline of no pollution control in 1970–1990, with resulting growth in pollution levels.*

Source: United States Environmental Protection Agency (1997), Table 20.

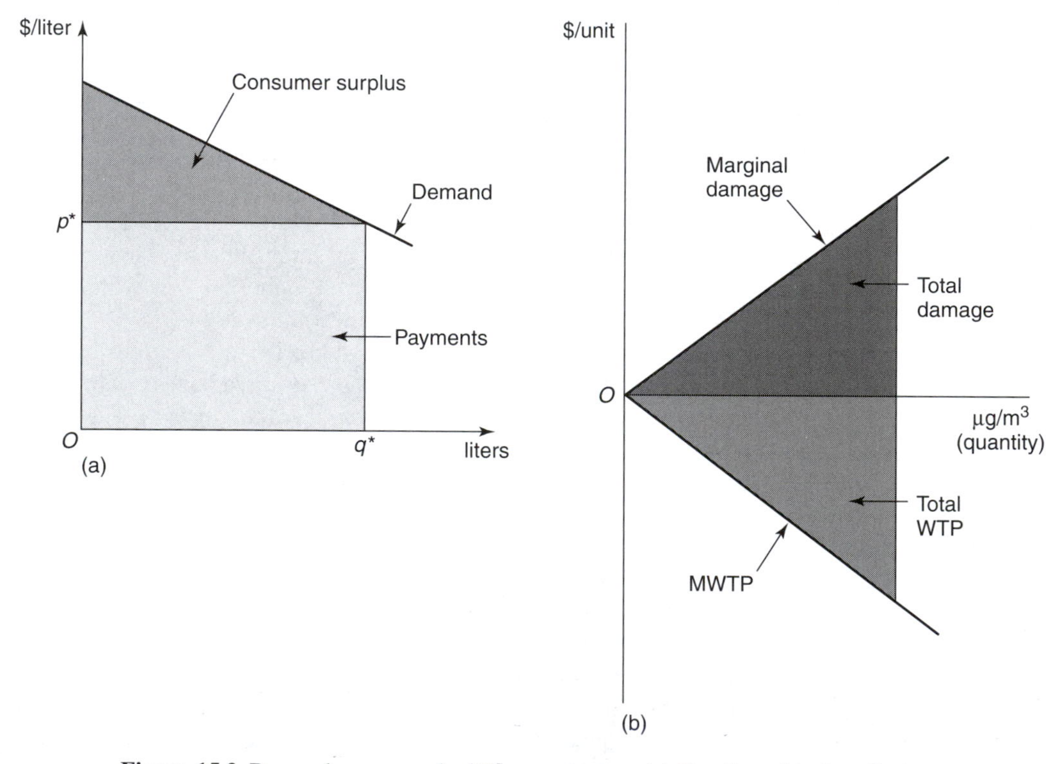

Figure 15.3 Demand curves and willingness to pay. (a) Gasoline; (b) air pollution.

Figure 15.3 shows hypothetical individual demand curves for two commodities: gasoline and air pollution. For gasoline, Figure 15.3a shows quantity demanded as a function of the price of gasoline. Remember "price" can be used interchangeably with "marginal willingness to pay." For air pollution, the lower part of Figure 15.3b shows the marginal willingness to pay for pollution (MWTP) as a function of the quantity of pollution. The top part of Figure 15.3b shows the marginal damage from pollution, which is precisely − MWTP.

Total willingness to pay to eliminate pollution involves adding up all of the marginal willingnesses to pay as pollution levels expand from zero to the current level. This is the area between the demand curve for pollution and the horizontal axis and is shown as the lighter shaded part of Figure 15.3b. Similarly, in Figure 15.3a, the shaded area shows the total willingness to pay for gasoline q^*. This shaded area consists of payments made (the rectangle) and consumer surplus (the triangle).

Similarly, for an environmental commodity, the area under the demand curve between two different quantities of the commodity gives the willingness to pay for the environmental commodity. Since no payment is made for the pollution (as there was for gasoline), this is also consumer surplus (though note that it is negative).

III. TYPES OF ENVIRONMENTAL GOODS

The environment is a very complex place. In understanding the value of the environment, it helps to use a classification scheme. We can classify goods based on the nature of the pollution: air quality is one good, water quality is another. Or we may classify goods based on the nature of the injured party: health effects of pollution, damage to agriculture, and damage to buildings and materials. Or we may categorize goods based on how people perceive the damages. This last categorization is a bit more subtle and has to do with whether consumers obtain utility from the environment by "using" the environment (e.g., hiking through the woods) or through more esoteric means (e.g., thinking about elephants roaming in Africa). Because this last dichotomy is somewhat unusual, we will examine it more carefully.

A. Use Value

Use value is the conventional notion of value associated with the consumption of a good. In the case of an environmental good, this could include current use ("I am currently visiting the park"), expected use ("I plan to visit the park later this year"), and possible use ("I might visit the park within the next 10 years"). Although this might be obvious, the point about possible use can be used to generate a positive value for a remote wilderness that few currently use, particularly if the very long term is taken into account.

There are several primary avenues whereby environmental goods impact humans.[3] One is through direct impact. This would include direct health effects of breathing polluted air (both mortality—death—and morbidity—sickness) as well as nonhealth effects such as bothersome odors, noise, or visual impacts. For instance, smoke from a power plant may result in a brown plume intercepting a vista you have come to enjoy. Although there may be no measurable physical impact on you from this plume, you find the view annoying and would be willing to pay some amount of money to be rid of it, should the opportunity arise. This is just as real an economic value as health effects. On the production side, pollution may enter directly into production functions, adversely affecting production processes. For instance, air pollution may make the manufacture of microchips (which require superclean conditions) more costly. Environmental goods may impact humans less directly, primarily through materials damage. Air pollution may damage or soil buildings requiring additional maintenance or cleaning.

A second avenue for environmental goods to impact humans is through damage to ecosystems. Agriculture, forestry, and fisheries are ecosystems from which humans directly derive economic benefit. Clearly pollution that degrades the performance of these ecosystems is undesirable. More subtle ecosystem effects relate to recreational use of ecosystems. Urbanization or pollution that may disrupt the ecosystem in a national park will have negative consequences for recreational users of that national park. Ecosystems are also a source of economic benefit through mechanisms such as providing genetic material for new pharmaceuticals.

B. Nonuse Value

Nonuse value is a controversial aspect of value. Nonuse value is a gain in a person's utility without the person actually using the good, using the word "using" very broadly. We may value ecosystems in remote parts of the world for reasons other than intending to visit the ecosystem or potentially obtaining something useful from the ecosystem. I may value the wilderness areas in the Sierra Nevada, not because I plan to make use of the wilderness but because others may, and that makes me feel good (gives me utility).

Three basic types of nonuse value are existence value, altruistic value, and bequest value. Existence value is the value a consumer attaches to *knowing* something exists (e.g., the African elephant example mentioned earlier). This would be in addition to any value associated with actual or potential "use." Altruistic value derives not from my own consumption but from the fact that I derive benefit when someone else gains utility. So if my neighbor derives benefit from my cleaning my front yard, I obtain utility from the fact that my neighbor is better-off. Bequest value is similar, though associated with the well-being of descendants. If I value passing a wilderness area on to the next generation, that wilderness has a bequest value to me, even if I never use it or intend to use it.

The purpose of dissecting value and placing it into these various categories is to understand the complexity whereby environmental goods confer value on consumers. Simply looking at the use value of an environmental asset may obscure much of the value. In fact some environmental assets have little use value but very significant nonuse value. We should point out that empirically it is often impossible to separately measure the components of value, for instance, how much value is altruistic vs. bequest. This is a bit like dividing the value of an ice-cream cone into texture and flavor. The categorization simply provides an intellectual framework to help ensure completeness when valuing environmental goods.

IV. MEASURING DEMAND

Because of the absence of markets for environmental goods, measuring demand is not straightforward. There are two basic approaches to measuring demand: revealed preference and stated preference. In revealed preference, we observe a real choice[4] in some market and cleverly infer information on the trade-off between money and the environmental good. For instance, we may notice that two communities are identical except that one has high housing prices and clean air and the other has lower housing prices and dirty air. We may infer that the difference in housing prices reflects the value people place on clean air. The second approach, stated preferences, basically involves asking people how much an environmental good is worth. Opinion polls and surveys are used to derive this information. This approach is controversial because of the absence of real choices; only real choices involve actual trades between money and environmental goods. Hypothetical choices lack a realism that may be necessary to obtain accurate information. There is a good deal of division within the economics profession on this point. There is also a gray region between revealed preferences and stated preferences, for example, experimental markets in which subjects (usually students) are given some money and asked to make choices (perhaps somewhat unusual choices) using that money. The decisions are real; the context is contrived.

Within the category of revealed preference, there are two basic approaches to measuring demand, hedonics and household production. In the hedonic approach, the goal is to see how the price of a conventional good (e.g., a house) varies as the amount of a closely related environmental good changes (e.g., the air quality in the vicinity of the house). This relationship is then used to infer value. The household production approach starts with the assumption that consumers will combine private goods with the environmental good to "produce" another good, which is the real source of utility. For instance, soundproofing may be combined with noise that impinges on the outside of a house to obtain particular indoor noise levels. Indoor noise is what is assumed to matter. Or a national park may be combined with expenditures to visit the park to generate a park visit, which is what really gives utility. In either case, by observing expenditures on the complementary private good, we obtain a lower bound on the value of the environmental good or bad. The outside noise must cause at least as much damage as the expenditures on soundproofing. The national park must be worth at least the travel expenditures.

In the arena of stated preferences, the dominant approach is termed contingent valuation. Contingent valuation relies on direct revelation of demand from consumers. The name literally means "value contingent on there being a market"—if there were a market, how much would you pay for the environmental good? These values are obtaining by directly questioning a sample of potential consumers of the environmental good.

Contingent valuation is a type of constructed market. In a constructed market, a researcher will take a situation in which no market exists and generate a market. Constructed markets can be hypothetical or real. For instance, in a study of duck hunting, Bishop et al. (1983), constructed a market for buying and selling duck hunting permits (the permits were normally distributed through a lottery). In this manner they were able to generate a demand curve for duck hunting permits. Laboratory experiments are also widely used to generate information on how people trade money for environmental goods. In a laboratory study of the value of risk reduction, Shogren et al. (1994) constructed an experiment in which students were given the option of eating food prepared subject to conventional safety standards (which entails some risk, though small) with food that had been prepared with even stricter standards. The researchers were interested in the behavior of demand for these risk reductions. Such laboratory experiments are useful because they enable the investigation of questions for which market data just is not available. These studies also involve real resource decisions ("real" money is on the line), which is an advantage over contingent valuation. Unfortunately, they are expensive to conduct, often involve non-randomly selected samples, typically graduate students in economics or related fields, and are difficult to generalize. Another type of constructed market is the official referendum. Some environmental good is to be provided using a tax on residents of a community. The proposal is put to the voters of a community. Although this is a very real choice situation, its usefulness is obviously limited to the environmental goods offered to communities through referenda.

V. CONSUMER DEMAND FOR MARKET GOODS

In this section we will review material that is covered for the most part in many intermediate microeconomics courses. Much of the theory can be applied to either consumers or producers, though the terminology is slightly different in the two cases. Because of the similarities, we focus on the consumer and do not consider the producer.[5]

It is important to emphasize that the basic unit of analysis is the individual demand curve, i.e., the demand curve for a specific individual. Economic theory tells us how to construct such an individual demand curve, based on utility maximization. The market or aggregate demand curve is the sum of individual demand curves. Economic theory has less to tell us about the nature of market demand.

The starting point for our analysis is a utility function for a consumer, $u(z,q)$ where z and q are two goods. This utility function gives the level of utility for various bundles of z and q.

A. Ordinary versus Compensated Demand

If we start with an individual's utility function, prices for z and q, and some money (income), the individual is assumed to choose how much to consume by maximizing utility. An ordinary demand function[6] gives the amount of a good a consumer chooses as a function of prices and income: $x_q(p_z,p_q,y)$. Figure 15.4 illustrates the graphic process of generating a demand curve for good q from indifference curves between goods z and q. As the price for q is changed, the slope of the budget lines (broken lines in Figure 15.4a) shifts; the utility maximizing choice of q also changes, tracing out a demand curve. Points A, B, and C are the choice points generated in Figure 15.4a. These are then represented in Figure 15.4b in price-quantity coordinates. This is an ordinary demand curve because income, rather than utility, is kept constant as the price of q changes. The curve shown in Figure 15.4b is the ordinary demand for q: $x_q(p_z,p_q,y)$.

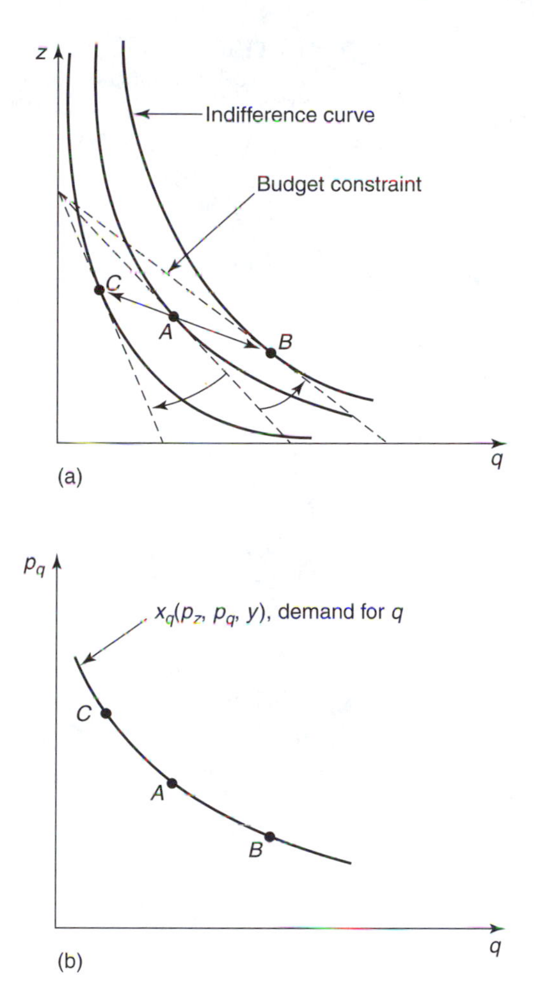

Figure 15.4 (a,b) Constructing an ordinary demand function for q. p_z, Price of z; p_q, price of q; y, income; $x_q(p_z,p_q,y)$, ordinary demand curve for q.

An alternate way to generate a demand curve is to keep utility constant as the price of q is changed. The only way to keep utility constant as prices change is to adjust income so that the consumer remains on the same indifference curve. The demand curve that traces out quantity demanded as a function of price, keeping utility constant, is called the compensated demand function[7]: $h_q(p_z,p_q,U)$ where U is a particular level of utility. This is shown in Figure 15.5. Figure 15.5a shows an indifference curve for utility level U_0. To trace out the demand curve, the price of q, p_q, is changed and income (y) adjusted so that utility stays at U_0 (on the same indifference curve). Figure 15.5b shows the resulting demand curve with points A, B, and C corresponding to the same points in Figure 15.5a.

Why should we worry about two different kinds of demand curves? The basic reason is that we are concerned with two aspects of demand: the price effect and the income effect. As the price of q goes up, people consume less, substituting into z. This is the substitution effect from a price change. But as prices go up, people also become poorer (their

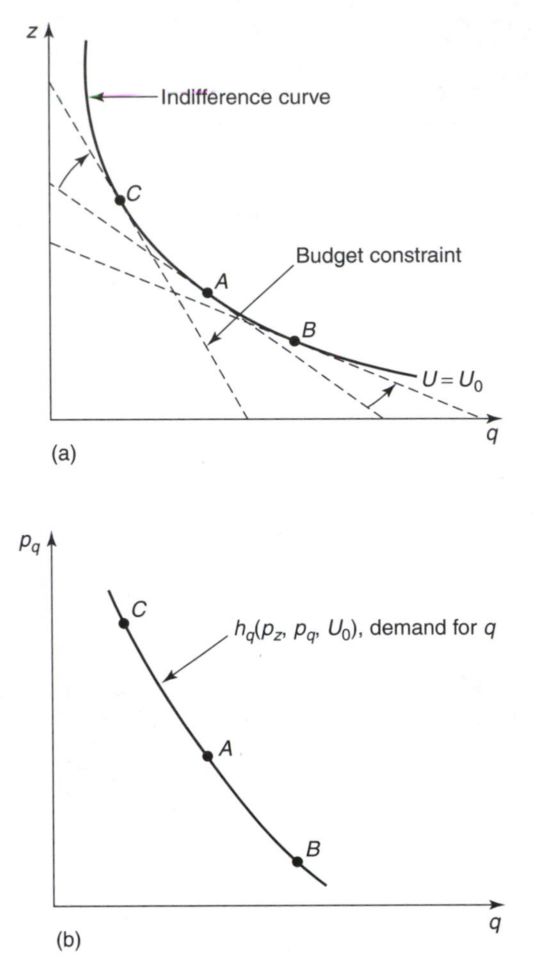

Figure 15.5 (a,b) Constructing a compensated demand curve for q. p_z, Price of z; p_q, price of q; U_0, utility (U) level; h_q (p_x, p_q, U_0), compensated demand function for q.

incomes do not go as far) relative to their situation before the price change. If they are poorer, they will consume less (typically). This is the income effect. A compensated demand curve shows only the price effect. Since utility is being held constant, we are not making people worse-off by a price change. Thus two points on a compensated demand curve represent two relative prices but the same level of utility. In contrast, two points on an ordinary demand curve represent two different prices *and* two different utility levels (since income is constant). Thus the price effect is bundled with the income effect.

We often wish to evaluate the effect of a government policy that changes relative prices. To evaluate such a policy, we want to examine the results of the price change only, not income effects (since income effects can be addressed through programs to redistribute income). In other words, we want to use the compensated demand function. In typical applications, however, it is the ordinary demand function that is measured, since we typically observe different prices and income, not different utility levels.[8]

We can connect the two demand functions conceptually. Figure 15.6 shows a plot of an ordinary demand curve (x_q) and two compensated demand curves (h_q), for different utility levels. One point shared by the ordinary and compensated demand curve is point A. At A on the ordinary demand curve, the consumer enjoys some level of utility, call it U_1. At this level of utility, we can trace out a compensated demand curve. The compensated demand curve must also pass through A, as is shown in Figure 15.6. A similar story would apply to point B.

B. The Expenditure Function

A useful way of rewriting the compensated demand functions is the following:

$$E(p_z,p_q,U) = p_z h_z(p_z,p_q,U) + p_q h_q(p_z,p_q,U) \tag{15.1}$$

This is the expenditure function that indicates how much income is necessary to achieve utility U, facing prices p_z and p_q. If we are facing those prices, we know how much of z and q we want and what the prices of z and q are. Equation (15.1) simply indicates how much money we need to make that purchase.

The expenditure function is useful for two reasons. First, its units are measurable dollars, so we can easily compare different values of the expenditure function. Second, the parameters of the expenditure function include prices, not quantities consumed—all choices the consumer has to make have already been factored in. Note that the expenditure function is very similar to the utility function, except that instead of the quantities of z and q as arguments, we have the prices of z and q and we assume the consumer is choosing the optimal quantities to achieve a certain level of utility. Also, the units are money, not unmeasurable utility.

There is a simple relationship between the expenditure function and a compensated demand function. An example will help illustrate this relationship. Suppose you have \$20 to buy groceries. You go to the grocery store and buy an optimal basket of goods, including a loaf of bread. Now let us conduct a little thought experiment. Suppose the price of bread goes up by \$1. How much more money do you need to be able to afford the

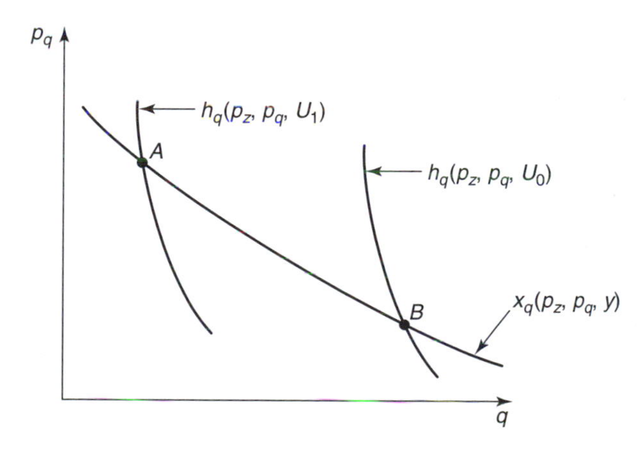

Figure 15.6 The connection between compensated and ordinary demand. p_z, p_q, Prices of z and q; U_0,U_1, utility levels; y, income; h_q, compensated demand for q; x_q, ordinary demand for q.

same basket (thus keeping utility approximately constant)? Precisely the number of loaves of bread in your basket times the price increase of $1:

$$\Delta E(p_z,p_q,U) = h_q(p_z,p_q,U)\ \Delta p_q \tag{15.2}$$

or

$$h_q(p_z,p_q,U) = \Delta E(p_z,p_q,U)/\Delta p_q \tag{15.3}$$

In other words, the ratio of the change in the expenditure function to a change in the price of a good is the same as the compensated demand for the good.[9]

Although the expenditure function may not be a familiar concept, it is a very similar to consumer surplus. Take a simple example. Suppose the price of z is p_z^*, resulting in consumption z^*. What is the surplus associated with consuming z^*? Figure 15.7 shows a demand function for good z—the line AB. Demand drops to zero at the "choke price" of p_z^c. If AB were an ordinary demand curve, the consumer surplus would be the area ABE, which is the area $ABCD$, less what is paid for z, $BCDE$. This is because all of the units of z consumed (except the last) are worth more than what they cost, and the difference between willingness to pay and cost, summed over all units consumed, is equal to consumer surplus (the shaded area).

If, however, AB is a *compensated* demand function, we do not get consumer surplus, though very similar logic applies. Referring to Figure 15.7, the first unit of z consumed is worth p_z^c to our consumer, giving surplus $(p_z^c - p_z^*)$; this must be subtracted from income to keep utility constant. As we expand consumption of z to z^*, the shaded area ABE represents the amount of income that must be taken away from the consumer to keep utility constant. So the area is the difference between (1) the income needed to yield utility U^* when the price of z is p_z^c and (2) the income necessary to yield utility U^* when the price of z is p_z^*: $E(p_z^c,p_q,U^*) - E(p_z^*,p_q,U^*)$. We see that the area under an ordinary demand curve corresponds to consumer surplus, whereas the area under a compensated demand curve corresponds to changes in constant utility expenditures. They are both welfare measures and, conceptually, very similar.

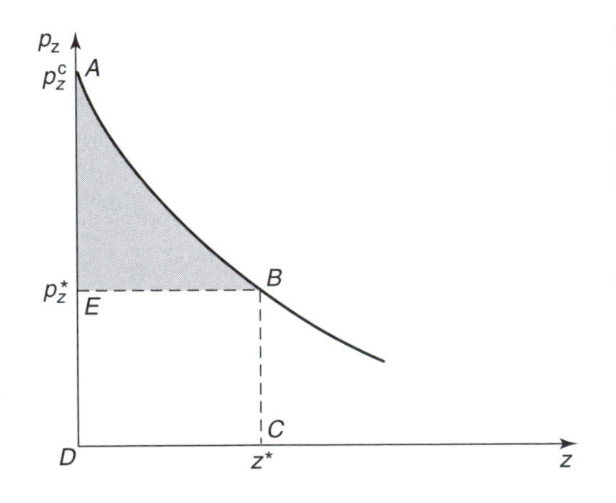

Figure 15.7 Surplus from ordinary or compensated demand. AB, Demand for z (ordinary or compensated); p_z^c, choke price for z; p_z^*, price of z; z^*, quantity of z consumed at price p_z^*; AEB, surplus.

C. Welfare Effects of a Price Change

Suppose the price of good z changes from p_z^0 to p_z^1. How do we evaluate the effect of this change on the consumer? In Chapter 4 we saw one way to do this. If p_z^0 results in consumption of z equal to z_0 and p_z^1 results in z_1, the welfare effect of the price change is the change in consumer surplus associated with changing from z_0 to z_1. And the consumer surplus from consuming some amount of z is the area under the ordinary demand curve, less any payment for the good.

But this measure of the welfare effect has one problem: income effects are mixed in with price effects. A more appropriate measure of the effect on consumers from the price change is obtained by looking at the area under the compensated demand curve, rather than the ordinary demand curve, which is done to obtain consumer surplus. And as we saw in the previous section, the change in area under the compensated demand curve, less any payments made for the good, is exactly equal to the change in the expenditure function associated with the price change.

But when a price changes the consumer moves from one level of utility to another. Since utility is constant for a compensated demand function, which compensated demand function (i.e., which utility level) do we use: the one associated with the initial price or the one associated with the final price? (Refer to Figure 15.6.) It turns out that we can use either one, though we get a slightly different answer in the two cases.

Think of the problem in terms of two possible questions about the significance of the price change. (1) "What amount of money could we give the consumer to compensate for the price change, i.e., to return the consumer to the original utility level?" We call this the *compensating variation* (CV). (2) "What level of money would the consumer be willing to pay to avoid the price rise?" We call this measure the *equivalent variation* (EV). In both cases we make the somewhat arbitrarily assumption that CV and EV are both negative for a price rise and both positive for a price decrease.[10]

These two terms can also be defined using the expenditure function. The expenditure function tells us how much income is necessary for a consumer to attain a given level of utility. We have two levels of utility in question, U_0 and U_1, associated with the two price levels for z. We can define $CV(p_z^0,p_z^1)$ and $EV(p_z^0,p_z^1)$ as the compensating and equivalent variations, respectively, of increasing the price of z from p_z^0 to p_z^1:

$$CV(p_z^0,p_z^1) = E(p_z^0,p_q,U_0) - E(p_z^1,p_q,U_0) \tag{15.4a}$$

$$EV(p_z^0,p_z^1) = E(p_z^0,p_q,U_1) - E(p_z^1,p_q,U_1) \tag{15.4b}$$

Take a moment to examine these definitions. We are making our consumer worse-off by raising prices. The consumer's nominal income before and after the price change is the same $[E(p_z^0,p_q,U_0) = E(p_z^1,p_q,U_1)]$. The compensating variation tells us how we can undo our negative action—compensate for the price change. The equivalent variation on the other hand converts our negative action into something easier to understand, an income loss. The equivalent variation indicates how much money we could take away from our consumer instead of changing the price and have the same effect as the price change. Thus we compare the income the consumer needs after the price change with the income the consumer would need without a price change but at the post-price-change utility level.

We can also visualize these terms graphically. Figure 15.8 shows several things. The basis of Figure 15.8 is an ordinary demand curve for z, EF. A price drop from p_z^0 to

p_z^1 causes the consumer (with constant income) to expand consumption of z from z_0 to z_1. Recalling Figure 15.6, utility is higher for our consumer after the price change; thus there are two compensated demand functions, one through E and one through F, representing utility before and after the price change. In Figure 15.8b the shaded area to the left of the ordinary demand curve is the consumer surplus associated with the price change; in Figure 15.8a the lightly shaded area to the left of the compensated demand curve EG is the compensating variation from the price change; and in Figure 15.8c the hatched area to the left of the compensated demand curve FD is the equivalent variation from the price change. Figure 15.8a, b, and c are superimposed in Figure 15.8d, in which we see that the consumer surplus appears to be bounded below and above by the compensating and equivalent variation. We also see that the areas are not too different. For small price changes we need not be concerned with the differences: EV, CV, and consumer surplus are approximately the same. For large price changes, the differences among EV, CV, and consumer

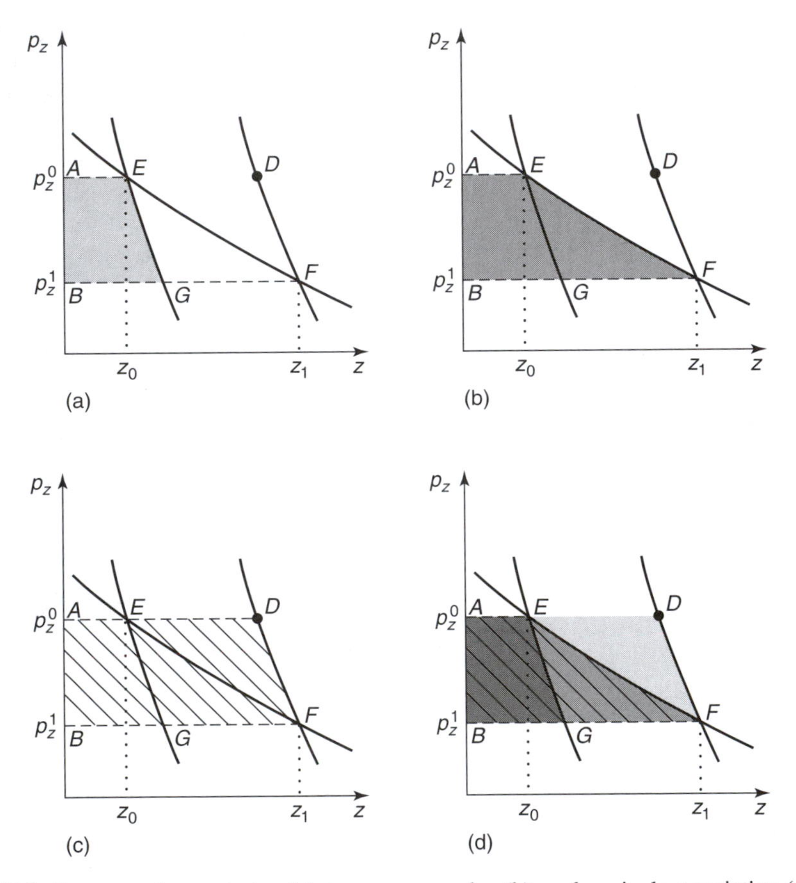

Figure 15.8 Compensating variation (a), consumer surplus (b), and equivalent variation (c) from a price change. (d) Superimposition of a, b, and c. Price of z decreases from p_z^0 to p_z^1. z_0, Initial quantity of z consumed; z_1, final quantity of z consumed; p_z^0, initial price of z; p_z^1, final price of z; EF, ordinary demand curve for z; EG,DF, compensated demand curves for z; $AEGB$, compensating variation; $AEFGB$, consumer surplus; $AEDFGB$, equivalent variation.

surplus may be large, though not always.[11] Also, as in clear from Figure 15.8, $CV \leq$ consumer surplus $\leq EV$ (provided the good is normal).

VI. CONSUMER DEMAND FOR ENVIRONMENTAL GOODS

A. Restricted Demand

So far, we have examined the demand for goods without regard to whether the good is environmental. In fact the discussion of the previous section is found in most rigorous intermediate microeconomics textbooks, and applies to goods generally. What happens with environmental goods, or public goods in general, is that the quantity is not chosen by the consumer. Suppose in our notation above that q is the environmental good. We can still talk about demand and expenditure functions, except that the price of q is an inappropriate parameter; rather we should deal simply with the quantity of q. We let z remain a conventional market good. Thus we can write the ordinary demand function for z as $x_z(p_z,q,y)$ and the compensated demand function as $h_z(p_z,q,U)$. We might, for instance, observe the demand for boat rentals (z), and see that it depends on the price of boat rentals and water quality (q).

The expenditure function in this case is straightforward:

$$E(p_z,q,U) = p_z\, h_z(p_z,q,U) \tag{15.5}$$

If there were more marketed goods, Eq. (15.5) would simply have additional terms representing the other market goods.

Notice that the demand functions and the expenditure functions are very similar to before, except that the quantity of q appears rather than the price of q. We call this type of demand and expenditure functions *restricted* demand and expenditure functions.

There is an interesting relationship between q and E that will tie the restricted expenditure function to the conventional expenditure and demand functions. A simple example will clarify the relationship. Let us start (again) with a grocery basket with a variety of groceries that you have chosen, including five loaves of bread priced at $1 each. The cost of the groceries is $35: $5 for the bread and $30 for everything else. Now suppose we do not observe the price of bread and see you are spending $30 for nonbread groceries when you have five loaves of bread in your basket. The question is, can we determine how much the bread is worth to you only by observing your demand for the other groceries? This is akin to observing demand for market goods when consuming a nonmarket environmental good; our problem is finding the value of the environmental good.

The answer is yes. We can determine the value of one more loaf of bread. Suppose we give you one more loaf of bread and observe that you do not need to spend quite as much on other groceries to keep utility constant. We observe from your compensated demand function for the other groceries how much income must be taken away to keep utility constant. This is your marginal valuation of a loaf of bread. If $1 has to be taken away to make up for the extra utility from the extra loaf, the value of that extra loaf is obviously $1.

Now consider the restricted expenditure function again [Eq. (15.5)]. What happens if we increase q by a little, say Δq? Since q is desirable, utility will go up unless we reduce income accordingly. The ratio of the change in the expenditure function to the change in q is the value of q *at the margin*. This is the marginal willingness to pay for q, which we denote p_q. If q were offered on the market at this price, and income were supplemented by $p_q\, q$, the amount chosen voluntarily would be exactly q. Thus we have a connection between compensated demand and the restricted expenditure function:

$$\Delta E(p_z,q,U)/\Delta q = -p_q \tag{15.6}$$

Think of Eq. (15.6) as having four unknowns, p_z, p_q, q, and U. Give me the value of three of these unknowns and I can give you the value of the fourth, using Eq. (15.6). The equation can be rearranged to yield q as a function of the remaining variables:

$$q = h_q(p_z,p_q,U) \tag{15.7}$$

which is the compensated demand function for q.

This can also be seen graphically. We start with a compensated demand curve for our market good, z: $h_z(p_z,q,U)$. Suppose the price of z happens to be p_z^* and the amount of q is q_0. This demand curve is shown in Figure 15.9, along with the quantity of z that will be consumed, z^*. Now suppose we increase q a bit, to $q_0 + \Delta q$. Since q is environmental quality, presumably it is desirable. Thus to keep utility constant, we need to take away some z when q is increased. The resulting compensated demand curve for $q_0 + \Delta q$ is also shown in Figure 15.9, slightly lower than the original one.

What is the compensating variation for this change in q? The shaded area in Figure 15.9 corresponds to the income equivalent of the change in q. Increasing q a little increases utility, so we need to take a bit of income away to keep utility constant. That is the shaded area in Figure 15.9. This can also be interpreted as the marginal willingness to pay for the change in q.[12]

It is important to emphasize the implications of this result. What we have shown is that if we measure demand for ordinary goods for different levels of the environmental good, we may be able to back out a demand curve for the environmental good, even though there is no market for that good!

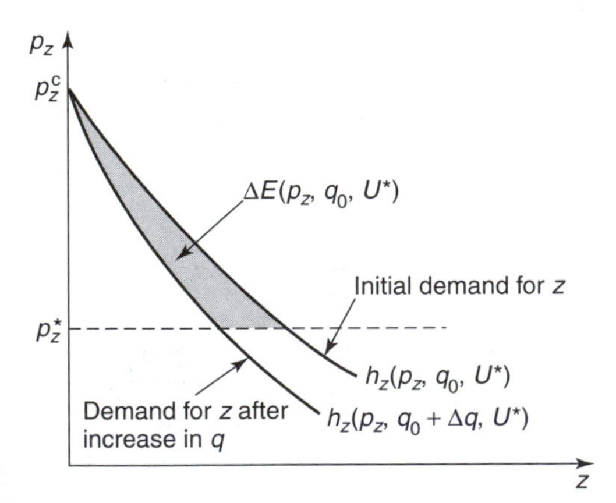

Figure 15.9 Marginal willingness to pay for an environmental good. p_z, Price of z; E, expenditure function; ΔE, change in expenditure function; U^*, utility level (fixed); h_z, compensated demand for z; p_z, price of z; p_z^*, fixed price of z; p_z^c, choke price for z; Δq, change in environmental good, q.

B. Measuring the Welfare Effects of Environmental Improvement

We can also evaluate the welfare effects of changes in environmental quality using compensating and equivalent variation. Start with a level of environmental quality, q_0. Suppose an individual consumes q_0 and has income of y available. Let us say that the resulting utility is U_0. We then change the amount of environmental quality consumed to q_1. The individual will now be at a new level of utility, U_1.

What is the value of the environmental change? One way of valuing the action that changed quantities is to determine the amount of money, in lieu of the action, that would have the same effect on utility. The amount of money that would keep the individual at the original level of utility *with the change* is called the *compensating surplus* and the amount of money that would move the individual to the new level of utility *without the change* is called the *equivalent surplus*. Conceptually, these two terms are very analogous to compensating and equivalent variation in the case of a market good. The reason for using the term "surplus" instead of "variation" is that the consumer is not free to vary the quantity of q.[13] In all other respects, the concepts are identical.

These two terms can be defined using the expenditure function as in Eq. (15.4). $CS(q_0,q_1)$ is the compensating surplus of moving from q_0 to q_1 and $ES(q_0,q_1)$ is the analogous equivalent surplus:

$$CS(q_0,q_1) = E(p_z,q_0,U_0) - E(p_z,q_1,U_0) \tag{15.8a}$$

$$ES(q_0,q_1) = E(p_z,q_0,U_1) - E(p_z,q_1,U_1) \tag{15.8b}$$

Take a moment to understand these two definitions. Consider first compensating surplus.

Let us make things a little less abstract by saying that the amount of q that our consumer consumes has been changed from q_0 to q_1, and think of the change as being for the worse. So we want to compensate consumers for the change in q—we want to give them an amount of money that brings them back to the utility level they were at before the change. This is the compensating surplus. The compensating surplus as defined in Eq. (15.8a) is the difference in the income needed to achieve the old level of utility at the old quantity $[E(p_z,q_0,U_0)]$ and the income needed to keep the same level of utility at the new quantity $[E(p_z,q_1,U_0)]$.

The equivalent surplus, on the other hand, is the change in income that would have the same ("equivalent") effect as the quantity change, without the quantity change actually occurring. In other words, we should compare the expenditures that yield utility U_1 at the original quantity with the expenditures necessary to yield utility U_0 at the original quantity:

$$ES(q_0,q_1,) = E(p_z,q_0,U_1) - E(p_z,q_0,U_0) \tag{15.8c}$$

which is actually the same as Eq. (15.8b) because $E(p_z,q_1,U_1) = E(p_z,q_0,U_0)$.

These are subtle definitions, perhaps difficult to understand. So consider a more concrete example. Kofi lives in a house close to the ocean with only a vacant field between him and the ocean. He has a nice view of the ocean. Now it turns out Lars owns the vacant field and wants to build a house that will block Kofi's view of the ocean. There are two ways to view the money equivalent of the damage to Kofi from the house built by Lars. One measure would be the maximum amount Kofi would pay Lars not to build the house. Clearly, Kofi would be willing to pay an amount such that his income after the payment and with a clear view of the ocean results in the same utility as his income without the payment and without the view. This is the equivalent surplus—a money measure that is equivalent to the loss of the view. Kofi actually will be indifferent between paying this amount to prevent the loss of the view and not paying but losing the view.

Another way of looking at the value would be if the town Kofi lives in has laws that prohibit construction that blocks another's view without permission of the viewing party. If Lars wants to build his house, he has to pay Kofi for that right. How much would he have to pay Kofi? He would have to pay the amount of money so that Kofi's utility with the payment but without the view is the same as with no payment but no blocked view—in other words, Kofi's original level of utility. This is the compensating surplus—Lars has to compensate Kofi for the loss of utility from taking away his view.

It is easy to see that these two measures should be roughly the same. If the item in question is relatively unimportant, the amount you would have to compensate me for giving it up (the *CS*) should be roughly the same as the amount I would be willing to pay you to obtain it (the *ES*). Any differences would be due to the income effect—you are richer when you own the item and are being compensated for it than when you must pay to obtain it. Income effects should be negligible when the item is of small value relative to your overall wealth.[14]

To close this discussion, we can examine *ES* and *CS* graphically. Figure 15.10 shows quantities of z and q that may be consumed. As before, assume z is a conventional private good and q is an environmental good, supplied directly. If the price of z is p_z, and income is y, the budget constraint is $y = p_z z$, a horizontal line in the $z - q$ plane as shown in Figure 15.10. This is because q is an environmental good that is supplied but for which payment is not made. At the initial level of q, q_0, utility is U_0. Suppose the quantity of the environmental good then increases to q_1, yielding utility U_1. The compensating surplus is the monetary value of z that would return to U_0, at the new quantity of the envi-

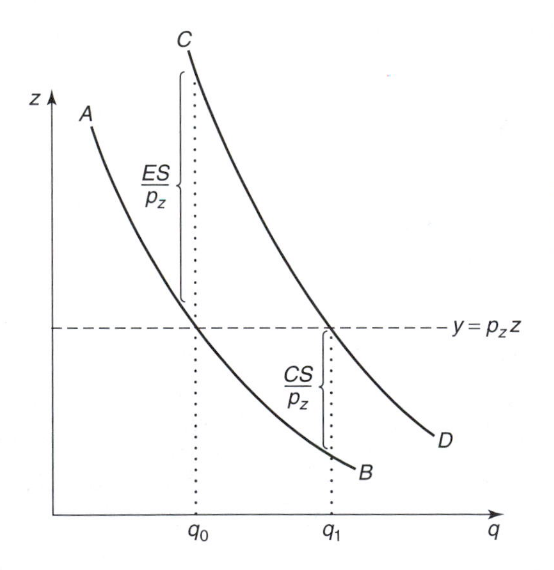

Figure 15.10 Surplus with quantity change. *AB*, Indifference curve, $U = U_0$; *CD*, indifference curve, $U = U_1$; *ES*, Equivalent surplus; *CS*, compensating surplus; p_z, price of z; y, income; q_0, initial quantity of environmental good; U_0, initial utility level; q_1, final quantity of environmental good; U_1, final utility level.

ronmental good, as shown in Figure 15.10. The equivalent surplus is the monetary value of z that would move to U_1, instead of changing q.

SUMMARY

1. Environmental goods provide special challenges to demand theory because of an absence of markets and a frequent lack of a cost of provision.

2. Environmental goods can be classified by their effect or manner in which they enter utility or production. Effects would include direct impact (e.g., health or materials damage) and ecosystem impact. Use and nonuse value refer to the way an environmental good is perceived by the consumer. Nonuse value includes existence value, altruistic value, and bequest value.

3. There are three basic approaches to measuring demand for environmental goods. Hedonic prices and household production are two methods based on revealed preferences. Contingent valuation is the third method, and is based on stated preferences. There are several hybrid methods, including experimental and constructed markets.

4. Compensated demand shows how quantity demanded depends on prices, keeping utility constant. Ordinary demand shows how quantity demanded depends on prices, keeping income constant.

5. Both compensated and ordinary demand functions conventionally have the price of all goods (or bads) as arguments. Restricted demand involves quantities instead of prices substituted for some goods, typically environmental goods.

6. The expenditure function indicates how much income is necessary to achieve a specified level of utility at a specified set of prices.

7. When the price of a good changes by a small amount, the change in expenditure function indicates the amount expenditures must change to keep utility constant. The ratio of the change in expenditures to the change in price is equal to compensated demand for that good.

8. A restricted expenditure function involves one or more goods (usually environmental goods) specified in quantity terms instead of prices. In this case in which quantity changes, the expenditures necessary to achieve the same utility change. The ratio of the expenditure drop to the quantity increase is equal to the marginal value of the good [Eq. (15.6)].

9. Suppose the quantity of an environmental good decreases. The compensating surplus measures the amount income would have to go up to return utility to the original level. The equivalent surplus measures the amount income would have to drop to have the same effect as the decline in environmental quality (but without the quality change). The concepts of compensating and equivalent surplus for quantity changes are very similar to the concepts of compensating and equivalent variation for price changes.

PROBLEMS

1. Anna was interested in Felix's preferences regarding air quality in Santiago. She reasoned that since air pollution could be washed off with soap, Felix should view air quality (A) and soap (S) as substitutes. Air quality is measured by an index, with larger values corresponding to higher quality. Holding annual soap consumption at five bars, she asked Felix to picture his utility with air quality at three different levels: 2, 4, and 5. Call these initial situations 1, 2, and 3, yielding utility U_1, U_2, and U_3. At each of these levels of air quality, she tried varying the amount of soap, asking how air quality should change to keep utility constant. During this mental experiment, she asked Felix to imagine his consumption of all other goods was fixed. She obtained the following data, showing values of A to keep utility constant at the three initial levels (U_1,U_2,U_3).

	Air quality (A) at		
S	U_1	U_2	U_3
1	10	20	25
2	5	10	12.5
3	3 1/3	6 2/3	8 1/3
4	2.5	5	6 1/4
5	2	4	5
6	1 2/3	3 1/3	4.17
7	1.43	2.86	3.57
8	1.25	2 1/2	3 1/8
9	1.11	2.22	2.78
10	1	2	2.5

Plot the three indifferences curves for soap and air quality implied by these data.

2. Using the data in Problem 1, suppose at air quality level 5, Felix chooses to buy four bars of soap when they are priced at \$1 per bar. Now air quality rises to 10. What are the compensating and equivalent variations associated with this change in air quality? (*Hint*: ignore expenditures on all goods except soap.)

3. We can generalize Problem 1 by stating that there are three commodities Felix consumes: air quality, soap, and a basket of other goods, denoted by x. Anna concludes from her research that Felix's utility function is $U(x,S,A) = x\,S\,A$. The price of x is always 1. Plot two restricted compensated demand functions for soap (price of soap on the vertical axis, quantity of soap on the horizontal axis). One demand function should be for $A = 4$ and the other for $A = 5$. In both cases, assume the utility level is 100.

4. In the Amoco Cadiz oil spill off of Northern France, fisheries, water fowl, and beaches were temporarily negatively impacted. List five types of consumers, identify the nature of goods these consumers demand that might be injured by the spill, and indicate the nature of demand for a contaminated environment (use or nonuse, bequest, altruistic, etc.).

5. Suppose housing (H) and air quality (A) are the only two things entering into Jose's utility and Jose's utility function is $U(H,A) = A \cdot H$. Suppose Jose's income is \$10 and the price of housing is \$2.

a. If air quality is $A = 2$, how much housing will Jose purchase? Plot Jose's indifference curve and budget constraint through this point, showing how much housing is consumed.

b. Suppose air quality rises to $A = 4$. Plot Jose's new indifference curve and new consumption point.

c. How much would Jose be willing to pay to increase air quality from 2 to 4?

6. Why would the simple plot of price vs. quantity in Figure 15.1 not be considered a demand curve for gasoline consumption in the United States? What additional information do you think you would need in order to be able to estimate the market demand for gasoline?

7. Using Figure 15.8, show that, for a normal good, the compensating variation is less than or equal to the equivalent variation, regardless of whether the price rises or falls.

NOTES

1. See Bergstrom et al. (1982) for an analysis of the demand for public schooling.
2. Table 15.1 shows a mean benefit of $23 trillion dollars. The same report shows the present value of costs of reaching that level of benefits of $0.53 trillion dollars. Keep in mind that this says nothing about the marginal benefits and marginal costs.
3. Freeman (1993) discusses some of these categories.
4. A real choice means a choice that involves a commitment of resources. Your response when I ask you how much you would pay for a ticket to a concert involves no real choice; actually buying a ticket for £10 is a real indication that the value of a concert ticket to you is at least £10.
5. Refer to any microeconomics text (e.g., Varian, 1996) for more information on these issues.
6. An ordinary demand function is also called a Marshallian demand function, after the British economist, Alfred Marshall. Also, the words "demand curve" and "demand function" are used synonymously.
7. A compensated demand function is also called a Hicksian demand function, after the British economist, John Hicks.
8. Fortunately, there are ways of obtaining the compensated demand function from an ordinary demand function, although the methods are too advanced for consideration here (see Hausman, 1981; Kolstad and Braden, 1991).
9. In calculus terms, the partial derivative of the expenditure function with respect to the price of a good is the same as the compensated demand for the good.
10. Although it might make more sense for the compensation for a price rise to be positive, it is equally intuitive that the income equivalence of a price rise (equivalent variation) should be negative. To focus on the magnitudes of these measures, rather than the sign, both are assumed negative for a price rise (undesirable to the consumer) and positive for a price drop (desirable for the consumer).
11. Willig (1979) elaborates on this point. See also Lankford (1988).
12. We have glossed over a technical point that really need not concern us. This argument that the change in the expenditure function fully captures the marginal willingness to pay really applies only if there is a close link between z and q, called "weak complementarity." Weak complementarity between z and q means that if z becomes so expensive that the consumer no longer consumes any of it, the consumer also stops caring about the environmental good. So z and q "go together"—they are complements. For instance, suppose a river is valued only for swimming and there is an admission charge to access the river. If the price of swimming is high enough, swimming will stop and river pollution becomes irrelevant. This is an example of weak

complementarity between swimming and pollution. Refer to Kolstad and Braden (1991) or Freeman (1979, 1993) for further information on this.

13. Lankford (1988) discusses this issue, though at an advanced level.
14. It is an empirical fact that compensating surplus/variation and equivalent surplus/variation do not appear to be the same for many environmental goods. Hanemann (1991) has argued that this is because the implicit value of many environmental goods is large; thus the income effect will also be significant. A significant income effect will drive a wedge between the *ES* and *CS*. In other words, how much we will pay to protect a major national park from commercial development may be considerably less than the amount we would willingly accept as compensation so that the park may be taken from the public and developed.

16 HEDONIC PRICE METHODS

This is the first of two chapters that are concerned with revealed preference methods of inferring demand for environmental goods and services. Recall that revealed preference means that we infer preferences for environmental goods from observed behavior in actual market transactions. In this chapter we will focus on making the inference by observing markets that are closely related to the pollution or other environmental good. For instance, if air quality varies throughout a city, what can the variation in property values in the city tell us about how people value clean air? Or since some occupations are more risky than others (e.g., working on high voltage lines), what can the wage premium associated with risky jobs tell us about how people value risk differences?

The primary empirical approach to measuring these values is what has become known as hedonic price methods. The basic idea is best conveyed through examples: measure the price of houses for a variety of different air pollution levels and attempt to see how the price changes when the air pollution changes, keeping everything else constant. Or for risky occupations, measure how wage rates change with risk and then infer the change that would be induced in wages from a small change in risk. In either case, we start with an hedonic price function that gives the price of houses or the wage rate as a function of air pollution or risk as well as other variables. Once we have quantified this relationship, the next step would be to infer the marginal value of a change in the air pollution or risk variable.

In the next section we will look at the general problem of relating pollution levels to property values, ignoring the labor market. We will then generalize this to examine the effect of pollution on property values and wages when producers and labor are mobile. We then turn to a detailed examination of hedonic theory before examining a classic study of the demand for air pollution, based on property values.

I. POLLUTION AND LAND RENTS

What determines the price of land? The price of a parcel of land is the value of the stream of services that the parcel can be *expected* to provide in the future, netted back to the present. Note the word expected. There is some uncertainty in the value of land since it is unclear exactly how useful it will be in the future. This is one reason land prices fluctuate over time.

Because expectations are unobservable, we often distinguish between the rental price of land and the asset price of land. The rental price of land is the price of renting the land for a short period of time (for example, a year). Expectations about the future play little or no role in the rental price. In contrast, the asset price of land is the price of taking title to the land in perpetuity. A farmer with agricultural land near a city may expect the land to be valuable for housing at some point in the future; this will increase the asset price of the land today. The rental price, however, will be totally related to the land's value during the rental period.

In a rural area, in which the land is used for agricultural purposes, the rental value of the land in any particular year is the expected value of output from the land, less the cost of labor, seed, and other nonland inputs necessary to obtain the output. If the land is poor, the land will yield less, and thus its value will be lower. Or if air pollution is severe, the yield will also be less and thus the rental value of the land will be lower.

If the land is used for housing, it provides services to the occupants of the land, including a place to locate a house and pleasant surroundings. In contrast to agricultural land, location can be very important in determining the rental value of land used for housing. Land in the middle of the Amazon may support a house with a nice view, but if people do not want to live in the Amazon, housing demand will be low and most of the land will be used for agricultural purposes. In contrast, land in downtown Tokyo is expensive because there are more uses for the land than there is land available. The value of land in an urban area has little to do with the agricultural productivity of the land and everything to do with the balance between the demand for and the supply of land for building. The classic economic model of a city consists of a downtown in which everyone works, surrounded by concentric rings of housing. In each ring, the price of land is the same since the commuting time (to the city center) is constant. Rings that are further from the center have lower land prices since commuting times are longer. Far from the center, the price of land drops so much that agriculture is a better use of the land than housing. This determines the extent of the city.[1]

The basic issue arises when air pollution varies from one location to another. Can we infer anything from the land value differences that might be induced by these variations in pollution levels? For instance, if land in a clean air area is priced higher than land in a dirty air area, can we assume that all of the difference is due to the difference in pollution levels? The problem is a bit more complex than it might at first appear. We will first consider the case in which the land is used for agricultural purposes and then turn to the slightly more complex case of land being used for housing.

To focus on the effects of pollution on land prices, we will assume that wages are unaffected by pollution. This might be the case if wages are set in a national or regional market and pollution varies over a smaller area. Thus wages will not vary from an area of high pollution to an area of low pollution. Later in this chapter we will explicitly introduce wages in addition to land rents and examine how both respond to differences in pollution levels. But for the time being, we assume wages are exogenous.

A. Agricultural Land

Suppose we have a valley devoted to agricultural production. Furthermore, the valley is divided into two regions, a clean region and a dirty region. This is illustrated in Figure 16.1. Crops do not grow as well on the land in the dirty region—productivity is lower than the clean region due to the pollution. This lower productivity will be reflected in lower rental prices for the land. We would expect the difference in value between the two regions to reflect the difference in productivity of the land in the two regions due to air pollution, since everything else is the same.[2]

Our primary question, however, is how much landowners in the dirty region would be willing to pay to clean up the pollution in the dirty region? Is it exactly the amount that land prices are depressed in the dirty region? If so, we just observe the land price differential and we have a measure of the willingness to pay for clean air. Whether this is true depends in large part on the nature of the agricultural markets for the produce grown on the land. If the market is purely local, the price of the agricultural output from the valley will be determined locally. Cleaning up the pollution will increase the supply of the crop and probably reduce the crop price, in both the dirty and the clean area. This results in a reduction in the land price in the clean area, in conjunction with an increase in the land price in the dirty area. Consumers as well as producers (landowners) have benefited from cleaner air (through lower-priced products). Thus the land value change will not fully capture the value of the cleaner air.

On the other hand, the market for many crops (e.g., wheat) is regional if not national or global. In such a case, the price of the product will be unaffected by the cleanup of the local air pollution. Thus the land price in the clean area will not change when pollution is eliminated in the dirty area. In this case, the land value change is a reasonable proxy for the value to agriculture of the environmental improvement.[3]

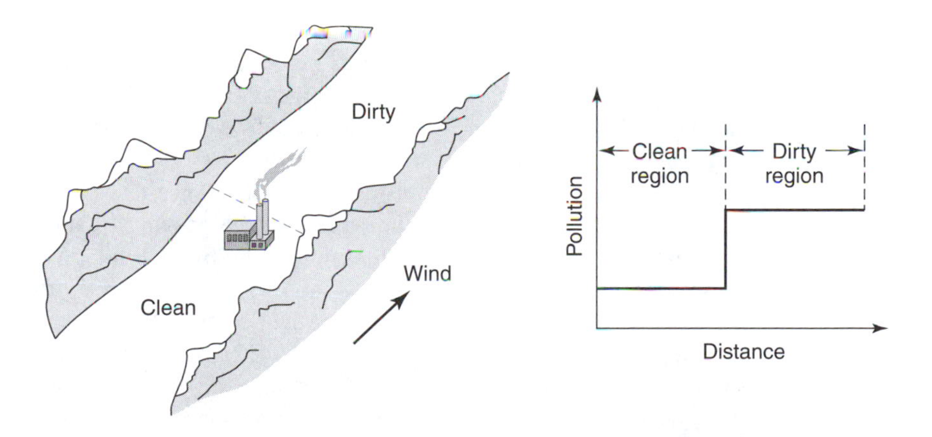

Figure 16.1 A valley divided into a clean region and a dirty region.

B. Urban Land

Now suppose we are dealing with land used for housing. In this case the value of the land does not reflect the value of the agricultural products produced from the land but rather the utility of the people residing on the land. Following the example above, suppose we have a city, part of which suffers from air pollution (the dirty part) and part of which is clean. There are many factors that determine the desirability of a parcel of land, including proximity to jobs. We will assume that everything is the same between the clean and dirty regions (such as commuting time and neighborhood characteristics), except for air pollution. We can ask the same question as we did for the agricultural example: Does the lower value of property in the dirty region fully capture the damage from the air pollution? We also assume that everyone in the city is the same, both in terms of their preferences and in terms of their incomes.

We consider two types of cities, a "closed city" and an "open city." The closed city assumes in- or out-migration is not possible—the population is fixed. In contrast, the open city is assumed to be one of many cities among which the population is free to move, locate, and relocate. The closed city may seem contrived—there is no city in which the population is prohibited from leaving or arriving. However, the closed city is a metaphor for a region (perhaps a country) in which mobility is less than perfect.

If we have an open city, we can conclude that the utility level of the typical person is the same, no matter where they live. With an open city, mobility among cities is costless; thus if there is any city that yields higher utility, people will move there, bidding up prices and depressing attainable utility until the utility difference has been erased. The same applies for different locations within a city. This means that people who live in the dirty part of the city must have the same utility as people who live in the clean part; otherwise, they would not put up with the dirty air. To compensate for the pollution, living is cheaper. Now if we clean up the pollution, the value of land will increase. But only landowners will be any better off. Why? Because the utility level is the same over all of the cities and a clean up in one portion of one city will not significantly change the overall level of utility. Thus when the air becomes cleaner, the land value will rise to compensate exactly for the increased quality of life. Put differently, the increased value of the land is a perfect measure of the value of the air quality improvement. Of course if we relax our assumption of homogeneity, and in fact consider a heterogeneous population (different tastes, different labor skills, different incomes), this conclusion must be tempered.[4] We will consider the interplay between wages and property values in the next section.

Now we turn to the case of the closed city. Since there is no mobility among cities in this case, the utility level of the typical citizen of our closed city is determined by internal factors, not a multicity equilibrium driven by mobility. We do know that the utility level within our city is the same for all residents. Thus the difference in property values between the clean and dirty parts of the city reflects the damage from air pollution. Now suppose we eliminate the pollution. Will the rise in property values in the dirty area reflect the value of the cleaner air? Generally, the answer is no. The reason is that the utility level of the typical resident of the city will rise as well as property values. The elimination of pollution has generated two effects, one observable (property value increases) and one unobservable (utility level increases).

What can we conclude from these various models of the effect of pollution on property values? The general conclusion is that in some but not all cases, the value of improving environmental quality will be reflected in increased property values. Situations in which this would not be the case involve secondary changes in utility or economic surplus brought about by the improved environmental quality, and not reflected in property values. This could be an increase in consumer surplus from lower agricultural prices or an increase in overall well-being brought about by the improvement and not bid away through higher land prices.[5]

C. The Effects of Climate on Agriculture

There is a recent application of these concepts for agriculture to the problem of global warming. As was discussed in Chapter 2, burning fossil fuels generates carbon dioxide and a great deal of fossil fuels have been burned worldwide since the industrial revolution began. The problem with carbon dioxide and similar gases in the atmosphere is that they make the atmosphere more opaque to outgoing infrared radiation (heat), which makes it more difficult for the earth to shed its incoming solar radiation. This is the greenhouse effect. The results are complex, varied, and not perfectly understood, but a major outcome is that the earth warms—"global warming." To complicate matters, much of the disruption that may be caused by emissions of greenhouse gases has yet to occur. It is emissions today and over the coming decades that may lead to real problems well into the twenty-first century. Policymakers must determine how much emissions of greenhouse gases should be cut back today to reduce effects many decades from now.

A major unknown in global warming policy is just how much we should worry about raising global temperatures by a few degrees Celsius. To help answer this, several economists at Yale University set out to measure the effect of climate change on agriculture in the United States (Mendelsohn et al., 1994). They noted that today there are many different climates in different parts of the United States; consequently, it should be possible to measure the effect of different climates on agricultural productivity. It is then easy to estimate how a change in climate would affect agriculture. Their approach is very similar to the examples presented earlier in this chapter regarding different pollution levels over an agricultural valley. Here we have the entire United States and instead of pollution, we have different climates. Just as we can estimate the willingness to pay to reduce pollution by comparing land values in parts of the valley with different pollution levels, we should be able to infer the willingness to pay in agriculture to avoid a 3°C temperature rise (for example) by examining two agricultural areas that are the same in all respects except that one has a climate on average 3°C warmer than the other.

The approach of Mendelsohn et al. (1994) was to try to explain the value of agricultural land in each of the approximately 3000 counties in the United States. They posited that land values depend on factors that determine agricultural productivity: climate (the long-run average weather), soil quality, altitude, slope (flat vs. hilly), latitude, and tendency to flood. A final factor that will contribute to land values but not agricultural productivity is proximity to urban areas. The equation to estimate is of the following form:

$$y = \alpha_0 + \alpha_c x_c + \alpha_l x_l + \alpha_u x_u + \epsilon \tag{16.1}$$

where y is the price per acre of agricultural land, x_c represents climate variables (e.g., average July temperature), x_l represents variables describing the quality of land from the point of view of agricultural productivity, and x_u represents variables relating to nonagricultural factors that might change the land price, such as proximity to urban areas. All variables are county averages, using readily available public data sources. The parameters α_0, α_c, α_u, and α_l are to be statistically estimated while ϵ is an error term. The authors statistically estimated Eq. (16.1) using 2933 observations on counties around the United States for a single year.

The last step of their analysis is to use the estimated Eq. (16.1) to compute what the loss in land value might be from a nationwide increase in average temperature of nearly 3°C coupled with an 8% increase in precipitation. They calculate that the loss in land value (using 1982 prices) would be $141 billion.[6] Because this is a change in land values, this is a one-time, not an annual loss. We can interpret this as the willingness to pay by agriculture to avoid this level of climate change. To place this in context, the total valuation of agricultural land in the United States in 1982 was $730 billion; 1982 farm income was $164 billion, with the value of all crops produced approximately half of this, at $72 billion (United States Bureau of the Census, 1987).

II. POLLUTION, LAND PRICES, AND WAGES

As we look at various communities around the world, we see that some are nicer to live in than others, for environmental as well as other reasons. In the previous section we examined the extent to which these differences are reflected in property values. But we all know that some of the worst places to live from the point of view of environmental quality have some of the highest property values (e.g., the center of many of the world's metropolises) and that many rural areas have high environmental quality and low property values. We also know that wages are often higher in cities than in rural areas. Do wages rise to compensate for deteriorating environmental quality? If this is the case, we must consider wage differences as well as property value differences in understanding the value of changes in environmental quality.

The basic question is, to what extent do higher levels of pollution depress property values but increase wages to compensate for the pollution? To understand the interplay among pollution levels, property values, and wages, we will construct a simple model involving multiple cities with competition among the cities for firms and residents. If pollution levels are different in the different cities, we expect to see differences in wages and/or property values.[7]

A. A Model of Wages, Land Prices, and Pollution

We start with a number of cities that differ in one respect: the level of pollution. Let the pollution level be denoted by p. In our example there is one composite produced good, call it X. Firms produce X using labor and land and may locate in any city. In a particular city, the wage rate is denoted by w and the land price or rent by r. These prices may of course vary from city to city. Firms seek to locate where X can be produced most cheaply. The price of the composite commodity is determined by the world market, outside of our analysis. We adjust the units for X so that the price of X is 1.

Consumers are identical in terms of preferences and use their wage income to purchase goods (X) and land for housing (L). Each consumer offers one unit of labor (such as a person-year) and receives a wage for that. If each consumer has the utility function $U(X,L,p)$, each consumer has the following utility maximization problem:

$$\max_{X,L} U(X,L,p) \tag{16.2a}$$

such that

$$w = X + rL \tag{16.2b}$$

This problem for the consumer results in some level of utility for each particular set of prices and pollution levels, w, r, and p. We can write the utility level that results from these prices and pollution as $V(w,r,p)$. Because we assume that people can move from one city to another easily and at no cost, it must be true that utility levels are the same for all cities. Call that constant utility level k. So $V(w,r,p) = k$, no matter what the prices or pollutin levels are.

Now consider the producer side of the market. With a constant cost industry,[8] the average cost of producing X must be equal to the product price, which is the same for all cities:

$$c(w,r,p) = 1 \tag{16.3}$$

Because each firm must sell its product for the same price, firms operating in different cities can use more of one input only if its price is less or there are cost savings in the use of the other input. If rents are higher in one city, wages must be lower to compensate, given the same level of pollution; otherwise, the firm would locate only in the lower cost city.[9]

Now how does pollution affect the firm? Obviously, if a clean environment is necessary for production, higher levels of pollution will make production more costly. In this case, we say pollution is "unproductive." Alternatively, high levels of pollution may be the result of lax pollution control regulations, in which case costs are lower for a polluting firm: pollution is "productive" in an indirect way. A third option is that pollution has no affect on the firm. Perhaps the firm is a law office that is unaffected by low or high levels of pollution. In this case we say that the pollution is "neutral."

Now let us consider two cities, both with firms producing X. One city has pollution level p_1, which is lower than the other city's pollution level, p_2. What can we say about land prices and wages? Remember that the utility level of a citizen of the first city must be the same as the utility level of a citizen of the second city. Because higher pollution is undesirable, some compensation is necessary to keep utility constant—in the high pollution city, either wages are higher or land prices lower or both. Figure 16.2 shows two lines of constant utility for the two cities, $V(w,r,p_1) = k$ and $V(w,r,p_2) = k$. Each line represents combinations of w and r that yield utility k for the particular level of pollution. Because pollution is undesirable to residents, the line associated with p_2 is lower than $V(w,r,p_1) = k$. Either lower land prices or higher wages

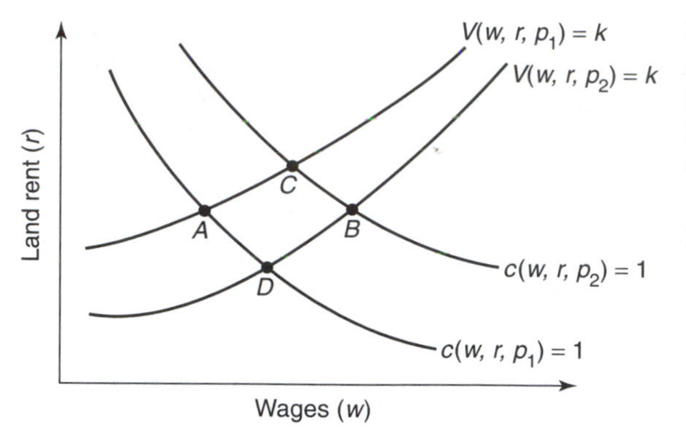

Figure 16.2 Effect of pollution on land rents and wages with productive pollution. A, Original equilibrium pollution $= p_1$; B, new equilbirium pollution $= p_2$; $c(w,r,p) = 1$, isocost line; $V(w,r,p) = k$, isoutility line. Note: $p_2 > p_1$.

are necessary to compensate for the increased level of pollution to keep utility constant at k.

Also shown in Figure 16.2 is Eq. (16.3) for the two different pollution levels. As shown, pollution is "productive" for the firm, meaning that higher pollution levels imply that the firm need spend less on pollution control. To keep costs constant, either land prices must rise or wages must rise to compensate the firm for the benefit of higher pollution levels, to ensure that Eq. (16.3) holds.

Now we can put these two perspectives together, comparing prices in the two cities. Point A is the level of wages and land prices with p_1 and point B is the level of wages and prices with p_2. It is clear from Figure 16.2 that when pollution is "productive," increased pollution levels raise wages while having an ambiguous effect on land prices. This is as expected. Higher levels of pollution drive away labor but attract firms. Higher wages in more polluted areas make up for the advantage to firms and tend to attract some people back. Higher land prices in the polluted area could also offset the cost advantage to firms, but higher land prices would not attract any workers. Thus wages must be higher in areas of higher pollution.

Pollution, of course, need not be "productive." If increased levels of pollution make it more expensive for a firm to operate, the two constant cost lines in Figure 16.2 would be reversed—the lower line will be associated with higher levels of pollution. In this case the equilibrium prices at low pollution levels will be point C while the prices for higher pollution levels will be point D. Pollution thus clearly depresses land prices but has an ambiguous effect on wages. Basically, pollution is bad for firms and workers. For a high pollution city to have any firms or residents, firm costs must be lower (lower wages or lower land prices) and working conditions must be better (lower land prices and/or higher wages). This is why land prices clearly decline as pollution increases but the effect on wages is indeterminate.

The third possible situation is that pollution has absolutely no effect on a firm. In this case, the two lines of constant cost in Figure 16.2 will be coincident. We can then conclude that increased pollution will decrease land prices and increase wages. The relative magnitude of each depends on the slope of the isocost line in Figure 16.2. A steep line means that land prices are not very important to the firm, in which case elevated pollution will be manifest almost totally in lower land prices. If the isocost line is fairly flat, a small increase in land prices must be offset by large wage decreases to keep costs constant: this means that land is much more important to the firm than labor. In this case, an increase in pollution is almost completely reflected in higher wages.

What we have seen from this analysis is that it is not possible to clearly conclude that higher pollution levels are totally reflected in either land prices or wages. In actuality, the extent to which pollution changes influence wages vs. land prices depends on how pollution influences the productive sector of a city's economy. The model here is very stylized, which makes it difficult for us to conclude exactly how higher pollution levels in a particular city will actually be manifest. The main thing we learned here is that both wages and land prices may be affected by pollution.

B. The Effect of Climate on Wages and Rents

One recent application of this wage–rent model is due to Nordhaus (1996), who examines the case of climate change. Earlier in this chapter we presented an analysis of Mendelsohn et al. (1994) valuing the effects of climate change on agriculture. William Nordhaus sought to determine the extent to which the amenity value of climate to individuals is reflected in wages and property values. This value is a final demand value, unrelated to production.

Nordhaus (1996) examined different counties in the United States (some 3000), expecting that wages and land prices would vary from county to county but that the price of traded goods would be equalized. Thinking of climate as sunshine, Figure 16.3 shows a possible relationship between a sunshine index (more is better), wages, and housing prices. This relationship is consistent with sunshine being neutral in production: more sunshine is desirable and thus drives down wages and drives up land prices. For higher levels of sunshine, housing prices rise and wages fall (termed nominal wages in Figure 16.3). Departing from the framework of our model in the previous section, Nordhaus avoids dealing with changes in two variables (housing prices and wages) by adjusting wages by the cost of living (including housing costs). These adjusted wages are denoted "real" wages in Figure 16.3. The higher the housing costs, the less income the consumer has available to spend on consumption. As the sunshine index increases, the real wages drop. Real wages are a purchasing power adjusted measure of income. Thus if I make $10,000 and

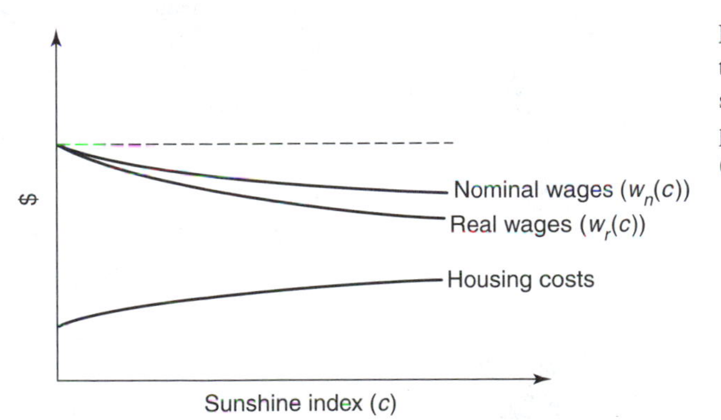

Figure 16.3 Hypothetical relationship among climate ("sunshine"), wages, and housing prices. Adapted from Nordhaus (1996).

you make \$5000, but the cost of living for me is twice what it is for you, our real wages are the same. It takes me \$2 to buy what you can buy for \$1. This adjustment allows the author to focus on how real wages change with climate.

The model proposed by Nordhaus relates real wages (w_r) to climate variables and other factors that determine real wages (such as education level of the population):

$$w_r = \alpha_0 + \alpha_C C + \alpha_Z Z + \epsilon \tag{16.4}$$

Equation (16.4) indicates that real wages are determined by climate (C) and nonclimate factors (Z) plus an error term (ϵ). The changes in the cost of living, including housing, have been netted out of the real wage in Eq. (16.4). Nordhaus (1996) statistically estimates Eq. (16.4) using a cross section of U.S. counties (approximately 3000 data points). In doing so, the climate includes many variables (such as mean annual temperature, summer temperature, precipitation) and Z includes many nonclimate variables that are thought to determine wages (e.g., population density, education level, ethnic background, unemployment rate).

Although Nordhaus presents a variety of results, Figure 16.4 presents his results for the mean annual temperature. The horizontal axis shows mean annual temperature for a county relative to the average over all counties. Thus zero would be the national average mean annual temperature. Wages are computed for average values of Z (the nonclimate variables). The vertical axis shows $[w_r(C) - w_0]/w_0$, where $w_r(C)$ is the real wage as a function of the mean annual temperature and w_0 is the wage at the national average mean annual temperature. Thus the vertical axis shows in percentage terms how climate drives real wages up or down. Note that a region with a slightly lower temperature would actually see higher real wages—the current national average does not result in the highest wages. Second, a region with a substantially higher temperature would expect real wages to be lower by as much as 5%. Because real wages involve both nominal wages and housing costs, when real wages decline, we do not know whether nominal wages are rising or falling or whether housing costs are rising or falling.

The importance of this study is in showing how different levels of an environmental good can influence wages and housing prices. Unfortunately, seeing how an environmental good influences equilibrium prices is not generally enough for us to compute de-

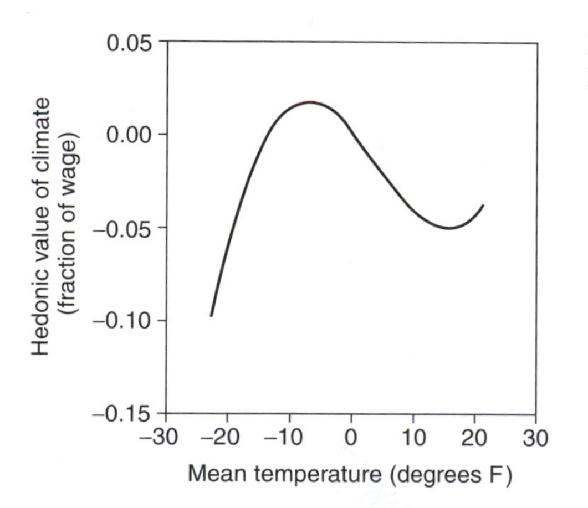

Figure 16.4 Real wage variation with climate. From Nordhaus (1996).

mand for the environmental good. To do that, we require more structure. We turn to that issue in the next section.

III. HEDONIC PRICE THEORY

We now consider the issue of measuring the demand for pollution based on observations of how pollution influences prices and markets. In the previous sections we noted the circumstances under which changes in pollution influence land markets and labor markets. This is only the beginning of being able to measure the demand for pollution.

In the "real world," we are often confronted with bundles of goods with a single price for the whole bundle. Yet, we are in fact interested in the price of an element of the bundle. For instance, we observe the prices of houses that are bought and sold. But a house consists of a bundle of characteristics such as the number of rooms, the neighborhood quality, and the surrounding environmental quality. We note that houses with fewer rooms fetch lower prices, as do houses with lower air quality. By observing the prices of many houses with differing characteristics, is it possible to back out the implicit value that is being placed on air quality (for instance)? Similarly, different occupations (jobs) have different characteristics, including the level of health or mortality risk. Typically, jobs with greater risks pay higher wages. By observing the wages associated with many different occupations, can we infer anything about how much workers value small changes in risk?

These examples suggest the focus of hedonic price theory: inferring the value placed on characteristics of goods based on the observed price of a bundle of goods.[10] It is somewhat more complicated that these examples suggest, in part because of the interaction between the supply of characteristics (building more houses with more rooms) and the demand for characteristics.[11] In this section we will develop this theory, followed by some empirical examples in subsequent sections.

We will consider the simplest possible structure, that of a good with a single characteristic. It is probably easiest to think of this as houses with different levels of a characteristic (such as the number of rooms or pollution levels). Although we are considering a single characteristic, it is not difficult but it is tedious to extend this to the case of multiple characteristics. We will examine two basic questions. The first involves determining the price of the good in the market. The second involves inferring consumer willingness to pay for changes in the characteristic—a demand curve.

The basic setup is a single homogeneous area, such as a city or part of a city, which can be considered a single market from the point of view of our good (a house). Each house can be characterized by a single variable, z, which we will assume represents air quality levels. We are interested in how the price of a house varies with the pollution level. In other words, we are interested in $p(z)$, the house price as a function of air quality levels. This price function is an equilibrium concept, resulting from the interaction of supply and demand. In much the same way that conventional supply and demand interact to produce a single market price, forces of supply and demand interact to produce a price function—a single house price for every pollution level.

To understand the nature of $p(z)$, we need to look at both the consumer side and the producer side of the market. We will assume the market is competitive, which means that producers take $p(z)$ as given and consumers take $p(z)$ as given (just as in a conventional market, producers and consumers take p as given). The way in which producers and consumers react to $p(z)$ will give us insight into how $p(z)$ emerges in the market.

A. The Consumer

Suppose a typical consumer has utility function U and income y. The consumer buys exactly one house. The consumer must decide how to allocate income between the house (choosing the level of air quality z) and ordinary goods, denoted by x (nominally priced at 1). Thus the consumer's problem is

$$\max_{x,z} \; U(x,z) \tag{16.5a}$$

such that

$$x + p(z) = y \tag{16.5b}$$

Equation (16.5a) states that the consumer should choose appropriate levels of x (ordinary goods) and z (housing characteristics) to maximize utility. Equation (16.5b) states that while doing this maximization, the consumer must be sure to stay within the budget, keeping expenditures on ordinary goods and expenditures on the house equal to income.

Another way of thinking about the problem is to determine, for particular values of z, the amount of x that needs to be consumed to achieve a particular level of utility, $\hat{U}$? This then determines how much money is spent on x and, in turn, how much income is available to spend on the house. Fixing z, we can solve for the value of x that satisfies $U(x,z) = \hat{U}$. This then defines a particular amount of income available for the house: $y - x = \theta$. Another way of looking at this would be to find the θ that satisfies

$$U(y - \theta,z) = \hat{U} \tag{16.6}$$

So given values of z, y, and $\hat{U}$, we determine how much money is available for the house, $\theta(y,z,\hat{U})$. We call this a bid function because it represents the amount of money the consumer may bid for the house with characteristics z, to keep utility at the level $\hat{U}$, assuming income y.

Figure 16.5 shows, for a single consumer, two bid functions for two different levels of utility. Each function shows how much the consumer would be willing to bid for different levels of air quality (z). Note that the lower bid function is associated with higher levels of utility. Figure 16.5 also shows the hedonic price function, $p(z)$, determined by the market. The consumer's problem is to determine what level of z to choose to maximize utility. For this consumer, the point of choice is the point at which the bid function is just tangent to the price function, $p(z)$—the point at which the bid function just "kisses" the price line. This gives maximum utility with the requirement that the amount the consumer is willing to pay (θ) must be equal to the price, $p(z)$. If we were to do this same exercise for another consumer or even the same consumer with a different income, we would get a different set of bid functions and a different choice point for z.

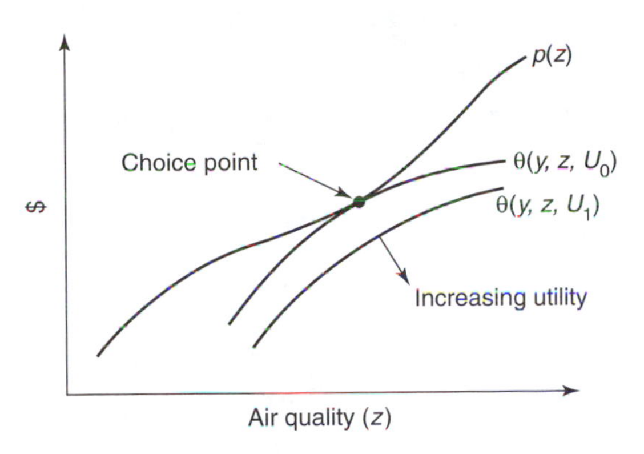

Figure 16.5 Consumer choice. $\theta(y,z,U)$, Amount of money avaiable to bid for house with air quality z, for a person with income y, obtaining utility U; $p(z)$, price of house with air quality level z.

B. The Producer

There are also producers of housing, and this side of the market is somewhat easier to characterize. Some producers are particularly good in building large up-market houses; other producers specialize in other segments of the housing market. Different producers will have different costs. Let us start with some particular producer with a given cost structure. We will assume there are constant returns to scale so that we may consider the costs of producing one house of characteristics z.[12] Suppose the producer faces input prices r; unit costs can be given by $c(r,z)$. If the producer offers a price of ϕ, then profits per house are given by

$$\Pi = \phi - c(r,z) \tag{16.7}$$

We can rewrite this as the price necessary to obtain a certain level of profit, given the level of the characteristic, z: $\phi(r,z,\Pi)$. We call this the offer function. The offer function indicates the price at which the producer will offer the house to obtain a particular profit level, Π, given a particular value of input prices, r, and a particular value of z. If the producer wants to sell a house, this offer curve has to intersect the price line. This is illustrated in Figure 16.6.

In Figure 16.6, the hedonic price function is shown, $p(z)$, as well as two offer functions for some particular firm, one function for each of two profit levels, Π_1 and Π_2. Note that higher curves are associated with higher profits. Thus the correct choice of z is the one at which an offer curve just "kisses" the hedonic price line. As with the case of consumers, different producers will have different sets of offer curves and thus different choice points along the hedonic price function.

C. Market Equilibrium

We stated at the beginning of this section that we were working backward to see how the hedonic price function is constructed. We are now in a position to see that every point along the hedonic price function corresponds to tangency between the bid function of some consumer and the offer function of some producer. This is shown in Figure 16.7. In Figure 16.7 we see three pairs of producers and consumers, each with tangency be-

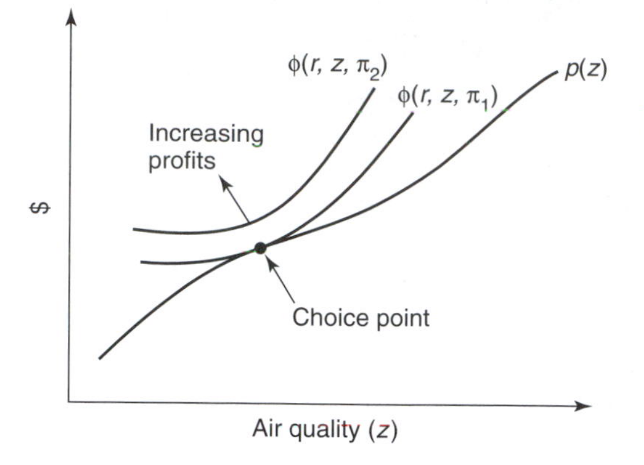

Figure 16.6 Producer choice. $\phi(r,z,\pi)$, House offer price when air quality level is z, input prices r, necessary to give profit π; $p(z)$, price of house with air quality level z.

tween their offer and bid functions. This tangency determines a choice of z, along with a price, $p(z)$. In this manner, the entire price line is constructed. Thus at each point along the hedonic price function, we conclude that there is some consumer's bid function and some producer's offer function both tangent to the price line.

This tangency is important. It means that the slope of the offer function (with respect to z) is equal to the slope of the bid function, which is in turn equal to the slope of the price function. The slope of the price function indicates how much the price goes up if the level of z goes up by one unit. This is the marginal price and is analogous to a conventional price for z. To see this, suppose we have a shopping cart full of groceries, including three loaves of bread. We are unsure about the price of bread, but we do know that the value of the entire shopping cart is £40. We also know that if we add one more loaf of bread to the cart the value of the cart goes up to £41. What must the price of bread be? The marginal price of the shopping cart with respect to the number of loaves of bread is £1/loaf. Clearly, the price of bread is £1. If we were unable to buy loaves of bread individually, but only as part of bundles of groceries, we could use the same mental exercise to conclude the marginal price of a loaf of bread is effectively £1 per loaf.

The fact that the slopes of the bid, offer, and price lines are identical at each choice point means that the producer, the consumer, and the market all have the same marginal

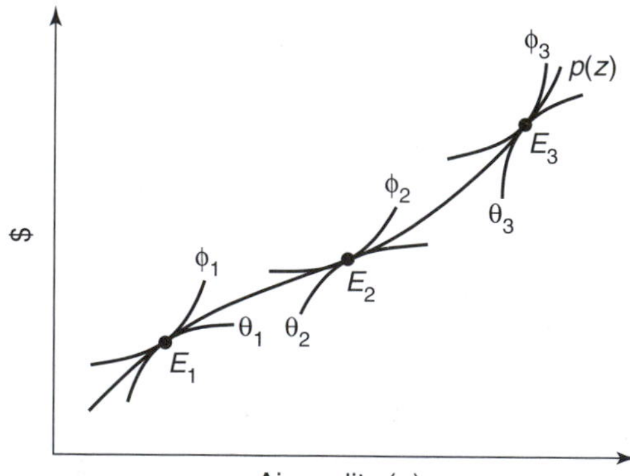

Figure 16.7 Equilibrium in an hedonic market. ϕ_i, Offer function for three producers, i = 1,2,3; ϕ_j, bid function for three consumers, j = 1,2,3; E_k, equilibrium between producer k and consumer k, k = 1,2,3; $p(z)$, price of house with air quality level z.

valuation on a unit of the characteristic. The slope of the offer function is the increased price necessary to compensate the producer for supplying one more unit of z. The slope of the bid function is the extra amount of money the consumer is willing to pay for one more unit of z. And in market equilibrium, these are equal and are also equal to the slope of the price function with respect to z.

D. Willingness to Pay

The previous sections described the origins of the hedonic price function. But what we are really interested in is a demand function or marginal willingness-to-pay function for particular characteristics. For instance, suppose air quality surrounding a house is the characteristic of interest. We have measured the hedonic price function and calculated its slope at various points. Thus we know the marginal price of air quality at different levels of air quality (z). But this is not demand—all we know is what the market price is of air quality. What we really need to know, for a particular individual, is what that individual's marginal willingness to pay is for one more unit of air pollution. The problem is that each individual chooses only one consumption point along the hedonic price function, not several consumption points. One point does not a demand function make.

The distinction between the marginal price of z as a function of z and the marginal willingness to pay for z as a function of z is illustrated in Figure 16.8. Shown is the slope of a hedonic price line. It generally trends downward in z. Also shown are two marginal willingness-to-pay (MWTP) functions for two different individuals. The point of intersection of these MWTP functions with the marginal price line gives us the choice of z for the individual. But how much each consumer is willing to pay for other quantities of z is obtained from the MWTP function, not the marginal price function. And these are different.

The solution to this is to assume the different individuals making choices along the hedonic price function are variants of the same person, simply with different incomes and characteristics (call these differences α). Thus we can translate the many observations on choice of z into a set of data on how marginal willingness to pay varies with z, and thus statistically infer a marginal willingness-to-pay function. We can do the same with pro-

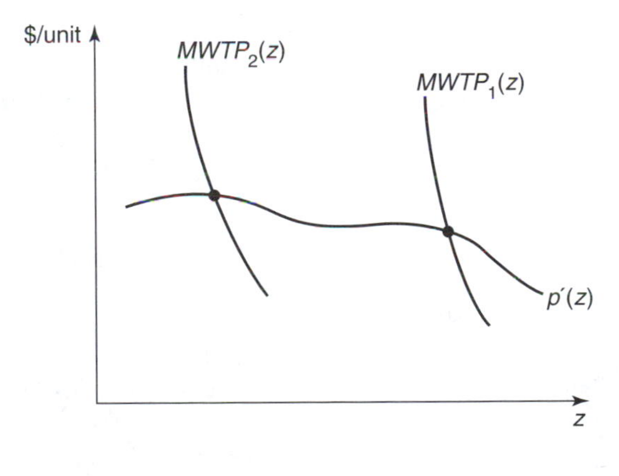

Figure 16.8 Marginal willingness to pay for z. $p'(z)$, Slope of hedonic price line with respect to z (marginal value of z); $MWTP_i(z)$, marginal willingness to pay for one more unit of z, consumers $i = 1,2$.

ducers denoting their differences by β. We thus have the following equations for supply and demand:

$$p'(z) = f(z,\alpha) \tag{16.8a}$$

$$p'(z) = g(z,\beta) \tag{16.8b}$$

In Eq. (16.8), $p'(z)$ refers to the slope of the price function, $p(z)$, with respect to z.[13] The right-hand side of Eq. (16.8a) is a marginal willingness to pay function. Recalling that the slope of the hedonic price function is analogous to the price of the characteristic, or marginal willingness to pay for the characteristic, Eq. (16.8a) states that the marginal willingness to pay depends on the level of the characteristic and other variables such as income. This is precisely how we would specify an inverse demand function: price is some function of quantity, income, and perhaps some other variables reflecting tastes. Similarly, Eq. (16.8b) is analogous to an inverse supply function, with the price related to the quantity supplied (in this case quantity of the characteristic) and factors that reflect the costs of the industry, β.

Although there are some substantial statistical problems associated with implementing this methodology,[14] the basic idea is that we first observe the hedonic price function in a market. Then we measure the slope. Then, using this slope as a price of the characteristic, we measure the demand for the characteristic. So if the characteristic is air quality and the good is housing, we first note how the prices of houses vary with air quality. We then compute how much those prices change when air quality changes by one unit. This gives us a "price" of air quality. We then note how people of different incomes react to different prices of air quality in choosing how much to consume. In the next section we discuss an implementation of this approach.

E. The Demand for Air Quality in Boston

Although there have been a number of analyses of demand for environmental quality using the hedonic method,[15] one of the classic studies is by Harrison and Rubinfeld (1978) for Boston, Massachusetts. They faced a very real and also typical environmental policy problem: Are the benefits of a proposed air pollution regulation sufficient to justify the costs? Their problem was having to quantify the benefits from tightening the automobile emissions control regulation for nitrogen oxides in Boston. They first estimated a household-level marginal willingness-to-pay function for nitrogen oxides. Using a meteorological model, they were able to calculate the change in pollution concentrations at different locations in Boston as a result of the new regulations. Then using their marginal willingness-to-pay function for nitrogen oxides, they calculated the benefits of the cleanup for an average household. They compute a value per average household ranging from \$59 to \$118 per year, in 1970\$.

The authors adopt the two-step procedure described in the previous section. The first step is to estimate the hedonic price equation for housing in Boston. Their goal in this step is to explain the price of houses. Their data are drawn from the decennial U.S. population census. Data for individuals are confidential and unavailable to the researchers. The smallest unit at which data are available is what is called the "census tract." A census tract consists of several city blocks, though this varies; it is basically the smallest unit that preserves anonymity of respondents and is thus ideal for public analysis.

For each census tract, the authors know the median price of owner-occupied houses, several structural variables (e.g., the average number of rooms in owner-occupied houses), several neighborhood variables (e.g., the crime rate), variables reflecting accessibility (e.g., distance to one of five employment centers), and an air pollution variable (annual average concentration of nitrogen oxide). In identifying this list of 13 explanatory variables, the authors tried to include anything that might be a determinant of the value of housing (and which was also available in the census data). The authors also had data on median household income in each census tract. Those data are not used to explain housing prices but are used to determine the marginal willingness to pay for air pollution.

Having specified the determinants of housing prices, the researchers used data from 506 census tracts in Boston to statistically estimate the housing value equation. The housing value equation is the hedonic price function and indicates how housing prices at the census tract level are determined by the 13 exogenous variables mentioned in the previous paragraph. They were able to fit the data to their equation fairly well, achieving an 81% fit.[16]

Remember that the hedonic price function is not a demand function—it simply indicates how housing prices change when characteristics change. And housing prices are determined by the interaction of supply and demand. But the housing price function does allow us to compute the *marginal price of nitrogen oxides* as reflected in housing prices. This is the slope of the housing price function with respect to nitrogen oxide concentrations. Thus the authors are able to compute this marginal price, denoted w. w then varies from one housing tract to another.

The second step the authors take is to estimate the demand for nitrogen oxides. Similar to Eq. (16.5a), they posit a demand equation of the form

$$\log(w) = a_0 + a_P \log(P) + a_Y \log(Y) \tag{16.9}$$

where P is pollution levels (nitrogen oxides), Y is household income, a_0, a_P, and a_Y are constants, to be statistically determined, and log represents the natural logarithm. Equation (16.9) is a classic demand function where the quantity demanded (P) depends on income (Y) and prices (w). Remember that w was computed from the first step of this process, the hedonic price function. The authors estimated Eq. (16.8) over the sample of 506 census tracts to obtain an equation for the marginal willingness to pay for air pollution. Three demand curves computed from this study are presented in Figure 15.2. These curves show the one-time payment consumers are willing to make for permanent changes in pollution levels.

We should mention that the study described here was conducted over two decades ago, before the statistical problems associated with estimating Eq. (16.9) were well known (see note 14). Consequently, the numerical values of the coefficients in Eq. (16.9) as presented by the researchers are not reliable. Nevertheless, the rigorous methodology presented by these authors and the clarity of their presentation make this study one of the most significant hedonic analyses of the demand for environmental goods.

The authors close their analysis by computing the value of the air pollution reduction associated with the new automobile emission regulations. As mentioned earlier, they compute the value of the expected pollution reduction to be on the order of $100 per year for the typical household.

F. The Value of a Statistical Life

One of the major applications of hedonic price analysis is in studying how much more workers are paid for riskier jobs. Different occupations involve different risks, and it is reasonable to expect that employers need to pay a wage premium to induce workers to undertake jobs involving higher risk. If we could measure this premium, we could obtain an estimate of the market value of small changes in fatality risk.

A number of authors in several countries have tried to make just such a calculation.[17] The general approach is to estimate a function of the form

$$w = f(\pi, z) + \epsilon \tag{16.10}$$

where w is the wage rate, π is the probability of death on the job, z is other characteristics of the job, and ϵ is an error term. As might be expected, there are problems in statistically estimating such an equation. One of the major problems is determining whose risk estimate should be used—the subjective risk as estimated by the worker or the objective risk as reflected in injury statistics. Another problem has to do with the age of the worker. Older workers have fewer remaining years left in their life; one would expect this to have some influence on their valuation of risk. Another problem is that there may be self-selection into high risk jobs. Workers who take such risks may place a very low value on their life, yet it is such workers who are setting the wage premium for riskier occupations. Nevertheless, a number of authors have estimated versions of Eq. (16.10). We will examine some of those estimates shortly.

The standard approach to using an hedonic price equation, such as Eq. (16.10), is to examine how the wage rate changes with changes in risk—the slope of the hedonic price equation. In the case of Eq. (16.10),

$$VSL = \frac{\Delta w}{\Delta \pi} = \frac{\Delta f(\pi, z)}{\Delta \pi} \tag{16.11}$$

Equation (16.11) indicates, for instance, how much the equilibrium wage rate increases when the fatality risk increases by a small amount. The slope of the wage-risk hedonic price function is called the *value of a statistical life* [*VSL* in Eq. (16.11)]. For instance, suppose we are at a risk level of 10^{-4} and we note that if the risk is increased to 2×10^{-4}, the annual wage goes up by $600. The slope of the hedonic wage function at this point would be $\$600/(2 \times 10^{-4} - 10^{-4}) = \$6{,}000{,}000$. One interpretation of this figure is that if we have 10,000 workers subject to an increased annual risk of 10^{-4}, we would expect, on average, one extra death to occur per year ($10^{-4} \times 10{,}000 = 1$). Workers would need to be paid $\$600 \times 10{,}000 = \$6{,}000{,}000$ in extra wages to willingly take this risk. If there are ways in which the employer can eliminate this increased risk for a sum less than $6,000,000, it would be in the employer's best interest to do so. Thus in a statistical sense, it is worth $6,000,000 to save a life.

This notion of placing a value on a statistical life sounds crass. But risks exist and it is not logical to spend enormous quantities of resources to reduce risks to the lowest

TABLE 16.1 Selected Empirical Estimates of the Value of a Statistical Life, Based on Hedonic Wage Studies

Year	Country	Average per capita income[a] (1990 US$)	Implicit value of statistical life (1990 US$ million)
1984	Australia	18,177	3.3
1986	Japan	34,989	7.6
1982	United States	19,194	16.2
1979	Canada	NA	3.6
1977	United States	28,000	10.0
1977	United Kingdom	11,287	2.8
1976	United States	17,640	6.5

[a]*Average income refers to study sample. NA = not available.*
Source: Viscusi (1993), Table 2.

possible level. In our everyday lives, we are constantly making trade-offs regarding risk, whether it involves using an old car with increased risk of mechanical failure but lower costs or the decision to take a plane trip (and risk a crash) to enjoy a vacation. In fact, the hedonic wage function indicates how workers willingly expose themselves to risk to have more of the resources income can buy.

Table 16.1 shows the results of selected hedonic wage studies involving risk–wage trade-offs, taken from a review of this literature by Kip Viscusi (1993). Although these studies were done in a number of different countries, the results do not vary too widely. In fact, Viscusi suggests that the several dozen studies he reviews for the most part find that the value of a statistical life is in the 2 to 7 million dollar range.

We should point out that estimating Eq. (16.10) gives information only on how the market wage depends on risk. We have not taken the second step in hedonic analysis of estimating an individual's willingness to pay for risk reduction. For the most part, empirical research has been content with measuring the hedonic price equation and has not taken the second and more difficult step of estimating the marginal willingness to pay for risk.

IV. CONCLUSIONS

The hedonic price method is one of the oldest approaches to determining the demand for nonmarket commodities, such as pollution. Its strength is that value is determined from actual market transactions. This is an important advantage; in fact, the lack of a basis in real market transactions is one of the primary criticisms of contingent valuation methods, which we will consider in a later chapter.

Of course there are shortcomings to the hedonic method. One is that it is necessary to find a market good whose value is influenced by the environmental commodity (e.g., housing prices are influenced by air quality). It is also important that all of the damage from the environmental commodity be reflected in the price of the market good. One shortcoming of property value studies is that residents may spend substantial portions of their day away from their home; the damage so incurred will not be reflected in property values.

The biggest problem is that many environmental problems do not directly influence observed market goods. What goods are affected by water pollution that may degrade recreation or possibly contaminate water supplies? What goods are affected by the loss of a species of plant? In those cases, other methods must be used, which we examine in detail in the next two chapters.

SUMMARY

1. In an agricultural region, the value of a change in air pollution in an area that constitutes a small part of the agricultural market will be reflected in the change in the price of land.

2. If a change in air pollution affects a substantial amount of land so that the price of agricultural products changes, land price changes will not reflect the value of the change in air pollution.

3. In an urban housing market, the value of changes in air pollution in a part of the market will be reflected in changes in property values if the region is open with free mobility in and out of the region, and if producers are unaffected by the change in pollution and property values. If the market is closed in that in- and out-migration are restricted, the value of pollution changes will not be fully reflected in land value changes.

4. If production is influenced by pollution or if property values are significant costs of production, a change in pollution will change both property values and wages. If pollution is "productive," in that more of it reduces costs for firms (perhaps because regulations are looser), an increase in pollution will raise wages and have an ambiguous effect on land prices. If pollution is "unproductive," in that more pollution increases production costs, an increase in pollution will depress land prices and have an ambiguous effect on wages. If pollution is "neutral" to the firm, an increase in pollution will decrease land prices and increase wages, although the amount of each depends on the relative importance of land and labor to production.

5. With hedonic price theory, we observe the price of a good as a function of certain characteristics of the good. For example, with housing we may observe that the price of a house varies with the air pollution level surrounding the house. From this we infer the marginal price of the characteristic as the slope of the price line with respect to the characteristic.

6. Hedonic price theory can further be used to estimate an individual demand curve. Having observed how the market values changes in the characteristic, we may estimate the demand or marginal willingness to pay for the characteristic using the marginal price of the characteristic, the level of the characteristic chosen by consumers, the income of consumers, and other factors that influence consumer preferences. There are serious econometric problems in estimating this demand.

7. In studies of wage–risk trade-offs, the marginal wage with respect to fatality risk has been called the value of a statistical life. The value of a statistical life is the ratio of the increase in the wage rate to the increase in the risk of fatality that induced the increase in wages. This is a market concept that is not the same as an average individual's marginal willingness to pay for risk reduction.

PROBLEMS

1. Suppose we have an agricultural valley 100 km by 50 km, with a polluting electricity power plant. The power plant causes pollution problems in a narrow strip of land downwind of the plant, 10 km in length and 500 m in width, but no problems outside of that area. The total effect of the pollution is to make crop land less productive. If we were to clean up the pollution, would we expect land prices to increase? Where? Would we expect wages to decline? Would the changes in land prices and/or wages fully reflect the benefits of cleaning up the pollution? Why or why not?

2. Oxana is conducting research on the demand for cleaner air in Novosibersk. She notices that the Ministry of Housing built many identical dwellings around the city approximately the same distance from the center. The only difference is the average level of air pollution. She notes the following rental prices and air pollution levels for these dwellings:

Pollution level ($\mu g/m^3$)	Rent (R/month)
30	497
50	492
70	485
90	475
120	455
150	430
200	375
250	305
300	219

   Plot the relationship between pollution levels (on the horizontal axis) and rent (on the vertical axis). On a separate graph, plot the relationship between pollution levels (on the horizontal axis) and the marginal valuation of pollution implicit in the relationship between rents and pollution levels (on the vertical axis).

3. After conducting her analysis of housing prices in Novosibersk, Oxana notes that everyone in her study has exactly the same income, graduated from the same school, and listens to the same music; in fact, everyone in town appears to have exactly the same tastes and income. Does Oxana have enough information to construct a typical marginal willingness-to-pay function for pollution in Novosibersk? Why?

4. Pierre is a software engineer, working in a heavily polluted city in northern France. He does not like the pollution, but the pay is good and housing is relatively cheap. He is considering taking a job with a different software company in Cannes, a lovely seaside community in southern France with very low pollution levels. Unfortunately, he discovers that housing is very expensive in Cannes; many people wish to live there and this has bid up the price of housing. Pierre tells his potential employer in Cannes that he will need a higher

wage to compensate for the higher cost of living. Is Pierre justified in making this request? Why or why not?

5. Figure 15.2 is drawn from Harrison and Rubinfeld's study of the demand for air pollution in Boston. Make a copy of Figure 15.2 and then estimate the total willingness to pay, for a household earning $15,000 per year, of reducing NO_x levels from 6 to 4 parts per hundred million.

6. An automobile maker is interested in the marginal willingness to pay by consumers for fuel economy in the cars the manufacturer produces. The automobile maker has asked you to conduct a hedonic analysis of automobile characterisitcs. Describe how you would carry out the analysis, being sure to discuss (a) which characteristics of cars you would measure, (b) how you would measure the hedonic price function, and (c) how you would translate this into a marginal willingness-to-pay function.

NOTES

1. This classic model is described more fully in Mills (1967).
2. As was mentioned earlier, it is really the rental price of land that should reflect the productivity differences. The asset (total) price of the land reflects future expectations, which are difficult to observe. For instance, people may expect air pollution regulations to be tightened in a few years, eliminating pollution problems. If so, the pollution will have only modest effect on the asset price of the land, even if it has substantial effect on the rental price of land.
3. This issue was the subject of much debate in the late 1960s and 1970s. See Freeman (1979) for a summary. The primary source papers are Lind (1973), Strotz (1968), and, more recently, Kanemoto (1988).
4. See Polinsky and Shavell (1976) for an analysis of a more general urban area.
5. There is a large empirical literature on using property values to infer the value of environmental improvement. The first such paper appears to be Ridker and Henning (1967).
6. The results presented by the authors are somewhat more ambiguous than this. They estimated two versions of Eq. (16.1), in one, placing more weight on the importance of the error for a county, depending on the agricultural significance of the county. When the weights were simply based on agricultural acreage in the county (more acres, more weight), the $141 billion loss resulted. When the weights were based on the value of agricultural output in the county, there was actually a gain associated with warming, on the order of $35 billion. The value-based weights tend to emphasize high-value crops such as vegetables grown in California or citrus in coastal areas, whereas the acreage-based weights tend to emphasize grains that have lower value but are planted widely in parts of the United States.
7. The model and results presented here are based on Roback (1982).
8. Recall that a constant cost industry has all firms operating at the bottom of their average cost curves. Further, there are no economies, diseconomies, or scarce factors that change industry prices or costs as the industry expands. Thus the long-run industry supply curve is horizontal, at a price equal to minimum average cost for any firm.
9. In the interest of full disclosure, we should point out that one of the primary reasons suggested as to why cities exist is agglomeration economies—economies associated with different firms being located in close proximity. For instance, there are may law offices in central London because of the ease of interaction between firms, access to courts, the use of common services, and similar reasons. Agglomeration economies violate our assumption of a constant cost industry.
10. One of the first empirical studies of this sort is Zvi Griliches' 1961 study of measuring quality change in automobiles (Griliches, 1971).

11. The modern theory of hedonic prices is due to Rosen (1974), although it has been expanded by a number of authors (see Palmquist, 1991).
12. Recall that constant returns to scale simply means that the cost of two identical houses is exactly twice the cost of one house. Thus we can focus on the cost per house—the unit cost function.
13. The slope of $p(z)$ with respect to z is of course the derivative of $p(z)$ with respect to z; the derivative is often denoted $p'(z)$.
14. The statistical problems are quite difficult to overcome. There are two basic statistical problems—identification and endogeneity. Identification means being able to disentangle the hedonic price function from the marginal willingness-to-pay function. Because the same data that are used to generate the hedonic price function are used to generate the demand function, it may be difficult to sort out the effect of exogenous variables on hedonic prices from the effect on willingness to pay. The second problem, endogeneity, has to do with the fact that the consumer really chooses both the marginal price of the characteristic and the quantity of the characteristic at the same time. In Eq. (16.7), the left-hand side is a function of z and z is also on the right-hand side. For standard least-squares statistical methods to be unbiased, the right-hand side variables must be truly exogenous. Palmquist (1991) and Freeman (1993) provide nice discussions of these problems. There are solutions to these problems in some cases, as discussed by Bartik (1987), Epple (1987), McConnell and Phipps (1987), and Epple and Sieg (forthcoming), among others. A third problem is that if more than one market is involved, the slope of the price line is insufficient information to characterize consumer and producer choice (Kolstad and Turnovsky, 1998). Deacon et al. (1998) contend that no empirical paper has yet overcome all of these difficulties.
15. For instance, Thaler and Rosen (1975) conducted one of the first analyses of the demand for risk reduction by examining how wages change with the riskiness of the job; Pommerehne (1988) conducts a hedonic analysis of property values in Basle, Switzerland to determine the demand for noise from airplanes and road vehicles; Quigley (1982) examines the effects of the presence of sanitation services on housing prices in El Salvador to determine the benefit of a public housing project. Bayless (1982) conducts a hedonic analysis of the salaries of university professors to infer the demand for air quality improvements. Clark and Nieves (1994) do a hedonic analysis of both wages and property values affected by noxious facilities, such as petroleum refineries. Smith and Huang (1993) survey hedonic air pollution studies, particularly those done in the United States.
16. The hedonic price function they estimated involved the logarithm of housing value on the left-hand side and a sum of the other terms or logs of the other terms on the right-hand side. They were able to explain 81% of the variation in housing prices over the sample (the R^2 was 0.81). The data used for the estimation are available on the World Wide Web for those interested in replicating the results (see Gilley and Pace, 1996).
17. Viscusi (1993) provides a very readable and relatively up-to-date review of studies of the value of risks to life based on hedonic wage studies and other methods.

17 HOUSEHOLD PRODUCTION

We turn now to another form of revealed preference for environmental commodities. The approach we consider involves environmental commodities whose effect on a person can be modified using market goods. If we are speaking of an environmental good (e.g., a national park), the use of market goods allows the individual to enjoy the good fully. If we are speaking of a bad (e.g., noise), market goods can be used to insulate the individual from the full damage from the bad.

Examples are instructive. First consider the case of a bad. If you are living in a house near a busy autobahn, the noise from traffic is a nuisance. However, what you may really be interested in is peace and quiet inside your house; at a cost you may insulate your house from noise and thus neutralize, at least partially, the damage from the noise. Although we as researchers cannot directly observe the damage to you from the noise, we can observe your efforts to neutralize the noise. These expenditures tell us something about how damaging the noise is to you. The more strongly you feel about the noise, the more you will spend on keeping the noise away. This is what is known as a defensive expenditure—money is spent to defend against the environmental bad. By observing defensive expenditures, we learn how individuals value the bad.

Now consider a good, such as a national park. To enjoy the national park fully, people must visit the park, and this costs money (for most people). The more utility a person gets from a visit to the park, the more money the person is willing to spend in visiting the park. Without visitation, the park provides little utility. For instance, people travel thousands of kilometers and spend thousands of francs to visit the game parks of Kenya—evidence that these visitors value these parks highly. In this example, the expenditure to enjoy the park is observable and related to the benefit an individual obtains from the park. By observing travel costs, we learn how individuals value the park.

Both of the cases are examples of what is called "household production." A household (or individual) combines an environmental good or bad with market goods to produce an "experience" that directly provides utility. In the first case, the indoor noise level is the produced commodity; in the second case it is a park visit. Insight as to the value of the environmental good or bad comes from observing consumption of the associated market goods, even though such insight may tell only part of the story.

In the next section we consider the case of bads—defensive expenditures. In the subsequent section we consider the case of goods—travel costs.

I. DEFENSIVE EXPENDITURES

The problem of demand for environmental bads is our first concern. The example above of noise pollution is typical. What expenditures defend against highway noise? In a house, you can install extra windows, use thicker walls (in new construction), and construct a noise barrier or wall (this is common). The household suffering the noise pollution will continue to invest in defensive measures until the marginal cost of additional measures exceeds the marginal benefit from the reduction in noise. It is easy to see in this case that defensive expenditures are a lower bound to the damage from the noise. It is likely that the noise has not been totally neutralized by the defensive expenditures and thus there is additional unaccounted for damage.

Air pollution is another example. What can be done to defend against air pollution? One damage from air pollution is soiling of property. An obvious measure is to use better paint on a house or paint it more frequently. Another step related to the health effects of air pollution is to install air purifiers and/or air conditioners in your house. Furthermore, additional medical care may offset some of the deleterious health effects of air pollution. Although air pollution cannot be totally neutralized in its effects, it can be reduced by appropriate expenditures.

Water pollution is our last example. It is not necessarily polluted groundwater that is damaging but rather the water consumed by the individual. There are several ways of defending against such pollution, including using bottled water, purchasing water purifiers, and increasing the monitoring of the well that is the source of drinking water. Once again, by observing these defensive expenditures, we learn about the significance of the pollution to the individual.

In all of these cases an individual combines a quantity of a public bad with a quantity of a market good to produce what actually gives utility—noise in the immediate surroundings, air quality inside the home, or water quality, as ingested. These expenditures are sometimes called defensive expenditures (defending against pollution) or averting expenditures (expenditures to avert the damage from pollution). Whichever term is used, the meaning is the same. We will first consider a simple economic model of defensive expenditures and then examine an application to the demand for improved air quality.

A. A Simple Model[1]

Consider the case of an individual subject to noise pollution (P) from a nearby road (higher Ps are worse). The individual, however, is really only interested in the level of quiet within the house, Q (higher Qs are better). Fortunately, the homeowner may buy noise insulation and other equipment to reduce the level of noise within the house. We call these purchases *defensive expenditures* and denote by $D(Q,P)$ the expenditures necessary to achieve indoor noise quality levels Q when the outside noise is P. We assume that these defensive expenditures provide no additional services other than to produce indoor environmental quality. Noise insulation would fit into this framework but an air conditioning system that reduces noise levels while improving the indoor climate would not.

Figure 17.1 shows a typical shape for this defensive expenditure function. As the desired indoor level of quiet (Q) increases, for a given outside level of noise, the defensive expenditures necessary increase. Furthermore, the marginal defensive expenditures increase as Q increases. This reflects the fact that it becomes more and more difficult (costly) to squeeze higher and higher levels of quiet from existing technology. Also shown in Figure 17.1 are two defensive expenditure functions, one for outdoor noise level P_1 and one for a slightly higher level of outdoor noise pollution, P_2. Note that as outdoor noise pollution levels increase, the quantity of defensive expenditures necessary to keep indoor levels of noise the same increases, and more so for higher levels of Q.

The individual is interested in consuming two commodities, quiet (Q) and other conventional goods (X). Only these two commodities enter into the individual's utility function. Outdoor noise does not enter explicitly, which is clearly a restrictive assumption in some cases. Assume the price of X is 1.[2] The consumer's problem is to choose X and Q to maximize utility, subject to a budget constraint that includes defensive expenditures:

$$\max_{X,Q} U(X,Q) \tag{17.1a}$$

such that

$$X + D(Q,P) = Y \tag{17.1b}$$

Equation (17.1a) indicates that it is necessary to adjust X and Q so as to maximize utility. Equation (17.1b) states that expenditures on X and on defensive measures must equal income (Y). In Eq. (17.1b), the variable P is outside of the control of the individual—the amount of road noise cannot be adjusted.

This consumer choice problem is represented graphically in Figure 17.2. (Remember that P is fixed.) Shown in Figure 17.2 is the budget constraint [Eq. (17.1b)]. In contrast to what we are used to for a budget constraint, in this case it is nonlinear. This is because of the presence of $D(Q,P)$ in Eq. (17.1b). Also shown in Figure 17.2 are two indifference curves involving X and Q. As is usual, higher curves represent higher levels

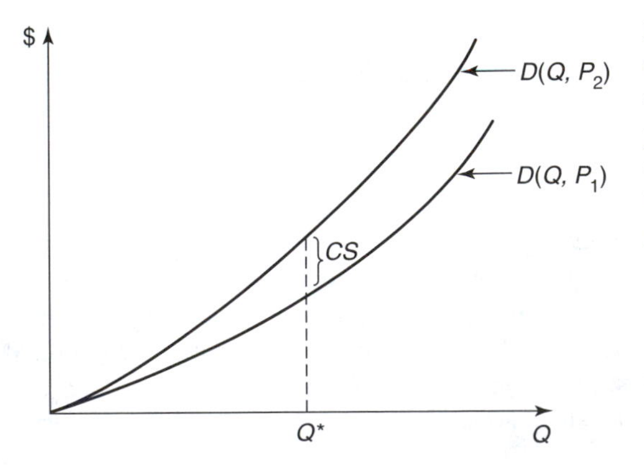

Figure 17.1 Defensive expenditures. Q, Quality of personal environment; P_1,P_2, outdoor pollution; CS, compensating surplus of change from P_1 to P_2 at Q^*; Q^*, optimal choice of personal environmental quality at P_1; $D(Q,P)$, defensive expenditures

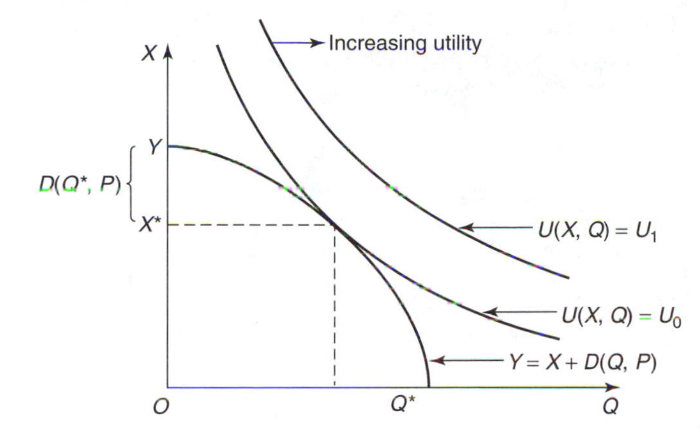

Figure 17.2 Consumer choice with defensive expenditures. $U(X,Q) = U_i$, Indifference curves; X, numeraire consumption good; Q, quality of personal environment; $Y = X + D(Q,P)$, budget constraint; $D(Q^*,P)$, defensive expenditures; Y, income; X^*,Q^*, optimal choice of X and Q.

of utility. The point of tangency between the budget constraint and an indifference curve represents the optimal consumer choice of (X^*,Q^*). Also shown in Figure 17.2 is the corresponding defensive expenditure, $D(Q^*,P)$, assuming that when $Q = 0$, defensive expenditures are zero.

Now suppose we increase the level of noise pollution slightly, say from P_1 to P_2. How does the observed change in defensive expenditures compare to the willingness to pay to avoid the pollution increase? If they are the same, observations of defensive expenditures will tell us something about the demand for noise pollution. If defensive expenditures are predictably higher or lower than willingness to pay, defensive expenditures may provide an upper or lower bound on willingness to pay.

It is simple to see what the marginal willingness to pay is for a slight change in pollution. Focusing on Eq. (17.1b), if P is increased slightly, defensive expenditures necessary to maintain Q will increase slightly. Consumers given enough income to pay for that increase in defensive expenditures will maintain the same level of utility. Using the terminology of Chapter 15, this is the compensating surplus (*CS*). Holding Q fixed, the amount D changes when P changes slightly is the marginal willingness to pay to compensate for the change in P. This is shown as *CS* in Figure 17.1 for Q^*.

But what is the total effect on observed defensive expenditures when P changes? Figure 17.3 shows how the budget constraint rotates downward when P increases. Furthermore, the increase in P results in a lower level of utility, shown as the indifference curve labeled U_2. The important thing to note is that the chosen Q drops as P increases. Q_1^* is chosen when $P = P_1$ and Q_2^* is chosen when $P = P_2$, with $Q_1^* > Q_2^*$. This is really quite intuitive. With an increase in outside noise, the cost of achieving a particular level of quiet indoors has increased. When something is more expensive, our normal response is to substitute other goods for that good. Thus with quiet a bit more expensive due to the increase in P, the consumer makes do with slightly less quiet in the house. This is what is shown in Figure 17.3. By cutting back on Q, the consumer is saving on defensive expenditures. This means that the observed defensive expenditures will be less than the marginal willingness to pay to avoid the pollution increase. This is important. This means that what we observe in the marketplace (changes in defensive expenditures) will understate the true marginal willingness to pay for pollution reductions.[3] Figure 17.3

is drawn so that the change in defensive expenditures is small. Recall from Figure 17.2 that the change in defensive expenditures is equal to the change in expenditures on X. In Figure 17.3 defensive expenditures could increase or decrease as pollution moves from P_1 to P_2.

This can also be seen algebraically. Let us start with indoor and outdoor noise levels Q and P. Suppose we change P slightly, by ΔP. Then let the consumer adjust the choice of Q and X. We would expect defensive expenditures to change by ΔD and indoor noise levels by ΔQ:

$$\Delta D = D(Q + \Delta Q, P + \Delta P) - D(Q,P) \quad (17.2a)$$

$$= D(Q + \Delta Q, P + \Delta P) - D(Q, P + \Delta P) + D(Q, P + \Delta P) - D(Q,P) \quad (17.2b)$$

$$= D_Q \, \Delta Q + D_P \, \Delta P \quad (17.2c)$$

where

$$D_Q = [D(Q + \Delta Q, P + \Delta P) - D(Q, P + \Delta P)]/\Delta Q \quad (17.2d)$$

and

$$D_P = [D(Q, P + \Delta P) - D(Q,P)]/\Delta P \quad (17.2e)$$

The interpretation of D_Q is the change in defensive expenditures associated with a slight change in indoor quiet levels, keeping outdoor noise levels constant. Similarly, the interpretation of D_P is the change in defensive expenditures from a slight change in P, keeping indoor noise levels constant.[4] As was argued earlier, D_P is precisely the marginal willingness to pay to avoid a change in P. Equation (17.2c) can be rewritten, dividing both sides by ΔP:

$$\Delta D/\Delta P = D_Q \, (\Delta Q/\Delta P) + D_P \quad (17.3)$$

This is the final result of our analysis. The last term on the right-hand side of Eq. (17.3) is the true marginal willingness to pay (compensating surplus) to avoid a small increase in P. The left-hand side of Eq. (17.3) is the observed change in defensive expenditures from a small change in P. Generally, we can show that for small increases in P,

$$\Delta D/\Delta P < D_P \quad (17.4)$$

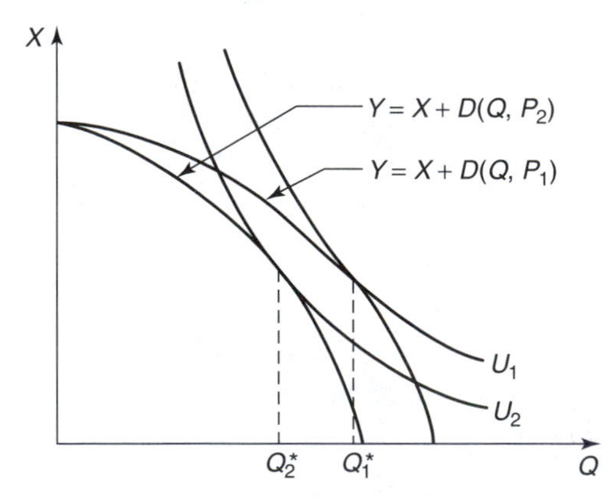

Figure 17.3 The effect of a change in pollution on defensive expenditures. U_1, U_2, Indifference curves; X, numeraire consumption good; Q, quality of personal environment; $Y = X + D(Q,P)$; budget constraint; P_1, P_2, two levels of outdoor pollution, $P_2 > P_1$; Q_1^*, Q_2^*, optimal choice of Q for P_1 and P_2.

This means that observed changes in defensive expenditures will understate the true willingness to pay to reduce pollution. Equation (17.4) holds because the first term on the right-hand side of Eq. (17.3) is negative. This is because D_Q is positive (by assumption—see Figure 17.1) and $\Delta Q/\Delta P$ is negative, as argued earlier (and as illustrated in Figure 17.3). An increase in P makes Q more expensive, which will result in less consumption of Q.

This result of course depends on the assumptions of our simple model. One significant extension of this model is to have the pollution level enter directly into the utility function. The most likely reason for this to occur is that it is not possible to defend against all of the deleterious effects of the pollution. For example, people live inside their house as well as outside, in their garden/yard.

B. An Example: Urban Ozone

There have been a number of empirical analyses of defensive expenditures, mostly in the health arena.[5] A study by Dickie and Gerking (1991) is a particularly interesting example of a defensive expenditures model being used to estimate the demand for ozone pollution in a major metropolitan area (Los Angeles, California). Ozone pollution in many cities is one of the most difficult air pollution problems to control; very significant resources are expended in trying to manage the ozone problem. Thus information on the benefits of reducing ozone levels is critical.

The study by Dickie and Gerking (1991) concerned two cities in the Los Angeles air basin, Burbank and Glendora. The distribution of daily peak ozone concentrations in these two cities in 1985 is shown in Figure 17.4. Until 1997 the U.S. standard for maxi-

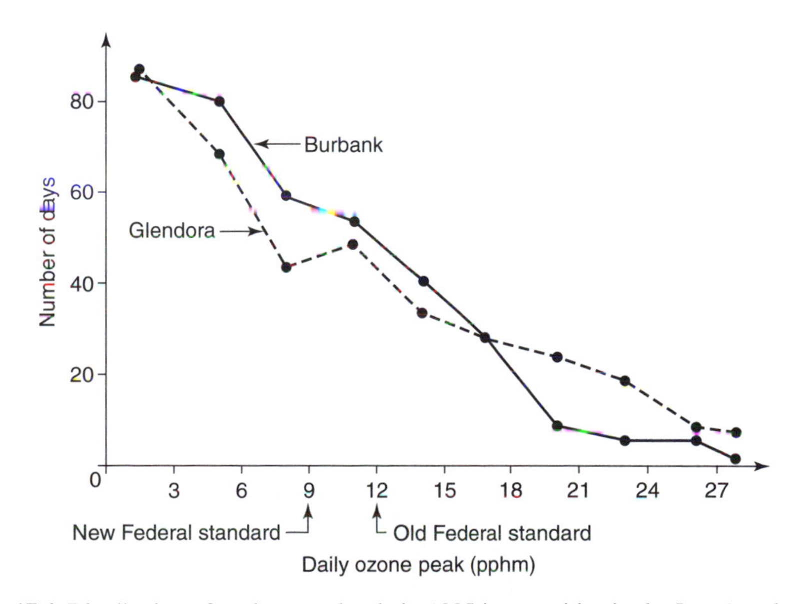

Figure 17.4 Distribution of peak ozone levels in 1985 in two cities in the Los Angeles air basin. Adapted from Dickie and Gerking (1991). Data are for intervals 3 pphm in width.

mum 1-hour ozone concentrations was 12 parts per hundred million (pphm); this is currently being revised downward to 9 pphm (which happens to be the existing California state standard). As can be seen from Figure 17.4, both these cities have a significant amount of air pollution, with many days during a year that exceed Federal standards. Glendora is the more polluted of the two cities. The authors ask the following policy question: What is the willingness to pay for an average resident of each city to reduce the maximum ozone level during a year to either 12 or 9 pphm?

The basis of their analysis is the willingness of people to purchase health care to compensate for the health-related damage from ozone pollution. They posit that when ozone causes coughing or other breathing difficulties, people spend significant amounts of money on medical care to alleviate these symptoms. By observing these expenditures, the authors hope to estimate the willingness to pay for reductions in ozone concentrations. This is a classic application of the defensive expenditures model of the previous section.

Start with a typical person with a utility function. Three variables enter the individual's utility function: health (H), air pollution (A), and market goods (X). Health is produced by the individual by combining medical care in the current period with air pollution. Health is also influenced by the individual's predisposition to sickness and the individual's previous health record and care. We can now say that defensive expenditures (M) will depend on health (H), air pollution (A), and the individual's own health record and experience (R). However, instead of writing an expression that gives defensive expenditures as a function of (H,A,R), we can write a health production function as

$$H = h(M,R,A) \tag{17.5}$$

which tells us how much health will result if M is spent on medical care when air pollution is A and the individual's predisposition to illness is R. This is simply another way of writing defensive expenditures.

The budget constraint for our individual will be a little unusual. We will assume the individual has a certain amount of time available to earn income (T). If our individual becomes ill, time and thus income is lost by either being in bed ill or by spending time visiting the doctor's office. It is important to recognize that visiting the doctor's office involves more than just out-of-pocket expenses. Furthermore, sickness causes a loss of time that could be better utilized in some other activity.[6] The consumer's problem is to maximize utility:

$$\max_{X,M} \; U(X,H,A) \tag{17.6a}$$

subject to

$$H = h(M,R,A) \tag{17.6b}$$

$$wT = q_X \, X + q_M \, M + w \, G(H) \tag{17.6c}$$

where w is the wage, T is the total time available, q_X is the price of X, q_M is the price of M (including time costs), and $G(H)$ is the lost time due to poor health. Thus the greater H (better health) is, the lower G will be. The consumer then maximizes Eq. (17.6a) in conjunction with Eq. (17.6b), subject to the budget constraint, Eq. (17.6c), yielding some maximum level of attainable utility that depends on prices, and the exogenous R and A:

$$v = V(w, q_X, q_M, R, A) \tag{17.7}$$

If we could observe utility in the population, we could statistically estimate Eq. (17.7); it would then be a relatively easy task to estimate the demand for A.[7]

Dickie and Gerking (1991) are creative in compensating for not observing utility; they use what is known as a random utility model. They note that what they observe is whether people visit the doctor. If a person visits the doctor, that is evidence that there must be a utility gain from visiting, as opposed to sitting at home and not visiting the doctor.

To see this, fix (w, q_X, q_M, R, A), let V_0 be the maximum attainable utility without any medical care ($M = 0$) and V_1 the maximum attainable utility with medical care ($M > 0$). Both V_1 and V_0 depend on prices, R and A. Although we do not observe utility in the population, we can observe whether people visit the doctor. If they visit the doctor, we can conclude that the utility of visiting the doctor is greater than the utility of not visiting the doctor ($V_1 - V_0 > 0$). Similarly, if they do not visit the doctor, we must conclude that utility with $M = 0$ is no lower than utility for some positive value of M ($V_1 - V_0 = 0$).

Although this might seem like relatively skimpy information, this framework is amenable to statistical estimation using what is know as the Probit model.[8] Since $V_1 - V_0$ is measured with error, the probability of visiting a doctor is equal to the probability that $V_1 - V_0 > 0$. This is why the model is known as a random utility model (RUM)—there is a random element to utility.

The authors convert Eq. (17.7) to one involving both V_1 and V_0:

$$V_1 - V_0 = W(w, q_X, q_M, R, A) \tag{17.8}$$

The authors specify a functional form for the right-hand side of Eq. (17.8) and couple it with the binary choice problem:

$$M > 0 \qquad \text{if } V_1 - V_0 > 0 \tag{17.9a}$$

$$M = 0 \qquad \text{if } V_1 - V_0 = 0 \tag{17.9b}$$

Whether M is zero or not (whether a doctor's visit occurs) is observed, even though $V_1 - V_0$ is not observed. This is sufficient to estimate Eq. (17.8) using the Probit.

Although we do not need to go through the details here, the authors of the study are able to estimate $V_1 - V_0$ and from that, to compute the demand for changes in A (air pollution). The researchers deal with a sample of 256 residents of the two cities mentioned earlier. They conducted detailed interviews of the respondents who were drawn from a prior study of respiratory disease.

After estimating the model, they are able to calculate (for an average individual) the total willingness to pay for a reduction in the maximum ozone concentration on any given day to the legal limit, either 9 or 12 pphm. They conclude that such an average individual would be willing to pay \$95–115 per year in Burbank to achieve the 12 pphm standard and \$171–205 to achieve the 9 pphm standard. The average citizen of Glendora is willing to pay more (not surprisingly since Glendora is more polluted): \$171–209 for the 12 pphm standard and \$261–314 for the 9 pphm standard. These are substantial figures, particularly when compared to figures for annual damage presented in other studies. For comparison, Dickie and Gerking (1991) cite two other studies that report total individual willingness to pay of less that \$4 per year per person.

The researchers also calculate the out-of-pocket medical expenses associated with elevated ozone in each of these two cities. Roughly speaking, the medical expenses are half the willingness-to-pay figures, suggesting that examining only the health costs of air pollution can significantly understate the damage.

II. TRAVEL COST

We now turn to the case of household production in which the consumer combines market goods with an environmental good (not a bad this time) to enhance enjoyment of the environmental good. In particular, we are referring to an environmental good that takes effort to enjoy, such as a national park. By observing the costly effort expended to enjoy the park, we can infer something about how the consumer values the park. When that effort consists of traveling to the park, what we are dealing with is the travel cost approach to valuing environmental goods. This method is most frequently applied to valuation of natural environments that people visit to appreciate. The most frequent application of travel cost to the demand for pollution is when pollution degrades a recreational experience, such as pollution affecting beach use. In this case, travel cost methods can be used to value the beach in its clean state and in its polluted state; the difference can be attributed to damage from the pollution.

The travel cost method is probably the oldest method of valuing environmental goods. Following World War II, the U.S. National Park Service solicited advice from a number of economists on methods for quantifying the value of specific park properties. Harold Hotelling responded in a two-page letter to the director of the Park Service, from which the following is excerpted:

> Let concentric zones be defined around each park so that the cost of travel to the park from all points in one of these zones is approximately constant. The persons entering the park in a year, or a suitable chosen sample of them, are to be listed according to the zone from which they come. The fact that they come means that the service of the park is at least worth the cost, and this cost can probably be estimated with fair accuracy. . . . The comparison of the cost of coming from a zone with the number of people who do come from it, together with a count of the population of the zone, enables us to plot one point for each zone on a demand curve for the service of the park. By a judicious process of fitting, it should be possible to get a good enough approximation to this demand curve to provide, through integration, a measure of the consumers' surplus resulting from the availability of the park.[9]

This seemingly straightforward prescription has generated a half century of research and the method is still far from perfected, though quite useful.

There are three major dimensions to travel cost analysis of the demand for an environmental good. One concerns how demand depends on quality of the good (for instance, water quality at a beach). A second is associated with the number and duration of trips during a period of time such as a year. A third concerns the treatment of substitute sites, such as when a visitor to a national park faces the choice of several parks.

We will start with a simple model of travel cost, one in which there is a single environmental good of varying quality. We will then extend this to the case of multiple goods. Finally, we will turn to issues associated with implementing the travel cost method.

A. A Simple Model of a Single Site

We start with a single consumer and a single environmental good, say a park. The park has a level of quality, q, that may be associated with congestion or air quality (or anything else that affects the quality of a visit to the park). Higher qs are better. Our consumer chooses two things—visits to the park (v) and a basket of market goods (x). We are interested in visits over the course of a fixed period of time, say a year. Assume that units in which x is measured are such that the price of x is unity. Let p_0 be the out-of-pocket expenses associated with a single trip to the park—automobile, train, or plane expenses, food, and admission charges. Suppose our consumer works for L hours at a wage w to earn a certain income. We are now in a position to write the consumer's utility maximization problem:

$$\max_{x,v} \; U(x,v,q) \tag{17.10a}$$

such that

$$wL = x + p_0 \, v \tag{17.10b}$$

The only problem here is that out-of-pocket expenses are not the only cost of visiting the park. Consumers must take time to travel to and to visit the park. And this time could be devoted to work in order to increase their income. They have chosen not to work, but to visit the park. But should we totally ignore their time commitments to the trip?

To answer this question, suppose our consumer has T hours of time available to devote to park visits and work. Whatever portion of T is devoted to work, the rest will be devoted to park visits, and vice versa. Denote by t_t and t_v, respectively, the travel time associated with a single round trip visit to the park and the on-site time associated with a single visit. The consumer then faces a time–budget constraint that must be appended to the utility maximization problem in Eqs. (17.10a) and (17.10b):

$$T = L + (t_t + t_v) \, v \tag{17.10c}$$

Equation (17.10c) can be substituted into Eq. (17.10b) to eliminate L and thus reduce the maximization problem to

$$\max_{x,v} \; U(x,v,q) \tag{17.11a}$$

such that

$$\begin{aligned} wT &= x + [p_0 + w\,(t_t + t_v)]\, v \\ &\equiv x + p_v \, v \end{aligned} \tag{17.11b}$$

where

$$p_v = p_0 + w(t_t + t_v) \tag{17.11c}$$

Note that Eq. (17.11) is a conventional utility maximization problem except that the price of a visit equals the out-of-pocket expenses plus the value of time devoted to the trip, with time valued at the wage rate. This is as expected. Since time spent traveling could alternatively be earning a wage, the opportunity cost of an hour of travel time is the wage, w.

As we will see later, this model is about as simple as possible. It is logical that consumers would rather be traveling than working on the assembly line. Thus they may not view leisure time as totally interchangeable with labor time. Further, they may not be able to earn additional money from wages, if their work hours per week are set by contract. How people value leisure time relative to their wage is an empirical question to which we will return later.

One can solve the maximization problem specified in Eq. (17.11) for a particular consumer. The result will be a demand function for visits to the park:

$$v = f(p_v, q, y) \tag{17.12}$$

where y is income (wT). Now let us focus on the quality variable, q. Is it possible, starting with Eq. (17.12), to measure the willingness to pay for a small change in q? The answer is yes. In fact, this is exactly the problem we considered in Chapter 15 in the context of restricted demand.

Figure 17.5 shows typical demand curves for park visits: the vertical axis is the full price of a park visit and the horizontal axis is number of visits (per year). Shown in Figure 17.5 are two demand curves, one for quality level q_1 and one for a slightly higher level of quality, $q_1 + \Delta q$. Clearly consumers like this quality increase since they are interested in consuming more park visits for higher qs. At a price of a park visit of p^*, the consumer demands v_1 visits at the lower quality level and $v_2 > v_1$ visits at the higher quality level. The consumer's willingness to pay for this increase in q is the increase in surplus associated with the quality increase—the shaded area ABC in Figure 17.5. If Δq is very small, the shaded area will be small; the ratio of the area of the shaded region to the change in q, Δq, is the marginal willingness to pay for increases in q. If this exercise is repeated for a variety of quality levels, the marginal willingness-to-pay function for quality will be generated.[10]

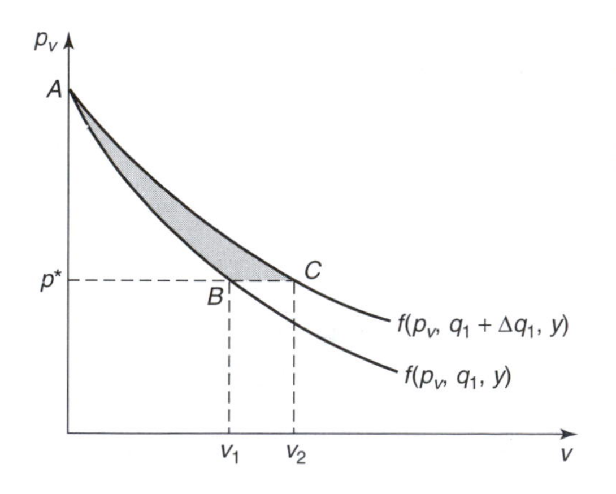

Figure 17.5 Surplus from a change in quality. v, Number of visits to the park; q, quality of park visit; p_v, price of a visit; p^*, actual price of park visit; $f(p_v,q,y)$, inverse demand for park visits; y, income; ABC, surplus gain from quality increase from q_1 to $q_1 + \Delta q_1$.

B. Multiple Sites

Now we turn to the case in which the consumer is choosing among multiple sites. For example, suppose our consumer wants to take a vacation at one of the game parks in Africa, but can visit only one. The consumer will examine the cost of visiting all of the parks and make a choice. How can we represent this very real process in our travel cost model?

This problem of taking into account the choices a consumer may have has been a significant subject of research in the travel cost literature over the last three decades. At the simplest level, we can modify Eq. (17.12) to include the price of substitute sites. Thus if there are three sites, A, B, and C, the demand for any one site (e.g., site A) will be a function of the prices of visiting the other sites as well as the quality of the other sites:

$$v_A = f_A(p_A, p_B, p_C, q_A, q_B, q_C, y) \quad (17.13)$$

Conceptually, this representation of the role of substitutes is straightforward. The problem is in empirically implementing a model such as Eq. (17.13): with many substitute sites the formulation becomes unwieldy. For this reason, there have been a number of approaches to simplify what amounts to the site choice problem for consumers. Freeman (1993) discusses a number of these approaches. We will focus on one, called the random utility model or RUM for short. The ozone model discussed in the section on defensive expenditures was a random utility model. A similar approach is even more common in travel cost analysis.

Suppose we have a consumer and a set of sites available for the consumer to visit, $i = 1, \ldots, I$. Each site is described by a set of attributes, which we will call quality, q_i, and an access price, p_i—the travel cost. The consumer is also consuming unrelated market goods, x, which we assume have a price of 1. Visiting site i is hypothesized to give a certain amount of utility, which depends on the price of visiting i (not the price of substitute sites) and the quality of site i:

$$u_i = f(\beta, p_i, q_i, y) + \epsilon_i \quad (17.14)$$

where β is a set of parameters to be statistically estimated and ϵ_i is an error term that represents factors that are unknown to the observing economist. This is where the name of the approach comes from—the expression for utility in Eq. (17.14) has a random component, ϵ_i.

If we could observe utility, we could statistically determine the value of β in Eq. (17.14). But we do not observe utility; we never observe utility. What we do observe is consumer choice. And when the consumer chooses site i over site j, that must mean that the utility from i is higher than the utility from j:

$$\text{Site } i \text{ chosen} \Rightarrow u_i \geq u_j, \qquad \text{for all } j \quad (17.15)$$

Different values of β will yield different values of u_i and u_j. Since, in fact, u_i and u_j are random variables [from Eq. 17.14)], different values of β will yield different probabilities of the event ($u_i \geq u_j$, for all j). One value of β may make that event highly likely, whereas another value of β might make that event highly unlikely. Statistical methods can be used to choose a β that makes the observed consumer choices most likely.[11]

Having statistically estimated β in Eq. (17.14), we can compute the demand for trips to site i as a function of quality of the site and the price of a visit. We can then examine how this demand changes when the quality of the site changes. The details of this calculation need not concern us here.[12]

C. Implementation

Real problems always arise when we attempt to use theoretical models for empirical analysis. The travel cost model is no exception. There are two basic approaches to estimating the demand for a site based on travel costs, other than the random utility model discussed above. One method is basically that suggested by Hotelling and is called the zonal travel cost model (ZTC), because of its use of geographic zones around the site as the basic units of observation. The second approach is more data intensive, and more recent, and is called the individual travel cost model (ITC), because of its reliance on observations of the behavior of individuals.

1. The Zonal Travel Cost Model. The zonal travel cost model follows directly from the original suggestion of Hotelling, quoted earlier in this chapter. However, the method was first applied and developed in detail by Marion Clawson and colleagues at Resources for the Future in the late 1950s and 1960s (Clawson and Knetsch, 1966). The only data that need be collected are a sample of visitors to the site in question, identifying the origin of those visitors' visits. Having identified the origin of visitors, it is possible to estimate the number of visitors per year from each origin "zone." Knowing the population of each origin zone gives a visitation rate for the zone (e.g., 1 person for every 1000 population). This visitation rate is explained by two things: the travel cost from the origin zone to the site (the "price" of visiting the site) and the demographic/income characteristics of the population of the origin zone. The researcher thereby generates a set of data, one data point for each zone, indicating the visitation rate, travel cost, and characteristics of the zone.

The next step is to statistically determine how the visitation rate is affected by travel costs and zonal characteristics. Having done that it is possible to estimate how visitation will change if travel costs change, for instance, by applying an admission fee for the site. Choosing several admission fees and the resulting total attendance at the site generates points on a demand curve for the site.

To be more precise about this, we start with a given site, call it a park, and assume the distribution of people and alternative sites is fixed in the geographic region from which the park draws visitors. Divide the region around the park into Z zones, with travel time roughly constant from any point within a zone to the park. Index these zones by $z = 1, \ldots, Z$. Let the population of each zone be P_z and the average income in each zone, y_z. Let w_z be other demographic characteristics of the zone, such as the fraction of young people in the population (anything that might influence making a trip to the park). Compute the travel cost for visiting the park for each zone, excluding any admission fee to the park, and call that travel cost π_z. Let the admission fee be f (which is the same for all zones). Of course, f may be zero. Finally, take a survey of park visitors during a year to estimate the total number of visitors in a year, S, and the number of visitors from each zone, s_z. Define the visitation rate as $v_z = s_z/P_z$.

The next step is to statistically estimate a visitation equation:

$$v_z = g(\pi_z + f, y_z, w_z) \tag{17.16}$$

This equation explains the visitation rate (akin to the probability that a randomly selected resident of the zone will visit the park) on the basis of the price of visiting the park ($\pi_z + f$), income (y_z), and other characteristics of the typical zone resident (w_z). Demand for park visits (q) can then be written in terms of the function g:

$$q\,(f) = \Sigma_z\, P_z\, g(\pi_z + f, y_z, w_z) \tag{17.17}$$

Although in statistically estimating g, the admission fee (f) was at its actual value, we can now vary f and see how $q(f)$ changes, using Eq. (17.17). This is the aggregate demand curve for the park.

There are many applications of this method. One is particularly interesting, though not fully documented. The case is the Amoco Cadiz oil spill off the coast of France in 1978. The oil tanker Amoco Cadiz lost its steering entering the English Channel and crashed on rocks off the coast of Brittany, spilling over 200 thousand tonnes of crude oil. The damage from the oil spill in terms of lost recreation was the subject of a travel cost analysis (Brown et al., 1983; Brown, undated).

The basic idea behind their analysis is to estimate the demand for Brittany beaches for two different quality levels, one clean and one spoiled by oil. To accomplish this, visit rates (visits per 1000 population) were calculated for Brittany from 21 different regions of France in 1978 (the spill year) and 1979 (after cleanup). Two variables were posited to influence the visit rate: the travel cost to Brittany and the travel cost to the nearest alternate coastal area. The data on visitation rates were obtained from a survey of Brittany visitors (identifying visitor origin) in each of the 2 years, conducted by the French Institut National de la Statistique et des Études Économiques. The basic equation estimated in each of the 2 years was

$$v_i = \alpha_0 + \alpha_1\, p_i + \alpha_2\, a_i \tag{17.18}$$

where v_i is the visitation rate for zone i, α_0, α_1, and α_2 are parameters to be statistically estimated, p_i is the travel cost from zone i, and a_i is the travel cost to the nearest substitute coastal area, for zone i. Equation (17.18) is then estimated for each of the two years using 21 data points, one for each zone. Applying this visitation rate to the population of all the zones generates a demand curve for Brittany beach visits.

Although the authors report some problems in interpreting the results of their estimated Eq. (17.18), they are able to calculate the change in consumer surplus between the 2 years (the shaded area in Figure 17.5). They compute a loss due to the oil spill of approximately 7 million francs, which is about 2% of the total surplus from beach visits in 1979.

2. The Individual Travel Cost Model. One of the strengths of the zonal travel cost model is that it is not data intensive. One of its shortcomings is assuming all residents of a zone are the same (since only average zonal characteristics are used). An alternative, is to collect more information on individual visitors (and nonvisitors) and use those data in estimating a model of demand instead of zonal averages. This is now a very common approach to implementing the travel cost method; however, there are two primary problems with this. One is that data are expensive to collect, and the more data that are collected from an individual, the more expensive it is.[13]

Another problem arises if the data used are generated by actual park visitors. The problem is bias introduced by the fact that the sample is self-selected, not a random sample of all possible visitors.[14] Although it is possible in principle to correct for this self-selection, the problem does not arise if the entire population of potential visitors is sampled. Unfortunately, it is generally too expensive to sample the entire population surrounding a park; a large sample size would be needed because most people will probably not visit the park.[15] Large samples are more expensive than small samples.

3. Computing Travel Costs. There are several problems associated with implementing the travel cost method. The most significant problem is in estimating the value of time. Although the simple model presented above suggests that the opportunity cost of travel time is the wage rate, that is generally not considered to be the case in actuality. Most people consider the opportunity cost of travel time to be substantially less than their wage rate. This may be because they like traveling more than working. Or it may be due to the fact that they receive a fixed wage and cannot adjust their hours worked, at the margin.

There have been many studies of the value of time spent in traveling. The primary use for such information is not for travel cost models of demand but rather for planning urban transportation systems. The typical approach to determining the value of travel time is to observe how individuals trade off their time with other expenses. For instance, if people willingly choose to save $1 by taking a bus rather than a train from point A to point B, thereby lengthening their journey by 15 minutes, we may conclude that they value their travel time at less than $4 per hour. The general consensus of the urban transportation economics literature appears to be that people value their travel time somewhere in the region of 20–50% of their gross wage rate (Bruzelius, 1979; Small, 1992).

Another problem with computing travel costs arises particularly in the case of automobiles. Is it appropriate to simply use the extra costs associated with the trip (which would typically exclude some depreciation, insurance, and taxes) or should average costs of operating the vehicle be used? How should one treat the fact that a person may have purchased an expensive four wheel drive vehicle or motor home specifically to visit a park or parks? Perhaps someone has chosen his or her home location so as to be closer to recreation areas, even though other costs, including the costs of commuting to work, may then be higher.[16] There are many other tough problems that confront the researcher when trying to implement the travel cost approach.

SUMMARY

1. The household production model involves a consumer combining a market good with a nonmarket good or bad to produce a third "synthetic" good that actually enters the consumer's utility function

2. In the case of combining market goods with nonmarket bads, the household production model is usually referred to as a defensive or averting expenditure model. The idea is that the consumer spends money to ameliorate the damaging effects of the bad.

By observing defensive expenditures, we obtain information on the consumer's willingness to pay to reduce the level of the bad. In fact, observed defensive expenditures are a lower bound on the willingness to pay to avoid the bad.

3. The most common case of combining market goods with a nonmarket good is recreation in which travel costs are combined with a distant recreation location to produce a visit to the recreation site. Although this method can be used to value the use of a recreation site, it can also be used to infer the value of improvements in the environmental quality in the vicinity of a recreation site.

4. There are three basic versions of the travel cost model: the zonal travel cost model, the individual travel cost model, and the random utility model.

5. Data requirements are modest for the zonal travel cost model. Based on a survey of visitors to a recreation area, visitation rates are determined for different zones around the area. These estimates are combined with the population distribution to generate a demand function for visits to the recreation area.

6. The individual travel cost model takes an approach similar to the zonal travel cost model except that information on individual characteristics and behavior is used. This is more data intensive but can be more reliable.

7. The random utility model is used most when individuals are making choices among competing recreation areas. It involves inferring the demand for a recreation area based on revealed choice of one area over another.

PROBLEMS

1. In the travel cost analysis of the losses from beach use in Brittany due to the Amoco Cadiz oil spill (sectin II.C.1), can you think of any improvements that might improve the travel cost analysis? Avoid the necessity of collecting additional primary data.

2. Consider a variant on the defensive expenditure model presented in this chapter. Suppose air pollution is the nonmarket bad and soap is the market good that can reduce the effects of air pollution—we bathe more frequently and wash our possessions more frequently. Our personal hygiene (H) is determined by the pollution level (P) and soap used (S). Suppose the price of soap is p_0. Set up the consumer's utility maximization problem for choosing the right amount of soap, other goods (x), and perhaps H and P. You do not need to solve the problem.

3. Consider Arturo, an individual with utility for a composite good (X) and indoor air quality (Q): $U = XQ$. Indoor air quality depends on pollution levels outside (P) and defensive expenditures (D): $D = PQ^2$. Let Arturo's income be Y.

   a. Write Arturo's problem of utility maximization subject to a budget constraint, mathematically. Draw Arturo's budget constraint and several indifference curves.

   b. Suppose $Y = 10$ and $P = 1$. What are the optimal amounts of X and Q for Arturo?

   c. What happens to the optimal X and Q (show your work) when (i) P increases to 2 and (ii) Y increases to 20?

   d. It is observed that low income individuals are exposed to more pollution. Is your answer to part (c) consistent with this?

4. We have discussed cases in which components of demand or willingness to pay for environmental goods and bads can be observed through their interaction with market goods. For

each of the following public bads, identify possible related and observable market goods, and comment on aspects of demand that are not closely related to any market behavior. For example, if an oil spill contaminates a coastal region for a period of time, lost beach visits are potentially measurable, whereas disutility from effects of the oil spill on marine life may not be inferable by observing consumption of any market good.

a. Degradation of wild and scenic areas of Kakadu National Park due to mining activity.

b. Global climate change.

c. Urban air pollution.

d. Stratospheric ozone depletion.

e. Groundwater contamination.

5. The text suggests that substitutability can affect demand for public goods and bads in the same way that it does for private goods and bads. To be more concrete, consider trying to estimate damages from an oil spill that caused the closure of several beaches along a coastal area. Why is it important to consider whether other local beaches were left uncontaminated and therefore did not close?

Individuals may also have the opportunity to substitute across *time* as well as *space* in their consumption of environmental goods. What do we mean by substitution across time and why is it important in estimating the damage in a case such as this one?

6. Consider the demand for park visitation, Eq. (17.12).

a. We discussed (and illustrated in Figure 17.5) the effect of a change in p_v on demand for visits, v. How would you expect the other variables, q and y, to effect v?

b. Given your answer to part (a), can we say what will happen to demand for park visits when the wage rate, w, increases? Keep in mind that y depends on w.

7. You have been commissioned to estimate the demand curve for admission to EuroLand, an amusement park in France. To do this, you spend a day surveying visitors to the park. Before doing this, you divide the area around the park into 10 zones, with the distance from the park approximately constant within a zone. You ask each person you interview where they come from. Based on that information, and figures on annual attendance at the park, you are able to calculate the annual number of visitors from each zone. Your data are shown below:

Zone	Distance from park (km)	Zonal population	Number of visitors
1	10	5,000	500
2	20	10,000	900
3	30	25,000	2000
4	40	10,000	700
5	50	100,000	6000
6	60	500,000	25,000
7	70	200,000	8,000
8	80	50,000	1500
9	90	100,000	2000
10	100	100,000	1000

You note that admission to EuroLand is 150 francs per person. You also calculate that transportation costs, including time costs, are approximately 2 francs per person per kilometer in each direction. Ignore the value of time in the park.

Calculate the demand curve for visits to EuroLand, showing visits as a function of the entry price. How many fewer visitors would you expect there to be if the management raised the admission fee to 200 francs?

***8.** This problem shows how we can develop lower and upper bounds on the value of a change in pollution, based on observed changes in defensive expenditures, assuming that pollution does not enter directly into the utility function, only the produced level of environmental quality. Consider the same model as discussed in the chapter: a consumer is subject to noise pollution P and chooses an indoor level of quiet, Q [incurring defensive expenditures $D(Q, P)$], and consumption of other goods, X, priced at 1. Suppose noise pollution levels decrease from P_0 to P_1 and as a result the consumer's choice of indoor environmental quality increases from Q_0 to Q_1 and utility increases from U_0 to U_1. The price of X will remain unchanged at 1; thus suppress the price of X and define the restricted expenditure functions $E(P,U)$ and $E(P,U;Q)$ as the expenditures necessary to achieve utility level U, when outdoor noise is P and, in the case of $E(P,U;Q)$, when indoor environmental quality is held at Q. Similarly, define restricted compensated demand for X, $X(P,U)$ and $X(P,U;Q)$.

a. Show that the observed change in defensive expenditures is less than or equal to $D(Q_1,P_0) - D(Q_1,P_1)$.

b. Show that $D(Q_0,P_0) - D(Q_0,P_1) = E(P_0,U_0) - E(P_1,U_0;Q_0)$

c. Show that the compensating surplus from moving from P_0 to P_1 is greater than or equal to $D(Q_0,P_0) - D(Q_0,P_1)$.

d. Show that the equivalent surplus from moving from P_0 to P_1 is less than or equal to $D(Q_1,P_0) - D(Q_1,P_1)$.

e. If the compensating surplus from a reduction in pollution is less than the equivalent surplus (as is typical), what result do you obtain about the lower and upper bounds on the welfare implications of the pollution change and its relation to observed changes in defensive expenditures?

NOTES

1. This model presented in this section is adapted from Bartik (1988) and Courant and Porter (1981). Some simplification has been necessary here; the interested reader is directed to either of these papers for a more rigorous development of the concept of defensive expenditures.
2. Since the units with which we measure X are arbitrary, there is no loss in generality in assuming the price is 1.
3. This result does not apply in all cases. Courant and Porter (1981), for instance, adopt very specific assumptions that reverse this conclusion. However, the result presented here generally applies.
4. Most readers will recognize that D_P and D_Q are simply partial derivatives of D with respect to P and Q, respectively. Clearly $D_Q(Q,P + \Delta P) = D_Q(Q,P)$ for ΔP infinitesimal.
5. Gerking and Stanley (1986) examine damage from air pollution in St. Louis (United States) by studying health care expenditures. See also Harford (1984) and Shibata and Winrich (1983).
6. If the authors of this study explicitly included leisure in their model, the choice among work, leisure, illness, and doctor's visits could be made more explicit. That would be a much more complicated model, however.

7. Estimating demand for A from the indirect utility function [Eq. (17.7)] is beyond the scope of this book. See Kolstad and Braden (1991) for a detailed though somewhat advanced treatment of this issue.
8. The Probit model is described in any good econometrics book (e.g., Kmenta, 1986). The idea is that we are interested in the relationship between some unobserved variable, y, and a number of exogenous variables, x: $y = x\beta + \epsilon$, where β is a vector of parameters and ϵ is an error term. We are interested in statistically estimating β. This is complicated by the fact that we do not observe y. But we do observe another variable, z, that takes on the value of 0 or 1, depending on the value of y. For instance, if $y > 0$, then $z = 1$; otherwise, $z = 0$. If we observe z, then the methods associated with the Probit allow us to estimate β.
9. Letter from Harold Hotelling to Director of National Park Service, dated June 18, 1947, reproduced in National Park Service (1949).
10. Figure 17.5 is comparable to Figure 15.9 except that in Figure 17.5 we are dealing with an ordinary demand curve, not a compensated demand curve.
11. Hausman et al. (1995) discuss the implementation of this in the context of a travel cost model of recreation demand. Kolstad and Braden (1991) discuss the general random utility model.
12. Refer to Bockstael et al. (1991) for an advanced treatment of this issue. Bockstael et al. (1987, 1989), Binkley and Hanemann (1978) apply this method to measuring the benefits of water quality improvement at beaches in Maryland and Massachusetts, respectively.
13. Some data are very difficult to collect on recreation surveys. Income is an example. Believe it or not, many people are unaware of their total income. Other people are part of a household; whose income should be used? Other people resent being asked about their income.
14. Bockstael et al. (1991) discuss this sample selection problem and some solutions in the context of travel cost models.
15. One prominent exception to this is the study done for the Exxon Corporation in support of their defense in the 1989 Exxon Valdez oil spill in Alaska (Hausman et al., 1995). In that study, a sample of the entire population of Alaska was used to estimate a travel cost model of recreation.
16. Randall (1994) discusses some of these problems and concludes that they present insurmountable obstacles to obtaining reliable information from the travel cost approach.

18 CONSTRUCTED MARKETS

The last two chapters were concerned with indirect, revealed preference methods for determining the demand for environmental quality. As meritorious as these methods are, they can be used to value only a small fraction of environmental goods. The demand for many environmental goods cannot be completely estimated using revealed preference methods. Existence value (see Chapter 15) is one example of a value that is very difficult to estimate with revealed preference methods.[1] If one values something (e.g., an arctic wilderness) for its mere existence, unconnected with the possible use, the consumer's behavior with regard to market goods will be unaffected by whether that something is available. If behavior is unaffected, revealed preference methods are difficult to apply. Additionally, there are many goods for which there is no logical market through which value can be reflected. The value of a scenic vista or climate change is difficult to completely determine using revealed preference methods.

It is for this reason that constructed market valuation tools have been developed. Richard Carson of the University of California uses the word "constructed market" to refer to the broad classes of valuation techniques that involve the construction of a market to be used to generate the value or demand for an environmental good. There are two basic types of constructed markets—hypothetical and experimental. Valuation through hypothetical constructed markets goes by assorted names such as "stated preference," "hypothetical valuation," or most commonly, "contingent valuation." Consumers are directly asked what they would pay for an environmental good, *if there were a market* (hence the word "contingent"). The second type of constructed market is an experimental market. In an experimental market, the researcher constructs all of the characteristics of a market, including trading money for a good. The researcher then observes behavior. Experimental markets can be in the context of a laboratory, perhaps with students as the subjects, in which case we call them laboratory experimental markets. Alternatively, experiments can involve the general population in a realistic setting; we would call these field experiments. The only requirement for an experimental market is that real money change hands in exchange for a real good or bad. In other words, it is a real market, except that it has been contrived by the experimenter. The ultimate field experiment is an official referendum in which the "experiment" is officially constructed and voters are asked to commit public money to provide some sort of environmental good.

I. STATED PREFERENCES

Stated preference methods of valuation involve finding an individual's willingness to pay for a good by posing a set of questions regarding preferences directly to the individual. Other methods of valuation may also involve surveys (such as the travel-cost approach of asking people who visit a park where they started their trip), but such surveys simply ask for recall of factual information. In a contingent valuation (CV) question, the individual is asked to imagine some situation that is typically outside the individual's experience and speculate on how he or she would act in such a situation.

For example, in a study of the value of improving air quality in Los Angeles, Brookshire et al. (1982) were interested in how much people would be willing to pay to increase the air quality in their community. The researchers generated photographs depicting three different levels of air quality—poor, fair, and good, defined in terms of visibility. Furthermore, the researchers generated a sample of 290 people scattered over nine communities with varying air quality. People in a community with poor air quality were shown the photos and asked about their willingness to pay to move to fair air quality and people with fair air quality were asked their willingness to pay to move to good air quality. The results ranged from $5.55 to $28.18 per month, amounts respondents were willing to pay for an increase in air quality.

Stated preference methods appear to date back to the early 1960s when Davis (1963) conducted a study of the value to recreators of the Maine (United States) woods. However, it was not for another decade that researchers picked up on the idea and began applying it widely to environmental valuation (Bohm, 1972; Randall et al., 1974). Since that time there have been thousands of contingent valuation studies of environmental goods, though most remain unpublished.[2]

During the 1980s and 1990s contingent valuation became very popular for valuing environmental goods. This was due to a number of factors, including the early 1980s movement in the United States toward conducting cost–benefit analyses of most environmental regulations, and the rise of what is known in the United States as "natural resource damage assessment." Several laws in the United States permitted the government to sue for damage to the environment from things such as hazardous waste accidents or oil spills (Portney, 1994; Kopp and Smith, 1989). To sue, it is necessary to compute the monetary equivalent of the damage. For many accidents, there is very little that can be done to place a value on damages, other than use contingent valuation.

CV has also found increased application in supporting government regulatory actions or provision of public goods. In Australia, the government debated whether to preserve or allow mining in an area known as the Kakadu Conservation Zone. In that debate, CV studies played a major role in steering public policy toward preservation (Carson et al., 1994).

Although it is quite easy to assemble a questionnaire, stand on a street corner, and administer the questionnaire to whomever walks by, it is quite another thing to conduct a careful, defensible CV study. There is a fairly well-developed protocol that is generally accepted by practitioners of CV. We will go through the details of the method, but first an example will serve to illustrate the approach.

A. An Application of Contingent Valuation: Drinking Water in Seoul

Korea has undergone dramatic economic development over the past few decades. But industrialization brings negative environmental side effects. In particular, several chemical spills into the Seoul water supply in 1991 caused great concern about the quality of the drinking water in that city of ten million inhabitants. Any water supply can be made safer, including that of Seoul, by, for instance, increasing the monitoring of the water supply and increasing storage to be used when contamination occurs. Of course, these improvements cost money. To ascertain the value of reducing the risk of drinking water contamination, Kwak and Russell (1994) undertook a CV study of households in Seoul. They considered revealed preference methods, but rejected them. There is no water quality variation across neighborhoods, so an hedonic method seemed inappropriate. There are defensive activities such as traveling to the mountains to obtain spring water. They concluded that the trip to the mountains involved more than simply picking up water; thus a household production analysis would be inappropriate. That left contingent valuation.

The first problem they faced was defining the "good" they would seek to value. They decided that the good would be making the drinking water safe enough that future chemical contamination would be very unlikely—defined as "the probability that you will be the next President of Korea." The next problem was in determining whom to ask. Obviously, they could not ask all ten million residents of Seoul, so they decided to sample. In selecting a sample it is important that it be random.[3] To randomize their sample of approximately 300 households, they scattered their sample geographically over the city. They then had to decide whom to question within the household.

The researchers constructed a questionnaire to use in interviews (Table 18.1). The questionnaire was designed to give background information on the problem, describe the "good" in question, elicit willingness to pay, and collect background information on the respondent, such as age, income, and number of people in the household. Asking about the willingness to pay for reducing the risk in the water supply is never easy. The researchers chose to use what is called a "payment card," which is a card with a number of figures, spanning the range of responses that might be expected. Each card has payment amounts along with several reference expenditure amounts. Thus a different card is used for individuals with different incomes. Figure 18.1 shows a card used in the Seoul study.

Finalizing the questionnaire to use was a multistep process. After developing a first draft with the advice of survey experts, the researchers pretested it on two individuals. Then they tested the questionnaire further by setting up "focus groups," groups of citizens who were asked to take the survey and then discuss it. After each step, the questionnaire was revised and refined. Finally, the survey was ready to be administered; they used a professional survey firm to contact the several hundred individuals in their sample.

Table 18.1 Excerpts from Seoul Questionnaire[a]

Section A — Introduction

Interviewer: Introduce yourself and explain the purpose of the survey. Indicate that the interview should take less than a half hour.

Section B — Background Attitudes and Information [6 questions]

1. What is your opinion about current tap water quality in Seoul?
   (a) very good; (b) good; (c) average; (d) bad; (e) very bad.
2. In the last five years, have you or your household taken any of the following actions to improve your drinking water?
   (a) installed water filter; (b) purchased bottled water; (c) boiled tap water regularly; (d) gone to a spring to obtain spring water.
3. What is the monthly combined income of your household after taxes, last year? [If refuses to answer, show card with ranges of income and ask to identify range of income]

Section C — Value of Water Quality Improvement

Interviewer: Two minute verbal description of major pollution accident in 1991 that contaminated the drinking water of Seoul. Because the consequences were not known immediately, many people drank contaminated water. [Visual props are used to describe the accident and its consequences.]

1. If the government takes no action, how many times do you think you might experience an accident similar to the Nak-Dong phenol accident in the next five years?

Interviewer: Give one minute verbal description of continuous monitoring system that would prevent introduction of pollutants into the city water supply. Indicate that with the system, the chance of drinking polluted water will be similar to the chance "that you will be a president of Korea." Chart is presented showing how taxes are currently apportioned among various uses (such as education, roads, sanitary services, and defense). Payment card for household's income is presented.

2. What is the most your household would pay in increased monthly taxes to reach the goal attainable with the new monitoring system?

[If response is zero, unusually high, or no response (refusal to answer), ask a follow-up question. If response is zero, ask reason for zero response. If refuses to answer, repeat question and try to get a response. If payment response is unusually high (higher than education expenditures), give respondent opportunity to revise his or her response.]

Section D — Background Information [7 questions]

1. What is your age?
2. What is the last grade of regular school that you and your spouse completed?
3. How many people live in your household? How many are under age 13?
4. What is your average monthly water and sewer bill?

[a]*Questionnaire follows an exact script that is paraphrased here.*

Source: Seung-Jun Kwak, personal communication, November 1997.

The next step is to analyze the data, developing a willingness-to-pay (WTP) function:

$$WTP(q_0,q_1) = f(P,q_0,q_1,Q,Y,T) \quad (18.1)$$

where q_0 is the initial risk in the water supply and q_1 is the risk after the water supply has been made "safe." P represents the prices of market goods, Q the quantity of other environmental goods, Y income, and T assorted characteristics of the individual, including

지불카드 5

-세후소득 월90~110만원 미만-

(정부의 다른 공공정책에 한 달 평균 지불하는 세금)

0		24,000	
100		26,000	
250		28,000	← 교육
500	← 위생	30,000	
750		35,000	
1,000		40,000	← 국방
1,500		45,000	
2,000		50,000	
3,000	← 보건	55,000	
4,000		60,000	
5,000		65,000	
6,000	← 도로	70,000	
7,000		75,000	
8,000		80,000	
9,000		85,000	
10,000		90,000	
12,000	← 주택	95,000	
14,000	← 사회보장 및 복지시설	100,000	
16,000		120,000	
18,000		140,000	
20,000		160,000	
22,000		200,000	

Payment Card 5

Monthly Household Income after Taxes
900,000–1,100,000 Won

(Average Monthly Amount in Taxes Paid for Some Public Programs)

0		24,000	
100		26,000	
250		28,000	← Education
500	← Sanitary Service	30,000	
750		35,000	
1,000		40,000	← Defense
1,500		45,000	
2,000		50,000	
3,000	← Health	55,000	
4,000		60,000	
5,000		65,000	
6,000	← Road & Highway	70,000	
7,000		75,000	
8,000		80,000	
9,000		85,000	
10,000		90,000	
12,000	← Housing	95,000	
14,000	← Social Security & Welfare	100,000	
16,000		120,000	
18,000		140,000	
20,000		160,000	
22,000		200,000	

Figure 18.1 Payment card from Seoul water-quality study.

tastes. In their analysis of their data, they asked respondents what they thought their current water quality was (see Table 18.1); the response was recorded as q_0. Variables representing individual characteristics included age, education level, number of children, and years resident in Seoul.

The average household's willingness to pay for cleaner water was 2560 Won ($3.28) per month, compared to the average water bill of 5000 Won per month. The researchers multiply the willingness to pay for the average household by the number of households to obtain an annual willingness to pay for Seoul of 83 billion Won ($106 million). The next logical step after constructing an individual willingness-to-pay equaton is to use it to construct an aggregate, population-wide measure of willingness to pay.[4]

B. Designing a Contingent Valuation Study

The Seoul water quality contingent valuation example above serves as a good introduction to the steps that should be taken in conducting a high-quality stated preference study. Some of what is now considered "best practice" has evolved over the years through research and experience.[5] However, in 1992, the U.S. Government decided that CV had become so important (and controversial) to its management of environmental resources that a high-level review was needed on the validity of the CV method. The U.S. National Oceanic and Atmospheric Administration (NOAA), the agency charged with writing regulations under the 1990 Oil Pollution Act, convened a panel of six distinguished economists and survey researchers, including two Nobel laureates, to evaluate the CV method. The NOAA panel concluded that CV could be useful, but certain practices would seem to be necessary to generate reliable estimates of willingness to pay.[6] These recommendations had a significant influence on what is considered "best practice" in conducting a CV survey. We discuss these findings below.

The issue is how to construct a "best practice" CV study to value some environmental good. Perhaps you want to find the willingness to pay for some nonmarginal change in the good, such as avoiding the extinction of a species. Perhaps you want to generate a demand curve/marginal willingness-to-pay function for some continuously varying commodity such as the peak daily ozone concentration in a community. To accomplish this, you must design, administer, and interpret a survey. A survey will typically consist of three parts: background information on the study and the environmental good, a section eliciting value, and a section asking background information on the respondent.

Carson (1991) describes six main components to a successful CV study:

- Define market scenario.
- Choose elicitation method.
- Design market administration.
- Design sampling.
- Design of experiment.
- Estimate willingness-to-pay function.

We will consider each of these in turn below.

1. Define Market Scenario. The market scenario is the information to be conveyed to a respondent (i.e., one of the people who will be asked about willingness to pay), to place the respondent in the right frame of mind to give meaningful responses to questions.

One of the first decisions to make is how to define the good to be valued. If a beach has been harmed by an oil spill, what environmental good should be valued? A day at the beach? The view of the beach? A degraded beach? Water pollution? Describing the good to be valued is only the beginning. It is also important to describe the context within which the good is supplied. If it is a day at the beach, how would payment be made? An admission fee? A parking fee?

It is important that the description of the market be realistic to the respondent and at the same time true to the eventual economic model that will use the data collected. If the market scenario is not understandable and plausible to the respondent, the data will be of questionable usefulness. Plausibility is important. For instance, if you ask respondents how much they would be willing to pay to make nuclear power completely safe, they may not believe that the goal is possible. If respondents do not understand the scenario or think it implausible, they may substitute an alternative scenario or simply not take the valuation exercise seriously.

An additional issue that needs attention in constructing the market scenario is the payment mechanism. Simply asking the willingness to pay to avoid some event, such as a fouled beach, is too abstract. The market scenario must be rooted in real-world experience, including the payment vehicle. Thus if the good is avoiding an oil spill at a beach, then a believable payment vehicle might be a tax on gasoline to hire additional inspectors for oil tankers. Of course it is also important to avoid a scenario that grates on the respondent's notion of right and wrong. Some people may feel it is the responsibility of the oil companies to pay for inspectors and may bridle at being asked to pay a tax for the service. In this case, it might be more appropriate to ask if they would be willing to require inspectors on oil tankers even if it meant raising the price of gasoline. The point is, considerable care must be taken in defining the good and the payment mechanism.[7]

Another issue, which is not easy to address, is providing the right context for the survey. For instance, suppose we are attempting to value cleaner groundwater (water in underground acquifers). Most people know very little about groundwater—what it is used for, how serious contamination is, what risks are associated with contaminated groundwater. To obtain informed responses to questions on this topic, some information will probably need to be provided to respondents. Obviously, time and other constraints allow very little education of respondents in the course of administering a survey. This raises the question of bias introduced by incompletely or inaccurately "educating" the respondent.

The NOAA Panel makes several recommendations regarding the market scenario. One point they make is that respondents should be reminded that substitutes exist. For instance, in ascertaining the value of a beach contaminated by an oil spill, the respondents should be reminded that there are many uncontaminated beaches a short distance away (if that is the case). The Panel suggests that the survey should be designed to avoid generating spurious emotions, such as a dislike of "big business." The Panel also urges that there be checks within the questionnaire to be sure that the respondent understands and accepts the information in the survey.

2. Choosing Elicitation Method. Having properly defined the market scenario, the next step is to decide how best to obtain the valuation response. This is obviously a very important part of the survey, and one of the most difficult to administer effectively. There are four primary ways of eliciting value: direct question, bidding game, payment card, and referendum choice.

The direct question is probably the most obvious. After the good has been described, respondents are simply asked their willingness to pay for the good. As Carson (1991) points out, one problem with this is that this direct valuation of a good is unfamiliar to most people. There are few real markets in which we are asked to generate our reservation price (highest price we would be willing to pay). People may not spend much effort in determining their willingness to pay, which may result in extreme responses (zeros and very large numbers).

The bidding game approach, used in the first "modern" CV study of an environmental good (Randall et al., 1974), starts with a WTP number and seeks a yes–no response. If the respondent replies yes, the amount is gradually increased until a "no" is received. Similarly, if the respondent replies "no," the amount is gradually decreased until a "yes" is received. The primary problem with a bidding game is what is known as *starting point bias*. If the good is unfamiliar to the respondent, the cues given by the surveyor may influence the response. If the surveyor asks about willingness to pay for groundwater cleanup and starts the bidding at $500 per year, the results will probably be different than if the bidding were started at $1 per year.

A third method, the payment card, was described in the context of the example of water quality in Seoul, presented in the last section. One problem with payment cards is that they cannot be used for telephone surveys. There may also be cues built into the range of values that are included in the card.

A fourth method is what is called referendum or discrete choice. Essentially a willingness-to-pay figure is offered to the respondent who is asked if he or she would be willing to pay that amount, "yes" or "no." Different respondents are offered different WTP figures, chosen to span the plausible range of willingnesses to pay. This approach is recommended by the NOAA Panel, because they thought it minimizes possible bias and is also familiar, in that people often vote yes/no on public referenda. One problem with referenda is that more data are needed to obtain statistically significant results, and this raises the cost of the survey.

3. Design Market Administration. Having designed the survey, it must be administered, i.e., respondents must complete the survey and their responses collated. There are three basic approaches to survey administration: mail, telephone, and in-person.

Mail surveys are the cheapest to administer, although they have problems. One problem is nonresponse. A mail survey is considered a success if only 30% of the surveys are not returned. But if there is some systematic reason for nonresponse, the results of the survey may be called into question. For instance, suppose people who feel less strongly about the environmental good tend not to respond. Another problem with mail surveys is that the respondent needs to understand the survey instrument. Therefore the survey must be relatively simple. Furthermore, there is no way to keep respondents from looking through the entire survey, which makes the control of the flow of the information in the survey difficult.

Telephone surveys are also relatively inexpensive to administer. However, telephones must be widely available within the population being surveyed. A telephone survey in Cairo (Egypt) would be most inappropriate because of the low penetration of telephones in the population. There may also be bias in terms of who answers the telephone. For instance, unemployed people may be more likely to be available. Another problem with telephone surveys is that visual cues cannot be used (e.g., photographs). This may lead to problems in eliciting values regarding changes in environmental quality, since photos are often used to convey the nature of the changed quality.

In-person surveys are the most expensive to administer but can be the most reliable. In fact, the NOAA Panel recommends in-person surveys whenever possible. One problem (other than costs) with an in-person survey is interviewer bias. It is difficult for an interviewer to always appear neutral when conducting an interview. Furthermore, since environmental goods are often perceived as desirable and socially "correct," respondents may be reluctant to reveal their unwillingness to pay if in fact they do not view the environment as very important (in the words of the NOAA panel, "social desirability bias"). Finally, if multiple interviewers are used, it is difficult to ensure that the results obtained by different interviewers are totally comparable.

Another issue in market administration is pretesting of the survey. The NOAA Panel emphasizes the importance of exhaustive pretesting of the survey instrument before the actual survey is conducted. This would include very careful analysis of the wording of each question and the organization of the survey. The survey can then be administered to test groups and adjusted based on feedback. The survey can also be administered to "focus groups" as was done in the case of the Seoul water quality survey. The NOAA Panel also urges that tests be done of interviewer and other biases.

4. Sample Design. There are two issues in choosing the people to answer the CV questionnaire. The first is to choose the group from which to draw the sample, and the second is to draw the random sample. The first issue is to determine the population of concern with regard to valuing the commodity in question. If we are dealing with visibility in Switzerland, should we target residents or visitors? If residents, what geographical area is of interest? Another issue is whether to examine individuals or households. Should children be included? How will these people/households be identified? By voting records or geographically as in the example of Seoul water quality? The answers to these questions define the *sampling frame*, the precisely defined population from which the sample is drawn. The appropriate sample is then drawn and the survey administered.

5. Experimental Design. The typical goal of a CV survey is to develop statistically significant estimates of willingness to pay for a particular environmental good or to test a hypothesis about the willingness to pay for the hypothetical good. Considering the cost of data collection, it is important to construct a survey carefully so that the appropriate information is collected in an efficient manner without unintentional biases. This is the process of experimental design. For example, in many referendum format CV surveys, each respondent is asked to vote yes or no on a different possible value for the environmental good. How those values are selected for presentation to the respondent is important for the ultimate usefulness of survey results. For instance, if the value is generated by the interviewer on the basis of the respondent's income, a bias may unintentionally be introduced into the data being collected. On the other hand, if the values are simply randomly chosen from a very wide range of possible values, it may be necessary to administer a very large number of surveys to obtain statistically significant results. The point is that design of the CV instrument, its administration, and its ultimate statistical analysis are very important steps that should not be taken lightly. This is what is known as experimental design.

6. Estimation of WTP Function. The last step is to take the survey results and correctly estimate the WTP function. This is obviously an important step and must be kept in mind as the survey instrument is developed. Sometimes this step is neglected until after the survey has been conducted, only to find that some vital piece of information is needed but was not collected on the survey. This outcome would suggest a flawed experimental design.

C. Problems with Contingent Valuation

Contingent valuation is highly controversial. The problem is complicated by possible conflicts of interest within the economics community. As mentioned earlier, CV has found significant application in the United States in assessing damage from environmental accidents, such as oil spills. CV has been used to generate very high estimates of damage to natural resources, primarily by estimating nonuse values. Many firms, including oil companies, would gain from discrediting the approach. A good deal of the research identifying flaws in CV has been funded by the Exxon Corporation.[8] A conflict of interest, of course, does not mean that the "anti-CV" work is distorted, just that there is an appearance of that possibility existing.[9] On the other side of the issue, academic careers have been built on the validity of CV.

There are many problems that have been identified with CV. A primary criticism is that the values elicited in CV surveys are not based on real resource decisions—they are hypothetical. How many of us have gone to a store thinking we would buy a new product, only to have second thoughts when faced with the prospect of relinquishing real money in exchange for the product? Many argue that without real resources at stake, the response to a WTP question is meaningless. Another way of looking at this is that there is no budget constraint in a hypothetical survey and without a budget constraint, choices are meaningless. The counter to that position is that there may be more scatter (variance) in the response to a hypothetical question than if real resources were at stake, but the average response will be unaffected by the hypothetical nature of the exercise.

Another issue that has been raised concerns ambiguity in what people are valuing. When you are asked how much you would be willing to pay to avoid the extinction of the lemurs in Madagascar, does the amount you are offering truly reflect your concern for the lemurs or are you simply "purchasing moral satisfaction" by responding that you would contribute substantially to their defense.[10] You get a "warm-glow" by saying that you would be willing to pay to provide an environmental good. Seip and Strand (1992) conducted a CV study in Norway, asking respondents if they would be willing to contribute 200 Norwegian Kroner to the leading Norwegian environmental organization (NNV) to protect Norwegian environmental resources. Of their sample 63% answered yes to that question. These people were then targeted by the NNV with a mail solicitation, asking for a contribution, with no reference to the survey. Fewer than 10% contributed, suggesting that the CV results were unreliable.

Another problem with CV has been called embedding. A typical problem for a CV survey is to determine the value of a specific natural resource, such as a particular park. However, there are usually substitute parks outside of the domain of the survey and there appear to be inconsistencies in how people value individual parks versus groups of parks. People may place the same value on cleaning up one lake as on cleaning up 10 lakes.

A related issue concerns existence value. Recall that existence value is value over and above use value. People may value the preservation of a wild Amazon river, even if they have no intention of visiting the river. There is some question about whether this type of value is valid, since existence value is never connected to a real payment. For instance, Desvousges et al. (1993) conducted a CV study of a contaminated lake in which a number of migratory water birds were killed. The birds were not endangered; thus the existence of the species was not in question, only a number of individual birds. The lake was hypothetical and thus not used by any of the respondents. The question was how much people would be willing to pay to eliminate the contaminated lake and thus save the birds. The willingness to pay was the same whether the number of birds at risk was 2000, 20,000, or 200,000, suggesting an embedding problem and possibly a fundamental problem with the idea of existence value.[11]

II. EXPERIMENTAL MARKETS

Probably the single greatest criticism of contingent valuation is the hypothetical nature of the exercise—no money is at stake when people articulate their willingness to pay. One solution to this is to construct a market where none existed before. There are two ways of doing this—field experiments and laboratory experiments.

A field experiment involves actually making a market in the "real world" where one has not previously existed. Typically, this is possible only where there is a governmental restriction that precludes market operation. An example would be markets in which goods are allocated with a lottery or on a first-come, first-served basis. An experimenter can obtain permission to substitute a market for such an allocation, and construct the experimental market in such a way as to learn about consumer behavior.

A laboratory experiment is much like the experiments psychologists have been conducting for decades. A group of volunteers is assembled, given real money to participate, and then faced with real decisions that involve giving up real money or receiving real money in exchange for the goods or bads the experimenter may present. We consider each of these types of experiments.

One characteristic of experiments is that their purpose is usually not to derive a value of a particular good but to test a theory in a controlled setting. This is in contrast to most other valuation methods in which the goal is to generate a marginal willingness-to-pay function that can be used for policy analysis or some other practical use. Experiments by their very nature apply in a very limited context. But because they are experiments, the conditions of the experiment are controllable by the experimenter. A major issue in valuation is the possible disparity between willingness to accept compensation (WTA) to give up a good vs. willingness to pay (WTP) to acquire a good. Theory suggests that there should be little difference between these two measures,[12] yet there is considerable evidence that they are not the same.[13] This issue is perfect for exploration using experiments because it is possible to control for extraneous factors that may drive a wedge between WTA and WTP.

A. Field Experiments

Field experiments are hard to construct because they generally require two basic conditions: (1) goods must be excludable (so that a market may operate) and probaly rival and (2) a market does not already exist. The problem is that most goods that are rival and excludable also have well developed markets.

One of the classic papers in valuation involving constructing an artificial market was done in the context of goose hunting in Wisconsin.[14] The good in question was access to a 10,000-hectare region in Wisconsin for the purpose of hunting geese. The good is not really a goose (for eating or stuffing) but rather the experience of getting out in the "wilderness" for a few days to hike around, possibly shooting a goose. Although many readers will perceive an incongruity between a wilderness experience and hunting, the fact is that many people enjoy hunting for the outdoors experience.

The interesting thing about this particular setting is that hunting permits are required to use the area and those hunting permits are in short supply. However, rather than use a market to issue them, the state of Wisconsin uses a lottery. The researchers conducted three analyses of the willingness to pay for goose hunting in this area: travel cost, contingent valuation, and experimental markets. The CV experiment involved asking a sample of holders of the permit how much they would be willing to accept as compensation in exchange for their permit. A similar sample was drawn prior to the lottery to ascertain willingness to pay for a permit. Thus estimates of willingness to pay and willingness to accept compensation were computed. The most interesting part of the experiment was that the researchers also drew a sample of 237 permit holders and sent them checks for random amounts between $1 and $200 asking recipients to do one of two things: (1) cash the check and return their permit or (2) return the check and keep the permit. Only one person chose a third route (cash the check and keep the permit).

The researchers found that the CV results were subject to a great deal of variability. Depending on the elicitation method, the mean response for willingness to accept compensation from the CV survey ranged from $67 to $101. In contrast, the mean response for the willingness to pay was $11 to $21, depending on how the question was posed. The mean of the field experimental market was $63. A travel cost estimate of the value of a permit was $32. The significance of the experimental market result is that real money was involved in the decisions made by the respondents. Although the market was artificial, it was not hypothetical.

B. Laboratory Experiments

In the example of goose hunting, it was fortuitous that what was essentially a private good (a hunting permit) was not allocated by a market but rather distributed by the state. This allowed the researchers to set up an artificial market. This is not always possible.

Another approach that has been used is what are called laboratory experiments. These involve groups of subjects (typically university students) who are paid a fixed amount to participate in an experiment for perhaps a few hours. These experiments involve the participants exchanging real money for experimental commodities. Payment is set up so that two criteria are met: (1) participants make real decisions regarding real resources and (2) the participants leave the experiment richer than when they began. The second requirement induces people to voluntarily participate.

One of the major uses of experimental economics in the environmental arena is to explore the divergence between WTP and WTA as well as the validity of contingent valuation. Both of these issues were addressed in the goose hunting study discussed in the previous section. There are two laboratory experiments that address these issues that are interesting and worth examining. One concerns the demand for consuming a foul-tasting liquid, important because of its connection to pollution. The other study concerns the existence value of an individual tree—also an issue of significant import in environmental economics.

1. Sucrose Octa-Acetate. In constructing an experiment in the demand for bads (as opposed to goods), the problem of not causing harm to an experimental subject is a significant constraint. It is most likely unethical, and in any event goes against most guidelines for experimenting on humans, to intentionally expose people to harm to determine their willingness to pay to avoid the harm. Psychologists have used a bitter and unpleasant tasting liquid, sucrose octa-acetate (SOA), for such purposes. SOA appears to be the only known laboratory substance that is very unpleasant and yet nontoxic. In an interesting use of SOA, Coursey et al. (1987) set out to answer several questions: How do hypothetical willingnesses to pay to avoid tasting SOA compare to "true" willingnesses to pay? And the same can be asked about willingnesses to accept payment to taste SOA. The second type of question is, how do WTA and WTP compare in an experimental market setting?

To answer these questions, they constructed two basic experiments, each consisting of several parts. One experiment concerned willingness to pay to avoid tasting SOA; the other experiment concerned willingness to accept payment to taste SOA. Each experiment involved eight economics students. Each experiment started by asking hypothetical valuation questions and by having subjects taste a few drops of SOA. Let us focus on the real experiments.

In the WTP experiment, each student was first asked about hypothetical values; after that the students participated in a series of auctions. Each student was asked to submit a bid as to how much he or she would pay to avoid tasting one ounce (approximately 30 ml) of SOA (tasting is defined as holding a one ounce cup of SOA in your mouth for 20 seconds). Bids were collected and ranked from highest to lowest. The fifth highest bid was announced to be the going price to avoid tasting SOA. The four individuals who bid highest would pay the going price (real money) to avoid the tasting; the other four individuals would have to taste the SOA. This procedure was repeated a number of times. The first four times it was done, no tasting occurred; these were considered practice runs. From that point on, however, real tasting did occur. The students were paid $10 to participate in each WTP auction. A similar procedure was adopted for WTA.

Two interesting things happened in these experiments. One result is that WTA starts very high when students have little experience with the SOA. After repeated trials, the WTA bids drop considerably. In contrast, WTP bids are much more stable, changing only slightly as auction trials are repeated. This suggests that WTP is a better measure of value than WTA, particularly when the goods are unfamiliar.

A second result concerns the validity of hypothetical values. In comparison to the "real" values that emerge from the experimental auctions, Hypothetical measures of WTA appear to be biased upward, whereas WTP measures are more consistent with experimental values.

2. Norfolk Pines. Another experiment that seeks to explain the divergence between WTA and WTP was done by Boyce et al. (1992). They explore the extent to which existence value may drive a wedge between the two measures. In particular, they suggest that if one "owns" an environmental good, ownership may be accompanied by a moral obligation of stewardship. This may not exist when the good is owned by someone else. Thus WTA will have to overcome a larger hurdle than WTP; we would expect WTA to be larger than WTP. Although this is a somewhat unusual theory that draws on unconventional arguments, the hypothesis can be tested with an experiment.

The experimenters utilized 115 staff members from the University of Colorado and conducted four basic experiments, two for WTA and two for WTP. Each person participated in only one of the four experiments and was paid $30–40 to participate. The commodity in question was a small potted plant, a Norfolk Island pine tree. These are commonly used as houseplants. In one of the four experiments, each subject was given a plant and was asked to submit a bid indicating the price at which they would sell their plant back to the experimenter (WTA). A price was then randomly generated and all offers below that price were accepted. A similar experiment was conducted in which the subjects were asked to submit a bid indicating how much they would pay to acquire such a tree (WTP). These bids are interpreted as representing the use-value of the Norfolk pines. The average WTA bid was 66% higher than the average WTP bid, a statistically significant difference.

So far, the experiment is quite standard. But the next step is to try to measure the existence value of the plant, over and above the use value. The experimenters do this by slightly modifying the experiment. They convince the subjects that if the plant remains with the experimenter (if the subject fails to purchase it or ends up selling it), the experimenter will kill the tree. The experimenters found that the death threat raised both WTP and WTA. WTP increased by 60% and WTA increased by 130%, and both of these increases are statistically significant. This suggests that the existence of the tree had value for the subjects, though that value may be a reflection of "duty" or moral obligation, which might not be the case in a more disconnected context. Furthermore, the fact that WTA was affected more significantly by the threat of tree death tends to reinforce the arguments about the importance of moral obligation, that it is higher when the subject has ownership of the tree.

This experiment does suggest the importance of existence value. But perhaps a more important contribution of the paper is to indicate the importance of experimental markets in exploring questions of valuation and even economic theory in the case of environmental goods where markets rarely exist.

III. REFERENDA

One type of constructed market is very official and not at all experimental—a voter referendum on the provision of some public good, often accompanied by a tax levy to pay for it. If we knew how individuals voted on such referenda, and the characteristics of the voters, we would be able statistically to estimate the demand for the particular public good. Unfortunately, balloting is usually secret. The best we can do is observe how groupings of voters vote, and utilize the characteristics of the grouping to explain the vote.

One of the earliest analyses of voting on environmental referenda is by Robert Deacon and Perry Shapiro (1975). They analyze two state-wide referenda in California, both related to the provision of a public good. One referendum concerned safeguarding the coastal area of the state from excessive development, providing an uncluttered coast as a public amenity. The authors examine a cross section of 334 California cities, comparing the percentage of the population voting in favor of the proposition with the characteristics of the city—income, education, political preference, and other variables. Although income did not turn out to be significant, other factors such as education, political orientation, and the fraction of the population in unskilled labor did turn out to be significant. Unfortunately, the data are not available to construct a demand curve for coastal protection, because information on the price is unavailable. Nevertheless, this approach gives important understanding of the demand for environmental goods.

The Deacon and Shapiro study was updated and expanded in a more recent analysis by Kahn and Matsusaka (1997). These authors also examine voting in California on environmentally oriented referenda, except that they are able to analyze 14 different such referenda. Furthermore, they use county-level voting data rather than city data. As with the Deacon and Shapiro analysis, they do not estimate a demand curve, since no prices are available for the environmental goods. They are able to measure how income affects demand and they reach some interesting conclusions. They find that most environmental goods are normal goods, meaning that demand increases as income increases. But some publicly provided environmental goods turn out to be inferior goods at higher income. It would seem that some goods, such as urban parks, are privately provided by high income people, which reduces their demand for public provision. The underlying goods of course may still be normal, even though when publicly provided they are inferior.

In summary, referenda can provide real insights on preferences for public good, though the approach is still not developed well enough to be able to generate demand curves for typical environmental goods.

SUMMARY

1. Stated preference methods of valuation are often termed contingent valuation. Contingent valuation is a method of valuing environmental goods that uses a survey of the population, with the survey directly asking willingess to pay for the environmental good. Thus the approach involves a hypothetical market with no real resource commitments behind the response.

2. Some consensus has developed regarding how to carefully conduct a contingent valuation study. The process of making a study more convincing and reliable almost always involves greater data collection expense.

3. The six basic steps in conducting a contingent valuation estimate of demand for an environmental good are define market scenario, choose elicitation method, design market administration, design sampling, design experiment, and estimate willingness-to-pay function.

4. The primary strength of contingent valuation is its applicability to virtually any good or bad. Criticism is usually based on either the lack of real resources being at stake or empirical evidence from existing studies showing logically inconsistent responses.

5. Experimental markets are another kind of constructed market. An experimental market involves defining a real market where none existed before. By real market we mean a situation in which real resources must be commited in association with decisions regarding the environmental good.

6. Experimental markets can involve artificial markets whereby a market is constructed in the population at large (field experiments) or they can involve laboratory markets in which a group of subjects is encouraged (usually by monetary payment) to participate, trading money for goods presented by the experimenter (laboratory experiments).

7. Referendum statements of value involve official votes on proposals to provide an environmental good, usually financed by tax payments. Such methods can be reliable but are not common.

PROBLEMS

1. A truck carrying insecticide has crashed, spilling hazardous chemicals into a creek, killing fish. Eventually the creek recovers with no permanent damage. You are charged with assessing the environmental damage. Identify one commodity that has been harmed and describe a CV study to assess the willingness to pay to avoid the damage.
2. In the Exxon Valdez legal case, several contingent valuation studies were part of estimates of liability damages owed by the oil company, Exxon. Damages were argued to include the value for all U.S. residents of an unpolluted Prince William Sound (where the spill occurred), bringing the issue of nonuse value to the front of the debate. Describe why a resident of the United States far from Alaska might associate real value (be willing to devote real resources) to a place he or she will never see or experience, such as coastal Alaska or the Amazon basin.
3. In conducting telephone surveys, some households always refuse to participate, saying they "don't have time" or "can't be bothered." If researchers accept these inevitable nonresponses and base their study on the other households, those who patiently answer each question, what biases are likely to be built into the sample? Can you think of a way to deal with this type of problem, either practical or statistical?
4. Consider being confronted with two possible scenarios while participating in a laboratory experiment:
   a. Someone offers to sell you a coffee mug. You have a reservation price for the mug (it is a nice one), and you bargain with the seller until you agree on a price that is less than or equal to your reservation price.
   b. Someone gives you a coffee mug, then begins to bargain with you to buy it back. You have a reservation selling price and you agree on a price greater than or equal to that reservation price.

   Is your reservation price in part (a) the same as your reservation selling price in (b)? Why or why not? Should such considerations matter in the design of experiments to elicit the value of public goods?

NOTES

1. To my knowledge, the only possible revealed preference method of inferring the existence value of a commodity is by observing voluntary provision of public goods. This approach has not been widely utilized, however.

2. Carson et al. (1995) provide a very detailed list of CV studies, through the early 1990s.
3. The sample can be intentionally constructed to emphasize certain segments of the population, as long as the manner in which the sample deviates from random is known. Kwak and Russell (1994) use such a "stratified random sample."
4. Unfortunately, some of the explanatory variables are derived from the survey and unknown for the wider population (such as the subjective assessment of current water quality). This makes aggregation more difficult.
5. Carson (1991) and Mitchell and Carson (1989) describe in detail the consensus on best practice in conducting CV studies.
6. The conclusions of the NOAA Panel were reported in January 1993 in the *Federal Register*, a U.S. Government publication of official notices and regulations (National Oceanic and Atmospheric Administration, 1993).
7. The NOAA Panel recommends that willingness to pay be measured, rather than willingness to accept compensation. In the example just given, it would be preferable to ask what people would be willing to pay to avoid an oil spill rather than what they would be willing to accept from an oil company for the right to spill oil. The laboratory experiments described later in this chapter provide support for focusing on willingness to pay.
8. Much of this work is reported in Hausman (1993) and was done in support of Exxon in the Exxon Valdez oil spill litigation. Despite the source of the funding, the economists conducting the research are very well respected and the work is of generally high quality (though many have found fault with aspects of the work).
9. A very interesting set of papers on CV appeared in the Fall 1994 *Journal of Economic Perspectives*. In that issue, Paul Portney, a member of the NOAA Panel set the stage for two papers representing two sides of the CV issue (Portney, 1994). Michael Hanemann from the University of California presented a defense of CV (Hanemann, 1994) whereas Peter Diamond and Jerry Hausman from MIT presented a negative critique of CV (Diamond and Hausman, 1994). Not coincidentally, Hanemann (among others) was employed by the state of Alaska to conduct CV surveys in support of its claim of damage in the Exxon Valdez oil spill. Hausman was employed by the Exxon Corporation in support of its defense in that case, part of which rested on discrediting the CV work done by the opposition.
10. This point has been made in a widely cited article by Kahneman and Knetsch (1992).
11. One possible alternative explanation is that respondents viewed the good being valued as "a world safe for these birds" rather than the size of the bird population.
12. The main theoretical reason why WTP and WTA should differ is if the good is substantial relative to income. Thus there will be income effects depending on whether one is giving up the good or gaining the good. If the value of the good is small relative to income, these differences disappear. Hanemann (1991) and Kolstad and Guzman (1999) explore additional theoretical reasons for a divergence between WTP and WTA, even when there is no income effect.
13. Cummings et al. (1986) review a number of studies that measure WTP and WTA and show a large disparity between the two measures.
14. Bishop et al. (1983).

References

Abrams, B. and M. Schmitz, "The 'Crowding-Out' Effect of Government Transfers on Private Charitable Contribution," *Public Choice*, **33**:29–41 (1978).

Adar, Zvi and James M. Griffin, "Uncertainty and the Choice of Pollution Control Instruments," *J. Environ. Econ. Mgmt.*, **3**:178–188 (1976).

Aivazian, Vadouj A. and Jeffrey L. Callen, "The Coase Theorem and the Empty Core," *J. Law Econ.*, **24**:175–181 (1981).

Akerlof, George, "The Market for Lemons: Quality Uncertainty and the Market Mechanism," *Quart. J. Econ.*, **84**:488–500 (1970).

Anderson, J. W., "The Kyoto Protocol on Climate Change: Background, Unresolved Issues and Next Steps," Resources for the Future Discussion Paper (Resources for the Future, Washington, DC, January 1998).

Angelini, Fabio, "Glossary of Terms and Events Relevant to European Community Environmental Law," *Ecol. Law Quart.*, **20**:177–196 (1993).

Arrhenius, Svente, "On the Influence of Carbonic Acid in the Air Upon the Temperature of the Ground," *London, Edinburgh, Dublin Philos. Mag. J. Sci.*, **41**:237–277 (1896).

Arrow, Kenneth J., *Social Choice and Individual Values* (Wiley, New York, 1951).

Arrow, Kenneth J. and Anthony C. Fisher, "Environmental Preservation, Uncertainty, and Irreversibility," *Quart. J. Econ.*, **88**:312–319 (1974).

Ashby, Eric and Mary Anderson, *The Politics of Clean Air* (Oxford University Press, Oxford, England, 1981).

Atkinson, Scott E., "Marketable Pollution Permits Acid Rain Externalities," *Can. J. Econ.*, **16**:704–722 (1983).

Atkinson, Scott E. and Donald H. Lewis, "A Cost-Effectiveness Analysis of Alternative Air Quality Control Strategies," *J. Environ. Econ. Mgmt.*, **1**:237–250 (1974).

Atkinson, Scott E. and T. H. Tietenberg, "The Empirical Properties of Two Classes of Designs for Transferable Discharge Permit Markets," *J. Environ. Econ. Mgmt.*, **9**:101–121 (1982).

Averchenkov, Alexander, Alexander Golub, Konstantin Gofman, and Vladimir Groshev, "The System of Environmental Funds in the Russian Federation," in OECD Centre for Co-Operation with the Economies in Transition, *Environmental Funds in Economies in Transition* (OECD, Paris, 1995).

Barnett, A. H., "The Pigouvian Tax Rule under Monopoly," *Am. Econ. Rev.*, **70**:1037–1041 (1980).

Barnett, Harold J. and Chandler Morse, *Scarcity and Growth* (Johns Hopkins University Press, Baltimore, 1963).

Baron, David P., "Design of Regulatory Mechanisms and Institutions," in R. Schmalensee and R. Willig (Eds.), *Handbook of Industrial Organization* (North-Holland, Amsterdam, 1989).

Baron, David P. and David Besanko, "Regulation, Asymmetric Information and Auditing," *Rand J. Econ.*, **15**:447–470 (1984).

Barrett, Scott, "Self-Enforcing International Environmental Agreements," *Oxford Econ. Papers*, **46**:878–894 (1994).

Barrett, Scott, "The Strategy of Trade Sanctions in International Environmental Agreements," *Res. Energy Econ.*, **19**:345–362 (1997).

Barrett, Scott, "Economic Analysis of International Environmental Agreements: Lessons for a Global Warming Treaty," in Organization for Economic Cooperation and Development, *Responding to Climate Change: Selected Economic Issues*, pp. 109–149 (OECD, Paris, 1991).

Bartik, Timothy J., "The Estimation of Demand Parameters in Hedonic Price Models," *J. Pol. Econ.*, **95**:81–88 (1987).

Bartik, Timothy J., "Evaluating the Benefits of Non-marginal Reductions in Pollution Using Information on Defensive Expenditures," *J. Environ. Econ. Mgmt.*, **15**:111–127 (1988).

Bator, F. M., "The Anatomy of Market Failure," *Quart. J. Econ.*, **72**: 351–379 (1958).

Baumol, William J. and David F. Bradford, "Detrimental Externalities and Nonconvexities of the Production Set," *Economica*, **39**:160–176 (1972).

Baumol, William J. and Wallace E. Oates, *The Theory of Environmental Policy*, 2nd Ed. (Cambridge University Press, Cambridge, England, 1988).

Bayless, Mark, "Measuring the Benefits of Air Quality Improvements: A Hedonic Salary Approach," *J. Environ. Econ. Mgmt.*, **9**:81–99 (1982).

Beard, T. Randolph, "Bankruptcy and Care Choice," *Rand J. Econ.*, **21**:626–634 (1990).

Bedell, Christine, "Feedlot Price: $1.82/Month," Santa Maria (California) *Times*, pp. A-1 and A-6 (June 28, 1996).

Bergson, A., "A Reformulation of Certain Aspects of Welfare Economics," *Quart. J. Econ.*, **52**:314–344 (1938).

Bergstrom, T. C., D. L. Rubinfeld, and P. Shapiro, "Micro-based Estimates of Demand Functions for Local School Expenditures," *Econometrica*, **50**:1183–1206 (1982).

Bergstrom, Theodore, Lawrence Blume, and Hal Varian, "On the Private Provision of Public Goods," *J. Public Econ.*, **29**:25–49 (1986).

Bhagwati, Jagdish and T. N. Srinivasan, "Trade and the Environment: Does Environmental Diversity Detract from the Case for Free Trade?", in J. Bhagwati and R. Hudec (Eds.), *Fair Trade and Harmonization* (MIT Press, Cambridge, MA., 1996).

Binkley, Clark S. and W. Michael Hanemann, "The Recreation Benefits of Water Quality Improvement: Analysis of Day Trips in an Urban Setting," U.S. Environmental Protection Agency Report EPA-600/5-78-010, Washington, DC (June 1978).

Bishop, Richard C., Thomas A. Heberlein, and Mary Jo Kealy, "Contingent Valuation of Environmental Assets: Comparisons with a Simulated Market," *Natural Res. J.*, **23**:619–633 (1983).

Bockstael, Nancy E., W. Michael Hanemann, and Cathy L. Kling, "Modeling Recreational Demand in a Multiple Site Framework," *Water Res. Res.*, **23**:951–960 (1987).

Bockstael, Nancy E., Kenneth E. McConnell, and Ivar E. Strand, "Benefits from Improvements in Chesapeake Bay Water Quality," Vol. III of "Benefit Analysis Using Indirect or Imputed Market Methods," U.S. Environmental Protection Agency Report EPA-230-10-89-070, Washington, DC (October 1989).

Bockstael, Nancy E., Kenneth E. McConnell, and Ivar Strand, "Recreation," in J.B. Braden and C.D. Kolstad (Eds.), *Measuring the Demand for Environmental Quality* (North-Holland, Amsterdam, 1991).

Bohm, Peter, "Estimating Demand for Public Goods: An Experiment," *Eur. Econ. Rev.*, **3**:111–130 (1972).

Bohn, Henning and Robert Deacon, "Ownership Risk, Investment, and the Use of Natural Resources," Department of Economics Working Paper, University of California, Santa Barbara (January 1997).

Bovenberg, A. Lans and Ruud A. de Mooij, "Environmental Levies and Distortionary Taxation," *Amer. Econ. Rev.*, **84**:1085–1089 (1994).

Bower, Blair T., "Mixed Implementation Incentive Systems for Water Quality Management in France, the Ruhr, and the United States," pp. 212–227 in P.B. Downing and K. Hanf (Eds.), *International Comparisons in Implementing Pollution Laws* (Kluwer-Nijhoff, Boston, 1983).

Boyce, Rebecca R., Thomas C. Brown, Gary H. McClelland, George L. Peterson, and William D. Schulze, "An Experimental Examination of Intrinsic Values as a Source of the WTA-WTP Disparity," *Am. Econ. Rev.*, **82**:1366–1373 (1992).

Braden, John B. and Stef Proost, "Economic Assessment of Policies for Combating Tropospheric Ozone in Europe and the United States," pp. 365–413 in J.B. Braden, H. Folmer, and T.S. Ulen (Eds.), *Environmental Policy with Political and Economic Integration: The European Union and the United States* (Edward Elgar, Cheltenham, England, 1996).

Brander, James A. and Barbara J. Spencer, "Export Subsidies and International Market Share Rivalry," *J. Int. Econ.*, **18**:83–100 (1985).

Bresnahan, Timothy F. and Dennis A. Yao, "The Nonpecuniary Costs of Automobile Emissions Standards," *Rand J. Econ.*, **16**:437–455 (1985).

Broadway, Robin W. and David E. Wildasin, *Public Sector Economics*, 2nd Ed. (Little, Brown, Boston, 1984).

Brookshire, David S., Mark A. Thayer, William D. Schulze, and Ralph d'Arge, "Valuing Public Goods: A Comparison of Survey and Hedonic Approaches," *Am. Econ. Rev.*, **72**:165–177 (1982).

Brown, Gardner, "Estimating Non-Market Economic Losses from Oil Spills: Amoco Cadiz, Steuart Transportation, Zoe Colocotroni," unpublished manuscript, Department of Economics, University of Washington, Seattle (undated).

Brown, Gardner M., Jr., Richard Congar, and Elizabeth A. Wilman, "Recreation: Tourists and Residents," Chapter 4 in U.S. National Ocean Service, *Assessing the Social Costs of Oil Spills: The Amoco Cadiz Case Study*, National Oceanic and Atmospheric Administration Report, NTIS PB84-100536, Washington, DC (July 1983).

Browning, Edgar K., "On the Marginal Welfare Cost of Taxation," *Am. Econ. Rev.*, **77**:11–23 (1987).

Bruzelius, Nils, *The Value of Travel Time: Theory and Measurement* (Croom Helm, London, 1979).

Buchanan, James M., "External Diseconomies, Corrective Taxes, and Market Structure," *Am. Econ. Rev.*, **59**:174–77 (1969).

Buchanan, James M. and William Craig Stubblebine, "Externality," *Economica*, **29**:371–384 (1962).

Buchanan, James M. and Gordon Tullock, "Polluters Profits and Political Response: Direct Controls Versus Taxes," *Am. Econ. Rev.*, **65**:139–147 (1975).

Bulckaen, Fabrizio, "Emissions Charge and Asymmetric Information: Consistently a Problem?," *J. Env. Econ. Mgmt.*, **34**:100–106 (1997).

Cabe, Richard and Joseph A. Herriges, "The Regulation of Non-Point-Source Pollution Under Imperfect and Asymmetric Information," *J. Environ. Econ. Mgmt.*, **22**:134–146 (1992).

Calabresi, Guido "Transaction Costs, Resource Allocation and Liability Rules—A Comment," *J. Law Econ.*, **11**:67–73 (1968).

Carlton, Dennis W. and Glenn C. Loury, "The Limitations of Piouvian Taxes as a Long-run Remedy for Externalities," *Quart. J. Econ.*, **95**:559–566 (1980).

Carlton, Dennis W. and Jeffrey M. Perloff, *Modern Industrial Organization*, 2nd Ed. (Harper Collins, New York, 1994).

Carson, Richard, "Constructed Markets," in J. B. Braden and C. D. Kolstad (Eds.), *Measuring the Demand for Environmental Quality* (North-Holland, Amsterdam, 1991).

Carson, R. T., J. Wright, N. J. Carson, A. Alberini, and N. Flores, "A Bibliography of Contingent Valuation Studies and Papers," Natural Resource Damage Assessment, Inc., La Jolla, CA (January 1995).

Carson, Richard T., Leanne Wilks, and David Imber, "Valuing the Preservation of Australia's Kakadu Conservation Zone," *Oxford Econ. Papers*, **46**:727–749 (1994).

Chichilnisky, Graciella, "North-South Trade and the Global Environment," *Am. Econ. Rev.*, **84**:851–874 (1994).

Christiansen, V., "Which Commodity Taxes Should Supplement the Income Tax?," *J. Public Econ.*, **24**:195–220 (1984).

Clark, David E. and Leslie A. Nievers, "An Interregional Hedonic Analysis of Noxious Facility Impacts on Local Wages and Property Values," *J. Environ. Econ. Mgmt.*, **27**:235–253 (1994).

Clawson, Marion and Jack L. Knetsch, *Economics of Outdoor Recreation* (Johns Hopkins University Press, Baltimore, 1966).

Coase, Ronald H. "The Problem of Social Cost," *J. Law Econ.*, **3**:1–44 (1960).

Coase, R. H., "The Coase Theorem and the Empty Core: A Comment," *J. Law Econ.*, **24**:183–187 (1981).

Conrad, Klaus and Catherine Morrison, "The Impact of Pollution Abatement Investment on Productivity Change: An Empirical Comparison of the U.S., Germany and Canada," *South. Econ. J.*, **55**:684–697 (1989).

Cook, Phillip J. and Daniel A. Graham, "The Demand for Insurance and Protection: The Case of Irreplaceable Commodities," *Quart. J. Econ.*, **91**:143–156 (1977).

Cooter, Robert D., "How the Law Circumvents Starrett's Nonconvexity," *J Econ. Theory*, **22**:499–504 (1980).

Cooter, Robert D., "Coase Theorem," in J. Eatwell, M. Milgate, and P. Newman (Eds.), *The New Palgrave: A Dictionary of Economics* (Macmillan, London, 1987).

Cooter, Robert and Thomas Ulen, *Law and Economics*, 2nd Ed. (Addison-Wesley-Longman, Reading, MA., 1997).

Copeland, Brian R., "Taxes versus Standards to Control Pollution in Imperfectly Competitive Markets," University of British Columbia Department of Economics Discussion Paper 91-03, Vancouver, BC, Canada (1992).

Cornes, Richard and Todd Sandler, *The Theory of Externalities, Public Goods and Club Goods*, 2nd Ed. (Cambridge University Press, Cambridge, 1996).

Costanza, Robert (Ed.), *Ecological Economics: The Science and Management of Sustainability* (Columbia University Press, New York, 1991).

Courant, Paul N. and Richard Porter, "Averting Expenditure and the Cost of Pollution," *J. Environ. Econ. Mgmt.*, **8**:321–329 (1981).

Coursey, Don L., John L. Hovis, and William D. Schulze, "The Disparity Between Willingness to Accept and Willingness to Pay Measures of Value," *Quart. J. Econ.*, **102**:679–690 (1987).

Cropper, Maureen L. and Wallace E. Oates, "Environmental Economics: A Survey," *J. Econ. Lit.*, **30**:675–740 (1992).

Cropper, Maureen L., Sema K. Aydede, and Paul R. Portney, "Rates of Time Preference for Saving Lives," *Am. Econ. Rev. Papers Proc.*, **82**(2):469–472 (May 1992).

Cropper, Maureen L., Sema K. Aydede, and Paul R. Portney, "Preferences for Life Saving Programs: How the Public Discounts Time and Age," *J. Risk Uncertainty*, **8**:243–265 (1994).

Crowards, Tom M., "Natural Resource Accounting: A Case Study of Zimbabwe," *Environ. Res. Econ.*, **7**:213–241 (1996).

Cummings, Ronald G., Davis S. Brookshire, and William D. Schulze (Eds.), *Valuing Environmental Goods: An Assessment of the Contingent Valuation Method* (Rowman & Allanheld, NJ, 1986).

d'Arge, Ralph C., William D. Schulze, and David S. Brookshire, "Carbon Dioxide and Intergenerational Choice," *Am. Econ. Rev. Papers Proc.*, **72**(2):251–256 (May 1982).

Dahlman, Carl J., "The Problem of Externality," *J. Law Econ.*, **22**(1):141–162 (1979).

Dasgupta, Partha, Peter Hammond, and Eric Maskin, "On Imperfect Information and Optimal Pollution Control," *Rev. Econ. Studies*, **47**:857–860 (1980).

Davis, R., "Recreation Planning as an Economic Problem," *Natural Res. J.*, **3**:239–249 (1963).

Deacon, Robert and Perry Shapiro, "Private Preference for Collective Goods Revealed Through Voting on Referenda," *Am. Econ. Rev.*, **65**:943–955 (1975).

Deacon, Robert T., David S. Brookshire, Anthony C. Fisher, Allen V. Kneese, Charles D. Kolstad, David Scrogin, V. Kerry Smith, Michael Ward, and James Wilen, "Research Trends and Opportunities in Environmental and Natural Resource Economics," *Environ. Res. Econ.*, **11**:383–397 (1998).

Des Jardins, Joseph R., *Environmental Ethics: An Introduction to Environmental Philosophy*, 2nd Ed. (Wadsworth, Belmont, CA, 1997).

Desvousges, William H., F. Reed Johnson, Richard W. Dunford, Sara P. Hudson, K. Nicole Wilson, and Kevin J. Boyle, "Measuring Natural Resource Damages with Contingent Valuation: Tests of Validity and Reliability," in J. A. Hausman (Ed.), *Contingent Valuation: An Assessment* (North-Holland, Amsterdam, 1993).

Diamond, Peter A. and Jerry A. Hausman, "Contingent Valuation: Is Some Number Better than No Number?," *J. Econ. Perspect.*, **8**(4):45–64 (1994).

Dickie, Mark and Shelby Gerking, "Willingness to Pay for Ozone Control: Inferences from the Demand for Medical Care," *J. Environ. Econ. Mgmt.*, **21**:1–16 (1991).

Downing, Paul and William Watson, "The Economics of Enforcing Air Pollution Controls," *J. Environ. Econ. Mgmt.*, **1**:219–236 (1974).

Epple, Dennis, "Hedonic Prices and Implicit Markets: Estimating Demand and Supply Functions for Differentiated Products," *J. Pol. Econ.*, **95**:59–80 (1987).

Epple, Dennis and Holger Sieg, "Estimating Equilibrium Models of Local Jurisdictions," *J. Pol. Econ.* (forthcoming).

Eskeland, Gunnar S., "A Presumptive Pigovian Tax on Gasoline: Complementing Regulation to Mimic an Emission Fee," *World Bank Econ. Rev.*, **8**:373–394 (1994).

Farber, Daniel A. and Robert E. Hudec, "GATT Legal Restraints on Domestic Environmental Regulations," in J. Bhagwati and R. Hudec (Eds.), *Fair Trade and Harmonization*, (MIT Press, Cambridge, MA, 1996).

Feshbach, Murray and Alfred Friendly, Jr., *Ecocide in the USSR* (Basic Book, New York, 1992).

Fisher, Anthony C. and W. Michael Hanemann, "Option Value and the Extinction of Species," pp. 169–190 in V. Kerry Smith (Ed.), *Advances in Applied Micro-Economics*, Vol. 4 (JAI Press, Greenwich, CT., 1986).

Folmer, Henk, Pierre van Mouche, and Shannon Ragland, "Interconnected Games and International Environmental Problems," *Environ. Res. Econ.*, **3**:313–335 (1993).

Foster, Vivien and Robert W. Hahn, "Designing More Efficient Markets: Lessons from Los Angeles Smog Control," *J. Law Econ.*, **38**:19–48 (1995).

Freeman, A. Myrick III, *The Benefits of Environmental Improvement* (Johns Hopkins University Press, Baltimore, 1979).

Freeman, A. Myrick III, "Depletable Externalities and Pigouvian Taxation," *J. Environ. Econ. Mgmt.*, **11**:173–179 (1984).

Freeman, A. Myrick III, *The Measurement of Environmental and Resource Values: Theory and Methods* (Resources for the Future, Washington, DC, 1993).

Freixas, X., R. Guesnerie, and J. Tirole, "Planning Under Incomplete Information and the Ratchet Effect," *Rev. Econ. Studies*, **52**:173–192 (1985).

Fullerton, Don and Thomas C. Kinnaman, "Household Responses to Pricing Garbage by the Bag," *Am. Econ. Rev.*, **86**:971–984 (1996).

Gerking, Shelby and Linda Stanley, "An Economic Analysis of Air Pollution and Health: The Case of St. Louis," *Rev. Econ. Stat.*, **68**:115–121 (1986).

Gillley, Otis W. and R. Kelley Pace, "On the Harrison and Rubinfeld Data," *J. Environ. Econ. Mgmt.*, **31**:403–405 (1996).

Goulder, Lawrence H., "Environmental Taxation and the Double Dividend: A Reader's Guide," *Intl. Tax and Public Finance*, **2**:157–183 (1995).

Goulder, Lawrence H., Ian W. H. Parry, and Dallas Burtraw, "Revenue-Raising Versus Other Approaches to Environmental Protection: The Critical Significance of Preexisting Tax Distortions," *Rand J. Econ.*, **28**:708–731 (1997).

Graham, Daniel A., "Cost-Benefit Analysis Under Uncertainty," *Am. Econ. Rev.*, **71**:715–725 (1981).

Gramlich, Edward M., *Benefit–Cost Analysis of Government Programs* (Prentice-Hall, Englewood Cliffs, NJ, 1981).

Griliches, Zvi, "Hedonic Price Indexes for Automobiles: An Econometric Analysis of Quality Change," in Zvi Griliches (Ed.), *Price Indexes and Quality Change: Studies in New Methods of Measurement* (Harvard University Press, Cambridge, MA, 1971).

Grossman, Gene M. and Alan B. Krueger, "Economic Growth and the Environment," *Quart. J. Econ.*, **112**:353–377 (1995).

Hahn, Robert W., "Economic Prescriptions for Environmental Problems: How the Patient Followed the Doctor's Orders," *J. Econ. Perspect.*, **3**:95–114 (1989).

Hahn, Robert W., "Market Power and Transferable Property Rights," *Quart. J. Econ.*, **99**:753–765 (1984).

Hahn, Robert W. and Gordon Hester, "Where Did All the Markets Go?: An Analysis of EPA's Emission Trading Program," *Yale J. Reg.*, **6**:109–153 (1989).

Hahn, Robert W. and Roger G. Noll, "Designing a Market for Tradeable Emissions Permits," Chapter 7 in W. Magat (Ed.), *Reform of Environmental Regulation* (Ballinger, Cambridge, MA, 1982).

Hanemann, W. Michael, *A Methodological and Empirical Study of the Recreation Benefits from Water Quality Improvement*, Ph.D. dissertation, Harvard University, Cambridge, MA (1978).

Hanemann, W. Michael, "Information and the Concept of Option Value," *J. Environ. Econ. Mgmt.*, **16**:23–37 (1989).

Hanemann, W. Michael, "Willingness to Pay and Willingness to Accept: How Much Can They Differ?," *Am. Econ. Rev.*, **81**:635–647 (1991).

Hanemann, W. Michael, "Valuing the Environment through Contingent Valuation," *J. Econ. Perspect.*, **8**(4):19–43 (1994).

Harford, Jon D., "Firm Behavior Under Imperfectly Enforceable Pollution Standards and Taxes," *J. Environ. Econ. Mgmt.*, **5**:26–43 (1978).

Harford, Jon, "Averting Behavior and the Benefits of Reduced Soiling," *J. Environ. Econ. Mgmt.*, **11**:296–302 (1984).

Harford, Jon D., "Self-Reporting of Pollution and the Firm's Behavior Under Imperfectly Enforceable Regulations," *J. Environ. Econ. Mgmt.*, **14**:293–303 (1987).

Harford, Jon D. and Winston Harrington, "A Reconsideration of Enforcement Leverage When Penalties Are Restricted," *J. Public Econ.*, **45**:391–395 (1991).

Harr, Jonathan, *A Civil Action* (Random House, New York, 1995).

Harrington, Winston, "Enforcement Leverage When Penalties Are Restricted," *J. Public Econ.*, **37**:29–53 (1988).

Harrison, David Jr. and Daniel L. Rubinfeld, "Hedonic Housing Prices and the Demand for Clean Air," *J. Environ. Econ. Mgmt.*, **5**:81–102 (1978).

Harvey, Charles M., "The Reasonableness of Non-Constant Discounting," *J. Public Econ.*, **53**:31–51 (1994).

Hausman, J. A. (Ed.), *Contingent Valuation: An Assessment* (North-Holland, Amsterdam, 1993).

Hausman, Jerry A., "Exact Consumer's Surplus and Deadweight Loss," *Amer. Econ. Rev.*, **71**:662–676 (1981).

Hausman, Jerry A., Gregory K. Leonard, and Daniel McFadden, "A Utility-Consistent, Combined Discrete Choice and Count Data Model Assessing Recreational Use Loses Due to Natural Resource Damage," *J. Public Econ.*, **56**:1–30 (1995).

Head, J. G., "Public Goods and Public Policy," *Public Finance*, **17**:197–219 (1962).

Heller, Walter P. and David A. Starrett, "On the Nature of Externalities," in Steven A. Y. Lin (Ed.), *Theory and Measurement of Economic Externalities* (Academic Press, New York, 1976).

Hershleifer, Jack, "From Weakest Link to Best Shot: The Voluntary Provision of Public Goods," *Public Choice*, **41**:371–386 (1983).

Heyes, A. G., "Cutting Environmental Penalties to Protect the Environment," *J. Public Econ.*, **60**:251–265 (1996).

Hicks, John R., "The Valuation of the Social Income," *Economica*, **7**:105–124 (1940).

Hoel, Michael, "International Environment Conventions: The Case of Uniform Reduction of Emissions," *Environ. Res. Econ.*, **2**:141–159 (1992).

Hoffman, Elizabeth and Matthew L. Spitzer, "The Coase Theorem: Some Experimental Tests," *J. Law Econ.*, **25**:73–98 (1982).

Holmström, Bengt, "Moral Hazard in Teams," *Bell J. Econ.*, **13**:324–340 (1982).

Hourcade, J. C. et al., "A Review of Mitigation Cost Studies," in J. Bruce, H. Lee, and E. Haites (Eds.), *Climate Change 1995: Economic and Social Dimensions of Climate Change* (Cambridge University Press for the Intergovernmental Panel on Climate Change, Cambridge, England, 1996).

Howe, Charles W., "Taxes versus Tradable Discharge Permits: A Review in the Light of the U.S. and European Experience," *Environ. Res. Econ.*, **4**:151–169 (1994).

Huppes, Gjalt and Robert A. Kagan, "Market-Oriented Regulation of Environmental Problems in the Netherlands," *Law Policy*, **11**:215–239 (1989).

Husted, Steven and Michael Melvin, *International Economics*, 4th Ed. (Addison-Wesley, Reading, MA, 1997).

Jaffe, Adam B. and Robert N. Stavins, "Dynamic Incentives of Environmental Regulations: The Effects of Alternative Policy Instruments on Technology Diffusion," *J. Env. Econ. Mgmt.*, **29**:S-43–S-63 (1995).

Jaffe, Adam B., Steven R. Peterson, Paul R. Portney, and Robert Stavins, "Environmental Regulation and the Competitiveness of U.S. Manufacturing," *J. Econ. Lit.*, **33**:132–163 (1995).

Jorgenson, Dale W., "Productivity and Postwar U.S. Economic Growth," *J. Econ. Perspect.*, **2**(4):23–42 (1988).

Jorgenson, Dale W. and Peter J. Wilcoxen, "Environmental Regulation and U.S. Economic Growth," *Rand J. Econ.*, **21**:314–340 (1990).

Just, Richard E., Darrell L. Hueth, and Andrew Schmitz, *Applied Welfare Economics and Public Policy* (Prentice-Hall, Englewood Cliffs, NJ, 1982).

Kahn, Mathew and John G. Matsusaka, "Demand for Environmental Goods: Evidence from Voting Patterns on California Initiatives," *J. Law Econ.*, **40**:137–173 (1997).

Kahneman, Daniel and Jack L. Knetsch, "Valuing Public Goods: The Purchase of Moral Satisfaction," *J. Environ. Econ. Mgmt.*, **22**:57–70 (1992).

Kaldor, Nicholas, "Welfare Propositions of Economics and Interpersonal Comparisons of Utility," *Econ. J.*, **49**:549–552 (1939).

Kanemoto, Y., "Hedonic Prices and the Benefits of Public Projects," *Econometrica*, **56**:981–989 (1988).

Katzman, Martin, "Pollution Liability Insurance and Catastrophic Environmental Risk," *J. Risk Uncertainty*, **55**:75–100 (1988).

Kelly, David L., "On Kuznets Curves Arising from Stock Externalities," Department of Economics Working Paper, University of California, Santa Barbara (October 1997).

Klaassen, Ger, "Trade-offs in Sulfur Emission Trading in Europe," *Environ. Res. Econ.*, **5**:191–219 (1995).

Kmenta, Jan, *Elements of Econometrics*, 2nd Ed. (Macmillan, New York, 1986).

Kneese, Allen V. and Blair T. Bower, *Managing Water Quality: Economics, Technology, Institutions* (Johns Hopkins University Press, Baltimore, 1968).

Kneese, Allen V. and William D. Schulze, "Ethics and Environmental Economics," Chapter 5 in A. V. Kneese and J. L. Sweeney (Eds.), *Handbook of Natural Resource and Energy Economics*, Vol. I (North-Holland, Amsterdam, 1985).

Kolstad, Charles D., "Empirical Properties of Economic Incentives and Command and Control Regulations for Air Pollution Control," *Land Econ.* **62**:250–268 (1986).

Kolstad, C. D., "Uniformity vs. Differentiation in Regulating Externalities," *J. Environ. Econ. Mgmt.*, **14**:386–399 (1987).

Kolstad, Charles D., "Clean Air and Energy: Reform of the PSD Provisions of the Clean Air Act," *J. Air Waste Manage. Assn.*, **40**:177–184 (1990).

Kolstad, Charles D., "Fundamental Irreversibilities in Stock Externalities," *J. Public Econ.*, **60**:221–233 (1996a).

Kolstad, Charles D., "Learning and Stock Effects in Environmental Regulation: The Case of Greenhouse Gas Emissions," *J. Environ. Econ. Mgmt.*, **31**:1–18 (1996b).

Kolstad, Charles D. and John B. Braden, "Environmental Demand Theory," in J. B. Braden and C. D. Kolstad (Eds.), *Measuring the Demand for Environmental Quality* (North-Holland, Amsterdam, 1991).

Kolstad, Charles D. and Rolando Guzman, "Information and the Divergence Between Willingness-to-Accept and Willingness-to-Pay," *J. Environ. Econ. Mgmt.* (1999).

Kolstad, Charles D. and Michelle A. L. Turnovsky, "Cost Functions and Nonlinear Prices: Estimating a Technology with Quality-Differentiated Inputs," *Rev. Econ. Stat.* **80**:444–453 (1998).

Kolstad, C. D., T. Ulen, and G. Johnson, "Ex Ante Regulation vs. Ex Post Liability for Harm: Substitutes or Complements?," *Am. Econ. Rev.*, **80**:888–901 (1990).

Kopp, Raymond J. and V. Kerry Smith, "Benefit Estimation Goes to Court: The Case of Natural Resource Damage Assessments," *J. Policy Anal. Manage.*, **8**:593–612 (1989).

Kozeltsev, Michael and Anil Markandya, "Pollution Charges in Russia: The Experience of 1990–1995," in R. Bluffstone and B. A. Larson (Eds.), *Controlling Pollution in Transition Economies: Theories and Methods* (Edward Elgar, Cheltenham, England, 1997).

Krämer, Ludwig, *Focus on European Environmental Law* (Sweet and Maxwell, London, 1992).

Kreps, David M., *A Course in Microeconomic Theory* (Princeton University Press, Princeton, NJ, 1990).

Kriström, Bengt and Pere Riera, "Is the Income Elasticity of Environmental Improvements Less Than One?," *Environ. Res. Econ.*, **7**:45–55 (1996).

Krutilla, John V., "Conservation Reconsidered," *Am. Econ. Rev.*, **57**:777–786 (1967).

Krutilla, John and Anthony Fisher, *The Economics of Natural Environments*, 2nd Ed. (Johns Hopkins University Press, Baltimore, 1985).

Kwak, Seung-Jun and Clifford S. Russell, "Contingent Valuation in Korean Environmental Planning: A Pilot Application to the Protection of Drinking Water in Seoul," *Environ. Res. Econ.*, **4**:511–526 (1994).

Kwerel, Evan, "To Tell the Truth: Imperfect Information and Optimal Pollution Control," *Rev. Econ. Studies*, **44**:595–601 (1977).

Laffont, Jean-Jacques, "More on Prices vs. Quantities," *Rev. Econ. Studies*, **44**:177–186 (1977).

Laffont, Jean-Jacques, *Fundamentals of Public Economics* (MIT Press, Cambridge, MA, 1989).

Laffont, Jean-Jacques and Jean Tirole, *A Theory of Incentives in Procurement and Regulation* (MIT Press, Cambridge, MA, 1993).

Laffont, Jean-Jacques and Jean Tirole, "Pollution Permits and Environmental Innovation," *J. Public Econ.*, **62**:127–140 (1996).

Lankford, R. Hamilton, "Measuring Welfare Changes in Settings with Imposed Quantities," *J. Environ. Econ. Mgmt.*, **15**:45–63 (1988).

Lenaerts, Koen, "The Principle of Subsidiarity and the Environment in the European Union: Keeping the Balance of Federalism," in F. Abraham, K. Deketelaere, and J. Stuyck (Eds.), *Recent Economic and Legal Developments in European Environmental Policy* (Leuven University Press, Leuven, Belgium, 1995).

Leopold, Aldo, *A Sand County Almanac* (Oxford University Press, New York, 1949).

Lerner, Abba, "The 1971 Report of the President's Council of Economic Advisors: Priorities and Efficiency," *Am. Econ. Rev.*, **61**:527–530 (1971).

Lewis, Tracy R., "Protecting the Environment When Costs and Benefits Are Privately Known," *Rand J. Econ.*, **27**:819–847 (1996).

Lichtenberg, E. and D. Zilberman, "Efficient Regulation of Environmental Health Risks," *Quart. J. Econ.*, **103**:167–178 (1988).

Lind, Robert C., "Spatial Equilibrium, the Theory of Rents, and the Measurement of Benefits from Public Programs," *Quart. J. Econ.*, **87**:188–207 (1973).

Lind, R. C., K. J. Arrow, G. R. Corey, P. Dasgupta, A. K. Sen, T. Stauffer, J. E. Stiglitz, J. A. Stockfisch, and R. Wilson, *Discounting for Time and Risk in Energy Policy* (Resources for the Future, Washington, DC, 1982).

Linklaters and Paines Solicitors, "United Kingdom," in M. Boes (Ed.), "Environmental Law," in R. Blanpain (Ed.), *International Encyclopaedia of Laws* (Kluwer Law International, Boston, 1992).

Low, Patrick and Alexander Yeats, "Do 'Dirty' Industries Migrate?," Ch. 6 in Patrick Low (Ed.), *International Trade and the Environment*, World Bank Discussion Papers #159 (World Bank, Washington, DC, 1992).

Luenberger, David, *Microeconomic Theory* (McGraw-Hill, New York, 1995).

MacMinn, Richard D. and Patrick L. Brockett, "Corporate Spin-offs as a Value Enhancing Technique When Faced with Legal Liability," *Insurance: Math. Econ.*, **16**:63–68 (1995).

Magat, Wesley A., Alan J. Krupnick, and Winston Harrington, *Rules in the Making: A Statistical Analysis of Regulatory Agency Behavior* (Resources for the Future, Washington, DC, 1986).

Malik, Arun S., "Markets for Pollution Control When Firms Are Noncompliant," *J. Environ. Econ. Mgmt.*, **18**:97–106 (1990).

Margulis, Sergio, "The Use of Economic Instruments in Environmental Policies: The Experience of Brazil, Mexico, Chile, and Argentina," in *Applying Economic Instruments to Environmental Policies in the OECD and Dynamic Non-Member Economies* (OECD, Paris, 1994).

Markusen, James R., Edward R. Morey and Nancy Olewiler, "Competition in Regional Environmental Polices with Plan Locations Are Endogenous," *J. Public Econ.*, **56**:55–77 (1995).

Martin, Robert E., "Externality Regulation and the Monopoly Firm," *J. Public Econ.*, **29**:347–362 (1986).

Mas-Colell, Andreu, Michael D. Whinston, and Jerry R. Green, *Microeconomic Theory* (Oxford University Press, New York, 1995).

McConnell, Kenneth E. and T. T. Phipps, "Identification of Preference Parameters in Hedonic Models: Consumer Demands with Nonlinear Budgets," *J. Urban Econ.*, **22**:35–52 (1987).

McGartland, Albert M. and Wallace E. Oates, "Marketable Permits for the Prevention of Environmental Deterioration," *J. Environ. Econ. Mgmt.*, **12**:207–228 (1985).

Meade, James, "External Economies and Diseconomies in a Competitive Situation," *Econ. J.*, **62**:54–67 (1952).

Meade, James E., *The Theory of Externalities* (AW Sijthoff, Leiden, The Netherlands, 1973).

Mehr, Robert I., *Fundamentals of Insurance*, 2nd Ed. (Irwin, Homewood, IL, 1986).

Mendelsohn, Robert, William D. Nordhaus, and Daigee Shaw, "The Impact of Global Warming on Agriculture: A Ricardian Study," *Am. Econ. Rev.*, **84**:753–771 (1994).

Mills, Edwin S., "An Aggregative Model of Resource Allocation in a Metropolitan Area," *Am. Econ. Rev. Papers Proc.*, **57**(2):197–210 (1967).

Mishan, E. J., "The Postwar Literature on Externalities: An Interpretive Essay," *J. Econ. Lit.*, **9**:1–28 (1971).

Mitchell, Robert C. and Richard T. Carson, *Using Surveys to Value Public Goods* (Johns Hopkins University Press, Baltimore, 1989).

Montgomery, W. David, "Markets in Licenses and Efficient Pollution Control Programs," *J. Econ. Theory*, **5**:395–418 (1972).

Mueller, Dennis C., *Public Choice II* (Cambridge University Press, Cambridge, England, 1989).

Murdoch, James C. and Todd Sandler, "The Voluntary Provision of a Pure Public Good: The Case of Reduced CFC Emissions and the Montreal Protocol," *J. Public Econ.*, **63**:331–349 (1997).

Musgrave, Richard A., *The Theory of Public Finance* (McGraw-Hill, New York, 1959).

Myles, Gareth D., *Public Economics* (Cambridge University Press, Cambridge, England, 1995).

Nash, Roderick Frazier, *The Rights of Nature: A History of Environmental Ethics* (University of Wisconsin Press, Madison, 1989).

National Oceanic and Atmospheric Administration, U.S. Department of Commerce, "Natural Resource Damage Assessments Under the Oil Pollution Act of 1990," *Federal Register*, 58:4601–4614 (January 15, 1993).

National Park Service, Land and Recreational Planning Division, "The Economics of Public Recreation: An Economic Study of The Monetary Evaluation of Recreation in the National Parks," U.S. Department of the Interior (Washington, DC, 1949).

Nordhaus, William D., *Managing the Global Commons: The Economics of Climate Change* (MIT Press, Cambridge, MA, 1994).

Nordhaus, William D., "Climate Amenities and Global Warming," in N. Nakicenovic, W. D. Nordhaus, R. Richels, and F. C. Toth (Eds.), *Climate Change: Integrating Science, Economics, and Policy*, Proceedings of Conference at International Institute of Applied Systems Analysis (IIASA), Laxenburg, Austria, IIASA Report CP-96-1 (December 1996).

Nordhaus, William D., "Climate Allowances Protocol (CAP): Comparison of Alternative Global Tradeable Emissions Regimes," Department of Economics Working Paper, Yale University, New Haven, CT (June 1997).

Oates, Wallace E. and R. M. Schwab, "Economic Competition among Jurisdictions: Efficiency Enhancing or Distortion Inducing?," *J. Public Econ.*, **35**:333–354 (1988).

Oates, Wallace and D. Strassmann, "The Use of Effluent Fees to Regulate Public-Sector Sources of Pollution: An Application of the Niskanen Model," *J. Environ. Econ. Mgmt.*, **5**:283–291 (1978).

Oates, Wallace E., Paul R. Portney, and Albert M. McGartland, "The *Net* Benefits of Incentive-Based Regulation: A Case Study of Environmental Standard Setting," *Am. Econ. Rev.*, **79**:1233–1242 (1989).

OECD Economic Outlook, Organization for Economic Cooperation and Development, Paris (December 1997).

OECD, *Economic Instruments for Environmental Protection*, Organization for Economic Cooperation and Development, Paris (1989).

OECD, *Environment and Taxation: The Cases of the Netherlands, Sweden and the United States*, Organization for Economic Cooperation and Development, Paris (1994a).

OECD, *Managing the Environment: The Role of Economic Instruments*, Organization for Economic Cooperation and Development, Paris (1994b).

OECD, *Environmental Taxes in OECD Countries,* Organisation for Economic Cooperation and Development, Paris (1995).

O'Ryan, Raul E., "Cost-Effective Policies to Improve Urban Air Quality in Santiago, Chile," *J. Environ. Econ. Mgmt.*, **31**:302–313 (1996).

Osborne, D. K., "Cartel Problems," *Am. Econ. Rev.*, **66**:835–844 (1976).

Page, Talbot, "Sustainability and the Problem of Valuation," Chapter 5 in Robert Costanza (Ed.), *Ecological Economics: The Science and Management of Sustainability* (Columbia University Press, New York, 1991).

Palmer, Karen, Wallace E. Oates, and Paul R. Portney, "Tightening Environmental Standards: The Benefit-Cost or the No-Cost Paradigm?," *J. Econ. Perspect.*, **9**(4):119–132 (1995).

Palmquist, Raymond B., "Hedonic Methods," in J.B. Braden and C.D. Kolstad (Eds.), *Measuring the Demand for Environmental Quality* (North-Holland, Amsterdam, 1991).

Parry, Ian W. H., "Pollution Taxes and Revenue Recycling," *J. Environ. Econ. Mgmt.*, **29**(No. 2, Part 2):S-64–S-77 (1995).

Pearce, David W. and R. Kerry Turner, *Economics of Natural Resources and the Environment* (Johns Hopkins University Press, Baltimore, 1990).

Peltzman, Samuel, "Toward a More General Theory of Regulation," *J. Law Econ.* **19**:211–240 (1976).

Pigou, Arthur C., *The Economics of Welfare*, 4th Ed. (Macmillan, London, 1962).

Plott, Charles R., "Externalities and Corrective Policies in Experimental Markets," *Economic J.*, **93**:106–127 (1983).

Polesetsky, Mathew, "Will a Market in Air Pollution Clean the Nation's Dirtiest Air? A Study of the South Coast Air Quality Management District's Regional Clean Air Incentives Market," *Ecol. Law Quart.*, **22**:359–411 (1995).

Polinsky, A. Mitchell "Notes on the Symmetry of Taxes and Subsidies in Pollution Control," *Can. J. Econ.*, **12**:75–83 (1979).

Polinsky, A. Mitchell and Steven Shavell, "Amenities and Property Values in a Model of an Urban Area," *J. Public Econ.*, **5**:119–129 (1976).

Pommerehne, Werner W., "Measuring Environmental Benefits: A Comparison of Hedonic Technique and Contingent Valuation," in D. Bös, M. Rose, and C. Seidl (Eds.), *Welfare and Efficiency in Public Economics* (Springer-Verlag, Berlin, 1988).

Porter, Michael E. and Claas van der Linde, "Toward a New Conception of the Environment-Competitiveness Relationship," *J. Econ. Perspect.*, **9**(4):97–118 (1995).

Porter, Richard C., "Michigan's Experience with Mandatory Deposits on Beverage Containers," *Land Econ.*, **59**:177–194 (1983).

Portney, Paul R., "The Contingent Valuation Debate: Why Economists Should Care," *J. Econ. Perspect.*, **8**(4):3–17 (1994).

Portney, Paul R. and Robert Stavins (Eds.), *Public Policies for Environmental Protection*, 2nd Ed. (Resources for the Future, Washington, DC, forthcoming).

Quigley, John M., "Nonlinear Budget Constraints and Consumer Demand: An Application to Public Programs for Residential Housing," *J. Urban Econ.*, **12**:177–201 (1982).

Randall, Alan, "The Problem of Market Failure," *Natural Res. J.*, **23**:131–148 (1983).

Randall, Alan, "A Difficulty with the Travel Cost Method," *Land Econ.*, **70**:88–96 (1994).

Randall, A., B. Ives and C. Eastman, "Bidding Games for Valuation of Aesthetic Environmental Improvements," *J. Environ. Econ. Mgmt.*, **1**:132–149 (1974).

Regan, Donald H., "The Problem of Social Cost Revisited," *J. Law Econ.*, **15**:427–437 (1972).

Rehbinder, Eckard, "Environmental Regulation Through Fiscal and Economic Incentives in a Federalist System," *Ecol. Law Quart.*, **20**:57–83 (1993).

Reisner, Marc, *Cadillac Desert* (Viking Press, New York, 1986).

Repetto, Robert, William Magrath, Michael Wells, Christine Beer, and Fabrizo Rossini, "Wasting Assets: Natural Resources in the National Income Accounts" (World Resources Institute, Washington, DC, June 1989).

Repetto, Robert, Dale Rothman, Paul Faeth, and Duncan Austin, "Has Environmental Protection Really Reduced Productivity Growth?" (World Resources Institute, Washington, DC, October 1996).

Ridker, Ronald G. and John A. Henning, "The Determinants of Residential Property Values with Special Reference to Air Pollution," *Rev. Econ. Stat.*, **49**:246–257 (1967).

Roback, Jennifer, "Wages, Rents and the Quality of Life," *J. Political Econ*, **90**:1257–1278 (1982).

Roberts, Marc J. and Michael Spence, "Effluent Charges and Licenses Under Uncertainty," *J. Public Econ.*, **5**:193–208 (1976).

Rosen, Sherwin, "Hedonic Prices and Implicit Markets: Product Differentiation in Perfect Competition," *J. Political Econ.*, **82**:34–55 (1974).

Russell, C. S., W. Harrington and W. J. Vaughn, Chapter 4 in "Economic Models of Monitoring and Enforcement," *Enforcing Pollution Control Laws* (Resources for the Future, Washington, DC, 1986).

Sagoff, Mark, "Should Preferences Count?" *Land Econ.*, **70**:127–144 (1994).

Samuelson, Paul, *Foundations of Economic Analysis* (Harvard University Press, Cambridge, MA, 1947).

Sappington, D. E. M. and J. E. Stiglitz, "Information and Regulation," in E. Baily (Ed.), *Public Regulation* (MIT Press, Cambridge, MA, 1987).

Schelling, Thomas C., "Intergenerational Discounting," *Energy Policy*, **23**:395–401 (1995).

Schmalensee, Richard, "Option Demand and Consumer's Surplus: Valuing Price Changes Under Uncertainty," *Amer. Econ. Rev.*, **62**:813–824 (1972).

Schmutzler, Armin and Lawrence H. Goulder, "The Choice Between Emission Taxes and Output Taxes under Imperfect Monitoring," *J. Environ. Econ. Mgmt.*, **32**:51–64 (1997).

Schoenbaum, T. J. and R. H. Rosenberg, *Environmental Policy Law* (The Foundation Press, Westbury, NY, 1991).

Schulze, William and Ralph C. d'Arge, "The Coase Proposition, Information Constraints and Long Run Equilibrium," *Am. Econ. Rev.*, **64**:763–772 (1974).

Segerson, Kathleen, "Uncertainty and Incentives for Nonpoint Pollution Control," *J. Environ. Econ. Mgmt.*, **15**:87–98 (1988).

Seip, Kalle and Jon Strand, "Willingness to Pay for Environmental Goods in Norway: A Contingent Valuation Study with Real Payment," *Environ. Res. Econ.*, **2**:91–106 (1992).

Seskin, Eugene P., Robert J. Anderson, Jr., and Robert O. Reid, "An Empirical Analysis of Economic Strategies for Controlling Air Pollution," *J. Environ. Econ. Mgmt.*, **10**:112–124 (1983).

Shafik, Nemat, "Economic Development and Environmental Quality: An Econometric Analysis," *Oxford Econ. Papers*, **46**:757–773 (1994).

Shafik, Nemat and Sushenjit Bandyopadhyay, "Economic Growth and Environmental Quality: Time-Series and Cross-Country Evidence," Policy Research Working Paper WPS 904, The World Bank, Washington, DC (June 1992).

Shavell, Steven, "A Model of the Optimal Use of Liability and Safety Regulation," *Rand J. Econ.*, **15**:271–280 (1984).

Shogren, Jason F., Seung Y. Shin, Dermot J. Hayes, and James B. Kliebenstein, "Resolving Differences in Willingness to Pay and Willingness to Accept," *Amer. Econ. Rev.*, **84**:255–270 (1994).

Shibata, H. and J. S. Winrich, "Control of Pollution When the Offended Defend Themselves," *Economica*, **50**:425–38 (1983).

Sigman, Hilary, "A Comparison of Public Policies for Lead Recycling," *Rand J. Econ.*, **26**:452–478 (1995).

Sigman, Hilary, "Midnight Dumping: Public Policies and Illegal Disposal of Used Oil," *Rand J. Econ.*, **29**:157–178 (1998).

Slade, Margaret E., "Trends in Natural Resource Commodity Pricing: An Analysis of the Time Domain," *J. Environ. Econ. Mgmt.*, **9**:122–137 (1982).

Slovic, Paul, Baruch Fischoff, and Sarah Lichtenstein, "Rating the Risks," *Environment*, **21**:14–20, 36–39 (April 1979). [Reprinted in T. Glickman and M. Gough (Eds.), *Readings in Risk* (Resources for the Future, Washington, DC, 1990.)]

Small, Kenneth A., *Urban Transportation Economics* (Harwood Academic Publishers, Reading, MA, 1992).

Smith, V. Kerry, and W. H. Desvousges, "An Empirical Analysis of the Economic Value of Risk Changes," *J. Political Econ.*, **95**:89–114 (1987).

Smith, V. K. and W. H. Desvousges, "The Valuation of Environmental Risks and Hazardous Waste Policy," *Land Econ.*, **64**:211–219 (1988).

Smith, V. Kerry and Ju Chin Huang, "Hedonic Models and Air Pollution: Twenty-Five Years and Counting," *Environ. Res. Econ.*, **3**:381–394 (1993).

Solow, Robert, "An Almost Practical Step Towards Sustainability," Invited Lecture, Resources for the Future, Washington, DC (October 8, 1992).

Spulber, Daniel F. "Effluent Regulation and Long-Run Optimality," *J. Environ. Econ. Mgmt.*, **12**:103–116 (1985).

Spulber, Daniel F., "Optimal Environmental Regulation Under Asymmetric Information," *J. Public Econ.*, **35**:163–181 (1988).

Spulber, Daniel F., *Regulation and Markets* (MIT Press, Cambridge, MA, 1989).

Squires, Dale, "Productivity Measurement in Common Property Resource Industries: An Application to the Pacific Coast Trawl Fishery," *Rand J. Econ.*, **23**:221–236 (1992).

Stanners, David and Philippe Bourdeau, *Europe's Environment: The Dobřiš Assessment* (European Environment Agency, Copenhagen, 1995).

Starrett, David, *Foundations of Public Economics* (Cambridge University Press, Cambridge, England, 1988).

Starrett, David and Richard Zeckhauser, "Treating External Diseconomies—Markets or Taxes?," in John W. Pratt (Ed.), *Statistical and Mathematical Aspects of Pollution Problems* (Marcel Dekker, New York, 1974).

Stavins, Robert N., "Transaction Costs and Tradeable Permits," *J. Environ. Econ. Mgmt.*, **29**:133–148 (1995).

Stigler, George, "The Economic Theory of Regulation," *Bell J. Econ.*, **2**:3–21 (1971).

Strotz, R. H., "The Use of Land Rent Changes to Measure the Welfare Benefits of Land Improvement," in J.E. Haring (Ed.), *The New Economics of Regulated Industries: Rate-Making in a Dynamic Economy*, Economics Research Center, Occidental College, Los Angeles (1968).

Thaler, Richard and Sherwin Rosen, "The Value of Saving a Life: Evidence from the Labor Market," in Nestor E. Terleckyj (Ed.), *Household Production and Consumption* (National Bureau of Economic Research, New York , 1975).

Thomas, Alban, "Regulating Pollution Under Asymmetric Information: The Case of Industrial Wastewater Treatment," *J. Environ. Econ. Mgmt.*, **28**:357–373 (1995).

Thompson, Peter and Laura A. Strohm, "Trade and Environmental Quality: A Review of the Evidence," *J. Environ. Dev.*, **5**:363–388 (1996).

Tideman, N. and G. Tullock, "A New and Superior Process for Making Social Choices," *J. Political Econ.*, **84**:1145–1159 (1976).

Tierney, John, "Betting the Planet," p. 53 *et seq*, *New York Times Magazine* (December 2, 1990).

Tietenberg, Thomas H., "Indivisible Toxic Torts: The Economics of Joint and Several Liability," *Land Econ.*, **65**:305–319 (1989).

Tobey, James A., "The Effects of Domestic Environmental Policies on Patterns of World Trade: An Empirical Test," *Kyklos*, **43**:191–209 (1990).

Toman, Michael A., "Economics and 'Sustainability: Balancing Trade-offs and Imperatives," *Land Econ.*, **70**:399–413 (1994).

Ulph, Alastair, "The Choice of Environmental Policy Instruments and Strategic International Trade," in Rudiger Pethig (Ed.), *Conflicts and Cooperation in Managing Environmental Resources* (Springer-Verlag, Berlin, 1992).

United Nations Environment Programme, *Environmental Data Report 1993–94* (Blackwell, Oxford, England, 1993).

United States Bureau of the Census, *Statistical Abstract of the United States: 1988*, 108th Ed., Washington, DC (1987).

United States Council on Environmental Quality, Executive Office of the President, *Environmental Quality: 25th Anniversary Report* (Government Printing Office, Washington, DC 1997).

United States Department of Commerce, Bureau of Economic Analysis, "Integrated Economic and Environmental Satellite Accounts," *Survey Current Business*, **74**(4):33–49 (1994a).

United States Department of Commerce, Bureau of Economic Analysis, "Accounting for Mineral Resources: Issues and BEA's Initial Estimates," *Survey Current Business*, **74**(4):50–72 (1994b).

United States Environmental Protection Agency, *Environmental Investments: The Cost of a Clean Environment* (U.S. Government Printing Office, Washington, DC, 1990).

United States Environmental Protection Agency, *The Benefits and Costs of the Clean Air Act: 1970 to 1990*, Office of Air and Radiation, U.S. Environmental Protection Agency, Washington, DC (October 1997).

Usategui, Jose M., "Uncertain Irreversibility, Information and Transformation Costs," *J. Environ. Econ. Mgmt.*, **19**:73–85 (1990).

Varian, Hal R., *Intermediate Microeconomics*, 4th Ed. (Norton, New York, 1996).

Viner, Jacob, "Cost Curves and Supply Curves," *Zeitschrift für Nationalökonomie*, **3**:23–46 (1931).

Viscusi, W. Kip, "The Value of Risks to Life and Health," *J. Econ. Lit.*, **31**:1912–1946 (1993).

Viscusi, W. Kip, "Economic Foundations of the Current Regulatory Reform Efforts," *J. Econ. Perspect.*, **10**(3):119–134 (1996).

Viscusi, W. Kip, John M. Vernon, and Joseph E. Harrington, Jr., *Economics of Regulation and Antitrust* (D.C. Heath, Lexington, MA, 1992).

Weimer, David L. and Aidan R. Vining, *Policy Analysis: Concepts and Practice*, 2nd Ed. (Prentice-Hall, Englewood Cliffs, NJ, 1992).

Weitzman, Martin L., "Prices vs. Quantities," *Rev. Econ. Studies*, **41**:477–491 (1974).

Weitzman, Martin L., "Optimal Rewards for Economic Regulation," *Am. Econ. Rev.*, **68**:683–691 (1978).

Weitzman, Martin L., "On the Environmental Discount Rate," *J. Env. Econ. Mgmt.*, **26**:200–209 (1994).

Wheeler, David and Ashoka Mody, "International Investment Location Decisions: The Case of US Firms," *J. Int. Econ.*, **33**:57–76 (1992).

Willig, Robert, "Consumer's Surplus Without Apology," *Amer Econ Rev*, **69**:469–474 (1979).

Wilson, Richard, "Analyzing the Daily Risks of Life," *Technology Review*, **81**(4):41–46 (1979). [Reprinted in T. Glickman and M. Gough (Eds.), *Readings in Risk* (Resources for the Future, Washington, DC, 1990).]

World Bank, *1998 World Development Indicators* (World Bank, Washington, DC, 1998).

World Bank, *World Development Report 1992: Development and the Environment* (Oxford University Press, New York, 1992).

World Resources Institute, *World Resources 1994–95* (Oxford University Press, New York, 1994).

World Resources Institute, *World Resources 1998–99* (Oxford University Press, New York, 1998).

Xepapadeas, A. P., "Environmental Policy under Imperfect Information: Incentives and Moral Hazard," *J. Environ. Econ. Mgmt.*, **20**:113–126 (1991).

Xing, Yuqing, and Charles D. Kolstad, "Do Lax Environmental Regulations Attract Foreign Investment?, Department of Economics Working Paper 6-95R, University of California, Santa Barbara (February 1997).

Yao, Dennis A., "Strategic Responses to Automobile Emissions Control: A Game-Theoretic Analysis," *J. Environ. Econ. Mgmt.*, **15**:419–438 (1988).

Zich, Arthur, "China's Three Gorges," *National Geographic*, **192**(3):2–33 (September 1997).

Author Index

Subject Index